Mark Friezer

PAUL HAM is a historian, specializing in twentieth-century conflict. A former journalist, he has worked for the Financial Times Group and was the Australia correspondent for *The Sunday Times* of London for fifteen years. He lives in Paris with his family.

ALSO BY PAUL HAM

Young Hitler

1914: The Year the World Ended

Sandakan: The Untold Story of the Sandakan Death Marches

Vietnam: The Australian War

Kokoda

HIROSHIMA NAGASAKI

The Real Story of the Atomic Bombings
and Their Aftermath

PAUL HAM

Picador

A Thomas Dunne Book
St. Martin's Press
New York

The author wishes to thank the Harry S. Truman Library and Museum in Independence, Missouri, for the presidential research grant which greatly assisted the completion of this book.

HIROSHIMA NAGASAKI. Copyright © 2014 by Paul Ham. All rights reserved. Printed in the United States of America. For information, address St. Martin's Press, 175 Fifth Avenue, New York, N.Y. 10010.

www.picadorusa.com
www.twitter.com/picadorusa • www.facebook.com/picadorusa
picadorbookroom.tumblr.com

Picador® is a U.S. registered trademark and is used by St. Martin's Press under license from Pan Books Limited.

For book club information, please visit www.facebook.com/picadorbookclub or e-mail marketing@picadorusa.com.

The Library of Congress has cataloged the St. Martin's Press edition as follows:

Ham, Paul.
    Hiroshima, Nagasaki : the real story of the atomic bombings and their aftermath / Paul Ham. — First U.S. edition.
        p. cm.
    Includes bibliographical references and index.
    ISBN 978-1-250-04711-3 (hardcover)
    ISBN 978-1-4668-4747-7 (e-book)
    1. Hiroshima-shi (Japan)—History—Bombardment, 1945—Moral and ethical aspects. 2. Nagasaki-shi (Japan)—History—Bombardment, 1945—Moral and ethical aspects. 3. World War, 1939–1945—Japan—Hiroshima-shi. 4. World War, 1939–1945—Japan—Nagasaki-shi. 5. Atomic bomb victims—Japan—Hiroshima-shi. 6. Atomic bomb victims—Japan—Nagasaki-shi. 7. Atomic bomb—Government policy—United States—History—20th century. I. Title.
    D767.25.H6H334 20124
    940.54'2521954—dc23

                                                                2014008489

Picador Paperback ISBN 978-1-250-07005-0

Picador books may be purchased for educational, business, or promotional use. For information on bulk purchases, please contact the Macmillan Corporate and Premium Sales Department at 1-800-221-7945, extension 5442, or write to specialmarkets@macmillan.com.

Originally published in Australia by HarperCollins Publishers, Australia Pty Limited

First published in the United States by Thomas Dunne Books, an imprint of St. Martin's Press

First Picador Edition: August 2015

10 9 8 7 6 5 4 3 2 1

Pour Marie-Morgane,
Mon amour, ma chérie:
La femme de mes rêves,
La sirène de ma vie ...

CONTENTS

NOTES ON STYLE

Metric and imperial measurements

This book uses a combination of metric and imperial measurements, depending on the source of information, and what was current usage at the time. Many quantities are commonly referred to in imperial measurements (for example, altitude and armaments) and some conversions have been made to assist readers. Exceptions were made when a measurement is in a direct quote or is an acceptably well-understood unit.

The general principle has been to present measurements where practical rounded in the metric scale.

1 inch = 25.4 millimetres	1 acre = 0.405 hectares
1 foot = 305 millimetres	1 pound = 0.454 kilograms
1 yard = 914 millimetres	1 gallon = 4.546 litres
1 mile = 1.6 kilometres	1 (US) ton = 0.9 tonnes

JAPANESE TERMS

Japanese names have been rendered in Western style of given name followed by family name. Place names have been expressed without using hyphens. The following glossary may be helpful:

-bashi	bridge
-cho	precinct
-gawa, -kawa	river
-gun	county
-hana, -bana	point
-ji, -dera	temple
-jinsha, -jinja	shrine
-ken	prefecture
-ko	port
-machi	township, precinct
-mura	township
-saki, -zaki, -mizaki	cape
-shi	municipality
-shima	island
-sone, -se	reef, shoal
-take, -dake	mountain
-ura	inlet
-yama, -san, -zan	mountain, hill

HIROSHIMA
NAGASAKI

CHAPTER 1

WINTER 1945

In Japan there is a philosophy of death and no philosophy of life.

Kiyoshi Kiyosawa, Japanese historian, January 1945

... there is a point beyond which we will not tolerate insult. If [the Russians] are convinced that we are afraid of them and can be bullied into submission, then indeed I should despair of the future relations with them and much else ...

Prime Minister Winston Churchill to President Franklin D. Roosevelt, March 1945

THE BIG THREE SMILED AT the world from the grounds of the Livadia Palace in the Crimean resort town of Yalta. It was February 1945. The chill blowing off the Black Sea pressed the leaders into greatcoats and fur hats: Prime Minister Winston Churchill, President Franklin D. Roosevelt and Marshal Josef Stalin were meeting here to carve up the old Continent devastated by war and decide the outline of the post-war world.

Peace in Europe was at hand. The destruction and unconditional surrender of Germany were imminent; Japan's defeat would assuredly follow. Roosevelt had honoured his agreement with Churchill to defeat 'Germany First', and the bulk of Allied troops were then in Europe. From the west, over the previous six months, General Eisenhower's armies had swept across northern France, freed Paris, defeated Germany's last stand at the Battle of the Bulge and reached the shores of the Rhine. From the east Soviet tanks, troops and artillery had rolled across the Baltic, smashed the Nazi grip on Poland and stood on the threshold of the Fatherland, 65 kilometres from Berlin. No conflict had matched in scale and fury the battle on the Eastern Front, where the Red Army and the Wehrmacht were locked in the vestigial shambles of total war; millions of troops had been killed or wounded and countless civilians slaughtered, raped or left homeless. From his Berlin bunker, the Führer continued hysterically to issue orders that imagined pristine armies on the march where there were only ragged columns of bleeding, hungry, broken men.

Winter kept them warm: the Big Three made a great show of friendship at Yalta, hosting banquets, raising toasts, joking. Photographs present Roosevelt, perhaps the greatest Democrat, now very sick, sitting up in his wheelchair wrapped in a black cape, evoking the patrician hauteur of a Roman tribune; Churchill, lounging about in his greatcoat like a breathless bulldog, radiating delight at the top table, cigar smoke trailing in the direction of his loquacious argument; and Stalin, small and sharp amid the gathering darkness, in his flashing eyes and faithless smile a fixity of purpose that seemed to concentrate the air of menace that preceded him like a personal storm.

In the closing stages of the conference Stalin offered an eloquent expression of goodwill tinged with a warning: 'It is not difficult to keep unity in time of war,' he toasted his comrades-in-arms, 'since there is a joint aim to defeat the common enemy ... The difficult talk will come after the war when diverse interests tend to divide the Allies. It is our duty to see that our relations in peacetime are as strong as they have been in war.'

Mutual distrust between Anglo-America and the Soviet Union simmered at Yalta. The Big Three brought deep suspicion and clandestine intent to the table. Several great issues threatened to destabilise, or possibly break, the West's alliance with Moscow: the question of German and Polish borders; the political status of Eastern

Europe; and the terms of Russia's involvement in the Pacific War. Long before Yalta the 'danger' of the Soviet Union had occupied anxious discussions in the State Department. For its part, Moscow was determined to reject any Anglo-American attempt to limit its hegemony over Eastern Europe. On both sides, anxious suspicions were about to flare into fierce disagreement.

•

A secret that would astonish the earth – had its contents been revealed – lingered over the Yalta talks: Roosevelt and Churchill arrived bound by a private agreement, signed on 19 September 1944 at the President's Hyde Park Estate in Washington, not to share with the Soviet Union or the world the development of an extraordinary new weapon that, in theory at least – it had not been tested – drew its power from an atomic chain reaction.* The British codenamed the weapon project 'Tube Alloys'; the American government dubbed it 'S-1'. The Hyde Park Agreement conceived of an Anglo-American duopoly over the development of an atomic bomb, ruled out any international controls over the new weapon and named, for the first time, its future target:

'The suggestion,' Churchill's one-page agreement with Roosevelt stated, 'that the world should be informed regarding tube alloys with a view to an international agreement regarding its use and control is not accepted. [The weapon] should continue to be regarded as of the utmost secrecy … but when a "bomb" is finally available, it should be used against the Japanese, who should be warned that this bombardment will be repeated until they surrender.'†

* The so-called Hyde Park Agreement was in effect a 'gentlemen's agreement', not an international treaty, or an official undertaking like the Quebec Agreement and Declaration of Trust, signed a year earlier; the Quebec Agreement bound Britain and America 'to control to the fullest extent practicable the supplies of uranium and thorium ores' and not to divulge any information about work on a tremendous new weapon (codenamed 'Tube Alloys') to third parties 'except by mutual consent'.

† Churchill softened this to read: '… but when a "bomb" is finally available, it might perhaps, after mature consideration, be used against the Japanese …', meaning either they seriously intended to debate the use of the weapon, or had changed the record to appease posterity, while fully intending to use it.

A handful of British and American officials were aware of Tube Alloys (or S-1): Churchill's and Roosevelt's closest cabinet colleagues and those entrusted with leading its construction. The then vice-president Harry Truman, most American and British politicians, and just about everyone else were ignorant of the project. Stalin and his top officials, via their spies (chief of whom was Klaus Fuchs, an exiled German physicist working in the US), were, however, already well informed. Indeed, at this time, Stalin knew more about work on the atomic bomb than virtually every US congressman.

One Washington insider was Henry Stimson, a conscientiously Christian, Ivy League alumnus who served as Roosevelt's Secretary of War. At 77 Stimson's long life bracketed the sabre and rifled musketry of the late 19th century, the machine guns of the Western Front and the recent firebombing of German cities. He now contended with the prospect of nuclear war. His outlook was Victorian; his morals, patrician. An 'unabashed elitist', Stimson believed 'richer and more intelligent citizens' should guide public policy, and that Anglo-Saxons were superior to the 'lesser breeds', as he was apt to say. He also dedicated his term as War Secretary to eradicating the nastier aspects of war: he detested the submarine; embraced the 1928 Kellogg–Briand Pact that called for the renunciation of war; and campaigned tirelessly for arms control, international co-operation and mutual trust. Indiscriminate slaughter vexed the conscience of this fastidious gentleman.

Stimson had no illusions that S-1 could be kept secret, and yet he believed sharing the secret of the new weapon with Russia should deliver something in return to America. Stimson knew the Russians 'were spying on [S-1]', as he recorded in his diary on 31 December 1944, and told the President so, 'but ... they had not yet got any real knowledge of it and that, while I was troubled about the possible effect of keeping from them [the work on the atomic bomb], I believed that it was essential not to take them into our confidence until we were sure to get a real *quid pro quo* from our frankness'. Roosevelt had said he agreed.

At Yalta, however, Roosevelt's mood changed. Notwithstanding Stimson's advice and the Hyde Park pact, the President felt tempted to divulge the atomic secret to the Russians. Circumstances had shifted: several French and Danish physicists knew of the bomb; and FDR pondered whether candour might remove the risk, and diplomatic uproar, of the French revealing it to Stalin first. Churchill was aghast: 'I was shocked,' he told Anthony Eden, the British Foreign Secretary, before Yalta, 'when the President … spoke of revealing the secret to Stalin on the grounds that de Gaulle, if he heard of it, would certainly double cross us with Russia.' Paramount in Churchill's mind was the preservation of Anglo-American control of atomic technology; only the British and Americans, Churchill believed, could be entrusted with it: 'You may be quite sure,' he told Eden, 'that any power that gets hold of the secret will try to make [the bomb] and that this touches the existence of human society.'

•

In the event, FDR kept the secret. Nobody spoke openly of the bomb at Yalta; the early atomic manoeuvring played out in private salons and the minds of men. The delegates' top priority was the division of Germany, whose defeat loomed as inevitable. The Big Three formulated an ultimatum to Berlin, which they announced on 11 February 1945: 'Nazi Germany is doomed,' it warned:

> It is our inflexible purpose to destroy German militarism and Nazism
> and to ensure that Germany will never again be able to disturb the
> peace of the world. We are determined to … wipe out the Nazi Party,
> Nazi laws, organizations and institutions, remove all Nazi and
> militarist influences from public office … It is not our purpose to
> destroy the people of Germany but only when Nazism and militarism
> have been extirpated will there be hope for a decent life for
> Germans …

The words may have been written for Tokyo: by extension, Roosevelt would accept nothing less than the 'unconditional surrender' of Japan. The popular phrase would prove a dangerous hostage to fortune. FDR first used it at Casablanca in January 1943, as an unintended *ad lib* at a press conference. A prior disagreement between two French generals reminded him of commanders Lee and Grant: 'We had a general called U.S. Grant,' he told reporters, 'he was called Unconditional Surrender Grant. The elimination of German, Japanese and Italian war power means [their] unconditional surrender.' The President's words surprised his listeners, appalled the State Department – which had not been informed and feared it would prolong the war and, initially at least, delighted Churchill: 'Perfect!' the British Prime Minister exclaimed. 'And I can just see how Goebbels and the rest of 'em'll squeal!' Under the terms of unconditional surrender, the Germans and Japanese would have to lay down their weapons, yield all territory won by conquest, and abandon the whole infrastructure and philosophy of militarism – or face annihilation. There would be no negotiation.

Churchill had gnawing doubts about the wisdom of the policy's extension to Japan; he well understood the Japanese people's fanatical devotion to their Emperor, and wondered whether the wording might be softened to encourage Tokyo to disarm. No doubt he continued to believe, as he told US Congress in May 1943, 'in the process, so necessary and desirable, of laying the cities ... of Japan in ashes, for in ashes they must surely lie before peace comes back to the world'. Yet might not a subtle relaxation of the surrender terms avoid further Allied losses, he wondered. And so at Yalta, alone among his American and Soviet colleagues, Churchill suggested that 'some mitigation [of the terms of surrender] would be worthwhile if it led to the saving of a year or a year and a half of war in which so much blood and treasure would be poured out'. The President dismissed Churchill's proposal. The Japanese would interpret leniency as weakness, Roosevelt argued. The American people would not tolerate peace negotiations with an enemy who had killed or maimed tens of thousands of American soldiers.

•

Japan had to be defeated before she would surrender. To this end, Churchill and Roosevelt openly sought Russian entry in the Pacific War. Whatever their personal view of the Soviet dictator – and Churchill loathed the tyrant George Orwell had recently described as a 'disgusting murderer temporarily on our side' who had packed off millions to the Siberian Gulag – the Americans and British needed Soviet help, not least because the Chinese, a spent force, had failed to defeat the Japanese occupying forces, who showed every sign of fighting to the last man. Hence London and Washington desired Russian entry 'at the earliest possible date', stated the US Joint Chiefs of Staff in November 1944. Russian military aid would, however, come at a cost, cautioned Roosevelt's right-hand man, the wily James Byrnes, on several occasions: the more America appealed for Soviet military assistance, the more Stalin would demand in return.

However, the Soviet position in the Pacific was complicated. Until Yalta, Stalin had trod a careful line between threatening and appeasing the Japanese. He initially praised the Soviet–Japanese Neutrality Pact, which had been ratified on 25 April 1941 at a lavish party in Moscow where Stalin had 'danced around like a performing bear', embraced and kissed the Japanese delegation and even toasted a '*Banzai!*' to the Emperor, according to witnesses. The pact stipulated 'peaceful and friendly' relations between the two countries until it expired in April 1946. Moscow, however, had had no intention of honouring the spirit of the agreement: in 1941 it bought time for Stalin to re-arm and confront the German threat, safe in the knowledge of a neutral Tokyo in his rear – hence the Generalissimo's performing bear act. In time, however, he resolved to turn his great armies east and avenge Russia's loss to Japan in the war of 1904–05. Indeed, Moscow's successive breaches of the spirit of the pact were drumbeats to the invasion of Japan.* In late 1944, Stalin ratcheted up the stakes in anticipation of striking a better deal at the peace. He fixed his hungry eye on the spoils of Japanese conquest. He wanted to be 'in at the kill', as he said, to recoup his down payment of men and materiel likely to be lost in the Pacific. His price included territory Japan had seized from Russia in 1905. In November 1944, Stalin reasserted his price for Pacific entry to Averell Harriman, the US ambassador in Moscow: the Kurils and South Sakhalin should be returned to Russia, along with leases on Port Arthur, Dairen and several railway lines; Outer Mongolia, which had in fact been under Soviet control since 1921, should remain 'independent'. Harriman saw no serious objection and Stalin's timely denunciation of Japan as 'an aggressor on a plane with Germany' cheered their relationship along.†

Stalin further pressed these demands on the table at Yalta. He aimed to cement the Soviet Union's strategic claim on Asia. Britain and America acquiesced. On 11 February 1945, Stalin, Roosevelt and Churchill signed a Top Secret 'protocol': 'The former rights of Russia,' it stated, 'violated by the treacherous attack of Japan in 1904' would be restored, and the islands and leases 'handed over' to the Soviet Union after the surrender of Japan.

---

* Stalin's contempt for the pact was demonstrated as early as December 1941, when, a week after Pearl Harbor, he told British Foreign Secretary Anthony Eden that the best way to get Russia into the Pacific War would be 'to induce Japan to violate the Neutrality Pact'. The Soviet Union would need four months to move the Red Army to the Far East, he added. Again, in October 1943, at a meeting of foreign ministers in Moscow, the Russians vowed to help defeat Japan if the Allies opened a second front against Germany to relieve the Red Army, then fighting 80 per cent of the German forces. 'When the Allies succeeded in defeating Germany, the Soviet Union would then join in defeating Japan,' Stalin had told US officials. After dinner, the Russian party showed a film about the Japanese invasion of Siberia, and drank to the day when American and Russian forces were 'fighting together against the Japs': 'Why not? Gladly – the time will come,' observed Molotov, the Soviet Foreign Minister, and 'downed the drink'. Their entertainment found formal expression at the Teheran Conference in November 1943, where Britain and America agreed to Stalin's demands and, with a photograph and a handshake, committed tens of thousands of troops to the Normandy landings in exchange – in part – for a Russian commitment to join the Pacific War.

† This 'bombshell' did not mean a hardening of Soviet policy towards Japan, Moscow brazenly assured a nervous Tokyo; in truth, however, Stalin had already started planning the invasion of Japanese-occupied Manchuria; it would involve stockpiling weapons and supplies for 30 additional divisions on the border, according to dispatches from Major General John Deane, chief of the US Military Mission in Moscow.

Roosevelt said nothing of this deal, struck the day after he was officially supposed to leave the Crimea (Stalin had personally asked the President to stay another day). The strictest secrecy prevailed; Byrnes, hitherto one of the President's most trusted advisers, later claimed he was unaware of the deal; Congress was not informed. The President believed the agreement fair: '[The Russians],' he said later, 'only want to get back that which has been taken from them.'*

•

The 'Russian Protocol' was a rare moment of unity at Yalta. In fact deep divisions flared over the break-up of Germany and pushed American and British relations with the Soviet Union to the point of collapse. The argument was technically over the nature of the political system in the recently liberated countries of Eastern Europe. The dispute, however, went to the heart of the question of whether liberal democracy or Soviet communism would prevail in the post-war world.

Stalin's brutal pragmatism outmatched Churchill's florid eloquence on the future of Poland. Replying to the British leader's rosy defence of Polish freedom, Stalin barked that 'twice in the last 25 years the "Polish door" had opened and let hordes of Germans overrun Russia … Russia was determined this time that it would not happen again'. Stalin proved the 'real boss' of proceedings at Yalta and a formidable adversary; his generals jumped to his elbow at the slightest nod.

Churchill and Roosevelt clung to a few brittle reeds in this Soviet gale: the Polish leaders in exile, they insisted, should be invited to

---

* Their secrecy showed Roosevelt and Churchill appreciated the sensitivity of the 'Russian Protocol'; but neither anticipated the future storm over the 'horrendous concessions' made to the Soviet Union, 'an enemy' who entered the Pacific War near the end. In fairness, at the time of signing, Russia was then an ally, had sustained huge casualties in Europe, and no one knew how long the Pacific War would last. Roosevelt sought to lash as many strings to his bow – including, if need be, the Red Army. He knew, of course, that the Russian concession would not play well in America, and on his return to Washington entrusted it to the care of his Chief of Staff, Admiral William Leahy, who locked America's copy of the protocol in the White House safe. There it sat until after the President's death.

participate in a post-war Polish democracy. Churchill had recognised the Polish émigré government in London at the start of the war; how could he face Parliament if he abandoned them? The London Poles were 'our' Poles: democratic, hostile to communism, with popular support; Poland's provisional government then resident in Lublin were Moscow's Poles, subservient to Soviet demands. Churchill 'objected violently' to any recognition of the Lublin Poles.

To no avail: the Soviet leader would impose his chosen regime on Poland, by stealth. On paper Stalin agreed at Yalta to 'guarantee' 'unfettered elections' by 'secret ballot', 'universal suffrage' and so on. In practice, Stalin immediately reneged on the Yalta declaration. By March 1945, the Soviet Union had torn up the script, bit by bit: Molotov obfuscated and delayed over the agreed terms, while the Red Army consolidated its grip on Warsaw. Poland soon fell behind the Soviet shadow, as Churchill had warned. With the prescience that had marked his wilderness years, the British leader perceived in Soviet actions a pattern of forceful acquisition, the unyielding nature of which had partly eluded Washington. 'The Russians,' he wrote to Roosevelt on 8 March 1945, 'have succeeded in establishing [in Eastern Europe] the rule of a communist minority by force and misrepresentation … which is absolutely contrary to all democratic ideas … Stalin has subscribed on paper to the principles of Yalta which are certainly being trampled down …'

Churchill's undying faith in the nostrums of freedom and democracy, his steadfastness when all seemed lost, profoundly moved Roosevelt. The British Prime Minister was the first to grasp the true nature of the beast behind Stalin's grey uniform and steely black eyes. 'We are in the presence of a great failure and utter breakdown of what was agreed at Yalta …' he reminded Roosevelt on 13 March 1945. In that doom-laden rhetoric lay the beginning of the end of the Soviet–American alliance, the fault lines of the Cold War. Britain and America stood helplessly aside – there was little they could do – as Moscow claimed a string of East European 'satellites' against the hated Germans. The Western leaders had blinked; then they shut their eyes.

The President's last letters to Churchill reveal a mind on the threshold of a new and terrifying world: 'Our peoples,' Roosevelt wrote on 27 March 1945, 'are watching with anxious hope the extent to which the decisions at the Crimea are being honestly carried forward.' They were not; on 3 April, negotiations with the Soviets were 'at breaking point', Averell Harriman, America's ambassador in Moscow, warned the President. Harriman dispatched a list of Soviet breaches of Yalta to Washington.

Soviet–American relations were now frigid and Roosevelt very ill. The President's famed diplomacy found no traction in the brutish new dialogue that ran roughshod over his patrician decency. James Byrnes, the man many saw as president-in-waiting, had a clearer grasp of the forces that engulfed them, and shared Churchill's glimpse of the coming darkness. Indeed, Byrnes already saw the dispute with the Soviet Union in military terms: 'Because of Russia's potential developments and strong army, [Byrnes] thinks [the] US should stay well-armed,' his private secretary observed.

•

While the Big Three wrestled over the future of Europe, fighting in the Pacific approached the shores of Japan. Here the Allies faced an enemy who, it appeared, was willing to fight to the last man, woman and child. As Joseph Grew, US ambassador to Japan for 10 years, had warned in 1943: 'I know Japan … I know the Japanese intimately. The Japanese will not crack. They will not crack morally or physically or economically, even when eventual defeat stares them in the face … Only by utter physical destruction or utter exhaustion of their men and materials can they be defeated.' The surprise attack on Pearl Harbor; the brutality of Japan's southern advance; the torture and butchery of prisoners and locals; the enslavement of nations – all alerted the Allies to a new kind of foe, and a new kind of war. They were not fighting an opponent who hoped to live – even the Germans surrendered when overwhelmed; nor an enemy who observed any recognisable moral or physical constraints; but rather a kind of unnatural spirit that seemed to glorify cruelty and death. That, at least, was how the Allies viewed 'the nips' – from the soldier at the frontline to the people and political leaders at home.

By early 1945 a darker vision of the Pacific enemy prevailed. Washington and London had gathered evidence of Japanese atrocities that portrayed an enemy intent not only on torturing and murdering prisoners and civilians – in contravention of all recognised rules of warfare (the Japanese had refused to sign the Geneva Convention of 1929, which prescribed the humane treatment of prisoners); but one that had adopted such brutality, in the name of racial conquest, as a policy of war. Allied war planners were aware of the facts, for example, of the 1937 'Rape of Nanking', during which the Japanese butchered tens of thousands of Chinese civilians and raped more than 20,000 women. More recent Japanese atrocities involved American soldiers: on the Bataan Death March, for example, 2330 American and 7000 Filipino prisoners died of starvation, sickness, torture and execution after General Douglas MacArthur's forces surrendered to the Japanese in the Philippines on 9 April 1942. 'To show [the prisoners] mercy is to prolong the war,' was how the *Japan Times* justified the general treatment of prisoners at the time.

They showed little mercy to the very end, as countless examples demonstrated: in Rabaul in 1942 Japanese troops tied 160 Australian prisoners to palm trees and bayoneted them to death, as practice, placing a sign, 'It took them a long time to die', beside the bodies; in Palawan, in December 1944, the Japanese commander ordered the elimination of 150 American prisoners, who were drenched in petrol, crammed into air shelters, and ignited; those who tried to escape were shot. Washington and London were aware of 'Unit 731', Japan's biological warfare unit, which was developing a 'bacillus' bomb that would spread lethal bacteria and poison enemy food and water supplies. The Imperial Army had used biological weapons, containing typhoid and cholera bacteria, against Chinese cities in 1942. In late 1943, Unit 731 reportedly planned to spread lethal bacteria over Burma, India, Australia and New Guinea; and in 1944, to drop biological weapons on the Americans at the Philippines and Saipan.* The Japanese forces' proposed 'cholera strategy', according to enemy documents captured on Luzon in March 1945, recommended spraying bacterial solutions by aeroplane; dropping bombs containing bacteria; dropping infected insects, animals and animal tissue; and leaving pathogenic organisms behind while retreating.†

Evidence of Japanese war crimes implanted in Allied minds a cold and unyielding hatred, which intensified the sense that they were fighting a retributive war – not only against the Japanese armed forces, but against the Japanese people. This thinking permeated the highest levels of Allied command. In Admiral Bill Halsey's eyes, the Japanese were 'bestial apes', a commonly held view; to the Australian commander, General Sir Thomas Blamey, the Japanese were 'not normal human beings' but 'something primitive': 'Our troops have the right view of Japs,' he said. 'They regard them as vermin.'

Indeed, American, British and Australian servicemen were trained to think of the Japanese as 'bloody little yellow swine', 'semi-educated baboons' and 'filthy monkeys' – some of the less vehement epithets used in troop training. Marines went into battle with the words 'Rodent Exterminator' stencilled on their helmets. *Leatherneck*, the marines' magazine, cast the enemy as a species of lice, *Louseous Japanicus*, that had reached epidemic levels: 'Before a complete cure may be effected the origin of the plague, the breeding grounds around the Tokyo area, must be completely annihilated'. This issue appeared

---

* The Allies would not learn the extent of Japan's biological warfare capability – that it planned the mass production of cholera, typhoid and paratyphoidal bacilli weapons, and had conducted experiments on human beings, mostly Chinese – until after the war.

† The Allies were not yet aware of the extent of the Imperial forces' inhumanity. Some of the grisliest Japanese crimes – the horrific torture of captured B-29 crews, medical experiments on live prisoners, the Sandakan death marches – did not emerge until after the war.

on 9 March 1945, the night of the firebombing of Tokyo. Frontline soldiers did not need to be told to hate 'the nip'. The evidence of their eyes decided the servicemen's feelings: the sight of their comrades' mutilated bodies and the suicidal fury of Japanese troops. In this sense, the soldiers' hatred was emotional – and understandable.

At home, the media routinely portrayed the Japanese as beneath contempt: cunning little rats; disfigured, slant-eyed freaks; a simian invasion, and so on. To the press, this was a racial war in all but name: 'In Europe,' wrote Ernie Pyle, the GIs' favourite war correspondent, 'we felt that our enemies, horrible and deadly as they were, were still people. But out here I soon gathered that the Japanese were looked upon as something subhuman or repulsive; the way some people feel about cockroaches or mice.' Upon seeing Japanese prisoners, Pyle wrote: 'They were wrestling and laughing and talking just like normal human beings. And yet they gave me the creeps and I wanted a mental bath after looking at them.' Serious magazines such as *Science Digest* and *Time* soberly examined whether the Japanese people were in some way genetically inferior: 'Why Americans Hate Japs More than Nazis' ran one headline in *Science Digest*. Hollywood never cast a good Japanese, but often a good German.

Western governments were ready to exploit and popularise the hatred of the Japanese. The Germans were a bit 'like us' – if deceived by an evil doctrine – but 'there was no such thing as a good Japanese; Japanese were all evil beyond redemption', was a typical view. Washington legislated to intern Japanese Americans, though not German or Italian ones, 'giving official imprimatur to the designation of the Japanese as a racial enemy'.

Nor did Allied leaders attempt to distinguish the Japanese armed forces from the Japanese people, and easily conflated the loathing of the military regime with a general contempt for the race. Roosevelt himself was not above making crude racist jokes about the Japanese, and commissioned a study of the 'scientific' evidence of the inferiority of the 'Asiatic' races and the effects of racial crossing, in which

Smithsonian Professor Hrdlicka remarked that the Japanese were 'utterly egotistic, tricky and ruthless'. Churchill, a grand old white supremacist, wanted those 'yellow dwarf slaves' dead in great numbers, as soon as possible; he never forgave them for humiliating Britain at Singapore. Australia's Labor government outdid them all, introducing a formal policy of racial hatred as an instrument of war, which broadcast advertisements that ended, 'We always did despise them anyhow'. A lone voice opposed the use of hatred as war policy: Robert Menzies, Australia's conservative Opposition leader, in one of his most dignified speeches, attacked this hysterical demonisation of a whole people.

•

Ordinary Japanese people were ignorant of, or wifully blind to, the atrocities being committed in their name. The truth of what their soldiers had done at Bataan and on dozens of coral atolls throughout the Pacific went unreported. The torture and murder of prisoners was not a feature of Japan's prison camps in previous wars; during the Russo–Japanese and Great War, the Japanese had treated their Russian and German prisoners with care and dignity.

The people, however, heard little from the frontline except news of real or imagined victories. In the winter of 1944–45, most Japanese refused to believe the portents of doom that filtered through after the American landing on the Philippines. They were too thoroughly immersed in the myth of Japanese supremacy to contemplate defeat; surrender was unthinkable.

A series of spectacular military triumphs had persuaded many ordinary Japanese of their sacred destiny – to rule the world. By 1945 this notion relied on a mystical faith in Japanese 'spirit', the residual delusion of four decades of unbeaten conquest. In 1894, the Meiji Emperor looked out from his headquarters in Hiroshima, the point of his troops' embarkation and triumphal return, flushed with pride after victory in the first modern war with China. Greater laurels awaited the armies of Nippon: only the fall of Singapore in 1942 would imbue the Imperial name with greater reverence than Japan's defeat of Russia in 1904–05. The astonished world witnessed in those offensives an undeveloped Asian nation, scarcely freed from the shackles of feudalism, crush the armies of the Tsar and the British Empire. Europe and America strained to comprehend how this little archipelago, so recently 'opened' to the West, managed to wipe out the Russians at Port Arthur, then deemed the world's most impregnable fortress; inflict 100,000 Russian casualties at the battle of Mukden in March 1905; and sink Russia's Baltic fleet, which had sailed halfway around the world to meet its dismal end on the seabed of the Tsushima Strait. To Russian anger, Japan seized under the Treaty of Portsmouth in 1905 a lease on the southern Liaotung – later the Kwantung – Peninsula; control of Russian railroads and other assets in Manchuria; a claim on the southern half of the island of Sakhalin; control of the strategically important cities of Dairen and Port Arthur; and a virtual protectorate over Korea – the 'dagger' pointing at the heart of Japan – which Tokyo formally annexed in 1910 as a bulwark against future Russian aggression. The wounded Russian bear crawled back into her cave; her pride never forgave the Japanese for the loss of so much blood and wealth. Indeed, Moscow's humiliation long outlived the Russian Revolution, and the new Soviet empire hankered for revenge.*

The defeat of Russia emboldened the Japanese to embrace a policy of conquest. The country's acute shortage of raw materials also impelled it to covet an Asian empire that would supplant the British, French and Dutch colonies in the Pacific. The wealth of the British Empire had not gone unnoticed in Tokyo. To this end Japan's rulers – in league with the armed forces – precipitated the occupation of Manchuria in

---

* In the short term Russia and Japan soothed their animosity with a series of agreements that divided Manchuria into 'spheres of influence', the south going to Japan and the north to Russia, separated by a military demarcation line; in 1916 the two countries signed a pact to support each other if a third power threatened either – a direct affront to America's plan to broaden trade with Manchuria. Russia stuck grudgingly to this deal, staying silent when Japan attempted to prise concessions from China with the 21 Demands, the aggressive intent of which was not lost on America and Britain.

1931 that led to the creation of the puppet state Manchukuo, infuriating Russia; withdrew sulkily from the League of Nations; provoked the Marco Polo Bridge Incident of 1937 in order to justify the full invasion of China; joined fellow pariahs Germany and Italy in the Tripartite Pact, signed on 27 September 1940; and occupied French Indochina, triggering US trade sanctions and the likelihood of war with America. In October 1937 President Roosevelt had demanded the quarantining of Japan and Germany for spreading an 'epidemic of world lawlessness'; in December 1941, without warning, Japan attacked Pearl Harbor.

Throughout Japan's military expansion, the Imperial forces claimed to be acting in the Emperor's name, or with the Emperor's tacit approval. Since the 1920s, the Japanese people had been taught to believe in the policy of military expansion as the divine right of Nippon, an expression of the Imperial Will. In the 1930s, Tokyo's newly minted propagandists dusted down the ancient idea of the Emperor's divinity. *The Essence of the Kokutai* (the Imperial state), published in 1937 by the Thought Bureau of the Ministry of Education, described the Emperor as a deity in whom the blood of all Japan ran, back to Jimmu and the Sun Goddess. 'Our country is a divine country,' stressed *The Essence*, 'governed by an Emperor who is a deity incarnate.' Belief in the *Kokutai* became orthodoxy.*

Hirohito, accordingly, despite his diminutive appearance, shrill voice and spectacles, embodied the power of the sun, 'the eternal essence of his subjects and the imperial land'. He existed at the heart of Japanese identity. The people worshipped him as *Tenno Heika*, the 'Son of Heaven', and a divine monarch. Their adoration of the Emperor cannot be understated: killing or removing him dismembered the body and soul of the nation; the rough equivalent of the crucifixion of Christ.

---

* The roots of this belief may be traced to the Meiji era, which promoted, using the tools of modern propaganda, a new version of Shinto that drew together earlier rituals into an 'ethical foundation' for the nation. An amalgam of mythology and expediency, the new state religion made Emperor worship 'compulsory and universal'.

While he had no official policy-making role, Hirohito held the title of Supreme Commander of the Imperial forces. Often he was the dupe of the commanders, who used his name to justify aggression. Until 1944, Hirohito approved in silence, or through courtiers, or other codified channels, Japan's policy of military expansion. He excused the elite Kwantung Army's crimes in Manchuria as 'excesses'; he approved the invasion of China; he formally ordered the capture of Nanking; and he frequently exhorted the troops to rise to the challenge. Often Hirohito gave the green light by saying nothing, or cued action with the faintest approbation: on the eve of Pearl Harbor, for example, he pointedly did not ask newly appointed Prime Minister Hideki Tojo to attempt to heal relations with Britain and America, as he had asked Tojo's predecessors. Or he simply failed to check aggressive interpretations of his words: when Tojo fell, in 1944, he pressed Japan's new leaders to continue to prosecute the war, which they interpreted as Imperial orders to destroy America and Britain.*

•

---

* Hirohito had various motives for supporting the militarists, chiefly economic. On his accession in 1926 (the dawn of Showa, the era of 'enlightened harmony'), he inherited a dysfunctional economy on the verge of collapse. The Great Depression accelerated the process. Conquest seemed the only hope of saving the Japanese economy and feeding his people. He thus promoted the Asian land grab that would deliver vital supplies of oil, food and coal. Another reason was fear for the lives of the Imperial Household. Military officers threatened anyone who questioned the Emperor's divine right to rule. The army, in particular, claimed the Emperor as its guiding spirit, regardless of what the Emperor said or did. In coup after coup, these self-styled 'soldiers of the gods' (*shimpeitai*) claimed to be acting in the Emperor's name, a delusion that culminated in the slaughter in February 1936 of the Lord Keeper of the Privy Seal, the Finance Minister and other high officials whom the soldiers accused of obstructing Japan's Imperial destiny. The perpetrators were shot, but several 'coup leaders' – usually hotheaded junior officers – got off lightly because their intentions were judged 'patriotic' or 'sincere', however violent their methods. Such 'sincerity' won public and judicial sympathy, and the army's lower ranks continued to rampage with impunity, dispensing summary justice with the sword fetish of their samurai forebears. Indeed, some came close to endangering the Imperial court; not that any would dare lay a finger on the Emperor's person. If Hirohito blinked, if he challenged the militarists' policy of conquest, junior officers were prone to attack or murder his advisers for 'poisoning' the Imperial mind. In this sense, Hirohito was, at least in part, a captive of the armed forces; thus, for a range of cultural reasons he often said what they wished to hear.

The Japanese regime promoted love of Emperor in tandem with hatred of the West. The people were exhorted to hate the enemy, and hate him to death. The Americans and their allies were cowards and monsters, morally depraved and barbaric, who sent skulls of Japanese boys home to America as souvenirs (as several Japanese newspapers claimed). Anti-American articles and posters appeared every day: 'If one considers the atrocities which [the Americans] have committed against the American Indians, the Negroes and the Chinese,' fumed the respectable economic newspaper *Nihon Sangyo Keizai*, on 5 August 1944, 'one is amazed at their presumption in wearing the mask of civilisation.'

'The demons and beasts are desperate in their all-out counter-offensive!' stated a leaflet issued by Japan Steel's Hiroshima Plant. 'We will wipe them out by increasing production! Now is the time to send letters of encouragement to our soldiers!'

Japanese children were prime targets of this propaganda. Throughout the 1940s, school posters urged children to 'Kill the American Devils'; boys and girls were instructed to attack images of Churchill and Roosevelt. The message found its mark. In February 1943 teachers asked schoolchildren in Aomori Prefecture to suggest ways of disposing of some 12,000 'blue-eyed sleeping dolls' donated to Japanese schools years earlier by American charities. Of 336 children in one school, 133 chose to burn the dolls; 89, to dismember them; 44 to send them back to America; 33 to throw them into the sea; 31 to exhibit and torture them; and five to drape them with a white flag. One child suggested they be used as models for identifying American spies.

The Japanese regime keenly implemented this program of demonisation on the country. In 1942, for example, an army PR man rebuked on national radio a Tokyo woman who had exclaimed, 'Poor fellows!' at a ragged line of American prisoners of war passing in the street; in April 1943 the army severely criticised a newspaper that had dared mention 'the existence of positive elements in the American heritage'. And older Japanese found it hard to shed their admiration for European and American culture and technology, which they had been taught to respect, even emulate, in earlier benign times.

In the 1940s, 'Thought Prosecutors' roamed the cities under the control of the Justice Ministry, ferreting out 'dangerous thinkers' – pacifists, leftists, journalists and Koreans. Meanwhile, Special Higher Police (*tokko ka*), deployed under the Peace Preservation Law, monitored the mind as well as the voice of Japan. That meant throttling the expression of both. In 1944, a *Mainichi* reporter thoughtfully asked in an article, 'Can Japan Defeat America with Bamboo Spears?' A furious Tojo had the miscreant dispatched to China. Persistent dissidents were tortured. But few challenged the censorship laws. Between 1928 and 1945, only 5000 people were found guilty of violating the Peace Preservation Law. In 1934, the peak year, 14,822 were arrested and 1285, prosecuted; in 1943, those figures were 159 and 52 respectively. Only two were sentenced to death: Richard Sorge, the Soviet master spy, and his accomplice Hotsumi Ozaki, for espionage.

By 1945, most Japanese had become compliant self-censurers who rallied around the war effort. State-approved intellectuals applauded the war as a sacred cause against 'Anglo-Saxon exploitation'. Poets eagerly volunteered to recite their haiku in factories and at the front. Newspaper editors exulted in news of victory and distorted evidence of looming defeat.

By suppressing the most obvious truth – that Japan was losing the war; yoking a brutal version of *Bushido*, the samurai code, to Japan's 'divine destiny'; and imposing a series of nihilistic slogans ('one hundred million hearts beating as one' – *ichioku isshin*; 'The eight corners of the world under one roof' – *hakko ichiu*) on the nation, Japan's more fanatical commanders hoped to lead the people in an act of national *seppuku* (ritual suicide), a blood sacrifice to the Emperor, rather than surrender. As 1945 opened, the regime seemed to be succeeding in binding the people to a mass-suicide pact: most Japanese showed a willingness to fight to the death, with bamboo spears, if need be.

Not all Japanese were completely fooled. Doubts stirred in the minds of the better-informed, or more intelligent. After the fall of Saipan, in late 1944, many people privately began to question whether they were winning the war. The struggle to survive seemed grossly at odds with the glad tidings of government propaganda. 'If we're winning the war why have we so little food?' people reasonably wondered. Government agents requisitioned all the rice they could carry, leaving middle-class families hungry, and poorer ones, starving. 'We were always told, "we're winning the war, we're winning the war",' remembers a then 20-year-old maths teacher. 'But everything was rationed. The shops were all closed.'

'We heard only good news,' said Kiyomi Igura, a young nurse from Nagasaki in 1944. The city held a lantern festival at news of the fall of Singapore and 'everybody walked about holding lanterns'. That was February 1942; in the winter of 1945, Nagasaki 'began to have doubts', she said, 'but no-one could bring themselves to say that Japan might lose.' Victory was assured, despite the food shortage: 'The mood of the time was very much that Japan would definitely win the war.'

A few brave citizens dared to criticise the government and challenge the Peace Preservation Law. Some broke the censorship rule that forbade the reading of pamphlets dropped from American planes; they read of terrible losses on distant battlefields – in New Guinea, Burma and the Philippines. The people grew dimly conscious of a coming trauma, of a creeping realisation that 'we were all going to be killed'.

Those are the words of a man who, at great personal risk, committed his thoughts to paper. The liberal historian Kiyoshi Kiyosawa wrote a 'diary of darkness' in which he charted the moral and spiritual degradation of Imperial Japan. A well-connected, cultured man, Kiyosawa struggled in vain to reconcile the obscenity of Japanese military rule with his rectitude and intellect. His was a voice of sanity in a world of madness: 'When I listen to the morning radio,' he wrote on 15 December 1943, 'I find it completely insulting to the intelligence. There is the attempt to make the entire nation listen to stuff that has descended to this ... Even if I do listen, I am enraged.'

Kiyosawa's calm, almost innocuous, words render his barbs sharper. He noted the effect of Tokyo's policies on ordinary people: soaring inflation, parasitic black marketers, cheated farmers and acute malnutrition ('Not one orange appears in the shops') were not incidental hardships, the tolerable sacrifices of war. They were signs of a nation on the brink of military and spiritual collapse, the miscarriage of bad leadership and public stupidity: '... the media world is still getting its ideas from divine inspiration. Is it possible to win the war in this way?'

He took a scythe to the regime:

I would like to mow down
the thick weeds of silly ideas and politics
which thickly surround us ...

The kamikazes were not heroic; they were blind, lost young men, a terrible waste: 'Outstanding youths are on the brink of ... complete destruction.'

'Spirit' alone would not win the war; the 'ghosts of our fathers' were dead: 'It goes without saying that the intellectual background of the Pacific War is based on extreme feudal ideas. The celebrity of ... the "Forty-seven Ronin" [the fable of 47 Hiroshima samurai who disembowelled themselves for their Lord] has never been as intense as it is at present.'

The US Navy blockade ensured Japan's steady exhaustion of raw materials. The shops' shelves were bare. 'Japan has finally come to an internal stalemate,' wrote Kiyosawa. The empty holes that had been shops in the Ginza, Tokyo's retail district, looked 'as if teeth had been extracted'. By August 1944, the 'Greater East Asian War' had 'robbed' Japan of all kinds of iron. The railings on bridges, the fences of cemeteries, and even the bronze statues of the Sojidera, a Zen Buddhist temple in Yokohama, no longer existed. All had been requisitioned, removed and melted down. Hiroshima, Nagasaki and other Japanese cities were similarly denuded; their temples and churches stripped of platinum, gold and other metals, to be turned into weapons.

Kiyosawa dared to ask for whom the bell tolls. It tolled for Japan, for Tojo and his cabinet, whom he judged the most stupid on record and who were forced to resign in July 1944, after the loss of Saipan. Tojo's resignation statement on 19 July disgusted him: 'I deeply regret the anxiety that [this loss] had caused to His Majesty ... But these developments [give us] the opportunity to smash the enemy and win the war. The time for the decisive battle has arrived.' Kiyosawa held Tojo and his cabinet personally responsible for 'plunging Japan into misery'. In January 1945 Kiyosawa lost faith in Japanese spirit and descended into his own dark place: 'In Japan there is a philosophy of death and no philosophy of life.'*

•

A short-lived duumvirate of General Kuniaki Koiso and Admiral Mitsumasa Yonai succeeded Tojo and continued, at least in public, to drum up enthusiasm for the war. In private, Yonai and other relatively moderate cabinet members dared to discuss how they might terminate it. By the end of the year most high officials knew the war was unwinnable.† Some ministers secretly contemplated open surrender.

---

* Four months later, Kiyosawa died of pneumonia brought on by malnutrition; he was 55. His *Diary of Darkness* was not published until 1948, when it became a Japanese classic. In 1945 his warning sat unread, the ghost of truth.

† There had been many internal warnings of the coming defeat: for example, cabinet ministers had read Admiral Sokichi Takagi's prescient analysis of February 1944, which concluded that Japan 'could not possibly win the war; therefore she must seek a compromise peace'; army and navy chiefs had seen Colonel Makoto Matsutani's 'Measures for the Termination of the Greater East Asian War' of spring 1944, which argued that, facing the choice of surrender or national suicide, Japan should press only for the preservation of the Emperor and yield all else (Tojo had Matsutani sent to China for his efforts).

In late 1944 Foreign Minister Shigenori Togo dared inform Marquis Koichi Kido, Lord Keeper of the Privy Seal and the Emperor's closest adviser, that 'unconditional surrender, may be unavoidable'. Kido, among the best informed of the elite, had earlier confided in his diary that 1944 looked 'precarious', and drew up a 'peace proposal', which he set aside for extraction at the appropriate time.

In early 1945 these men came together as a loose gathering that would be called the 'peace party', a clandestine group of top officials who secretly believed that Japan had lost the war; that a way must be found to open negotiations with America; and that any peace deal must preserve the Imperial line. They deeply distrusted the armed forces, and lived in constant fear of assassination. Foreign minister Togo, the most consistent 'dove', became their unofficial leader. With his round glasses, caterpillar moustache and thoughtful demeanour, Togo fitted the traditional Japanese mould of the intellectual public servant. Born in 1882, in Satsuma, to a samurai family, he had risen rapidly through the foreign service to become ambassador to Germany in 1938. Having lived in 'the west', Togo felt he understood it. He had also warned against militarism. His essay, 'A Foreign Policy for Japan Following Withdrawal from the League of Nations', of 27 March 1933, had urged Japan to consolidate in Manchuria and advance no further into Asia: 'It is essential that … we avoid conflicts with other countries, unless conflict be forced upon us,' Togo had cautioned. 'The basic policy towards the United States should be … to prevent war.' His quiet demurral fell like a snowflake on a gathering firestorm.

Twelve years later, in January 1945, Togo, together with Kido, Yonai and other less consistent moderates, found themselves contemplating the imminent destruction of Japan. Circumstances forced them to consider not *whether*, but *how* to surrender. Their long debates hinged on three questions: (1) How to persuade the Japanese forces to lay down their weapons? (2) What were the most favourable peace terms Japan could hope for? (3) Would the Emperor, if shown the gravity of the situation, be prepared 'to wager the future of his throne' and intervene to end the war? They knew that the armed forces 'could only be controlled through the Emperor, whose influence could deal a crushing blow to any would-be opposition' to ending the conflict.

Meanwhile, the American forces drew closer to the Japanese main islands. In early 1945 the Tokyo regime deemed the cities on the southern island of Kyushu the frontline of a planned American invasion. The inhabitants were warned to prepare themselves. Every day the people scanned the horizon and strained their ears for the sight and sound of enemy aircraft. The airfields of Saipan, now in the hands of the US, were within striking distance of Tokyo, and from December 1944 there were nightly air raids on the capital, Osaka and other cities.

The moderates ensured the Imperial Household was kept aware of the rising threat to the homeland. The Emperor heeded the warnings, and started to distance himself from the militarists. From late 1944 Hirohito sensed the war was lost and requested an assessment of the military outlook from Prince Fumimaro Konoe, a member of one of Japan's most prestigious families. Konoe, an intelligent observer of events, had, like Togo, advised against going to war with America. On 14 February 1945, as American forces invaded the Philippines and the Big Three met at Yalta, Konoe delivered his verdict: 'I regret to say,' he told the Emperor, 'that Japan's defeat is inevitable.'

CHAPTER 2

TWO CITIES

Annihilate America and England, one, two, three!
Radio chant during morning exercises in Hiroshima and other Japanese cities

My friends and I thought that Japanese Christians had ties with
the British and Americans and that was why Nagasaki had been
spared bombing until then.
Teruo Ideguchi, Nagasaki schoolboy, aged 15 (in early 1945)

THE TWO CHILDREN KNELT ON the *tatami* mat by the fire with bowed heads and outstretched hands, a girl aged five and a boy 10. If they dropped a grain of rice grandpa got 'really mad', the girl remembered. Sometimes he hit them with his bamboo pipe. The children were more hungry than afraid and waited every day, like two Japanese Oliver Twists, for their daily rice ration. This morning grandpa looked down on their little heads, monk-like in the wartime haircuts, and gravely placed a rice ball in each palm. 'We were beaten many times,' the girl said.

Their grandfather, Zenchiki Hiraki, was a tall, lean, stooped man who shuffled about on his cane in a cloud of tobacco smoke. He had lived here, on a small farm near Hiroshima, for most of his 80 years. In 1945 little had changed since the Meiji Restoration of 1868. The earthen kitchen, the drenched fields, the grinding poverty, were the same; and still no gas or running water. And what use was electricity in the nightly blackouts? As then, a coal brazier heated the house; a deep well supplied water. His grandchildren gathered firewood in the forests and bathed in the neighbours' germ-filled wooden tub, queuing in freezing winters, just as he had done.

Every morning the old man exercised, inspected his fields and clapped awake the spirits in the little family shrine. He would gaze at the little altar in silence. On family occasions he appeared in his best kimono, emblazoned with the family crest. His dignified appearance disguised a mean character. *Zenchiki* means 'very good'; when he died, his neighbours were quick to say, 'Zenchiki was not *zenchiki* at all'. He displayed in his intermittent rage and submission a cast of mind common among the Japanese peasantry: in public, he prostrated himself before whoever happened to control his patch of *padi*; at home, he ruled with the iron fist of a miniature *daimyo* (warlord).

Zenchiki was a bitter man; fate had cheated him. His ancestors had been prominent wealthy people, by village standards. They called the village bridge Hiraki Bridge; the village ruins, Hiraki Ruins; the village itself, Hiraki … Did that not count for something? Had he not upheld the family name? Yet here he was, the ailing leaseholder of the shadow of a farm near Hiroshima that had flourished during the Meiji era until a succession of crushing taxes and poor harvests, culminating in the great famine of 1934, left it in this gnarled and parlous state. Zenchiki leased four *tan* – barely an acre – growing mostly rice, with small plots of barley, potatoes, *nashi* pears and *daikon* radishes; 'every corner,' noted one witness of the region's farms, 'diked and leveled off, even though the growing surface is less than a man's shirt; every field soaked with manure and worked and reworked … nothing thrown away, nothing let go, nothing wasted.' Unable to afford workers, the old man relied on relatives and other villagers – nine or ten families. When the harvest fell short of the government's wartime rice quota, which it often did, he pleaded with neighbours to make up the shortfall, an ordeal he regarded as less humiliating than failing to meet the wartime food demands of Hiroshima and the nation. The lifeblood of the war effort was oil and rice – rice that may, for all he knew, be destined for his son's battalion.

Pre-war Showa – in the early reign of Hirohito – did little to ease the lot of ordinary Japanese such as the Hiraki family. At the very top of society were 19 samurai families – the future *zaibatsu*, Japan's family-controlled industrial empires – each of which received at least one million yen annually. At the bottom some 2,232,000 families scraped together a living on a yearly income of about 200 yen. In 1930, 84 per cent of Japanese held half the nation's household income. The decade after the Great Depression saw the collapse of agricultural prices and the silk industry. The divide between the 'two Japans' to which US ambassador Joseph Grew had referred in the late 1930s, deepened on several levels: soldier versus civilian, farmer versus city, peasant versus landlord – manifestations of a grossly inequitable society whose leaders saw fit to blame foreigners, communists, colonialists, Chinese or Americans for their country's peculiarly grim economic lot. If a kind of social harmony existed, in the sense of *giri* – the code of reciprocal social duty – it proved a brittle veneer in times of poor harvests and excessive taxation when furious farmers begged or rioted. From the terrible famine of 1787 to the utter destitution portrayed in *The Soil*, Takashi Nagatsuka's novel of 1910 based on true accounts, little had changed in the lives of the Japanese poor: 'Once the farmers had paid the rents … they were lucky to have enough left over to sustain them through the winter,' he wrote. During the worst famines, the peasants starved; and cannibalism and infanticide were common. Even as late as the 1930s, starving peasants killed unwanted children – another mouth to feed – echoing the practice of 100 years earlier when infanticide was 'widespread in the Inland Sea region', according to social scientist Nobuhiro Sato, '… but there the children are killed before their birth, thus making it appear that there is no infanticide'. Sato advocated military expansion to counter Japan's destitution. Leaders in Tokyo agreed. The acute shortage of raw materials, chiefly oil and coal, must be relieved by force, they decreed; a repeat of the terrible famine of 1934 was intolerable. Such were the excuses for the subjugation of Manchuria and China, which triggered US trade sanctions; such were the arguments for the invasion of Indochina – fed to a people moulded by force and propaganda into complicit pawns of the Pacific War effort.

•

The bright, expectant faces of the schoolboys and girls who spilled into Hiroshima Station in early 1945 belied this grim reality and lent a little hope to the city's grave wartime brow. From here they took streetcars or walked to school, some past Hijiyama, a hill just south of the station near the Red Cross Hospital. The summit of Hijiyama offered a complete view of this floating city of some 300,000 people: from here, in winter, Hiroshima looked leaden blue; in spring the cherry blossom fell in flakes, dusting Hijiyama in 'a drift of pink snow'; in summer the 'islands' and bridges swayed in the mirage of warm air that hovered over the floodplain like a shroud.

Hiroshiman schoolchildren usually crossed a river on their way to school: would anyone dare leap off a bridge? In early 1945, about five bridges spanned the Enko River, one of seven fingers of the Ota River, on the easterly approach to the city. The Ota Delta partitions the city; the centre of town lay, as it does today, at the junction of the Honkawa and Motoyasu tributaries where the Aioi Bridge forms a T-shape clearly visible from the air (see map, picture section 2). Most schools stood within a kilometre's radius of this riverine confluence – the biggest being the Fukuromachi National Elementary, the Honkawa National Elementary and the Hiroshima First Prefectural Girls' School. In 1945 they were the daily destinations of thousands of children. In January, the threat of air raids had not yet disrupted lessons; nor were students over 12 aware that within a few months a new law would compel them to work in the city centres as 'mobilised labour'.* For now, they were still permitted to swim at high tide and kick balls around at low tide on the banks of the Motoyasu.

The people of Hiroshima wore three badges of pride: the quality of their schools; the purity of their water; and the city's historic role as a castle town and military barracks. Water, as a trade route, natural form of protection and food source – as well as flood threat – soaked the city's history. Locals spoke fondly of their 'water metropolis'. The Ota River's water was reputed to be the purest in Japan, cleansed as it flowed south through the alluvial sieve of sand and stones. Later in the war, in the absence of *sake*, soldiers' families saw nothing improper in raising cups of local water to their sons and husbands. At any time of year dinghies, little flat craft and broader vessels plied the Ota, bumping in and out of the piers, under the many bridges, and around the 'islands' between the tributaries. These white stretches of alluvium gave Hiroshima its name: in 1591 Terumoto Mori, the *daimyo* who ruled the shores of the Inland Sea, established a foothold on the Ota Delta. Here he built a new fiefdom. One of his first acts was to rename the five villages huddled by the river, 'Hiroshima': 'hiro' meaning 'wide' (after his kinsman, Hiromoto); and 'shima', meaning 'island' (after his retainer, Fukushima Masanori, who would oversee the building of his castle): hence, the 'City of Wide Islands'.

The heart of town, the Nakajima Honmachi District, crowded onto the long 'island' where the Honkawa and Motoyasu tributaries meet (the site of today's Peace Park). Pre-war Nakajima was the city's entertainment and business district, packed with shops, tea rooms, temples, shrines, geisha houses, Kabuki and Noh theatres, the Honmachi shopping arcade, a public swimming pool; nearby were the Taishoya Kimono Shop and Sumitomo Bank. Festivals were regularly held there: in 1913 Nakajima 'appears to be lively', wrote a prudish military Hiroshiman commander, 'but the people are full of frivolity and fraud ... blinded by immediate profits'. He blamed their decadence

---

* In April 1938 the Diet (parliament) enacted the National Mobilisation Law, which empowered the government to control manpower, production, prices and wages; in 1944 and 1945 it was extended to all children over 12.

on the influence of Western customs. Such baleful foreign influences were less apparent in 1945, with the bars and businesses shut, the shelves empty and and the economy stalled.

Nearby, on the eastern bank of the Motoyasu River, just beside the 'T' formed by the Aioi Bridge, rose a grand 19th-century European structure of red brick and stone with a copper-green dome. Designed by Czech architect Jan Letzel, the city's exhibition centre opened in 1915. Hiroshimans took the building to heart, as a symbol of progress and enlightenment. In 1933 they renamed it Hiroshima Prefectural Industrial Promotion Hall, a proscenium for local producers. In 1945 it contained government offices. Nearby stood the Shima Hospital; a little beyond, the new Fukuya Department Store, painted dark brown in 1945 and virtually empty.

Clinging to the edges of Nakajima Honmachi were thousands of little family homes made of local cedar, paper screens and *tatami* mats, all grouped into wards bound by narrow streets in such profusion the houses seemed to tumble down the banks and into the river. The residents' lives whirred with clockwork precision: the war effort prescribed a rigid daily routine starting with morning exercise at 6am (often around shared radios to the chant 'Annihilate America and England, one, two, three') and prayer before the household shrine. Then to work: tending plots of sweet potatoes and radishes, gathering firewood, digging bomb shelters, clearing fire lanes, and loading buckets of human excrement onto fertiliser carts. On weekends neighbourhood associations (*tonarigumi* or *rimpohan*) organised self-defence and combat classes.

In 1945 Hiroshima faced the same, severe conditions as any other Japanese city. The war grimly impinged on how people ate and dressed. In January the average daily ration fell below 1500 calories, 65 per cent of the minimum required in Japan to sustain basic health.* Cases of night blindness due to malnutrition were common. Thoughts of

---

* The Ministry of Health and Welfare's nutritional standard for an adult male doing 'medium hard-labour' was 2400 calories a day and 80 grams of protein, notes Saburo Ienaga.

food occupied every waking hour: one child evacuee remembers 'inhaling tooth powder in order to withstand hunger pangs'. Most people supplemented the paltry ration with sweet potatoes and whatever they could scavenge or steal. In the rural areas, families were forced to beg after the government took the harvest. Mothers went from house to house pleading for salt and bean curd for their children; in time, they resorted to eating lily roots, mulberries, boiled snakes and crabs. As conditions deteriorated in the cities, their residents poured into the countryside in the hope of finding food, driving black market prices for fruit and vegetables to exorbitant levels; on average, the black market charged 4200 per cent more than official prices. The armed forces and political elite were well fed, however: 'Only the fools queue up,' Kiyoshi Kiyosawa wrote. 'Everything goes to the military, the black marketers and the big shots.'

Neither sailor suits beloved of Japanese schoolgirls nor bright kimonos were permitted under the government's 'lifestyle reform' policy. 'Extravagance is the enemy!' placards warned the ostentatious, who persisted in wearing traditional dress. Instead, the people were given coupons to buy drab *monpe*, the baggy grey uniforms of the National Defence Corps, and *geta*, wooden clogs. The poorest – such as Zenchiki's grandchildren – made do with *waraji*, straw sandals. 'Anything with a sense of elegance was forbidden,' said one kimono merchant. 'Kimono sleeves had to be cut short … They were supposed to look more gallant that way.' Hunger conquered female vanity, in any case. Women grimly traded their kimonos for rice balls, ignoring a 1940 ruling that forbade the sale of velvet, chiffon, lace (and other 'Western' fabrics), jewels and silverware. By 1945 the stores were empty; the food coupons useless. Hiroshiman women were forced to wear *monpe* by law and economic necessity: 'If I wasn't wearing my *monpe*,' one woman recalled, 'the military police would come along and give me a warning. It was very strict. We had no freedom at all.'

•

Military installations mingled with Hiroshima's schools, hospitals and theatres. Since 1888, Hiroshima Castle, a moated white tower set in gardens just north of the town centre, had been home to the 5th Division of the Imperial Army and its locally famous 11th Regiment. The Meiji Emperor had made the castle his headquarters during the first war with China, in 1894. In the 1940s, it served as a focal point for local recruits to the Imperial Army whose presence breathed life through the town; at any time, 20,000 to 40,000 reserve troops would parade on the castle's drill grounds prior to their departure for the Gaisenkan, the Hall of Triumphal Return, at the mouth of the Ota. This was the last point on the mainland from which millions of Japanese troops would depart for the killing fields of China, Russia and the Pacific – at least until late 1944, when the US naval blockade terminated Hiroshima's military function. By early 1945, with defeat looming, Hiroshima had lost its critical role as the army's embarkation point.

Hiroshima Castle, the home of the local warlord or *daimyo*, rose as a multi-tiered structure of stone parapets surrounded by a network of moats. An outer ring of palaces and ramparts protected the inner palace (*oku goten*, where the *daimyo* lived) and the immense keep (*donjon*). Dazzling tiled patterns bore the *daimyo*'s logo – in Hiroshima's case, a sacred carp.

In 1619 the shogun appointed Asano Nagakira as the new *daimyo* of the Hiroshima fiefdom, and granted him 420,000 *koku* (one *koku* equalled about five bushels), levied on local farmers, for loyalty in battle. Enriched, Asano and his heirs imposed upon Hiroshima a rough order that persisted for 250 years until the collapse of the shogunate in 1867. Under successive Asano lords, land reclamation and flood controls strengthened the city as the hub of regional power. The main beneficiaries were the samurai clans living in the castle grounds.

The spirit of the samurai persisted in Japan well into the 20th century, and nowhere as strongly as in Hiroshima. Under Asano rule, the city developed a rich samurai tradition, symbolised by the story of the 47 *Ronin*. Every local schoolchild knew by heart the tale of these lordless warriors who, in 1701, avenged the death of their master, committed *seppuku* and were deified as national heroes. The wife and son of their leader, Kuranosuke Oishi, were buried with the Asano family in Hiroshima.

The samurai thrived on war; in peace, they performed no productive work and lived leech-like off the grunt and sweat of the peasant. In Hiroshima, as in other castle towns, a samurai pursued a parasitic existence. Special laws set him apart from the common people; he could kill and maim with impunity. Corrupt samurai applied the privilege on a whim, as though lopping off a hand or leg or head were an entertaining blood sport. The samurai's end swiftly followed the Shoguns' demise. Japanese knights had no place among the military strategists and bowler-hatted clerks of the Meiji era. His beloved swords were no match for gunpowder; his topknots, more absurd than warrior-like.

After the return of Imperial rule, the samurai legend persisted in woodcuts and stories: the 47 *Ronin* of Hiroshima became a source of 'endless plots for plays'. Lesser samurai made little effort to obviate their redundancy; they sloughed off their duties and faded away, estranged warriors in perpetual limbo. Not all fell into drunken, libidinous, provincial obscurity; many retired with honour – or got a job. Some old soldiers managed the transition from warrior to bureaucrat and businessman with ease. The great houses of Mitsui, Mitsubishi, Sumitomo, Yodoya and Mazda (a powerful wartime business in Hiroshima) were borne of samurai families. Others entered politics: in the 1940s, most senior government ministers came from old samurai families. In 1945, as they girded their people for a possible invasion, these grim old men could hardly have descended from a more determined training ground than the warrior tradition of their ancestral past.

•

Open and outward-looking, a subtropical entrepot, and, in some ways, as exotic to the Japanese visitor as the foreigner … such were the attributes of Nagasaki, 'the port of myriad goods and strange objects', according to Sorai Ogyu, an 18th-century Confucian scholar. Green mountains cradle the city's lush harbour, the destination of traders and buccaneers over centuries. Puccini set *Madam Butterfly* here, a 'town of stone roads, mud walls, old temples, cemeteries and giant trees', wrote Japanese novelist Kafu Nagai, visiting from Tokyo in 1911. 'The colours of the forest trees appear fuller and brighter than those of Tokyo. The chirping of the cicadas is also quite different, falling over the town and its forests like a spring shower … the traveller feels that in Nagasaki he is in a place far from Japan …' The *minminzemi* cicada was said 'to chant like a buddhist priest reciting the kyo'.

Long before America's black ships entered Tokyo Bay in 1853, Nagasaki served as Japan's window to the West. In 1640 a Dutch trading party was allowed to stay there after the expulsion of the Spanish and Portuguese. The 'Hollanders' assured their hosts of the relative pliancy of their brand of Christianity, demonstrating their good Protestant faith by firing a few shells at the Japanese Catholics huddled in Hara Castle.

The Dutch aimed to profit, not to proselytise. In return for a financial monopoly, the two dozen or so Dutch traders agreed to confine themselves to the little island of Dejima, 180 by 60 metres, like quarantined animals behind high fences. A stone bridge linked them to the city. Their ships brought Chinese silk, tin, lead, pelts, clocks, mirrors and other curios, in return for gold and silver bullion, copper, jewellery, porcelain and lacquerware. The Dutch made the round trip from Holland to Java and Nagasaki 116 times between 1633 and 1850, when the trade ceased. On their occasional visits to Edo (as Tokyo was then known) the shogun was known to treat them like performing monkeys: 'They made us jump, dance, play gambols and walk together,' wrote Engelbert Kaempfer, a doctor who accompanied the Dutch ship to Edo in 1691. 'Then they made us kiss one another, like man and wife, which the ladies particularly shew'd by their laughter to be well pleas'd with ... After this farce was over we were order'd to take off our cloaks, to come near the skreen one by one ...'

Fascination with and deep suspicion of the West met in Nagasaki. News of the great convulsions of Western power – the French Revolution, the American War of Independence – came ashore like the tidal pulse of a distant explosion. Pernicious ideas, of 'freedom' and 'equality' and the 'rights of man', fluttered to life, and the names of Napoleon and Washington conjured great Western warlords in the Japanese mind. The study of European languages, science and medicine followed: Dutch doctor Philipp F. von Siebold, the 'surgeon-general' of Dejima, was allowed to establish a medical school in Nagasaki in the 1820s, where he taught local doctors. The Scottish merchant Thomas Glover ran a coal and weapons trading empire from the city. These exceptions did not mean the shogunate tolerated Europeans, all of whom seemed to worship the same god, dress similarly, behave in a slovenly manner, write horizontally – and yet claimed to come from different countries. The shogun and his warlords tended to bundle Europeans together as an homogenous evil, the progeny of a single white superpower.

Most were refused entry. In 1825 Edo issued an edict expelling white foreigners on sight, on the grounds of their religion: 'All Southern Barbarians and Westerners, not only the English, worship Christianity, that wicked cult prohibited in our land. Henceforth, whenever a foreign ship is sighted approaching any point on our coast, all persons on hand shall fire and drive it off ... If the foreigners force their way ashore you may capture and incarcerate them ... have no compunctions about firing on [the Dutch] by mistake ...' Only such extreme measures would banish the Europeans who were 'gathering like flies to a bowl of rice'.

Notwithstanding the prohibitions, Nagasaki continued to offer a point of foreign contact with this closed country until 1853, when American ships arrived to force open the door. That year, Commodore Matthew Perry entered Edo (Tokyo) Bay at the head of four dark-hulled vessels bearing 967 troops, on a mission to wring trade concessions from the Japanese on behalf of the American president. The 'opening' of Japan was both humiliating and enlightening: the last Tokugawa shogun conceded Japan's backwardness, at least in technology, and embraced the foreign devil with the slogan, 'Western science, eastern morals'. The edict of Keiki Tokugawa, to drive out all foreigners by 1863, was more a public posture – an example of Japanese sincerity of intent – than a serious policy. The study of 'barbarian' books, the voracious interest in Western culture, the huge trade and diplomatic expeditions that set sail from Nagasaki Harbour – all served the late Tokugawan and early Meiji policy of studying the West the better to challenge, emulate and oppose it. Economic and military power was the new priority; hence the Meiji slogan, 'Enrich the country, strengthen the army' (*fukoku kyohei*) with which Japan entered the 20th century.

•

'Eastern morals' were always seen as tainted in Nagasaki by the presence there of a virulent belief in the Western god. No matter how determinedly Tokyo sought to stamp out Christianity, it resisted centuries of hostility and prevailed in this strange city like a peculiarly stubborn infection. Twelve thousand Catholics lived in Nagasaki in 1945, the largest Christian community in Japan. They worshipped at Urakami Cathedral, then the biggest in Asia, built on the site of former Christian persecution a few kilometres north of the city centre: 33 years in the building, the cathedral accommodated almost 2000 worshippers.

The survival of Japanese Christianity in Nagasaki was a triumph of faith over experience. The city's Catholic history began in the mid-16th century, when Portuguese ships bore the first Jesuit missionaries. French, Spanish and Italians followed. In 1549 Francis Xavier arrived in Japan, and travelled the country he hoped to convert to Christ. The missionaries aimed their arts of suasion at the feudal lords, calculating that the commoners would emulate their superiors' example. The calculation worked. The elite were receptive; the *daimyo* and his lords shared a deep hostility toward 'Buddhist sectarians' – lapsed followers of Buddha – and helped to solder the new faith to the Japanese elite. Later Franciscan, Augustinian and Dominican missionaries preached directly to the common people.

In 1563 the *daimyo* Omura Sumitada and 25 acolytes in the Nagasaki fiefdom were baptised. Among them was Sumikage Jinzaemon, lord of the Nagasaki harbour area; 1500 of his subjects duly followed. They worshipped in the Todos os Santos church, formerly the Shotoku Buddhist temple. About this time, Omura ordered the construction of a port on the bay to enrich his fiefdom. Nagasaki Harbour opened in 1571 and admitted Dutch and French trading vessels. The Jesuits installed a printing press that disseminated Western literature as diverse as Aesop's Fables and Thomas a Kempis' *Imitation of Christ*. Christianity grew so quickly that by the end of the 16th century Catholic converts, according to historian Marius Jansen, 'may have neared 2% of the population [of Japan], a higher percentage than are Christian in Japan today'. The greatest concentration has always lived in Nagasaki.

The backlash began in 1587, under an edict from a taiko, Hideyoshi Toyotomi, which banned 'this pernicious doctrine' from the 'land of the gods'. The next year Edo authorities took direct control of Nagasaki and confiscated all church property. The Christians were driven into hiding. Occasionally they broke into open rebellion and were repressed. Over the ensuing three centuries Japanese Christians were constantly, viciously persecuted. 'No area,' wrote Jansen, 'had been more evangelised than the rugged Kyushu countryside around Nagasaki … No area was more immediately subject to dragnet searches and tortures designed to force repudiation of faith in Christianity.' On 5 February 1597, 26 Christians – six European missionaries, three Japanese Jesuits and 17 Japanese worshippers – were crucified in Nagasaki, 'their bodies left to rot on their crosses'. (All were canonised by Rome in 1862.) The great martyrdom – Great Genna – of 55 Nagasaki priests and laity followed on 10 September 1622. Successive edicts forced Japanese Christians to worship at Buddhist temples.

During the 17th century, the shogunate sought to stamp out Nagasaki's strange sect for good: Christian samurai were banished as *ronin*; captured priests were subject to torture 'so ingenious and fiendish' that six European priests renounced their faith at the hands of the Tokugawa 'Inquisitor'. By 1637 most of the 300,000 converts had been hunted down, tortured, executed or forced to recant. In response, the Japanese *Kirishitan* on the Shimabara Peninsula rose in revolt. The few survivors fought 'with the desperation of people who had nothing to lose'. Besieged within Hara Castle, they soon ran out of food and weapons and the Togukawan forces slaughtered all survivors. The next year the shogun closed the country to foreigners, outlawed the faith and exiled all foreign and mixed-blood missionaries. Those conditions prevailed for more than two centuries.

Yet the faith survived in hiding. Several thousand Japanese Catholics – the *Kakure Kirishitan*, or Hidden Christians – worshipped in caves and safe houses, symbols of the triumph of the missionaries' persistence. The faith's notoriety, perhaps more than its religious precepts, drew the young and romantic. Still, the persecution continued; if captured, the accused were forced to trample on Christian images and icons (the *fumie*) before being tortured, interned or executed. In 1865, the opening of the Oura church built in memory of the 26 martyrs tempted hundreds of Nagasaki's Hidden Christians out of hiding, to the astonishment of French missionary Father Bernard Petitjean. The people had mistaken the new church – built for foreign worshippers since the forced opening to the West – as an official reprieve and dared to pray in public for the first time in 200 years. The government responded with crushing efficiency: the *Kirishitan* were promptly seized and banished. Two years later, the Fall of Urakami terminated any hope of leniency: Nagasaki troops stormed the Christian ghetto, rounded up the 3414 local Catholics, including children and the elderly, and 'resettled' them in 19 detention centres in distant prefectures. Those who persisted in their faith were 'tortured, starved and put into slavery'. The new Meiji rulers behaved no less malignly than the shoguns who preceded them, and approved the exile of the *Kirishitan* and 'cleansing' of Urakami.

Foreign outrage, however, forced Tokyo to reconsider this policy. In 1873, the Meiji Emperor relented and withdrew the ban on Christianity, and in 1880 Bishop Petitjean transferred the seat of the southern vicariate from Osaka to Nagasaki, reflecting the latter's status as the spiritual home of Japanese Christianity. Exiled Catholics were allowed to return, shedding tears of joy as their ships entered the harbour. They celebrated Easter and Christmas in uniquely Japanese style, and their numbers increased. By 1940 scattered around the country there was almost double the number of Christians there had been in 1637.

The 20th-century revival of Emperor worship renewed Japan's ancient hostility to Christianity. Tokyo's samurai leaders considered Japanese Catholicism an unspeakable wart on the face of the *Kokutai* and Japanese Catholics no less than traitors. When the war began, relations between the faiths further deteriorated. 'There was little harmony with other religions at the time,' recalled Kazuhiro Hamaguchi, from a family of local Catholics. Catholics were humiliated, refused jobs and their children bullied – chiefly during the period of the sensational Japanese victories of 1941 and 1942. Yet the military regime grudgingly acknowledged the Christian presence. Rather than snuff out this odd community, Tokyo saw an opportunity to exploit the faith's inspirational value: Jesus Christ would be yoked to the war effort.* On 3 May 1940 Japanese Catholics received official

---

* While believers tried to reconcile their love of Christ with faith in the Emperor, the modern Japanese state found ingenious methods of assisting them. Some Japanese scholars interpreted Christianity as an extension of Confucianism. Others found elements of *Bushido* – the samurai creed – in the Christian willingness to die for his or her beliefs: that Christianity offered 'a goal worth dying for' later impressed the writer Yukio Mishima. The marriage of Catholicism and Shinto was masterfully demonstrated at a meeting of the Catholic Japanese hierarchy in April 1935. Christians 'may show reverence at Shinto shrines', they declared, in answer to a Ministry of Education edict that 'such reverence is merely an expression of patriotism and loyalty'.

recognition as the *Nippon Katorikku Kyodan* (the Japanese Catholic Religious Body); on Christmas Day 1941, three weeks after Pearl Harbor, Japanese priests were ordered to insert a 'prayer for victory' in all Christmas masses. Thereafter churches, monasteries and convents were regularly forced to pray for victory.

By November 1943 the war had emptied the pantry of the Eucharist: the supply of altar bread and foreign wine was exhausted. At Holy Communion the *Katorikku* resorted to bread made from cassava flour and vinegary wine matured in Tajimi. Platinum articles and all metal objects – bells and incense holders – were confiscated and melted down into weapons and bullets. War brought another exigency to Nagasaki that ran foul of the Catholic minority: demand for the services of a great many prostitutes and 'low-class geishas'. 'If such people did not live here,' observed the diarist Kiyoshi Kiyosawa, 'the "productive warriors" [factory workers] would not settle down here.'

•

A recent Catholic convert was Takashi Nagai, the son of an untrained 'herbalist' from a village in Shimane Prefecture whose ancestors had served as managers of the medicinal herb garden of the Matsue clan. Nagai's family had been strident followers of Shinto, devoted to the 'glory of the Imperial family ... as espoused by the grand shrine of Izumo in Shimane Prefecture'.

Nagai rejected Shintoism and moved to Nagasaki in 1931, a professed atheist. He shared the communist ideals of his fellow medical students and scorned the local Christians as 'slaves of Westerners, hoodwinked into clinging to an obsolete faith'. Hope for humanity, he believed, lay in science and Marxism. His views gradually changed while he was a lodger with the Moriyama family at their home in Urakami, where he met their only daughter, Midori. The Moriyamas were among the earliest of Nagasaki's Catholic converts. Their ancestors had led the Hidden Christians and witnessed the crucifixion of the 26 martyrs. Nagai found himself immersed in the stories and struggles of Japanese Christianity, the curious rituals of hymn and prayer, of going 'down on the knees', of mass and confession and the strange singsong voice with which his hospital attendant said the rosary half-aloud.

Nagai graduated as a radiologist and threw himself at the relatively new sciences of X-ray and radiotherapy. The invasion of Manchuria interrupted his work: he was conscripted into the 11th Hiroshima Regiment and served at the front in 1933–34. While abroad he read a copy of the catechism Midori had given him. It survived the military censor, who told Nagai, 'If you have time to read useless stuff about Western gods you had better know your Soldier's Manual!' On his return to Japan he was received into the Catholic Church and baptised. His conversion followed the death of his mother, whose dying eyes revealed, he claimed, 'a soul that may leave the body but endures for all eternity'. Pascal's *Les Pensées* had convinced him of the existence of the human soul. In 1934 he married Midori and she gave birth to the first of their four children, a boy, the next year.

Nagai practised in the hospital near Urakami Cathedral, whose bells measured out his day. War intervened again: in 1937 he found himself sailing to China as chief surgeon in the 5th Division Medical Corps. This time he saw through the propaganda of generals and politicians, and witnessed the awful reality of war – at one point, his commander ordered him to set alight the mattresses of his wounded patients if attacked – Japanese soldiers were not allowed to fall into enemy hands alive. Nagai meant to resist the order; the crisis passed. He served with distinction and was highly decorated. On his return to Nagasaki, he became a professor of radiology and applied X-ray therapy as a way of detecting tuberculosis (Japan, at the time, had one of the highest tuberculosis rates in the developed world). His faith hardened into a kind of spiritual armour and he embraced medical science as the revelation of God's handiwork: 'The size of planet earth is to an apple what an apple is to an atom! Will X-rays make it possible for us to see this microscopic world?' he wrote. Recalling Nagai's earlier attachment to Marxism, his colleagues dismissed the young doctor's zeal as 'theatrical' and fickle, a weathervane blown towards his latest obsession. In time, as his disciples later acknowledged, Nagai's faith would survive the toughest tests which God had set before Job.

•

Like Hiroshima, Nagasaki made education a priority. The city enjoyed a high rate of literacy and school attendance, and Urakami was an educational beehive: 11 schools stood within a kilometre of the cathedral, on the hillsides and valley floor.* Almost 10,000 schoolchildren and university students jostled for food, board and space around Urakami in 1945. They continued to attend classes despite the frequent air-raid alarms. And they were very hungry. Since 1944, the calorific value of rice and wheat had fallen by about a third. 'I clearly remember eating grass, roots and berries when the food ran out,' a Urakami resident, Kazuhiro Hamaguchi, recently recalled.

One sharp-minded schoolboy was Tsuruji Matsuzoe, 15, a second-year student at Nagasaki Normal School, a teacher-training college in Ohashi, just north of Urakami. Tsuruji hoped to become a junior high school teacher when he completed his six-year course. In early 1945 he lived with a family who owned a small bar downtown near the southern stop on the tramline. Every morning he took the streetcar to Ohashi, a 30-minute journey to the northern end of the line. The streetcar rolled up the eastern shore of Nagasaki Harbour; to the west, the green brow of Mount Inasa jutted over the bay which narrowed to a river as Tsuruji's streetcar turned north, up the Urakami Valley, past Hamaguchi and Matsuyama, terminating at Ohashi.

---

* To the west were Shiroyama National School (500 metres away), Chinzei Junior High (600 metres) and Keiho Junior High (800 metres); to the north, Yamazato National School (600 metres), the School for the Deaf and Blind (500 metres) and Nagasaki Industrial School (800 metres); on the valley floor, the Mitsubishi Industrial School for Boys (600 metres) and the Josei Girls' Technical School (600 metres); and to the east, Nagasaki Medical College and the Departments of Medicine and Pharmacology (500 to 700 metres).

Tsuruji's family were traditional Buddhists. Before the war they grew silkworm trees and tea on land near the village of Nakayamago, about 65 kilometres north of Nagasaki. His father, Yoshiro, 50, and mother, Hayo, 40, now spent their days growing rice and rye; the government took the rice; the family lived on the rye. The children did weeding and small jobs and went to school in Nagasaki. Of Tsuruji's six brothers, one called Seiji joined the navy as a 'child soldier' aged 17, and was killed off the coast of the Philippines in 1944.

'We found out about his death three or four months later,' Tsuruji recalls, 'when the navy sent a message. My father did not often show his feelings, even when his son was killed. My mother was much more vocal in her sadness. But they felt that his death was justified.'

In early 1945, Tsuruji and his classmates heard that their daily routine was about to change: as in other cities, school lessons would cease under the National Mobilisation Law. All children aged 12 and up would work in arms factories or on demolition teams. In Nagasaki, teachers instructed their students to prepare for labour in Nagasaki's Mitsubishi weapons factories; in Ohashi, where Tsuruji went to school, Mitsubishi operated an underground torpedo plant.

•

Meanwhile, the city's medical students – some of whom worked in Professor Nagai's radiology department – prepared for the mass casualties of air raids expected day and night. The vast Mitsubishi shipyards on the bay, the arms factories along the river valley, and the torpedo works in the hills, were obvious targets. In early 1945 few of these factories produced anything useful but that did not diminish their perceived value as air-raid targets.

And yet, strangely, as late as March 1945, Nagasaki – and Hiroshima – had not been heavily bombed. Their pristine condition seemed an affront to the wreckage of dozens of other cities. Neither Nagasaki nor Hiroshima had experienced the new incendiary bombs, horrific reports of which were emanating from Tokyo and Osaka. And with every passing day, hopeful rumours played on Hiroshima's and Nagasaki's sense of exception: perhaps the Americans were preserving us for occupation, went one story. Or, Nagasaki's residents hoped, perhaps the Cross would protect them? Many Catholic Japanese saw the hand of God in the city's eerie preservation, and fondly imagined their shared faith had restrained the Americans: surely the enemy knew of Nagasaki's historic links with Christianity, they quietly fancied.

Children echoed their parents' speculation: 'My friends and I thought that Japanese Christians had ties with the British and Americans,' said Teruo Ideguchi, the 15-year-old youngest son of a Buddhist family, then living in Urakami, 'and that was why Nagasaki had been spared bombing until then.' A halo of wishful thinking hovered over the Christian quarter, whose presence reassured local Buddhists – some of whom dared to hope that perhaps the existence in Japan of this strange, irrepressible faith, whose believers worshipped a man-god nailed to a tree, had deterred the Americans and saved the city.

Their hopes faded when, on 26 April 1945, a single B-29 bombed Nagasaki Station, killing 90 and wounding 170. A few days later 29 bombers destroyed ships in the harbour and, in late July, 32 planes attacked the Mitsubishi shipbuilding plant, putting the shipyard out of action. The world's greatest naval yard had built 47 battleships, cruisers and aircraft carriers, including the battleship *Musashi*, a 69,000-ton (62,500-tonne) monster completed in August 1942, sister to the more famous *Yamato* – the culmination of a nationwide ship-building frenzy that slipped 1,600,000 tons (1.4 million tonnes) in 1944 before 'both ships and shipyards were annihilated'. Nagasaki's residents knew little of these behemoths that had been raised under their noses; the ships were constructed behind giant screens and slipped in secret. The *Musashi*, for example, 'was launched while air-raid drills kept all residents inside their homes, with storm windows and curtains shut'; army surveillance units were posted outside homes with harbour views. In early 1945, however, the Mitsubishi shipyard produced virtually nothing; it had few resources, largely due to the US naval blockade, and served as a workhouse for Korean slaves and about 500 British, Dutch, American and Australian prisoners of war.

CHAPTER 3

FEUERSTURM

The destruction of Dresden remains a serious query against the conduct of Allied bombing ... I feel the need for more precise concentration upon military objectives ... rather than on mere acts of terror and wanton destruction, however impressive.

Winston Churchill, British Prime Minister, after the bombing of Dresden, 1945

AS SPRING OF 1945 APPROACHED, the Japanese people toiled in darkness, ignorant of the dimensions of their worsening predicament. Early in the year Washington had approved a new air offensive that would place millions of Japanese civilians in the cross hairs of a campaign of air-borne obliteration, the scale and concentration of which has no parallel in the history of war.

In 1918 the controversial US General William 'Billy' Mitchell – the founding commander of the US Air Force – envisaged aerial bombardment as the future of human conflict, possibly rendering the carnage of the trenches obsolete, he declared, even as it ushered in a darker era of destruction. In 1926, as air-war enthusiasts were gaining influence in Washington, Mitchell laid before Congress the concept of the 'strategic' air raid: it flew over the exhausted, sodden infantry thrashing about on the ground and struck deep in the heart of the enemy nation. It smashed factories and homes, killed women and children and, in theory, broke the will of the people to resist; the soldiers, conscious of their loved ones being slaughtered in the rear, would lose the will to fight. Mitchell already had in mind the Japanese, America's ally in World War I and now the emerging Pacific power whom he named as America's future enemy. For years he extolled strategic air power, and 'waves of long-range bombers', as the best means of defeating Japan – by burning their paper cities.

The world's first civilian victims of 'strategic bombing' were the few Belgians who died during the German Zeppelin raid on Liege in August 1914. The next was three-year-old Elsie Leggatt, of London's East End, killed by a bomb dropped from a German airship in 1915. German Zeppelins and Gotha heavy bombers subsequently unloaded 9000 bombs on Britain in 84 raids during the Great War, killing 1413 and wounding 3408 people. The lessons of this were terrifyingly vivid: in a future war involving long-range bombers, the victor would be the first to deliver a knockout blow that destroyed as many people, created as much chaos, and levelled as many homes and factories as possible. The point was to compel the surviving citizens to surrender, or rise in terror against their government and force it to surrender.

Mitchell's concept chimed with the concurrent ideas of a little book published in 1921 by Giulio Douhet, an Italian general, called *The Command of the Air*. Its perfect timing, on the cusp of the realisation of the importance of air supremacy, assured the book's international influence. Like Mitchell, the Italian envisaged a 'new form of war' – that of mass slaughter. 'To gain command of the air,' Douhet remarked, during the Great War, 'was to render the enemy harmless.' He prescribed an air strategy in which the victor would launch spectacular, pre-emptive strikes before the enemy had a chance to move, far less retaliate: 'A complete breakdown of the social structure,' he wrote, 'cannot but take place in a country subjected to ... merciless pounding from the air.' To end the horror and suffering, the people 'would rise up and demand an end to war'. In such circumstances, he asked, would not 'the sight of a single enemy plane be enough to stampede the population into panic?' In such a war, 'the decisive blows will be directed at civilians', and the victor would be the side that 'first succeeds in breaking down the … resistance of the other'. It would be 'an inhuman, an atrocious performance,' Douhet conceded. Wars of the future would make no distinction between combatant and civilian, he predicted. All would apply the 'most powerful and terrifying means, such as poison gas and other things, against the civilian population'. Attacking civilians was inevitable, that is, 'logically destined', because in future the women making shells, farmers harvesting wheat and scientists in their labs would all be deemed combatants. In such a war, the safest place 'may be in the trenches'. That black joke scarcely registered in such a book.

•

The Japanese and Germans were the first to apply a Douhet-style knockout blow: 5400 Chinese nationals died when Japanese aircraft dropped incendiaries on Chongqing, China, in 1939. The Luftwaffe's destruction of Guernica during the Spanish Civil War presaged air wars on a scale more terrible than Douhet had conceived: it ushered in the blitzkrieg, or 'lightning strike', which Goering applied with merciless efficiency on Poland, Belgium, France, the Netherlands and Britain. The Luftwaffe's raid on central Rotterdam in May 1940 killed more than 1000 civilians and wounded many thousands more. The Blitz rained bombs on London for 76 consecutive nights from 7 September 1940, leaving more than 40,000 civilians dead and more than 375,000 people homeless.

In late 1941 Winston Churchill, on the recommendation of Sir Charles Portal, Chief of the Air Staff, considered a new policy of 'area' raids on German cities. Precision bombing – of factories and refineries, for example – had failed, it was felt. Of those aircraft recorded as attacking their targets only one in three got within 8 kilometres, and only one in four did so over Germany, due largely to inadequate equipment, underdeveloped radar systems and poor training, according to the Butt Report of August 1941 on the effectiveness of precision bombing, compiled by David Bensusan-Butt, a civil servant in the War Cabinet Secretariat. The conclusions were a shock. Rather than attempt to amend the RAF's deficiencies, the British government chose to adopt a new policy, and gave their pilots a new mission: to render German cities 'physically uninhabitable', and 'the people conscious of constant personal danger', through Douhet-prescribed air raids on civilians. The RAF's new aims were brutally simple: (1) to achieve utter destruction; and (2) to incite the fear of death in the people.

Portal occupied himself with calculating the likely effect of dropping 1.25 million tons (1.1 tonnes) of bombs on German towns: they would destroy six million homes; leave 25 million people homeless; and kill 900,000 civilians and seriously injure one million. The first raids awaited further research. Indeed, studying the effect of destroying enemy homes became a kind of obsession for Lord Cherwell, formerly Dr Adolphus Frederick Lindemann, the Prime Minister's German-born scientific adviser. 'Having one's house demolished is most damaging to morale,' he told Churchill in March 1942 in his 'Dehousing Memorandum', after surveying the devastation of Hull and Birmingham. 'People seem to mind it more than having their friends or relatives killed.' Perhaps this projected Cherwell's fear of homelessness rather than the quality of friendship, or value of homes, in Hull. Regardless, nothing dissuaded him from his grim task: 'We should be able to do ten times as much harm to each of the 58 principal German towns. There seems little doubt that this would break the spirit of the people.' Portal agreed: the loss of the German's home, if not his family, would break enemy morale, he advised Churchill in November 1942.

Francis Vivian Drake, considered an expert on air war, seized the initiative in his book *Vertical Warfare* (1943). He claimed that area bombing would 'bring Germany to her knees' within six months. Drake recommended dropping 240,000 tons (218,000 tonnes) of bombs in that time, at a cost, he calculated, of 1660 Allied planes and the lives of 20,000 airmen. The crews' lives would not be wasted: they would win the war, he argued, because, 'It is outside the realm of possibility that the population of any country, no matter how determined or how desperate, could withstand anything like such a terrible tonnage ...' In the event, the Allies dropped 2.7 million tons (2.4 million tonnes) on Germany during World War II – 260,000 tons in March 1943 – for the loss of more than 160,000 airmen.

Air Marshal Arthur Harris, appointed commander in chief of Britain's Bomber Command in February 1942, wholeheartedly supported the policy he inherited – at least at the start of the air-raid 'experiment', as he called it: 'Germany ... will make a most interesting subject for the initial experiment. Japan can be used to provide the confirmation.' Portal masterminded the coming air war on German cities; Harris executed it. Unwilling to risk aircraft trying to pinpoint 'non-civilian' targets using poor radar, 'Bomber' Harris (also dubbed 'Chopper' and 'Butcher' by the RAF) argued that 'dehousing' the Germans would more effectively destroy their will to fight. This was pure Mitchell and Douhet. Until early 1942, air raids on Germany had neither specifically targeted civilians nor dropped the new incendiary (or jellied petroleum) weapons – early versions of napalm – to a substantial degree. That changed on the night of 30 May 1942 with the first 'thousand bomber' raid on Cologne, hitherto the most devastating air attack in war. 'Area' bombardment of dozens of German cities would follow – the intent of which was to destroy homes and civilian lives.

On 25 July 1943, Harris achieved, in Operation Gomorrah, the obliteration of most of the city of Hamburg, Germany's second most populous. 'The total destruction of this city,' stated Most Secret Operation Order No. 173 before the raid, '... together with the effect on German morale ... would play a very important part in shortening and in winning the war.' To complete 'the process of elimination' would require at least 10,000 tons (9000 tonnes) of bombs. The methods and results were unprecedented in the history of war: soon after midnight, 728 RAF aircraft dropped thousands of incendiary clusters and high explosives on Hamburg's urban areas. Within an hour, roaring fires and smoke covered 10 square kilometres of residential Hamburg. The city's fire department conjured a new word for the effect: a *Feuersturm* – firestorm – a phenomenon rare in nature. Perhaps a volcanic eruption over a forest, or a multitude of flaming geysers, or a band of arsonists in the bush on a hot summer's day would deliver the same result:

'Small fires united into conflagrations in the shortest time,' reported a secret German document of the time,

and those in turn led to the fire storms. To comprehend these ...
one can only analyse them from a physical, meteorological angle.
Through the union of a number of fires, the air gets so hot ... which
causes other surrounding air to be sucked towards the centre.
By that suction, combined with the enormous differences in
temperature (600-1000 degrees centigrade) tempests are caused ...
In a built-up area ... the overheated air stormed through the street
with immense force taking along not only sparks but burning timbers
and roof beams ... developing in a short time into a fire typhoon such
as was never before witnessed, against which every human
resistance was quite useless.

'Self-energised dislocation' was how the RAF described the scene in Hamburg, using a euphemism as callous as it was inexact, suggesting the bombers had merely ignited small fires that had mysteriously self-energised into a raging inferno, which had, of itself, dislocated the public. 'Terror-bombing' – a phrase coined by German Propaganda Minister Goebbels – more accurately described the most efficient way yet discovered of killing human beings: 'It would be ironical,' stated the British military historian Basil Liddell Hart, 'if the defenders of civilisation depend for victory upon the most barbaric and unskilled way of winning a war the world has seen.' That was his private view, expressed in a diary. In public the British rejoiced at the success, as 'the Hun' burned. After three nights of this torment, half of Hamburg ceased to exist. More than 30,000 civilians perished in the inferno. Harris later justified this air strategy as 'humane' compared with the British blockade of Germany, which reportedly would kill 800,000 civilians, according to a British White Paper. The word 'humane' had no place in this hypothetical debate; neither Harris nor civil servants in Whitehall could accurately calculate how many people the blockade would kill, directly or indirectly.

Over the course of the war, Bomber Command terror-bombed 70 German cities, of which 69 suffered the destruction of at least 50 per cent of their urban (industrial and residential) areas. The preponderance of war factories in the Ruhr Valley validated those cities as military targets, although most of the victims were factory workers. Elsewhere civilians were the main casualties: of 2,638,000 tons of bombs dropped on Germany and German-held territory, 48,000 tons – less than 2 per cent – fell on war-related factories, while 640,000 tons landed on 'industrial areas' – largely workers' homes. Indiscriminate terror strikes on residential areas accounted for most of the rest. Not all were as 'effective' as Hamburg; Harris lost 1047 bombers in his failed attempt to burn Berlin (in which American aircraft also took part).

In February 1945 Harris turned his attention to Dresden. Churchill and his advisers had selected the target while they were at Yalta, as part of Operation Thunderclap, a demonstration to Stalin of the Western Allies' resolve to strike deep in Germany territory. Dresden, the 'Florence on the Elbe', the paragon of baroque architecture, was not a military target by any stretch of the definition. In fact, it did not appear on Bomber Command's list of targeted German cities drawn up by Harris's second-in-command, Air Marshal Sir Robert Saundby. No doubt Dresden had an important post office and a railway marshalling yard. A local factory made gas masks. And 8 kilometres north of town an old disused arsenal produced soap, baby powder, toothpaste and items rumoured to be aircraft navigation instruments and bombsights. Whatever it made, the arsenal was outside the RAF's target area.

By the night of 13 February, Dresden's population had swelled to more than one million people, including 400,000 refugees – German and third-country civilians – fleeing the Soviet tank invasion. These people had nowhere to live. They huddled beneath the rococo angels and baroque eaves and flying buttresses of buildings like religious pilgrims in search of sanctuary.

That night, 796 RAF Lancaster bombers in two waves unloaded 650,000 incendiary bombs over Dresden. The aircraft met no ground fire; the city lay undefended. The pilots, some of whom felt affronted, even ashamed, by this lack of opposition, flew in low. The first wave dropped 4000-pound high explosives that broke open the roofs of buildings like the tops of eggshells; 750-pound clusters of incendiaries followed. The second wave encountered not a city but a raging furnace. Billowing clouds of smoke and flame obscured the aiming points. So they firebombed the fireball: 'There was a sea of fire covering ... 40 square miles [100 square kilometres],' a crew member of the last Lancaster over Dresden later said. 'We were so aghast at the awesome blaze that ... we flew about in a stand-off position for many minutes before turning home, quite subdued by our imagination of the horror that must be below. We could still see the glare of the holocaust thirty minutes after leaving.'

About noon the next day 311 US bombers joined the RAF over Dresden, the first US aircraft to participate in a civilian terror strike. It was a superfluous act of overkill. The pilots believed they were attacking a railway terminal. Instead, they pulverised whatever remained of the inner city. The rubble danced and the corpses fell to dust. Then, lest any sign of life dare show itself, scores of low-flying Mustang fighters strafed the smouldering ruin and mowed down dishevelled crowds on the river banks and in the gardens where a remnant of the Kreuzkirche children's choir and some British prisoners of war had sought refuge.

Kurt Vonnegut was in Dresden that night as an American prisoner of war. The author of *Slaughterhouse V* could not forget the sight of rows of asphyxiated people sitting up in a shelter, 'like a streetcar full of people who'd simultaneously had heart failure ... Those in underground shelters said they heard a strange howling sound, unlike any they'd heard, overhead: the sound of a tornado of flames.'

At least 100,000 civilians lost their lives in Dresden in a single night (upper estimates put the number of dead at 135,000). By comparison, 568 civilians died in the Coventry attack and the London Blitz claimed 40,000 lives. Dresden's dead included two trainloads of evacuee children aged between 12 and 14. Tens of thousands of body parts were unidentifiable. The Central Bureau of Missing Persons resorted to collecting wedding rings – it was the German custom to engrave the married couple's names inside the band – to aid identification; some 20,000 wedding rings were salvaged from unknown corpses. Soviet troops then trampled the piles of the dead into stacks and burned them in the Altmarkt (Old Market).

The story of Dresden swiftly reached London and Washington. Press reports of 'terror-bombings of German population centres' alarmed General Eisenhower and US Secretary of War Henry Stimson. General Carl Spaatz, Commander of US Strategic Air Forces in Europe, assured Eisenhower that the targets had been purely military and apprised Stimson of Dresden's importance as a transport centre. No evidence could be found to justify the claim by US General George Marshall that Russia had requested the 'neutralisation' of Dresden. The controversy dissolved behind the doors of officialdom. Barbs of doubt, however, lodged in the mind of Churchill, who had initially championed the attack and was fully informed of the execution of the air war. He now recoiled from the spectacle, adroitly shifted responsibility to Bomber Command, more specifically to Harris, and recast his own role for posterity:

'It seems to me,' the British Prime Minister wrote, in a famous minute of 28 March 1945, 'that the moment has come when the question of bombing German cities simply for the sake of increasing the terror … should be reviewed. Otherwise we shall come into control of an utterly ruined land … The destruction of Dresden remains a serious query against the conduct of Allied bombing … I feel the need for more precise concentration upon military objectives … rather than on mere acts of terror and wanton destruction, however impressive.'

Bomber Command greeted Churchill's remark with muted fury: most distressing, Saundby alleged, was the insinuation that Bomber Command had been terror-bombing Germany on its own initiative – when the orders clearly came from the War Cabinet. The Chiefs of Staff were similarly aghast, and would not be held responsible for the decisions of their political masters. The Chiefs and air commanders compelled Churchill to rephrase the minute: 'We must see to it,' ran Churchill's revised version, 'that our attacks do not do more harm to ourselves in the long run than they do to the enemy's immediate war effort.'

•

Until 1944 the US Air Force had not widely targeted civilians in the European theatre, preserving their ordnance for high-altitude precision strikes on bridges, factories, railways and military bases. General Carl Spaatz, then commanding the US Eighth Air Force, had at first refused to countenance terror-bombing. This culture of restraint was founded less on humanitarian considerations than on military pragmatism; however, it echoed Roosevelt's earlier revulsion at the 'inhuman barbarism' of German and Japanese civilian bombing, which, he told Congress on 1 September 1939, had 'sickened the hearts of every civilized man and woman, and has profoundly shocked the conscience of humanity.' Even so, the US Air Force kept its options open: terror-bombing had been a longstanding contingency, drawing on the recommendations of Mitchell in the 1920s. In November 1943 aircrews tested a new kind of incendiary weapon on a mock Japanese town built in the Dugway Proving Ground in Utah, an exact replica in miniature of a Japanese paper suburb, 'down to the books on the shelves and the matting on the floor', observed General Curtis LeMay. A fire brigade even simulated its hapless Tokyo counterpart. The testing resulted in the development of the new napalm-based M69 incendiary bomb, which came on stream at the end of 1944.

Roosevelt's revulsion faded as Germany and Japan set the terms of total war. By late 1944, driven by the exigencies of mounting casualties and enemy atrocities, Washington determined to take the war to the Japanese civilian. The first American incendiaries fell on Japanese-occupied Hankow, China, in December 1944 in an experimental raid. This breached the terms of the 1907 Hague Declaration prohibiting the 'discharge of explosives from balloons' and the General Rules of Aerial Warfare, which outlawed 'aerial bombardment for the purpose of terrorising the civilian population'. Legally, however, the US had no case to answer, having refused to observe the terms of the 1907 law; conveniently, the Hague's 'General Rules' of air combat were never ratified. Not that old rules of war restrained any of the combatants of World War II. Shortly the US Air Force brought forward plans for its first massive proto-napalm (jellied petroleum) strike on the Japanese mainland. The first target would be Tokyo.

•

General Curtis LeMay, commanding XXI Bomber Command, led America's strategic air offensive against the Japanese home islands and earned the cold respect, if not the affection, of the pilots in his charge. He had flown, with courage and skill, several air raids against the Germans in 1943; he was willing to do so over Japan, and would have done so had not his knowledge of S-1 – the atomic bomb development project – grounded him at the US air base in Saipan; his superiors could not risk the secret's extraction under torture. Instructed by his superior, General Hap Arnold, Commanding General of the US Army Air Forces, to give priority to cities, not factories, LeMay worried that low-level incendiary raids would put his aircrews dangerously at risk from ground fire. On the eve of XXI Bomber Command's first incendiary attack on Tokyo – scheduled for 9 March 1945 – he feared the loss of 300 planes and 3000 airmen. But he reassured himself that conventional air raids had severely weakened Tokyo's air defences, and the US naval blockade that surrounded Japan had denied the enemy any hope of reinforcements.

Aircrews in the packed Quonset hut on Saipan listened in awed silence to their pre-flight briefings: more than 300 Superfortresses (hitherto a maximum of 150 had been deployed in a single raid) packed with incendiaries would strike Tokyo at altitudes of just 5000 to 8000 feet (1500 to 2000 metres). Their mission, to burn the city, would involve dropping thousands of cylinders of napalm on Tokyo's most congested residential areas. Anxious crews were told to jettison guns and ammunition – and thus risk flying over enemy territory without retaliatory fire – in order to accommodate more M69 bomb clusters. Each cluster contained 38 incendiary cylinders, or bombs, of jellied petroleum. A single plane would carry about 40 clusters, making a total of 1520 incendiary bombs per plane. LeMay's instructions dismayed his airmen: most thought the operation impossible; surely their planes would be cut to pieces by anti-aircraft fire? The 'pall of a suicide mission' hung over the pilots, LeMay recalled. If the mission horrified US pilots, how would the Japanese react to a low-altitude incendiary raid? Shock tactics were LeMay's answer; the enemy had had no experience of a low-level mass incendiary strike.

•

The people of Tokyo heard the long, dreary wail of an air-raid warning at 10.30pm. They were accustomed to warnings and little troubled:

Tokyo, a city of 4.3 million, had thus far lost about 1000 people to air raids. They sat in their darkened homes awaiting the second decisive siren that confirmed an attack. A violent gale rattled the shutters of their flimsy paper-and-wood homes. For a while nothing happened. Then, a little after midnight, coast watchers detected the silver bellies – for a year, there had been no need to camouflage them, as they usually flew beyond the range of groundfire – of the first B-29 Superfortresses flying low over the water. Nicknamed 'Bikko' or 'B-san' in Japan, the Superforts reached the city at eight minutes past midnight. The raid seemed a feint, as the aircraft dropped few bombs and 'looked as though they were escaping towards the south of Boso Peninsula', observed one witness.

At 12.15am, coastal radio alerted Tokyo to more enemy aircraft, a whole formation in fact, also flying at unusually low altitude. They approached east Tokyo, the city's most densely populated area, whose scattering of small factories and cottage industries confirmed – in US Air Force public relations parlance – its designation as a 'military target'. The second air-raid sirens wailed. Hundreds of thousands of people scurried from their homes; some wore air-raid hoods and lugged buckets and wet towels; fathers carried sleeping mats and food; mothers bore children in their arms or on their backs. They ran towards the few concrete shelters. If these were full – and Tokyo's concrete shelters in total had room for only 5000 people – they resorted to shallow trenches, covered holes, anything underground. Expecting high explosives, they hoped to shield themselves from shrapnel and flying debris. Then, above the roar of the planes came a strange, new whizzing sound unknown to the people of Joto, the first targeted area on Tokyo's eastern plain, and the most densely populated area of the world.

Weather conditions were perfect for igniting a paper city: a cold, moonlit night with fierce northerly gales that would act as giant bellows to the storm. The incendiary canisters burst on impact. The 4-pound bombs bounced across the parks and rooftops, spewing flaming jellied petroleum onto homes, attics, alleys, schools, hospitals, temples and factories. The high winds fanned these spot fires into a fireball that sucked in the surrounding oxygen. What followed was a firestorm more terrible than anything seen in Germany.

The flat plain of Tokyo's *shitamachi* (downtown) residential area, where up to 84,000 people per square kilometre lived in a crush of little paper-and-wood dwellings, was the kindling for a hurricane of flames: 'The scattered fires came together into a single huge flame and 40% of the capital was burned to the ground,' the Japanese Home Affairs Ministry blankly reported. In his memoirs, LeMay chose a biblical metaphor: 'It was as though Tokyo had dropped through the floor of the world and into the mouth of hell.'

The second wave of aircraft 'saw a glow on the horizon like the sun rising,' pilot Robert Ramer recalled. 'The whole city of Tokyo was below us ... ablaze in one enormous fire with yet more fountains of flame pouring down ...' The pilots flew into clouds of black smoke and huge updraughts that buffeted the planes 'like embers over a campfire', and threw up 'the horrible smell of human flesh'.

On the ground, as the spot fires ignited and spread, the official policy was, 'Fight, don't run.' The neighbourhood associations, armed with mops, buckets and sandbags, rallied their wan tribes beneath an escarpment of fire. Those who stayed to fight were burned. Millions chose to flee the flames that chased them through the city like furies. The firestorm flung ahead gigantic cinders – burning beams, joists, palings – which smashed to the ground, or into buildings, lighting new spot fires that fed the advancing inferno. Homes and people, like trees in the path of a bushfire, burst into flames; families, the elderly, mothers and children went mad with pain and terror; victims rolled about on the molten streets unable to douse the jelly that burned to the bone. The people headed for the parks or along the train lines or rushed to the river and hurled themselves in. Coils of flame surrounded and ensnared the weak or slow or overburdened, who caught fire and fell, unhelped by the fleeing populace; others gave up and knelt at prayer in the direction of the Imperial Palace as the conflagration swept over them. No structures were safe or sacred: hospitals crashed down, their patients incinerated where they lay; temples collapsed on the bowed heads inside; schools, mercifully deserted at night, were ash by dawn.

The city sounded the 'all-clear' at 3.20am. In those few hours, 325 American Superforts had dropped almost half a million incendiary cylinders on the people of Tokyo. Twelve planes were lost, and anti-aircraft fire damaged 42 – such was the hopeless state of Japan's air defences. The homes of 372,108 families and about 4000 hectares of property were destroyed. Scores of temples, shrines, churches and convents burned. More than 1.15 million people fled the city. Nobody knows the exact number of dead, but close to 100,000 is the generally accepted figure, mostly of burns and asphyxiation. The US Strategic Bombing Survey (set up by Roosevelt in early 1945 to assess the air-inflicted damage to Germany and Japan) calculated 93,000 deaths but acknowledged that many bodies were uncounted. Japanese sources claimed that 72,000 bodies 'or more' had been removed and cremated by 15 March, most in hastily dug mass graves in public parks. The historian Mark Selden put the casualties at far greater, in a teeming city with 'ludicrously inadequate' firefighting measures fanned by the hurricane-force winds. This, he concluded, combined with LeMay's insistence that Tokyo be 'burned down, wiped off the map ... to shorten the war', surely killed tens of thousands more.

The US Air Force judged the first firebombing of Tokyo – several raids would follow – a great success, as measured by the scale of destruction and loss of life. General Arnold praised LeMay's brilliant planning and execution, and the courage of his crews. 'Under reasonably favourable conditions,' Arnold added, the US Air Force 'should have the capacity to destroy whole industrial cities.'

That is what they did. LeMay meant to take the war to the Japanese people with every weapon in his arsenal: 'Bomb and burn them until they quit,' was the general's guiding principle. In the following weeks

LeMay's XXI Bomber Command firebombed the urban areas of every major Japanese city, dropping almost five million incendiaries (98,466 tons/89,327 tonnes) – one-third of which fell in July 1945 – burning more than two million properties. Tokyo, Nagoya, Yokohama, Osaka, Kobe and Kawasaki were the worst hit, sustaining 315,922 casualties (of whom 126,762 were killed) and the loss of 1,439,115 properties covering 270 square kilometres. US pilots dropped millions of pamphlets a few days in advance of the attacks. One stated, 'America, which stands for humanity, does not wish to injure the innocent people, so you had better evacuate these cities.' Japanese military police ordered people not to read the pamphlets; in any case, half the leafleted cities were bombed within a few days of the warning.

Tokyo's leaders responded with mere propaganda – appeals to Japanese spirit – as city after city was laid to waste. Living in the bomb shelters was 'an adventurous and manly life,' two high-ranking Home Ministry officials assured the nation, 'but we cannot deny that it lacks stability from the standpoint of public order.' The ministry noted an 'insufficient' number of bomb shelters but lacked the materials or manpower to build better ones. On 1 June the regime secretly discussed moving the government, but publicly declared its determination 'to stay in the Metropolis, even if the Metropolis is reduced to ashes'.

•

The American press – with the exception of a few religious journals – delighted in the burning of Japanese cities. A euphoric media hailed the incendiary campaign as a brilliant strategy that would win the war and bring the boys home. In fact the media 'demanded more bombing of civilian targets' and criticised the earlier policy that had restricted US aircraft to targeting military and industrial facilities. *Time* magazine reported the Tokyo air raid as 'a dream come true … properly kindled, Japanese cities will burn like autumn leaves'. LeMay boasted to a press conference on 30 May 1945 that firebombing had killed a million Japanese. The American people similarly applauded the 'area bombardment' of 'arsenal cities'.

Some high US officials, however, demurred. A few days after LeMay's remarks, US War Secretary Henry Stimson privately feared the United States would 'get the reputation for outdoing Hitler in atrocities'. Admiral William Leahy, Roosevelt's Chief of Staff, and General Douglas MacArthur were similarly disturbed by what they saw as the utter barbarity of the air campaign. Yet Washington did nothing to curb the bombing – 'only LeMay's tongue'.

While publicly it claimed the targets were 'military', privately the US Air Force from March 1945 had abandoned any pretence that they were attacking military targets; they made no distinction between civilians and combatants: 'The entire population of Japan is a proper Military Target,' declared one US Air Force intelligence report. 'THERE ARE NO CIVILIANS IN JAPAN.' It was a belief shared by Japan's leaders, for whom the 'hundred million' Japanese people would assuredly fight to defend the homeland. Understanding that assuaged any qualms about terror-bombing, if they existed, in the minds of US air commanders.

•

Terror-bombing failed. The firestorms caused immense loss of life and property but failed to break the enemy's war machine and the people's will to resist. Douhet, Mitchell, Goering, Harris and LeMay had underestimated both the astonishing resilience of a people under siege, and the steadfastness – and callous disregard – of the regimes in charge. There would be no Douhet knockout blow; no domestic uprising; no surrender, in the aftermath. This is not the wisdom of hindsight; the London Blitz offered a valuable lesson: the German air raids hardened rather than weakened the British will to fight on. Terror-bombing could not defeat a country with strong air defences and high morale – or a totalitarian regime in charge – a lesson ignored in the firebombing of Germany and Japan, whose people similarly refused, or were ordered not, to capitulate.

Nor did the Allied campaign take into account the fact that Germany and Japan deployed the largest cohort of slave labour ever assembled: the Nazis herded more than five million Europeans off to work to rebuild the mines, rail and road networks, and factories of the Reich; Albert Speer, Germany's Minister for Armaments and War Production, was determined to keep them working at the frontlines and simply replaced the casualties with new workers. The Japanese similarly used Koreans, Chinese and prisoners of war as slaves – and as 'shields' against air attack – to work in arms and other war-related factories in 'frontline' cities. There were more than 140,000 white prisoners of war in the Japanese Empire, many of whom worked in mines and factories in the homeland in or near the bombed cities.

Bombing slaves and prisoners in residential areas did little damage to the German war effort. Civilian areas were 'unprofitable' targets; targeting them drained air resources from profitable targets and had little impact on Germany's productive capacity; in fact, German tank and fighter production rose in 1944.

As a memorandum to the US Secretary of War in June 1945 on the 'effectiveness' of the bomber offensive dryly observed: 'In contrast to the offensive against oil and transportation, there is considerable evidence that the attacks upon German cities, although extremely heavy, had a relatively indecisive effect upon German war production …' In conclusion, it noted, in the clipped tone of an afterthought, 'The Germans were far more concerned about air attacks on any one of their basic industries and services, such as oil and chemicals, steel, power and transportation, than they were about attacks on finished armament capacity or on city areas … The attacks on oil and transport were the decisive ones.' One cannot help but be struck by the devastating consequences for civilian life of failing to apprehend this before the act.

Bomber Harris himself reached the same conclusion after the war: 'Almost every German officer who knew anything about the subject' knew this: that the conventional destruction of factories, communication posts and transport lines inflicted much greater damage on the war effort than civilian 'area raids'. D-day succeeded largely because Allied aircraft destroyed the railway lines in northwest Europe: 'We had forty ... reserve divisions,' said one German officer, 'Your effective bombing ... made it impossible for us to remove our troops rapidly, if at all.' Without the aerial destruction of 'our lines of communication and transportation', said another, 'your invasion ships and barges would have been sunk or driven out to sea, and the invasion would have been a dismal failure'. Harris applauded the precision bombing of vital factories and rail lines in German-occupied France that preceded D-day. He cited, for example, the Lancasters' demolition of 'a small but very important needle-bearing factory consisting of only two buildings ... almost entirely hidden in cloud'. Indeed, by 1944 British and American aircrews had the skills and technology to deliver heavy concentrated night attacks with real precision. Their commanders decided instead, in 1945, to ratchet up the bombardment of civilians to a level of unprecedented ferocity.

Similarly, in Japan: LeMay's concentration on civilian destruction preserved much of the nation's war infrastructure: the visible rail network, the Kokura arsenal and vital coal ferry between Hokkaido and Honshu were still operating in mid-1945. So too were several major industrial centres. Their 'strangulation' would have defeated Japan 'more efficiently' than 'individually destroying Japan's cities', according to the US Strategic Bombing Survey. LeMay was ordered not to do so, in line with his personal mission to destroy Japanese civilian morale. In the broader picture, the US naval blockade as well as Fleet Admiral William 'Bull' Halsey's carrier aircraft – which attacked Japanese military targets with withering accuracy in July 1945 – destroyed Japan's capacity to wage war more effectively than

LeMay's indiscriminate air offensive. That offensive may be judged a moral and military failure.

The clearest manifestation of its failure was the people's resistance. They did not revolt. Insurrection was unthinkable to hungry, bombed civilians. The assumption that 'civilian hardship' produced public anger and political opposition 'did not stand up'. 'Counter-civilian coercion' merely hurt or killed ordinary people 'for no good purpose', concluded the historian Robert Pape; it was 'wasteful and immoral'. Major Alexander de Seversky, an air-war expert, put it rather more brutally in an appropriately named chapter: 'The Fallacy of Killing People': 'The dead can't revolt.'

No one could describe Major de Seversky as a bleeding heart. A Russian naval aviator who lost his right leg in combat against Germany in 1915, he went on to fly 57 combat missions over the Baltic Sea, downed 13 German planes and won every decoration within his government's gift. He defected to America after the Russian Revolution and later designed and built the first bombsight. His book *Victory Through Air Power* drew the attention of leading aviators and was made into an animated Walt Disney film, which Churchill and Roosevelt viewed at the Quebec Conference in 1943. De Seversky's opinion of civilian terror-bombing stands as its most clear-headed denunciation: 'In air battle, *killing is incidental to the strategic purpose* [my emphasis].'

Bomber Harris belatedly acknowledged this truth. The idea that bombing German cities would break the enemy's morale 'proved to be wholly unsound', he wrote in his memoirs after the war.

So too, in Japan: even without adequate air defences, the nation refused to yield. Undoubtedly air bombardment weakened Japanese morale, yet they would not surrender: 'The workers would still go to work or be forced to go ...' Most students, farmers and factory workers made every effort to stay on the job; in any case, the Japanese military police (*Kempeitai*), like the Gestapo, were on hand to compel them.

To the people of Hiroshima and Nagasaki it remained a mystery why they were spared from America's aerial onslaught. They remained unmolested in April 1945, at a time when most cities shrank under waves of Superfortresses. 'We heard about the destruction of Tokyo on the news,' Miyoko Watanabe, a Hiroshima resident, recalled. 'Everyone was in a state of panic. After the bombing of Tokyo more people started to evacuate their children in response.' The expected attack did not come. The reason lay in a cable LeMay received on 17 May 1945 at his HQ on Tinian Island, from where the atomic bombs would be flown to Japan:

TELECON MESSAGE – G-15-11: TOP SECRET
SUBJECT: RESERVED AREAS
TO: COMCENBOMCOM 21
FROM: COMAF 20
(LEMAY EYES ONLY)
THIS IS TOP SECRET IT IS DIRECTED THAT NO BOMBING ATTACKS
BE MADE AGAINST THE FOLLOWING TARGETS WITHOUT SPECIFIC
AUTHORIZATION FROM THIS HEADQUARTERS. THESE TARGETS
ARE THE CITIES OF HIROSHIMA, KYOTO AND NIIGATA. THIS
DIRECTIVE SHOULD RECEIVE MINIMUM DISTRIBUTION AND BE
ACCOMPLISHED WITHOUT PUBLICITY AND WITHOUT EMPHASIS ON
THE RESTRICTION AS THESE CITIES ARE TENTATIVELY
ESTABLISHED AS INITIAL TARGETS FOR THE 509TH COMPOSITE
GROUP.

Soon, the Pentagon would add Nagasaki to the target list for the atomic bomb.

CHAPTER 4

PRESIDENT

Our blows will not cease until the Japanese military and naval forces lay down their arms in unconditional surrender.

President Harry Truman, announcing the defeat of Germany, 8 May 1945

'I HAVE A TERRIFIC HEADACHE,' a very ill Roosevelt complained to his physician one afternoon in April. The President sat by the fireplace in the 'Little White House', his holiday retreat atop Pine Mountain in Warm Springs, Georgia. He slumped forward. Arthur Prettyman, his faithful servant – 'Negro valet', the press called him – and a Filipino mess boy carried the President to his bedroom. He fell unconscious and died in bed at 3.45pm (Warm Springs time), 12 April 1945, of a cerebral haemorrhage. The news broke at 5.47pm.

That evening Eleanor Roosevelt received Harry Truman in her rooms on the second floor of the White House. 'What can I do?' the anguished Vice President asked the widow of the man who had led America for 12 years. 'Tell us what we can do,' Mrs Roosevelt replied sympathetically. 'Is there any way we can help *you?*' She watched with near pity as the weight of the awesome responsibility that had fallen on this odd, unknown little man, whom many thought unable to carry it, registered.

At 7.09pm, a dazed Harry Shipp Truman was sworn in as America's 33rd president. The ceremony, in the West Wing of the White House, took four minutes. Truman, then 60, wore a blue speckled bow tie with a matching spotted handkerchief that seemed a little inappropriate beside the dour black suit of Chief Justice Harlan F. Stone. The new President looked wan and nervous, and had trouble repeating the opening lines of the oath of office. 'So help me God,' he concluded, clutching a Gideon's fetched from the drawer of Roosevelt's head usher. 'A faint, sad smile' lingered briefly on his face, reported the *Washington Post*, as he shook Stone's hand. Then came the milling and damp-eyed congratulations of colleagues, commanders and his family – his wife, Bess, and daughter, Margaret.

Surely few men have inherited a greater burden than Truman did that day. Short, compact, usually smiling, Truman looked, blinking out of his thick, round glasses, at a grieving world of high protocol, deep distrust and flashing bulbs. Few believed he could succeed, much less fill Roosevelt's shoes. He took command of the most powerful nation, then fighting a world war on two vast fronts; 16 million men and women in uniform; the world's largest navy; and 'more planes, tanks, guns, money and technology than ever marshalled by one nation in all history,' as his biographer, David McCullough, observed. He faced the twin challenges of carrying on the tough negotiations with the Soviets and reconverting the American war economy in anticipation of the end of the bloodiest conflict the world had known.

Truman was seen as a nobody. A provincial farmer, failed haberdasher and all-round 'nice man', according to *New Republic*, he appeared out of his depth in almost every way. He knew little of US foreign policy. He was not a military strategist. He had not read the Yalta minutes. He had received patchy information on something called the Manhattan Project, but had little idea of its purpose. Truman was a stranger outside his home state of Missouri. 'He didn't know the right people. He didn't know Harriman [the US ambassador in Moscow],' wrote his biographer. Many congressmen dismissed him as a failure before he started, labelling him the 'Missouri compromise' and the 'mousy little man from Missouri'.

Roosevelt had 'consciously excluded' his Vice President from the detail of military and foreign policy. The extent of Truman's ignorance, of which he soon became aware, might have unnerved a less confident man, but Truman swiftly rose to the challenge. Within an hour of his inauguration, the new President held a cabinet meeting; he strove to entrench an impression of continuity. He made three decisions that reassured the nation: the San Francisco Conference on the establishment of the United Nations, a new security organisation on which hopes for world peace depended, would go ahead that month as planned; all cabinet members would remain in their posts; and he would meet chief military commanders next morning to talk about winning the war 'at top speed'.

Shortly after that meeting, Henry Stimson drew Truman aside. The War Secretary quietly informed the new President of 'an immense project looking to the development of an explosive of almost unbelievable power'. The revelation jogged Truman's memory: as chairman of the Truman Committee, set up to unearth waste and profligacy in the armed forces, he had encountered on three occasions dead ends, at which his inspectors' inquiries were blocked. Back in June 1943, for example, then Senator Truman had been asked by Stimson not to inquire further into the cost of building a series of mysterious factories around America. 'I am one of the group of two or three men in the world who know about it,' Stimson had said. 'It's part of a very important secret development.'

'That's all I need to know,' Truman had replied. 'You don't need to tell me anything else.' Truman pursued it no further, but grasped something of the secret, as he revealed in a letter to his friend Lewis Schwellenbach, a federal district judge in Spokane. Schwellenbach had been concerned about immense earthworks on the banks of the Columbia River at Hanford, Washington. 'I know something about that tremendous real estate deal,' Truman wrote, in a gross breach of security, '… it is for the construction of a plant to make a terrific explosion for a secret weapon that will be a wonder. I hope it works.'

Later that year Stimson blocked another Truman Committee investigation of a proposed $500–$600 million factory in the Tennessee Valley: 'It's very secret and very, very dangerous,' Stimson said.

'A military secret?' Truman's inspector asked.

'It's the most dangerous one I have. That's all I can tell you ...'

'... You can tell me this about it,' Truman's investigator persisted, 'whether or not ... you might be able to utilize whatever you are doing in this war?'

'Oh yes,' said Stimson. 'It's to match possible dangers of the same kind, novel kind, from other countries. It's a race ... Some day after the war is over ... I can sit down with you over a fire and tell you things that will make your hair stand on end.'

On a third occasion, Truman's investigator had followed a trail of auditing discrepancies at the Hanford site and threatened Stimson 'with dire consequences' if the War Secretary refused to reveal the nature of the project to Congress. Stimson's patience was exhausted. He recorded in his diary: 'Truman is a nuisance and a pretty untrustworthy man. He talks smoothly but he acts meanly.' This was unfair. The members of the Truman Committee were simply doing their job – so well that they ran aground on the shoals of the atomic secret.

•

Roosevelt's remains were transferred to the East Room, where they lay in a casket of gunmetal grey, draped in flags, guarded by a soldier, sailor, airman and marine. Wreaths banked up despite Eleanor Roosevelt's plea to mourners not to send flowers. 'A small coloured lad, helping bring in the wreaths, was all but hidden behind a floral offering of white calla lilies,' wrote one reporter. The late President's empty wheelchair stood nearby. On the 15th they buried him at his Hyde Park estate in an austere military service without eulogies. Private contemplation seemed the appropriate way to remember one of America's finest leaders, and only the crunching footfall of West Point cadets and a 21-gun salute broke the silence.

Roosevelt had guided the country from the depths of Depression to the brink of victory. In his last year, however, he had deluded the people and himself about the likely nature of the post-war world. He 'juggled with balls of dynamite whose nature he failed to understand', Anthony Eden, the British Foreign Secretary observed. Roosevelt's misplaced faith in Stalin was regarded as his gravest error of judgment. In his ailing months he failed to confront the world as it was; to attempt more effectively to resist Russian demands and prepare America for the enormous trials ahead. Understandably, he was sick and exhausted. His old-world certainties, his faith in political reason and the power of diplomacy, seemed to expire with him; he died 'Micawber-like, still hoping for post-war co-operation', concluded the historian Wilson Miscamble.

Truman found himself shoved under those balls of dynamite, each of which he had to catch, understand and relaunch into the air: the unravelling of Yalta; the Soviet claims on Eastern Europe; the war with Japan; the meaning of 'unconditional surrender'; and now the greatest secret of all – the atomic bomb – and how and when it might be used. The task ahead plainly awed him. 'The world fell in on me,' he wrote to his sister-in-law on 12 April, the night of his inauguration. How could he succeed a man 'they all practically worshipped' and assume the 'terrible responsibility'? 'Boys,' he told the media the day after his swearing in, 'if you ever pray, pray for me now. I don't know whether you fellows ever had a load of hay fall on you, but when they told me yesterday what had happened I felt like the moon, the stars and all the planets had fallen on me.'

The President met the challenge head on, in characteristic fashion. Those who dismissed him underestimated his energy, adaptability and skill at finding practical solutions to complex problems. His great strength was his decisiveness. Nothing precipitate or whimsical governed his decision-making; once in possession of the facts and the views of his colleagues, Truman acted, firmly and decisively, and wholly without 'that most enfeebling of emotions, regret', as Dean Acheson, then Assistant Secretary of State, later observed of his friend.

Truman delivered his first congressional address to a traumatised nation on 16 April 1945. It came at an especially difficult time: that morning he had read the casualty figures in the papers. Over the course of the war, America had suffered 899,390 casualties, of whom 196,999 servicemen were dead, and the rest wounded, missing or had been taken prisoner; the names of 6481 had been added in the past week. Many more casualties were expected: on 1 April 183,000 American troops had invaded Okinawa, the first time foreign forces had set foot on the Japanese homeland. Yet most civilians still supported the war; polls revealed a determination to exact revenge on Japan for the American losses inflicted since Pearl Harbor. Amid this injured atmosphere Truman rose to give his first speech as President. After humbly acknowledging the greatness of his predecessor, he laid before Congress the two words that had so delighted Churchill at Casablanca and vexed the British leader at Yalta. Unlike Roosevelt, Truman was not a man to dissemble. He meant what he said:

'*We must carry on,*' he vowed. '… both Germany and Japan can be certain, beyond any shadow of a doubt, that America will continue the fight for freedom until no vestige of resistance remains! … America will never become party to any plan of partial victory! To settle for merely another temporary respite would surely jeopardize the future security of all the world. Our demand has been and it *remains*: Unconditional Surrender!' He banged the podium with a characteristic chop of his hand. Japan and Germany, he declared, had 'violated … the laws of God and of man'. He wanted 'the entire world' to know that America's direction 'must and will remain – UNCHANGED AND UNHAMPERED!' (Truman's emphases). He ended with a prayer to 'Almighty God', invoking the words of King Solomon, 'Give therefore thy servant an understanding heart to judge thy people, that

I may discern between good and bad: for who is able to judge this …?'
His ovation was long and sincere.

•

April 24, 1945
Dear Mr President [Stimson wrote],
I think it is very important that I should have a talk with you as soon
as possible on a highly secret matter. I mentioned it to you shortly
after you took office but have not mentioned it since on account of
the pressure you have been under …

Truman met Stimson the next day and heard the full story of a new
weapon so powerful it could 'end civilisation'. Reading from a long
memo, Stimson divulged the details of a secret organisation larger
than the biggest US corporation; of tens of thousands working on an
enterprise, the purpose of which they were ignorant; of huge factories
and laboratories situated on mesas, deserts and valleys; of swathes of
American businesses given over to developing new and untested
processes; of immense resources, deadly substances and remarkable
scientific advances; and of the cost to US taxpayers: upwards of
US$2 billion (US$24 billion in 2010).

'Within four months,' Stimson continued, 'we shall in all probability
have completed the most terrible weapon ever known in human
history, one bomb of which could destroy a whole city.' Britain had
contributed technical know-how, but the US controlled the resources
and processes used in the construction, a position of global dominance
it expected to hold for several years.

'The world … would be eventually at the mercy of such a weapon,'
Stimson said. 'With its aid even a very powerful unsuspecting nation
might be conquered within a very few days by a very much smaller
one … Modern civilization might be completely destroyed.' Control of
the 'menace' of atomic power 'would involve such thorough-going rights

of inspection and internal controls which we have never heretofore contemplated ... The question of sharing it with other nations and ... on what terms, becomes a primary question of foreign relations.'

The United States 'had a certain moral responsibility' to control the weapon and avoid the disaster to civilization, Stimson added, reprising his deep personal misgivings about the nature of modern warfare; on the other hand, if properly used, nuclear power might afford America the opportunity to bring peace to the world and 'save our civilization'.

Truman studied the memo with composure; he did not wish to appear alarmed. In the anteroom, the man in overall charge of building the bomb awaited his turn to speak. A large figure of imposing authority, General Leslie Groves had been ushered in via the back door to escape the attention of the press. Groves sat with the President and War Secretary and ran through the details of the operation. Soon he reached the production schedule: the first gun-type (uranium) bomb should be ready for use about 1 August 1945; the first implosion-type (plutonium) bomb should be ready for testing in early July, Groves minuted. Their talks placed 'a great deal of emphasis on the Russian situation', in the context of the arrival of nuclear power, Groves later noted. Stimson and Groves proposed that a committee be created to oversee the development and use of the atomic weapon; Truman agreed, and approved the 'Interim Committee' on 1 May.

That afternoon the War Secretary drove to Woodley, his Washington home, to dine alone. Shortly afterwards he heard startling news from a Pentagon staffer, interrupting his hopes of an afternoon nap: 'This active President of ours,' Stimson later wrote in his diary, has been 'wandering at large' in the Pentagon, and had made a phone call to London. In the investigative spirit of his old Truman Committee days, the President had embarked on a private fact-finding mission to gather more information on the bomb.

•

Once firmly in office, Truman swept through the White House like a whirling dervish, issuing words of encouragement, reassuring staff and dropping in on surprised officials unused to the backwoods bonhomie of this smiling Missourian, whose easy style contrasted happily with the aloofness of their previous boss. The corridors were full of strangers and whispers: 'The situation continues confused,' wrote Eben Ayers, a White House press officer, in his diary. 'There seem to be all sorts of strange people coming and going. Missourians are most in evidence and there is a feeling of an attempt by the "gang" to move in.' Truman naturally rewarded his loyal supporters with jobs, but the office gossip turned to Truman's background association with Missouri's corrupt Pendergast dynasty, whose family bosses had launched his political career.

The media quickly warmed to the new President's hoe-and-shovel humour. His meetings with reporters were light-hearted affairs, stripped of the gravitas of high office. He spoke rapidly and to the point, chopping the air with his hand for emphasis: Molotov, the Soviet Foreign Minister, was 'going to stop by' [on his way to the San Francisco Conference],' he revealed at his first press conference on 17 April, the largest hitherto convened in the White House, such was the fascination with this easygoing leader: 348 reporters and observers packed the Oval Office and hall outside to hear Truman announce that he would be 'happy to meet Marshal Stalin and Prime Minister Churchill' to further the peace talks. He confirmed that he had no plans yet to lift the wartime ban on horseracing, the brown-out and the curfew. 'Let's wait till V-E Day,' he added, smiling. The reporters loved it.

The President prepared well for his meetings with Molotov, scheduled for 22–23 April. Secretary of State Edward Stettinius had sent two long memoranda on Roosevelt's key foreign policy initiatives. Truman read with dismay of the rapid deterioration of relations after Yalta, Stalin's 'firm and uncompromising position' on every aspect of negotiations, Soviet intransigence on Poland, and the totalitarian conditions on the ground in Eastern Europe. He also read James

Byrnes' handwritten notes from Yalta. Byrnes, who shared Churchill's loathing for Stalin, presented a disturbing portrait of a Soviet regime hungry for conquest ruled by a dictator who displayed not the slightest intention of matching his words with deeds. Truman thus inherited – he did not trigger or engineer – Washington's deep distrust of the Soviet Union. Stalin had called the shots over the ailing Roosevelt, it appeared, handing Truman a difficult and dangerous choice: whether to continue to acquiesce before the Russians – and some observers felt Roosevelt had vacillated – or to adopt a much tougher line.

Several stark warnings buttressed his decision. A letter Roosevelt received before he died caught Truman's eye. It was from Stalin: 'Matters on the Polish question have really reached a dead end,' the dictator wrote. The Soviet government insisted on the appointment of Polish leaders who were 'friendly', in recognition of the blood Soviet troops had 'abundantly shed for the liberation of Poland'.

Stalin's letter reinforced the grim conclusions of a secret US Office of Strategic Services (OSS) intelligence report Roosevelt had commissioned before his death, and which Truman now read: 'Russia,' it stated, 'will emerge from the present conflict as by far the strongest nation in Europe and Asia – strong enough, if the United States should stand aside, to dominate Europe and … establish her hegemony over Asia. Russia's natural resources and manpower are so great that within relatively few years she can be much more powerful than either Germany or Japan has ever been …' If a Russian policy of expansion should succeed, 'she would become a menace more formidable to the United States than any yet known'.

In light of these and other warnings, Truman chose to adopt a tougher line with Moscow; his resolve sharpened after hearing Secretary of State Stettinius' naive analysis of why US relations with the Soviets had deteriorated: a moderate Uncle Joe, apparently, had been forced to renege on the Yalta deal to appease anti-Western sentiment in Russia. Truman replied with thin-lipped contempt for this analysis: 'We must stand up to the Russians,' he said, revealing a glint of steel beneath the bonhomie. 'We must not be too easy with them.' The implication was clear: Roosevelt and Stettinius had been exactly that at Yalta, appeasing the dictator despite evidence of Stalin's reign of terror and aggressive designs on Europe.

'Molly' – Vyacheslav Molotov – was a grim, balding man with a little moustache. He appeared at the White House in a dark blue suit and pince-nez. Despite, or perhaps because of, his officiousness (Lenin nicknamed him 'Comrade Filing Cabinet') this bland Bolshevik rose to be Stalin's most trusted deputy, in which capacity he oversaw the Stalinist Terror, approved the forced collectivisation of Soviet agriculture, and stood by as tens of thousands died in the consequent famine in Ukraine. Molotov later emerged as a signatory to the massacre of 22,000 Poles, including 8000 Polish officers, in Katyn Forest in 1940.

Truman received Molotov twice. At the second meeting, the President made clear his deep displeasure at Russia's failure to honour the Yalta agreements. Molotov replied truculently so Truman pressed him further. 'I told him in no uncertain terms that agreements [such as over Poland] must be kept [and] that our relations with Russia would not consist of being told what we could and could not do.' Co-operation 'was not a one-way street'.

'I have never been talked to like that by any foreign power,' Molotov snapped, according to witnesses.

'Carry out your agreements and you won't get talked to like that,' Truman replied. Years later the President wrote of the meeting, 'Molly understood me.'

Truman's tough line gave Molotov an excuse to tell Stalin that US policy had dangerously changed emphasis; the President, however, had simply deployed a stronger style, not a different strategy, in trying to check Soviet expansionism where softer techniques had failed. There was some truth, nevertheless, in Chief of Staff William Leahy's later assertion that Truman's 'great single political problem' would be 'getting along with the Soviets'.

Jubilation at the German surrender warmed this chilling atmosphere. On 2 May Truman revealed to the American press the death of Hitler; six days later he announced V-E Day, Victory in Europe. The President used the occasion to reissue the ultimatum to Japan: '*Our blows will not cease until the Japanese military and naval forces lay down their arms in unconditional surrender* [Truman's emphasis]'. What did 'unconditional surrender' mean for the Japanese people, Truman asked. It meant 'the termination of the influence of the military leaders' who had brought Japan 'to the brink of disaster' and the return of the Japanese armed forces to their homeland. It did not mean 'the extermination and enslavement of the Japanese people'.

The terms of surrender had thus fundamentally changed since Roosevelt's ultimatum to the Japanese nation: 'unconditional surrender' was now limited strictly to the Japanese armed forces; the people were to be spared. Hirohito's status, however – the critical stumbling block – remained unclear. Truman's intent had been to soften Japanese fears of a threat to their Emperor without appearing to weaken before the American people, 33 per cent of whom believed Hirohito should be executed while just 3 per cent thought he should be used as a puppet to rule Japan, according to a poll conducted on 29 May 1945. In a later Gallup poll of 29 June 1945, 70 per cent of Americans supported the execution or harsh punishment of Hirohito.

The President's change of emphasis reflected a secret Washington line that cautioned against the destruction of Hirohito. Hanging or dethroning the Emperor would be 'comparable to the crucifixion of Christ to us', concluded a gloomy study by General MacArthur's South West Pacific Command. 'All would fight to die like ants. The position of the gangster militarists would be strengthened immeasurably. The war would be unduly prolonged; our losses heavier than otherwise would be necessary.' Attacking the Emperor, warned an Office of War Information (OWI) directive, would provide

'Japanese propagandists with excellent material for unifying the people behind the militarists and for whipping up their fighting spirit'. James Forrestal, the Secretary of the Navy, approved the OWI proposal that modified the surrender formula. The British Foreign Office, for its part, 'believed the rigid demand for unconditional surrender would prolong the war'.

Some of the press picked up Truman's meaning; some urged him to go further: 'Japan should be told her fate immediately,' declared a *Washington Post* editorial on 9 May, 'so that she can be encouraged to throw up the sponge ... What we are suggesting, to be sure, is conditional surrender. What of it? Unconditional surrender was never an ideal formula.' But reducing the punishment of the regime responsible for shedding so much American blood was a fraught political exercise.

The Japanese people heard nothing of this; only their rulers in Tokyo had access to American press reports and US announcements. The old samurai puzzled over and swiftly rejected the new terms. Truman had not mentioned the Emperor by name, they noted; surely Hirohito, as supreme commander, was among those 'military influences' whose 'termination' could only mean death? The Japanese thus pressed on with their resolve never to capitulate. Nor did Germany's defeat sway them: Foreign Minister Shigenori Togo, nominally of the 'peace party', declared that Germany's surrender would 'not affect' the Japanese Empire's determination to continue the war against the United States and Britain.

•

Just who is Harry Truman, Americans wanted to know of this beaming Midwesterner who stood up to Russia and Japan and presumed to fill Roosevelt's shoes? A few weeks after being sworn in, Truman went to church alone – he relished his anonymity – and slipped into a rear pew 'without attracting any notice whatsoever. Don't think over six people recognized me', he recounted. And there he prayed.

Harry Truman was a kind, consciously humble farmer from Independence, Missouri, who loved poker, bourbon (though never to excess) and fashionable clothes – colourful bow ties, two-toned shoes and sharp double-breasted suits. 'There is one thing I notice about the President,' wrote press secretary Eben Ayers. 'He … seems to be one of those men who always wears matching combinations of socks, tie and handkerchief for his breast pocket. Perhaps it's a hang-over from the days when he was in the clothing business.' At first glance his wardrobe ill-suited him; he seemed dressed to compensate for something – perhaps his 'orneryness'. But if he looked as though he were about to sell you something dubious, his obvious sincerity removed those suspicions.

Truman appeared extraordinarily ordinary, an impression he indulged: '[I am] just a common everyday man whose instincts are to be ornery; who's anxious to be right,' he wrote as a young man in a love letter to Bess Wallace, his future wife. He grew up in harsh times, working long days on the family farm. At school he had been the swot, the unpopular kid who 'ran from a fight' and was 'blind as a bat' when he lost his glasses.

Farm work and the army toughened him; as an artillery officer in World War I he showed great mettle under fire. The men under his command in Battery D came to love this kind officer who shared the same background as many, and showed such fortitude in battle; his unit would remain a close-knit group for years after the war.

Truman overcame the deep prejudices of his childhood. His grandparents had been slave owners, as were most white folk on the 'nigger-hating' side of the Kansas–Missouri border. He had grown up 'in a family of negro haters', he wrote – the table conversation peppered with references to niggers, Chinks, Japs, wogs and kikes (Jews).

Truman came to see the value of people for what they did and said, not according to how they looked or to whom they prayed. The black workers at a nearby oil refinery contributed no more or less to society than he did, as a farm labourer, he observed. In 1924, the 38-year-old political aspirant faced down a Ku Klux Klan meeting, telling 1000 Klansmen that anybody who had to work under a sheet was 'off the beam', after which, he later recalled, he got down off the podium and strode through the parting crowd. He adopted a courageous civil liberties policy that dismayed many of his colleagues. 'I believe,' he told a mostly white audience in 1940 'in the brotherhood of man; not merely the brotherhood of white men, but the brotherhood of all men before the law … The majority of our Negro people find but cold comfort in shanties and tenements. Surely, as freemen, they are entitled to something better than this.' It was as radical as it got in Missouri in 1940 and won few local votes.

Unlike Roosevelt, 'who had been given things all his life – houses, furniture, servants, travels abroad', Truman had been 'given almost nothing', as he recorded. It amazed him, as much as those around him, that a man of such humble origin could rise to the office of Vice President; but to inhabit the White House seemed altogether unnatural. He ascribed his early success to luck rather than personal skill: 'No one was ever luckier than I've been since becoming [President],' he wrote on 27 May. 'Things have gone so well that I can't understand it – except to attribute it to God. He guides me, I think.' Truman was not the first American leader to claim divine guidance. Yet it was not idolatrous or presumptuous; rather a sincere expression of faith.

'There was nothing of Uriah Heep about him,' wrote the author Merle Miller, who interviewed Truman at length. He carried an air of the perennial underdog, and always seemed slightly put-upon; self-doubting. His happy reception from the American people was due to the fact that he was one of them, hopeful, fearful, down to earth … always *trying*. He had none of the grandeur of Roosevelt; nor the judicial steel of James Byrnes or the seductive idealism of Henry Wallace, the Secretary of Commerce, both of whom had been Truman's rivals for the vice presidency. He used the common touch, the open smile, the slap on the back, to remarkable effect. He seemed friendly, human, *normal*, in a world with its share of dilettantes and phonies.

Truman read deeply of history, chiefly the history of war: war was preventable only if you understood the causes, he believed. Historians rarely agreed, he realised – an observation drawn from his study of the Gospels: 'Those fellas saw the same things in a different manner,' and all presumed to be telling the truth. Only action would clear a path through the confusion that divided scholars, disciples and political advisers. A leader must decide, then act – and to hell with his detractors. Beside this paragraph in his heavily thumbed copy of Marcus Aurelius' *Meditations* – 'When another blames or hates you, or when men say injurious things about you, approach their poor souls, penetrate within, and see what kind of men they are. You will discover that there is no reason to take trouble that these men have a good opinion of you …' – Truman had scrawled 'True! True! True!'

The story of mankind was a moral continuum in Truman's mind, and the great political truths he read in 'old Plutarch' were as applicable now as then. Nothing really changed in human history, only the methods and the names of those who used them, according to this thinking. Good and evil, crime and punishment, hubris and nemesis, were ineluctable laws of the moral universe. There were bad men and good men, bad states and good states, and if the power to destroy wickedness should fall in the lap of the good men and good states, then they should use it without restraint. Alliances with states considered less wicked were a necessary evil, he believed – that is, until the war was over.

Truman's personal philosophy boiled down to his faith in the fundamental goodness and moral destiny of mankind. He was a straight shooter at a time when 'honest men took honest stands against unmistakable evil'. To the American people, at his best, he represented an era of family communities, honest dealing, and traditional values.*

---

* Truman was not always popular – at one point, during his later years in power, his approval rating slumped to 23 per cent, lower even than Nixon's 24 per cent.

Yet this comforting, if rather crude, delineation of good and evil led Truman to utter some awful clangers. When Germany launched a surprise attack on Russia in June 1941, then Senator Truman said, half in jest: 'If we see that Germany is winning we ought to help Russia; and if Russia is winning we ought to help Germany; and that way let them kill as many as possible.' To Washington's elite the remark sounded worse than boorish; it was unforgivably parochial.

•

May 1945 concentrated the President's mind on the Soviet question and the war with Japan. That month Soviet–American relations during the war reached a nadir. After Truman's missile to Molotov, Stalin accused the President of colluding with Churchill and dictating terms to the Soviet Union. Behind the scenes a knot of senior officials discussed the atomic question and its effect on Soviet–American relations. In top-secret talks on 14 and 15 May, Stimson made clear to Assistant Secretary of War John McCloy that possession of the bomb would give Washington a great advantage in post-war negotiations with the Russians. If the bomb worked, 'we really held all the cards' in a 'royal straight flush' in dealing with Moscow, he said. The problem, however, was whether the bomb would be ready before the peace negotiations with Stalin and Churchill in the Berlin suburb of Potsdam, scheduled to begin on 1 July 1945: 'We [probably] will not know until after … that meeting, whether this is a weapon in our hands or not,' Stimson said. 'It seems a terrible thing to gamble with such big stakes in diplomacy without having your mastercard in your hands.' The bomb had already acquired a 'diplomatic' role in Washington's relationship with Moscow, in the minds of Stimson and Truman.

Indeed, the President was determined to go to Potsdam fully armed. In a meeting with Joseph Davies, an influential Washington lawyer and former US ambassador to the Soviet Union, on 21 May, Truman intimated that the meeting with Stalin should be delayed until the bomb was ready: 'The president did not want to meet [Stalin] until July,' Davies later wrote, because the 'atomic bomb experiment in Nevada' had been postponed from June until then.

The next day Truman wrote in his diary of 'our deteriorating relations with Russia'. He put his faith in the forthcoming face-to-face meeting with Stalin, and held out the hope that the Potsdam meeting might bridge the differences between the two superpowers. Harry Hopkins, a trusted diplomat and old Soviet hand, was to take a message to Moscow the next day. Truman rather naively told Hopkins that he wanted – 'and I intend to fight to get it' – peace for the world 'for 90 years'. That included 'free elections' for Poland, which was on the point of disappearing behind the Iron Curtain. Hopkins could use diplomatic language 'or a baseball bat', whichever he felt the best way of handling Stalin, Truman granted.

Between 26 May and 6 June, Hopkins had six long discussions with Stalin, impressing upon the Soviet leader that the President expected him 'to carry his agreement [at Yalta] out to the letter'. The outcome was inconclusive: Stalin obfuscated and fogged, as usual, but at least the Soviets were in a negotiable frame of mind, the US ambassador to Moscow Averell Harriman, who attended the meetings, informed Truman; Harriman was not easily moved: he had described Soviet hegemony as 'a barbarian invasion of Europe'. This time, Stalin impressed upon the Americans his enthusiastic willingness to attend the Three Power Conference in Potsdam to discuss the occupation of Germany and the war with Japan. Thanks to Hopkins' visit, the Russians seemed amenable to modifying their stridency at Yalta; a far better atmosphere for Soviet–American relations prevailed in advance of the conference. Hopkins had opened the door.

Truman reciprocated with a gentler approach. He tempered the hard line used in his bruising encounter with Molotov. He refused, for example, Churchill's offer to meet privately before they travelled to Berlin to avoid aggravating the Soviet leader's suspicions. 'Stalin already has an opinion we're ganging up on him,' Truman wrote on

7 June. 'To have a lasting peace, the three great powers must be able to trust each other. And they must themselves honestly want it.' While he loathed the communist system, the President saw no reason why the two countries could not 'be friends'. Truman urged patience in order 'to try to understand that form [of government] and their views'. He would meet Stalin in July not without hope but also with deep distrust of the nation that many in Washington and the Pentagon – General Groves and leading Republicans – already saw as America's most dangerous future adversary.

•

Shadowing these great proceedings was the prospect – as yet incomplete and untested – of the atomic bomb, which drew diverse reactions in the minds of the American leadership. While both Truman and Stimson saw S-1, should it work, as a weapon of mass destruction and as a diplomatic lever, Truman had a pragmatic view of the bomb's utility and eschewed Stimson's 'virtual obsession' with the weapon's spectacular power. The President had little patience with apocalyptic images and biblical metaphors. He considered – at first – the bomb as simply another weapon, albeit a rather large one. A decorated artilleryman, he cast a gunner's eye over the new technology: it would be, in effect, just another shell, he reflected – though one with the power of millions of the shells he had fired at the Germans in 1918. In May, his view became more complex – and indeed, positive: the bomb would change the world if it succeeded, Truman thought – but not necessarily in the way Stimson feared. The President envisaged a nuclear-armed United States with the power to stare down the regimes that threatened it, and lead the world to a more peaceful place, secure in the knowledge that the nuclear secret was safe in American hands.

CHAPTER 5

ATOM

*It is conceivable that extremely powerful bombs of a new type may
thus be constructed. A single bomb of this type, carried by boat and
exploded in a port might very well destroy the whole port together
with some of the surrounding territory.*

Albert Einstein to President Roosevelt, 2 August 1939

THE REVELATION OF THE SCALE and history of the project he had
inherited astonished President Truman. The new world revealed itself
through committees and memos, private briefings and intelligence
reports. The size of the enterprise, the brainpower it tapped, the tens
of thousands of people it employed, awed and, at times, bewildered
him. Truman heard of teams of Nobel-prizewinning scientists hidden
away on mesas; of secret factories bigger than several football fields; of
discoveries in nuclear science that transformed the constitution of
matter with devastating consequences for humanity.

The atomic bomb project predated the war, and the momentum
swept Truman along in a direction, and towards conclusions, over
which he had nominal executive control but little real authority. He
fitted the model of the man in charge that Washington and the
military-industrial complex had prepared for him; in truth he was an
essential auxiliary in a process pre-decided in a netherworld of official
secrecy. There were secret uranium deals, gentlemen's agreements and

trans-Atlantic loose ends to absorb and manage; Roosevelt's hand on every one. And there was the expenditure – nearing $2 billion. All of this made a profound impression on Truman, who accepted the prevailing wisdom that the bomb, if it worked, should be used: it bore the stamp of a foregone conclusion. The decision of 'whether' had been made for him; the 'who' was Japan, now that Germany had surrendered. The remaining questions were: when, how and where (that is, which Japanese targets). Truman was expected not to meddle in or obstruct the process; rather to listen, understand and wave the juggernaut on, a role he performed as exactingly as the military-industrial complex expected of the man who had led the Truman Committee. 'The final decision,' the President later wrote in his memoirs, 'of where and when to use the atomic bomb was up to me. Let there be no mistake about it. I regarded the bomb as a military weapon and never had any doubt that it should be used.'

How mankind arrived at the creation of such a weapon is an epic scientific and industrial story, which began with a hypothesis about the constitution of matter, by an ancient Greek 2500 years ago. The philosopher Democritus coined the word *atomos* to describe solid, indivisible, unchanging particles that, he supposed, constituted the building blocks of matter. More than two millennia passed before someone refined the concept: in 1776, John Dalton, an English chemist, conceived of the atom as the basic counting tool of the physical world, and took the shape of a solid sphere, rather like a billiard ball. Daltonian and Democritian theories survived until the late 19th century when fresh discoveries discredited the notion of a round, unchanging unit: atoms were not solid – they consisted of sub-atomic particles that moved in space. Nor were atoms indivisible – they, or rather their nuclei, were separable; atoms were not unchanging – some of them existed in a state of flux and random decay: 'If left to themselves they change into something else; one calls this change a "decay". The decay occurs according to the laws of chance,' wrote the nuclear physicist Robert Oppenheimer.

Nobody had actually 'proved' the existence of the atom, however, or determined its basic structure. Their explanation eluded scientists until the English physicist Joseph Thomson demonstrated, in 1897, that the mysterious beams inside a cathode ray tube were in fact negatively charged 'corpuscles'. The corpuscles quickly became known as 'electrons'.

Atomic particles not only moved about in space, they contained a peculiar energy, a near-magical quality, as the German physicist Wilhelm Roentgen found when he placed his hand in the path of an electron beam streaming from a heated metal cathode. An image of his bones, in silhouette, appeared on a screen. He later captured, to her shock, the skeleton of his wife's hand (and wedding ring) on a photographic plate. The 'X-ray' – so named because nobody knew the nature of the invisible ray – appeared to share the properties of the strange glow emanating from a mineral called uranium, first observed by the French scientist Henri Becquerel and later collected by husband and wife Marie and Pierre Curie for use in their pioneering experiments with radiation.

Uranium occurs naturally in pitchblende, an apparently useless black metal, which is the waste product of silver mining. Uranium is the heaviest and one of the densest elements, and emits a strange fluorescence. The Curies called this effect *radio-actif* – of, or relating to, rays. With great effort and risk to their health, the Curies boiled away tons of sacks of pitchblende from which they distilled two other intensively radioactive elements: polonium and radium (a phial of which Marie kept by her bedside). Their critical finding was that some atoms were *inherently* radioactive and emitted radiation.

And some were more radioactive than others. A few days' exposure to 'several centigrams of a radium salt in one's pocket' will leave 'a sore which will be very difficult to heal,' Pierre Curie warned in his acceptance speech for the Nobel Prize in 1903, which the couple shared. Prolonged exposure may 'lead to paralysis and death'. Pierre Curie recommended 'a thick box of lead' for transporting radium.

Three years later he died in a road accident, but his wife Marie's overexposure to radiation probably contributed to her death from leukaemia in 1934. The US Radium Corporation ignored the Curies' warning: more than 100 female employees lost their teeth and died of mouth cancers in the 1920s and 1930s as result of repeatedly licking the tips of brushes dipped in glowing radioactive paint used on watch faces. The legal case of the 'Radium Girls' prompted new regulations to protect workers from radiation and other workplace hazards and ensured that American industrialists were well aware of the effects of radiation a decade before the use of the atomic bomb.

●

'All things are made of atoms,' concluded Richard Feynman, one of the 20th century's finest theoretical physicists, and all atoms contain sub-nuclear particles 'that move around in perpetual motion, attracting each other when they are a little distance apart, but repelling each other upon being squeezed together'. Nobody could actually see or prod or photograph the components. But the scientists knew they existed, through experiment and mathematical deduction.

Among the most gifted of experimental physicists was Ernest Rutherford, who worked at Cambridge's Cavendish Laboratory and the University of Manchester in the first decades of the 20th century. The gruff, hard-working son of New Zealand farmers, Rutherford quickly stamped his authority on the new field of atomic physics. 'We're playing with marbles,' was how he casually described his work. He had a tendency to sing 'Onward Christian Soldiers' around the lab, oblivious to the signs pleading for 'Silence'. His students found him an inspiration.

One of Rutherford's earliest and most important discoveries was the existence of various intensities and electromagnetic properties of radiation. His experiments were paeans of elegance and simplicity. Simply by wrapping uranium in aluminium foil, he discerned two types of radiation: alpha rays (or particles), which the foil readily absorbed; and beta rays (or particles), which penetrated the foil. (A Frenchman, Paul Ulrich, later discovered a third, far more penetrative ray – or particle – the gamma, which, it seemed, could pass through concrete and steel.)

Rutherford's triumph was the discovery of the fundamental composition of the atom, made up of a core – the nucleus – of positively charged protons (the neutron had not yet been discovered), orbited by negatively charged electrons set in 'empty' space. In another experiment, he and his colleagues were astonished to find that when they fired a beam of alpha particles through gold foil some bounced back and dispersed, as shown by scintillations on a screen. For Rutherford, this was 'quite the most incredible event that has ever happened to me in my life ... almost as incredible as if you fired a 15-inch shell at a piece of tissue paper and it came back and hit you'. The alpha particles had struck the core of matter, the nucleus, which, relative to the size of the atom, was 'like a gnat in the Albert Hall'. Electrons swirled around the hall in little orbits that mysteriously refused to collapse into the positively charged nucleus. Why this happened remained a mystery. Of one thing, Rutherford was certain: the atom no longer resembled plum puddings or billiard balls or marbles or any other solid metaphor. It was a little cosmos whose Saturnian bands of electrons orbited the nuclear sun.

In further experiments, Rutherford and his then protégé, a young Oxford physicist called Fred Soddy, detected a strange gas – argon – emanating from the radioactive element thorium. They were at a loss to explain it. Later, working with nitrogen, they observed the spontaneous transmutation, or disintegration, of one element to yield a separate, elemental by-product. The experiment revealed the strange propensity of atoms of one substance, in certain conditions, to 'decay' into atoms of another. Rutherford had not 'split' the atom, as reporters widely misreported. He had demonstrated the transmutation of the atom.

Atomic transmutation would be seen as one of the finest achievements of 20th-century physics. At the time, however, Rutherford and Soddy risked exposing themselves to charges of quackery and ridicule, like the scorn heaped on 17th-century alchemists who claimed to be able to turn base metal into gold. 'Don't call it transmutation,' Rutherford warned Soddy, 'they'll have our heads off as alchemists.' Indeed, Rutherford's colleagues – including Marie Curie – were somewhat sceptical. But later experiments by Rutherford and a string of physicists confirmed the case for alchemy: nature was indeed mutable; some elements were in a state of constant flux.

Rutherford was said to hear the 'whisper' of nature; but his discoveries provoked more whispers than they answered. How did atoms of one element disintegrate, or transform, into atoms of another? Did a separate particle drive the decay? How much energy would nuclear disintegration release? Soddy had attempted to answer the second question in a paper in 1903. The process of nuclear disintegration, he declared, would yield tremendous energy: '… the total … radiation [released] during the disintegration of one gram of radium cannot be less than 10 [that is, 100,000,000] gram-calories.' This was at least 20,000 times, and some thought a million times greater, than the energy released in a molecular change (that is, outside the atom). 'It was just conceivable,' the eminent reporter Sir William Dampier Whetham quoted Rutherford as saying in 1903, 'that a wave of atomic disintegration might … make this world vanish in smoke.'

From the start – and to assure their future funding – the scientists were drawn more to the military than the civil application of the new source of energy. The next year Soddy told a conference in Britain that if the energy latent in the atom 'could be tapped and controlled, what an agent it would be in shaping the world's destiny'. Whoever held the lever that released it 'would possess a weapon by which he could destroy the earth if he chose'. It sounded like a science-fiction fantasy, fit only for the readers of H.G. Wells' novel *The World Set Free*, which prophesied nuclear Armageddon unless the energy was controlled.

Indeed, Rutherford later quipped, 'some fool in a laboratory might blow up the universe unawares', a scenario he scorned as impossible.

•

The great Danish physicist Niels Bohr, a large, soft-spoken man with a great craggy brow, appeared a parody of the absent-minded professor. To his friends, he seemed unable to grasp the simplest concepts: obvious plot twists in a Hollywood western, for example. On at least one occasion he tried to light his pipe without noticing that its bowl had fallen off. And often he failed to complete his sentences. In lectures, he elided complex thoughts with 'but' and 'and', leaving gaps where there should have been words or, his students hoped, an explanation.

'My first thought,' observed Bohr's biographer Alexander Pais when he met the Dane, 'was what a gloomy face.' Bohr's somewhat dismal countenance belied a cheerful family man and father of six, capable of sudden flights of agility. He was an accomplished skier, cyclist and yachtsman who, as a young man, sat on the bench as reserve goal-keeper for Denmark's Olympic soccer team. When Bohr spoke, Pais' first impression vanished, 'never to return'. This was the man in whom nature had invested the mental capacity to *imagine* the processes of the universe, and then prove his imagination correct. He drew on conceptual thoughts, on models from the natural world – things one could see and touch. His scientific grammar perfectly suited his purpose: to devise a solution to a scientific problem that yielded nothing to classical physics or indeed to any existing idea of how the universe worked. Essentially, he perceived a deep paradox at the heart of matter: 'There are forces in nature of a kind completely different from the usual mechanical sort,' he wrote, 'properties of bodies impossible to explain ...' In this, during the 1920s and 1930s, he radically departed from the rational, determinist world of Rutherford and Einstein.

Bohr explained the electron's orbit in terms of a miniature solar system, as Rutherford had done; yet there the metaphor ended. The electrons, or planets, described bizarre orbits. They vibrated in a stable, or 'ground state', in orbiting bands, or 'shells'. When they absorbed external energy they were prone to leap or jump from one orbit, or shell, to another, and fall back down again, emitting energy or light quanta – a red-hot or white-hot glow. Hence, the 'quantum leap' – the tiniest transmigration and hardly the great leap forward suggested by the term in political discourse.

Each electron followed its own orbit: none shared another's shell, according to the 'exclusion principle' of Bohr's eccentric friend Wolfgang Pauli. They existed in their own eternal microscopic revolutions. The mystery, wondered Rutherford – the scientists inhabited a very small circle, and all knew of one another – was how an electron 'decided' at which frequency it vibrated when passing from one orbit to another. The answer, it appeared, was pure chance. The electron 'fell' into a frequency. This element of chance in the conception of the atom was critical: it would radically alter everything scientists then knew about the nature of cause and effect. The magic question it provoked was whether the light quanta – electrons – behaved as waves or particles. That depended on the conditions; sometimes they behaved as both. Einstein had foreshadowed in 1909 the fusion of the wave and particle theories of light: they were 'not to be considered as mutually incompatible', he wrote.

Bohr's work on the electron and the nature of light atoms tethered the arguments of nuclear physics to a stormy intellectual summit upon which only the more brilliant physicists dared set foot. Rutherford, Einstein, Bohr, Pauli, Max Planck, Max Born, Werner Heisenberg, Erwin Schrödinger, Enrico Fermi, Paul Dirac, Hans Bethe, Ernest Lawrence, Oppenheimer and Feynman were the star mountaineers on the learning curve of quantum mechanics. It was a steep one. Most were or would become Nobel laureates.

These scientists, and others, more or less shared Einstein's definition of the scientist as a detached inquirer who 'sold himself body and soul' to physics. They were observers, uninvolved; they probed and mixed and detonated – and stood back to watch the result. They were not creators, but rather nature's messengers, bearing news of the revelation of her wonders and perversities. For them, self-interest and political motivation were beneath contempt and banished from the lab. The possibility of objectivity, of pure empiricism, of the glory of the scientific method, animated their great experiments and sprawling equations.

•

In 1927 these physicists and others met at a hotel in Brussels – courtesy of the Belgian industrialist Ernest Solvay – to discuss the nature of matter. Never had so many of the world's greatest scientists met in such concentrated circumstances. The question facing the Solvay Conference was how the constituents of the 'sub-nuclear zoo', in Oppenheimer's ingenious conception, actually behaved, and whether such behaviour was observable and hence understandable.

One of the more colourful delegates was the Viennese physicist Erwin Shrödinger, a well-known philanderer who pursued beauty and sexual pleasure as intensely as he pursued the electron. A love affair with a young woman in a Swiss sanatorium in 1925 unified these consuming passions. She left him with the exquisite idea that the electron behaved rather like the strings of a violin: wave-like motions that oscillated to a point. Schrödinger's alpine epiphany – he had in fact encountered the wave theory of matter formulated by Louis de Broglie in 1924 – met fierce resistance at the Solvay Conference. Did it matter how electrons behaved if we have no way of examining them, argued the German physicist and mathematician Werner Heisenberg.* 'The more precisely the position of [a particle] is determined,' Heisenberg declared, 'the less precisely the momentum is known …

---

* Their disagreements were later reconciled through the discoveries of the phenomenally shy English genius, Paul Dirac.

and vice versa … the path [of a particle] comes into existence only when we observe it.' In other words, if we observe a particle's position in space, we know nothing about its momentum; if we measure a particle's momentum, we know nothing of its position in space. In so many words Heisenberg drew down the curtain on Einsteinian certainty – that matter behaved according to pre-determined properties or laws.

Of course, it was not as straightforward as this: argument and counter-argument seized the delegates for days. When the rancour settled, Heisenberg's Uncertainty Principle was left standing – the single sustainable theory that subsumed the rest: the atom, and by definition the universe, were inherently unknowable, uncertain. The *actual* behaviour of sub-nuclear matter was incomprehensible, because any attempt to measure it failed to *see it*. Rutherford's scintillations and Thomson's cathode ray beam were reflections, residual sprays of light quanta, and manifestly not the real thing. No doubt they were indirect evidence of the real thing, but the very core of matter eluded direct observation. Nature seemed a random event, an arbitrary occurrence. Contrary to Einstein's remark a fortnight later that 'God does not throw dice', God (if He existed) showed all the symptoms of being a compulsive gambler.*

•

Radioactive, divisible, transmutable, uncertain and, in suitable conditions, likely to release enormous amounts of energy: such were the properties of atomic matter known to physicists in the early 1930s. Could that energy be harnessed, they wondered. What form would the release of such energy assume?

James Chadwick, the son of an impoverished English couple, spent most of the Great War in a German prison near Berlin, where his gaolers permitted him to set up a laboratory in the stables. After the

---

* Einstein attacked quantum theory 'just as irrationally as his opponents had argued against relativity theory' – according to his old friend, the physicist Paul Ehrenfest.

armistice he resumed his promising career as a physicist and became a Rutherford protégé at Cavendish Laboratory in Cambridge.

In February 1932, Chadwick performed the experiments that demonstrated the existence of the neutron – so named for its neutral charge. The neutron, he discovered, binds with the proton to form the nucleus. To say Chadwick 'discovered' the neutron would be slightly misleading; in an odd kind of way, the neutron discovered Chadwick, revealing itself in conditions the physicist inherited and applied. Rutherford had guessed at the neutron's existence, and experiments by the Curies laid the groundwork. It struck Chadwick that a different number of neutrons sometimes appeared in atoms of the same element; these variants were called 'isotopes' – for example, the uranium isotopes U235 and U238, whose 'mass number' equalled the total number of protons (in uranium, always 92) and neutrons (respectively 143 and 146) in the nucleus. Heavy elements, such as uranium, have more nuclear particles than lighter ones, for example helium with two protons and two neutrons. The neutron, Chadwick showed, has no electrical charge and may break away from the nucleus, passing unimpeded into the nucleus of another atom. It was the 'magic bullet' that would break down the atom and solve the mystery of atomic transmutation.

In April 1932, Rutherford's team achieved another breakthrough, which was contingent upon Chadwick's work: the physicists John Cockcroft and Ernest Walton bombarded the nuclei of lithium with artificially accelerated protons, and succeeded in transmuting them into atoms of helium. It was the first demonstration of atomic transmutation, which the press inaccurately hailed as 'splitting the atom'.

In September 1933, Lord Rutherford (as he was then titled) appeared before the British Association for the Advancement of Science to present a paper on his team's findings at the frontiers of atomic science. Drawing chiefly on Chadwick's, Cockcroft's and Walton's work, which revealed the behaviour of the mysterious new particle, the tall, bewhiskered antipodean enthralled his audience with a masterful description of this protean world. Smashing neutrons into nuclei might one day transform all basic elements, Rutherford declared – yet it would be a very poor way of producing energy: 'We might in these processes obtain very much more energy than the proton supplied, but on the average we could not expect to obtain energy in this way ... and anyone who looked for a source of power in the transformation of the atoms was talking moonshine.'

•

On the morning of 12 September 1933 a restless, self-absorbed man sat reading an article in *The Times* in the lobby of the Imperial Hotel in Russell Square, London. It contained a report on Rutherford's speech entitled, 'Breaking Down the Atom ... Transformation of Elements'. Leo Szilard was a Hungarian émigré by way of Germany, where he had studied with Einstein and Planck, later moving to London to escape the rising tide of Nazism. He was a gifted physicist who understood quantum mechanics. Rutherford's dismissal of nuclear energy irritated him, because it seemed to condemn any future attempt to release the energy inside the atom. He wondered, indeed, whether Rutherford himself was 'talking moonshine'. Receptive to creative interpretations of atomic science, Szilard intimated to friends that he had read as prophecy rather than science fiction Wells' novel *The World Set Free*. Indeed, Wells' novels, which he had keenly read as a youth, would inspire many of his ideas. For Szilard, the novel anticipated the liberation of atomic energy, the development of atomic weapons and a war in which 'the major cities of the world are all destroyed by atomic bombs'.

Szilard picked up his newspaper and set off on his daily peregrination around London. In a now-famous sequence, he stopped at a red light on Southampton Row when the thought struck him 'whether Lord Rutherford might be wrong'. The light turned green. Halfway across the street, 'it suddenly occurred to me that if we could find an element which is split by neutrons and which would emit two neutrons when it absorbed one neutron, such an element, if assembled in sufficiently large mass, could sustain a nuclear chain reaction' – and thus produce enough energy to warm or destroy the world. The collision of one neutron into a nucleus would release two neutrons; two would release four; and within 'millionths of a second, billions of atoms would split, and as they tore apart, the energy that held them together would be released', he reasoned. A 'nuclear chain reaction' would, however, require 'critical mass' to sustain it, Szilard continued to muse. He reached the other side of the street; the traffic resumed. He walked on, resolving to alert American scientists to his idea.

·

Chadwick's discovery of the neutron, and Cockcroft and Walton's 'splitting' of the atom, energised the world's physicists. Matter was revealing its secrets with dazzling speed. In August 1932, at Caltech in California, Carl Anderson, another future Nobel laureate, discovered the existence of a new particle, the positron – the positively charged electron. A month later the physicist Ernest Lawrence and his colleagues at Berkeley, California, using their new invention – the cyclotron, an electromagnetic particle accelerator – were busy smashing apart assorted nuclei but were unable to sustain a chain reaction.

Robert Oppenheimer, later inaccurately described as the 'father of the bomb', worked at Caltech at the time. 'Lawrence's things are going very well,' Oppenheimer wrote to his brother Frank, also a scientist, around September that year (his letter is dated 'Fall'): 'He has been disintegrating all manner of nuclei, apparently with anything at all that has an energy of a million volts.' At that time, Oppenheimer was preoccupied with 'a study of radiation', 'trying to make some order out of the great chaos' and 'worrying about the neutron'.

Lawrence's atom smasher did not trigger a self-sustaining chain reaction because the energy expended in the effort exceeded the energy yielded. Lawrence's 'bullets' were positively charged protons, deflected at the last moment by other positively charged protons within the nucleus. It was like 'trying to shoot birds in the dark in a country where there were not many birds in the sky', wrote Einstein, when he heard of these developments.

Despite such intense efforts to disintegrate matter, nobody had achieved a nuclear chain reaction; throughout most of the 1930s it remained theoretical. Bohr, in a celebrated speech on 'compound nucleus theory' to the Danish Academy on 27 January 1936, described its likelihood as fabulous, and the exploitation of nuclear energy for military or industrial purposes a distant dream: 'Indeed, the more our knowledge of nuclear reaction advances the remoter this goal seems to become.'

In the late 1930s, with the prospect of war looming over Europe, fierce competition gripped the world's scientific communities and pressed events forward. Einstein, Szilard and other émigrés – whose 'Jewish physics' Hitler had reviled – feared that Germany might discover the means of sustaining a nuclear chain reaction before America. The nightmare of a nuclear-armed Nazi State compelled Szilard and fellow Hungarian physicist Eugene Wigner to visit Einstein one hot July day in New York in 1939 (Szilard was then working at Columbia University in Manhattan). The 60-year-old father of relativity, who had been living in the US since 1933 – when the Nazis came to power in his native Germany – received the Hungarians at his Peconic, Long Island, cottage in a white singlet and rolled-up white trousers; he had been out sailing. They sipped iced tea on the porch by the lawn. The Hungarians warned Einstein of the risk of Germany achieving a fast nuclear chain reaction first: '*Daran habe ich gar nicht gedacht*,' said Einstein slowly. 'I haven't thought of that at all.' They persuaded him to sign a letter to Roosevelt – pre-written by Szilard – about the risk of Germany developing a nuclear weapon and the need for immediate American action.

·

German scientists were further advanced than Szilard had imagined. In 1938 a 'benign, mild and kindly little man', German Professor Otto Hahn, and his collaborator, the Viennese physicist Professor Lise Meitner, produced solid evidence for the chain reaction Szilard had imagined. The couple, working at the Kaiser Wilhelm Institute of Chemistry in Berlin, improved on Ernest Lawrence's experiments by using the neutron, not the proton, as the 'bullet' to break up the nucleus. The neutral particle, they asserted, would escape electromagnetic deflection and sail unimpeded into the targeted nuclei – liberating other neutrons and thus triggering a chain reaction. Their assertions drew on the work of Szilard and Chadwick, and the extraordinary experiments of the renowned Italian physicist Enrico Fermi, who had achieved a nuclear reaction without realising it in 1934. Fermi's experiments actually split the uranium nucleus to produce a new element, but his findings were inconclusive and he mistook the identity of the new element.

In Germany, however, Meitner's sex and religion were twin handicaps to progress. Had she been a man, the Berlin laboratory's reading room would have admitted her; had she been Aryan, and not Jewish, Germany would have welcomed her. Hahn, fearful of German police chief Heinrich Himmler, agreed to have her removed from the laboratory and would later fail to acknowledge Meitner in his success. In 1944 Hahn would collect the Nobel Prize for the discovery of fission; the citation offered no mention of Meitner. Nonetheless, Meitner's interpretative genius sealed her place in the history of the ensuing events.

In the autumn of 1938, as Hitler consolidated his grip on Austria and prepared for the invasion of Czechoslovakia, Hahn and the physicist Fritz Strassmann conducted a series of experiments that would transform the world. They bombarded highly fissile uranium atoms with neutrons; the consequent disintegration of the nuclei yielded a strange substance with the atomic weight of barium (56 protons). In fact it was barium. Uranium had transmuted or 'decayed' into an entirely different element. Rutherford's alchemic predictions were again proved correct. Hahn described the event in *Naturwissenschaften*, on 6 January 1939. For a brief period these discoveries remained too arcane even for governments to censor the scientists, and Meitner, now in exile in Stockholm, was free to read the account. She asked the crucial question – where had the 36 liberated protons gone? And answered herself: the uranium nucleus had split into atoms of barium and an inert gas known as krypton.

'Very gradually we realised,' wrote Meitner's nephew, a physicist called Otto Frisch (who worked with Bohr in Denmark), 'that the breaking up of a nucleus into two almost equal parts was a process so different [from any results hitherto achieved] that it had to be pictured in quite a different way.' His aunt described the split in terms of a bacteria multiplying: at the point of impact, each atom, like a cell, stretched out, formed a waist, and divided – a process to which she gave the biological term 'fission'. *Nature* reported the experiment on 11 February 1939: 'Disintegration of Uranium by Neutrons: A New Type of Nuclear Reaction'.

Remarkably, the combined weight of the two liberated nuclei was lighter than the nucleus from which they were freed (by about one-fifth of the mass of a proton). Put simply, something had escaped in the process: in fact, the destruction of the residual mass had released a great deal of energy – enough *per atom* to make a grain of sand jump, which, in relative terms, was a spectacular little explosion. 'The picture,' Frisch wrote, 'was of two ... nuclei flying apart with an energy of nearly two hundred million electron volts ...' In short, the neutrons behaved as Szilard had imagined: liberated in an exponential frenzy from the nucleus, they flew off and smashed apart neighbouring nuclei until the available 'fissile material' – for example, the enriched uranium – had disintegrated; the reaction would release tremendous amounts of gamma (neutron) radiation, explosive power and heat. The scientific community quickly realised that if an atom of enriched uranium could toss a grain of sand, then a gram of enriched uranium had the latent energy of 3 tonnes of coal ... and a kilogram of the stuff would in all probability destroy a city. Otto Hahn was so disturbed by the implications for mankind that he considered suicide.

Frisch told Bohr of his aunt's and Hahn's discovery of fission a fortnight before it appeared in *Nature*. Barely able to contain his excitement – 'Oh, what idiots we have all been! But this is wonderful! This is just as it must be!' – Bohr clumsily revealed the secret on the ship bearing him to America, where he was to address the Fifth Washington Conference on Theoretical Physics. During Bohr's session, a colleague enlarged upon the leak, and the American delegates rushed to prove it for themselves: '... several experimentalists immediately went to their laboratories ... before Bohr had finished speaking!' Frisch recalled. Within two days the Carnegie Institution of Washington, Johns Hopkins University and the University of California had 'split' the uranium atom. On 5 February, Oppenheimer, then working at Caltech, wrote to a colleague: 'I think it really not too improbable that a ten cm cube of uranium ... might very well blow itself to hell.'

Nuclear fission was not the 'discovery' of one individual; it realised the cumulative efforts of Einstein, Rutherford, Bohr, Chadwick, Szilard, Lawrence, Oppenheimer, Fermi, Hahn, Frisch, Meitner and their collaborating scientists (H.G. Wells played a part, too). The possibility of atomic energy and the creation of atomic bombs were no longer 'moonshine'; the experiments demonstrated the principle of nuclear fission, but still nobody had yet created a *self-sustaining* chain reaction that would produce an atomic explosion. As the Nazis prepared to invade Poland, in July 1939, this small group of (mostly) émigré physicists urged Washington to act. On 2 August 1939 Einstein sent Szilard's letter, which he had signed, to Roosevelt via the economist Alexander Sachs, an official adviser to the President. It warned of the need for 'quick action': '... it appears almost certain,' Szilard wrote under Einstein's name, that 'a nuclear chain reaction in a large mass of uranium ... could be achieved in the immediate future.'

'It is conceivable,' the Einstein letter warned the President, 'that extremely powerful bombs of a new type may thus be constructed. A single bomb of this type, carried by boat and exploded in a port [he doubted atomic bombs could be airlifted] might very well destroy the whole port together with some of the surrounding territory.' The United States must move to secure the sources and supply of world uranium, to prevent Belgian Congo's large stock of uranium from falling into German hands. Germany, he warned, 'has actually stopped the sale of uranium from the Czechoslovakian mines which she has taken over'. Roosevelt immediately set in motion the powers for establishing a properly funded nuclear-weapons project, which would shortly assume the dimensions, in secret, of one of the largest corporations and absorb the time and brainpower of the smartest scientists.

CHAPTER 6

THE MANHATTAN PROJECT

*... This cloud of radioactive material will kill everybody within
a strip estimated to be several miles long. If it rained, the danger
would be even worse because active material would be carried down
to the ground and stick to it, and persons entering the contaminated
area would be subjected to dangerous radiations even after days ...
the bomb could probably not be used without killing large numbers
of civilians ...*

The Frisch–Peierls Memorandum, the first official document to describe with scientific conviction
the means of making an atomic bomb and its effects

THE AMERICAN ATOM BOMB PROJECT exploited discoveries made
elsewhere – in Denmark, Britain, Germany and France. They included
the influential 'Paris Group', whose members (some of whom later
worked with the Resistance) made the crucial finding in 1939 that
fission in a uranium nucleus liberates the two or three extra neutrons
necessary to sustain a chain reaction. And in the late 1930s European
scientific advances culminated in two admirably clear reports: the
Frisch–Peierls Memorandum of 1940, produced by Otto Frisch and
Rudolf Peierls, both of whom had immigrated to Britain and were
working at Birmingham University under the Australian scientist and
Rutherford protégé, Professor Mark Oliphant; and the Maud Report

of 1941, put together by the Maud Committee, a secret scientific-military group headed by James Chadwick.

These developments were complex, close-knit, fast moving and always communicated in strictest secrecy. Every document was marked 'Eyes Only' or 'Top Secret'. Taxpayers, of course, who would finance the project, knew nothing of the great events that led to the creation of the first atomic bomb. The Frisch–Peierls Memorandum was the first statement, in any country, to describe with scientific conviction the practical means of making the weapon. On receiving it, Oliphant, highly impressed, informed Sir Henry Tizard, doyen of the British scientific community and chairman of the Aeronautical Research Committee; in turn Tizard passed it to Britain's Scientific Survey of Air Defence, which recommended the creation of a special committee to develop the atomic bomb. This would be known as the Maud Committee, so named after a cable sent by Lise Meitner in June 1940 to England, addressed to Cockcroft and 'Maud Ray Kent', which subsequent inquiries identified as Bohr's childhood governess. 'Maud' became Meitner's code for the great man, then working in Britain. The Maud Committee was the predecessor to Tube Alloys, the British atomic project, created in 1941, which subsumed all these discoveries – and scientists – under Churchill's watchful eye. In time, like a larger fish gobbling up the smaller, America would devour the British program and digest the trans-Atlantic effort as part of the soon-to-be-created Manhattan Project.

The Frisch–Peierls Memorandum probably did more than any other study to spur on this train of events. Both physicists drew heavily on Niels Bohr's work in concluding that an atomic explosion would require a highly fissile – that is (in the right circumstances) an extremely divisible – substance such as the uranium isotope U235: 'A moderate amount of U235 would indeed constitute an extremely efficient explosive,' their memorandum stated. Five kilograms would liberate energy equivalent to several thousand tonnes of dynamite. The Frisch–Peierls Memorandum also warned of the horrors of radiation likely to result from an atomic detonation, the first official document to do so:

The radiations [sic] would be fatal to living beings even a long time after the explosion ... This cloud of radioactive material will kill everybody within a strip estimated to be several miles long. If it rained, the danger would be even worse because active material would be carried down to the ground and stick to it, and persons entering the contaminated area would be subjected to dangerous radiations even after days ... the bomb could probably not be used without killing large numbers of civilians ...

It concluded: 'It is quite conceivable that Germany is, in fact, developing this weapon.'

The Maud Report expanded on the work of Frisch and Peierls. It was the first Anglo-American governmental paper to explain *how* an atomic weapon might work, and it recommended Anglo-American collaboration in its construction. Subtitled, 'Atomic Energy ... for Military Purposes', Maud recognised that the cost and difficulty of extracting U235 placed the job beyond the reach of the British, already overwhelmed by the expense of the war effort. The extraction of 1 kilogram of fissile uranium per day would cost £5 million. Collaboration between the Americans and British was thus 'the highest priority'.

The Maud Report explained the bomb's detonation mechanism in terms that politicians could understand: A chain reaction would occur when U235 reached 'critical mass' – that is, the point at which neutrons were released from their nuclei – 'resulting in an explosion of unprecedented violence'. Detonation would result when two pieces of the fissile material – 'each less than the critical size but which when in contact form a mass exceeding it' – were smashed together at high velocity, using 'charges of ordinary explosive in a form of double gun'. Maud envisaged a nuclear blast equivalent in force to 1800 tons of TNT, releasing large quantities of radiation that would make the area 'dangerous to human life for a long period'.

The great difficulty Maud identified was this: how to extract U235 from common uranium ore? British scientists in Cambridge and

Birmingham had gone 'nearly as far as we can on a laboratory scale'. Only America had the resources to extract supplies of fissile uranium on an industrial scale. It concluded with a sweetener: an atomic bomb would prove cheaper than conventional explosives, 'when reckoned in terms of energy released and damage done ...'.

•

Meanwhile, the Americans were making great advances towards an atomic bomb but chose not to share these with their British allies. Roosevelt had approved the development of a bomb in 1939, after receiving Einstein's letter, and delegated Vannevar Bush and James Conant to drive 'general policy in regard to the program of bombs of extraordinary power'. Bush, an engineer who directed the Office of Scientific Research and Development, and Conant, an organic chemist who headed the National Defense Research Committee, together oversaw the process that eventually subsumed Britain's Tube Alloys – and Maud Committee – into S-1, America's atomic bomb project, and rendered the construction of the weapon an exclusively American enterprise.

It was not surprising that some American experts were loath to admit their debt to European science. On receipt of his copy of the Maud Report, the 'inarticulate and unimpressive' Lyman Briggs (as Oliphant described him), chairman of America's Uranium Committee – another secret agency in a web of official US bodies charged with identifying and securing the world's supply of uranium for military use – promptly locked it in his safe without showing it to the committee. Oliphant, a full-throated champion of US–British collaboration, flew to the US in August 1941 in an unheated Hercules bomber – such was the urgency of his mission – to investigate the fate of Maud, which no US official had responded to for weeks. On his arrival, the Australian was 'amazed and distressed' at Washington's failure to act. Oliphant persuaded Ernest Lawrence, the Nobel Prizewinning Berkeley physicist, to press the report on the US government. He also secured an audience with Conant and Bush. Aloof men conscious of their possession of great power, they were initially disdainful of this zealous emissary, and dismissed Oliphant's importunate appeal as 'gossip among nuclear physicists on forbidden subjects'. Conant nonetheless saw the need to act. On 3 October 1941, he obtained a complete copy of Maud and handed it to Bush; on 9 October, Bush placed it under Roosevelt's nose. Conant later cited Oliphant's persuasiveness as the 'most important reason' the US nuclear program changed direction from an uncertain experiment into an all-out effort to build a bomb.

•

In 1942 Conant and Bush worried that progress on the bomb was far too slow. They feared at this time that Berlin would get there first – a view shared by the émigré physicists, who despaired of the prospect of a nuclear-armed Germany.

On 13 June that year Conant and Bush wrote to War Secretary Stimson, Army Chief of Staff General George Marshall, and then Vice President Henry Wallace, urging them to hasten the decision to establish a secret organisation to co-ordinate the development of S-1. Their letter, intended ultimately for the President, conveyed the unanimous belief of several top American physicists that the production of an atomic bomb was possible, and outlined several methods of producing the fissile material (U235), all of which 'seemed feasible'. They envisaged the production of 'a bomb a month' by 1 July 1944 'with an uncertainty either way of several months'.

They recommended the physicists Arthur Compton, Ernest Lawrence and Harold Urey as scientific leaders of the operation. All were Nobel laureates, and all were immersed in nuclear physics. A deeply religious man, Compton defined science as 'the glimpse of God's purpose in nature', of which the atom and radiation were manifest signs – he won the Nobel Prize in 1927 for his work on X-rays; the exuberant Lawrence received the prize in 1939 for his invention of the cyclotron atom smasher; Urey, an intensely hard-working man who discovered deuterium, received the Swedish gong in 1934 for his pioneering work on the separation of isotopes. The whole project, Bush and Conant estimated, would cost roughly $85 million (another huge miscalculation). Stimson, Marshall and Wallace sent the recommendation to the White House, from where it was promptly returned, marked 'OK–FDR'.

The biggest and most expensive industrial operation hitherto launched in America rolled into production. Construction companies were to build, from scratch, enormous factories using industrial processes that had never been attempted; engineers to take new scientific discoveries onto production lines; and scientists to create and test complex mechanisms using chalkboard designs ready for factory application within months.

The 'Manhattan Engineer District [MED]', the project's official title – chosen for its innocuous ambiguity – was always strictly a military operation. It subsumed all civilian research and experimentation in atomic science. The man appointed, on 16 August 1942, to head MED operations was an army engineer, Colonel Leslie Richard Groves. Initially reluctant to accept the job – Groves had hoped for a combat role – once persuaded, the colonel pursued it with characteristic gusto. 'If you do this job right, it will win the war,' Lieutenant General Brehon Somervell, commanding general of US Army Service Forces, told Groves. If further persuasion were needed, Major General Wilhelm Styer, a member of the Military Policy Committee that oversaw the development of atomic energy, reassured Groves that his appointment would transform the war effort.

If they meant to fan his ego, they misunderstood the burly engineer. To a man of Groves' self-worth, success was not the issue – 'If I can't do the job, no one man can,' he later confided in his memoirs. In fact, Groves wanted assurances that the bomb *could* be built before he accepted the job: the budget was a mere $85 million; the science seemed vague and unformed; and his rank as colonel did not exert the necessary authority. Groves knew more about the Project than he let on, of course, due to his excellent contacts in the Army Construction Division, where he had overseen the building of the Pentagon. He was haggling over terms. The army met his needs immediately: Styer promoted him on the spot to brigadier general, and Groves' other key demands – top-level security clearance, a virtually unlimited budget, and total operational control – were promptly delivered.

•

The portrait of Groves as a great brute of a man, tyrannical and unyielding, has been widely received; his qualities less so. Groves' mastery of industrial engineering, his iron self-discipline and extraordinary administrative and organisational skills qualified him as possibly the only man willing or able to attempt to build an atomic bomb in the time available. His working hours (around the clock), dismissal of anyone not up to the job, and pachydermal indifference to criticism – as his colleagues often remarked – further recommended him. His superiors wanted a man able to withstand the pressure of surely one of the most difficult jobs of the war effort. To his detractors, Groves seemed more machine than man; to his admirers, such as his deputy and chief engineer Colonel Ken Nichols, he was 'outstanding' – 'extremely intelligent' and 'the most egotistical man I know'.

The third son of four children to an austere Presbyterian army chaplain, 'for whom thrift was one of the godliest of virtues', and a gentle, sickly woman worn down by the weight of constant travel and hard work, Leslie was a good little boy, by all accounts, whom family and friends preferred to call 'Dick'. Biblical remonstrations pursued the boy's youth and Sunday Sabbaths were strictly observed. The family was constantly on the move following Chaplain Groves' service itinerary. Illness stalked their peripatetic rounds of the nation: the chaplain suffered recurring bouts of malaria, which he had caught in Cuba on a visit in 1898; Groves' crippled sister undertook strenuous exercises to straighten her hunched spine; and his mother, Grace, suffered from a chronic heart condition. The Groves home had the atmosphere of a 'family hospital', noted the general's biographer.

The boy grew into a tall, athletic young man who, as a student at Massachusetts Institute of Technology, kept his own counsel; he was self-absorbed, diligent and friendless. The premature death, from pneumonia, of his elder brother Allen, his parents' favourite, stung young Leslie to action. He defied his father's wishes and enrolled in the US Military Academy at West Point in 1916, where his arrogance won him few friends. He had inherited his father's tight-fistedness and refused to pay for his laundry, earning him the nickname 'Greasy', an appellation that pursued him through life. Lonely and unpopular, he rose through sheer grit and intelligence unrelieved by the personal charm or good humour that eased the advancement of less talented men.

He never shouted or swore. He led by quiet intimidation; those who angered him received 'the silent treatment'. As a young captain, he 'gave the impression of a man of great latent power, who was biding his time,' observed the historian Robert Norris. His baleful stare and his snap decisions left a trail of anxiety.

The general travelled tirelessly around his secret empire, which grew rapidly during 1942–44 to embrace city offices, university laboratories and secret factories on remote prairies, from the Midwest to Manhattan. Weekly, his private train shot through the dark fields of the Midwest towards another trembling recipient of his wrath. Everything he did was shorn of clutter, time-wasters and verbiage; all fine-tuned to the task ahead. He had no time for small talk or pleasantries. His memos were brief and abrupt, regardless of the seniority of the recipient; he closed meetings when he decided, even with superiors. At his first meeting with President Roosevelt, he impertinently announced that he had to leave early. An example of the Groves style was the following memorandum:

GEN GROVES TO DR V BUSH, REAR ADMIRAL WR PURNELL, MAJ-
GEN WD STYER: On 11 May 1943, the MED entered into a fixed-fee
($1.00) cost contract with Union Mines Dev Corp to determine the
world resources of uranium and, to the extent possible, bring such
resources under the control of the US Government.

Groves thus commandeered the world's available supplies of yellowcake.

If knowledge is power, Groves was among the most powerful people in America at this time. His security clearance matched the highest levels of the military and political establishment. He knew the fate of the earth – insofar as nuclear weapons might decide it – ahead of senior politicians and military commanders. The State Department was unaware of the bomb until February 1945 when Groves decided to let them in on the secret; General Douglas MacArthur was not officially informed until mid-1945; and Fleet Admirals Ernest King and Chester Nimitz, respectively Commander in Chief United States Fleet, and Commander in Chief Pacific Fleet, were only informed, on Groves' recommendation, in early 1945.*

By then, Groves had been working on the bomb for three years. His power over the weapon was complete. Anyone who opposed the general – and by extension, the bomb's purpose – found themselves mysteriously removed (irritating scientists, notably Szilard, were smoothly sidelined). Few questioned the intent of the project; their very awareness of it assured their approval. And Groves had powerful champions, chief among them Bush, Conant and, of course, the new President. His thinking was thoroughly in line with that of Truman and Byrnes, who believed that Japan was the immediate target, but

* On 27 January 1945 King wrote to Nimitz about 'a new weapon' that 'will be ready in August of this year for use against Japan', and to prepare accordingly. The next month Groves' personal emissary, Fred Ashworth, flew to Guam with a letter for Nimitz signed by Admiral King. Ashworth was convinced Groves had written it: 'What the letter said was that there would be a thing called the atomic bomb in his theater about the first of August,' Ashworth later wrote. Nimitz received Ashworth in private, read the letter and asked: 'Don't those people realize we're fighting a war out here? This is February and you're talking about the first of August.' Ashworth replied, 'Well, this is just to let you know what's happening.' Nimitz looked out the window, turned and said: 'Well, thank you very much, Commander. I guess I was just born about twenty years too soon.'

Russia was America's ultimate future enemy: 'There was never ... any illusion on my part but that Russia was our enemy, and the [Manhattan] Project was conducted on that basis.'

In time, he would exert a disproportionate influence over senior politicians; his unique position and mastery of the project's detail made him indispensable to its success, and they bowed to his demands. It was on Groves' insistence, for example, that in 1943 Byrnes, then head of the Office of War Mobilization, was told to abandon an inquiry into the $2 billion spent on the Manhattan Project because it might damage security. The President quashed the Byrnes review; and while Byrnes would become a staunch supporter of the bomb, at the time he was anxious to know why, and on what, so much public money was being spent. Stimson, too, felt the shadow of this granite presence: Groves surreptitiously drew attention to the War Secretary's doom mongering and negativity, and would gradually help to marginalise him.

Groves' first actions were eminently practical. In 1942 he moved to secure control of the world's largest uranium supply, in the Belgian Congo, through negotiations with Edgar Sengier, the managing director of the mining giant Union Minière du Haut Katanga. Sengier, a discreet Belgian banker and engineer, was relieved: the State Department, then under Edward Stettinius, had previously misunderstood the importance of uranium and rebuffed his approaches. Groves understood all too well: in a meeting with Colonel Nichols, the chief engineer of the Manhattan Project, on 18 September 1942, Sengier agreed to supply 1250 tons (1134 tonnes) of ore that had been imported from the Congo and was stored in some 2000 steel drums in a Staten Island warehouse. More would follow.

An equally pressing priority after his appointment was personnel: the general needed better than the best; failure was not an option. If the Project fails, General Somervell had joked, he and Groves might as well buy houses next to the Capitol, 'because you and I are going to live out our lives before Congressional committees'. Most importantly, Groves needed a lab leader. The exacting résumé narrowed the field to a handful: Ernest Lawrence, however, could not be spared from his work on the cyclotron at Berkeley and Arthur Compton was inseparable from the Metallurgical Laboratory (a codename) of Chicago University, where he co-led the work on nuclear fission with Fermi and Szilard. That left Robert Oppenheimer.

•

Few people have experienced a more controversial rise and fall in the public eye than Julius Robert Oppenheimer. A tall, thin, lanky man with blue eyes and a shock of dark hair, he walked, or rather advanced, on the balls of his feet, giving the impression that he floated by. His baggy suits and porkpie hat created a faintly clown-like effect. Oppenheimer was a Jew; his former girlfriend, Jean Tatlock, a communist; his brother, Frank, a Jewish communist. Those facts placed Oppenheimer outside the East Coast Anglican establishment. His communist associations deeply compromised him, decided the FBI and the security services, which initially refused to clear him for work on the Manhattan Project.

Oppenheimer's exceptional intellectual and, as it proved, administrative, gifts overrode the security risk, Groves believed. That he was related to, or slept with, communists did not mean he shared their beliefs. Groves, who had read everything he could about 'Oppie', saw a distinction that eluded the secret services. On 20 July 1943, the general confirmed the scientist as 'absolutely essential to the Project'. Thus began one of the oddest and most effective working partnerships in American history: Groves, immense, boorish and demanding; Oppenheimer, frail, cultured and intellectual.

Oppenheimer was born in 1904 to a wealthy, liberal New York family, and his early correspondence conveys the impression of an extremely clever young man in the thrall of his transcendent intellectual gifts. At Harvard (where he attended Bohr's lectures in 1923) and, as a Rhodes scholar at Oxford, he applied his prodigious mind as a tool for rapid self-advancement. At Harvard he explained his desire to jump straight to advanced physics on the grounds that he received 'a 96' in his physics entrance exam, grade As in all his subjects, and his 'partial' reading then involved four volumes on kinetic theory, thermodynamics, statistical mechanics and quantum theory; James Crowther's *Molecular Physics*; Henri Poincaré's *La Physique Moderne* and *Thermodynamique*; James Walker's *Introduction to Physical Chemistry*; Wilhelm Ostwald's *Solutions*; J. Willard Gibbs' *On the Equilibrium of Heterogenous Substances*; and Walther Nernst's *Theoretische Chemie*. These were advanced texts; Oppenheimer read them in English, French and German. He also read Ancient Greek and Latin. A few months earlier he had completed his freshman year, with top grades in French prose and poetry (Corneille through to Zola), the history of philosophy, and courses in maths, chemistry and physics. 'Whatever reading or work you may advise, I shall be glad to do ...' he added. He was 19.

The young Oppenheimer styled himself a philosopher-aesthete who affected an interest in literature ahead of science. *A La Recherche du Temps Perdu* by Marcel Proust left a deep impression on him. He quoted a favourite passage from memory: 'Perhaps she would not have considered evil to be so rare, so extraordinary, so estranging a state, to which it was so restful to emigrate, had she been able to discern in herself, as in everyone, that indifference to the sufferings one causes, an indifference which, whatever other names one may give it, is the terrible and permanent form of cruelty.'

He would be an occasional poet, as this example of his student juvenilia (1923) intimated:

... When the celestial saffron
Is faded and grown colourless,
And the sun
Gone sterile, and the growing fire
Stirs us to waken,

We find ourselves again
Each in his separate prison
Ready, hopeless
For negotiation
With other men.

And an art critic: 'The three things you sent me,' he wrote that year of sketches by a friend, '... show a good deal more care and inspiration in their design than the van Dyke and Giotto things you defend – I like immensely your abstract, particularly, I think, for its obvious but skillful repetition of color and texture, and the corresponding dramatic rapidity. Best of all, though, I like the nude, in spite of its technical scraggliness ...'

Harvard's recognition of his talents did not surprise him. He drolly wrote at the end of another superb semester, 'The work goes much as before: frantic, bad and graded A.' He graduated with distinction – summa cum laude – within three years, and celebrated privately with two friends and a bottle of laboratory alcohol: 'Robert, I think, only took one drink and retired,' remarked his friend, Fred Bernheim.

In his mid-20s he suffered from depression, hallucinations and suicidal feelings – 'a tremendous inner turmoil'. He self-diagnosed a schizoid personality and seems to have recovered through sheer hard work and strength of mind. At one point, in London, he dismissed his Harley Street psychiatrist as 'too stupid': '[Robert] knew more about his troubles than [the doctor] did,' wrote a friend. 'Robert had this ability to ... figure out what his trouble was, and to deal with it.'

There was nothing rebellious or dissipated in the young Oppenheimer; rather something joyless ... a space of pure intellect in his 'separate prison' from other men. 'Conrad's *Youth*,' he wrote, dismissively, 'is a beautiful novelette on the futility of youthful courage and idealism.' Perhaps, but were not the futile pursuits of ideals a rite of passage in the young? Oppenheimer, on the contrary, strove for perfection: of what use were ideals if one failed in their pursuit? His overriding psychological impulse was a fear of failure. In this he had something in common with Groves. 'Ambitious' is too crude, too obvious, a term for such complex men; they acted in defiance of, or in spite of, the voices in the wings as surely as they pursued the laurels of success.

On 17 October 1931, Oppenheimer's mother died. 'I am the loneliest man in the world,' he wrote somewhat disingenuously, as he had long felt estranged from her and compensated for this with excessive displays of somewhat contrived affection. After her death he immersed himself in his work: 'On the Stability of Stellar Neutron Cores', 'On Massive Neutron Cores' and 'Behavior of High Energy Electrons in Cosmic Radiation' were among his (co-written) scientific papers of the 1930s, unimpressive alongside the output of Lawrence, Compton and Fermi. Yet Oppenheimer was an intellectual dilettante, a modern Rennaissance man. He took up the study of Sanskrit: 'I have been reading the Bhagavad Gita with ... two other Sanskritists,' he told his brother in 1933. If any single mind has attempted to reconcile Eastern and Western cultures and bridge the Sciences-Arts divide, it was his.

The precocious student grew into a loyal friend, excellent teacher and inspiring leader. Along the way, at Harvard and in the 1920s at Cambridge and Göttingen in Germany, he met the greatest physicists of the day and formed close relations with Max Born, his professor at the University of Göttingen with whom he wrote 'On the Quantum Theory of Molecules', his most cited work. In the 1930s in California, as a professor at Berkeley University and later at Caltech, his students became his admiring disciples. The physicist Hans Bethe, one of his most staunch friends – who would work with him on the atomic bomb – said of him: 'Probably the most important ingredient he brought to his teaching was his exquisite taste. He always knew what were the important problems, as shown by his choice of subjects. He truly lived with those problems, struggling for a solution, and he communicated his concern to the group ... He was interested in everything, and in one afternoon they might discuss quantum electrodynamics, cosmic rays, electron pair production and nuclear physics.' His mature attributes – the rare fusion of scientific excellence, self-discipline and organisational flair – marked him for leadership of the Manhattan Project.

Another lure was his deep yearning to belong; to be inside the tent. A longing to know the world and the people who presumed to know and control him possessed Oppenheimer. He wanted not so much to share the nest of the American establishment as to feel able to share it, and then take or leave it. His family's wealth could buy yachts, ranches and horses but not this – the coming and going of a welcome insider. His sheer brilliance would force admission to the gilded enclave: it convinced Groves, and Groves persuaded the innermost sanctum of American power.

•

In the late 1930s Oppenheimer worked closely with Ernest Lawrence and the cyclotron pioneers at Berkeley, ushering him into the orbit of work on the bomb project. On 21 October 1941, at a meeting at General Electric's Schenectady laboratories, New York, Oppenheimer calculated the critical mass of U235 that would be required for a chain reaction – and an effective weapon. Conant, who had been one of Oppenheimer's lecturers, invited him to conduct further work on 'fast neutron' calculations that were critical to a chain reaction, and that year Oppenheimer opened a 'summer school' in bomb theory at Berkeley. His appointment to the Manhattan Project briskly followed in September 1942.

Oppenheimer and Groves chose Los Alamos, New Mexico, as the site for the bomb's laboratory. An unlikely horseman, Oppenheimer knew the terrain; he owned a ranch in the nearby Sangre de Cristo Range. The site sat atop a red-earth mesa, the Pajarito Plateau, near White Rock Canyon and the Valles Caldera, within an hour's drive of Santa Fe and Albuquerque, within easy reach of water, and near the train line and flight paths to major cities. To the cosmopolitan scientists housed in an erstwhile boys' boarding school, which the US government had bought, it seemed impossibly remote.

Here the finest physicists and chemists came willingly, or were persuaded to come. Oppenheimer plundered the nuclear physics departments of the most prestigious universities in the country, and the world – to those institutions' intense irritation. Most of the scientists were in their 20s and 30s; some had just finished their degrees. They gathered on the 'Hill', as they named Los Alamos, rather like Swift's Laputans on their floating island. Their ambitions went further than extracting sunbeams from cucumbers, however. They were attempting 'a far deeper interference in the natural course of events than anything ever before attempted', Bohr told Churchill on 22 May 1944 (Bohr had been smuggled out of German-occupied Denmark in 1943). Even at this stage Bohr consistently warned of an impending arms race and urged openness with the Russians. This little impressed Churchill, who scolded the physicist at their May meeting ('I did not like that man with his hair all over his head,' the Prime Minister later told Lord Cherwell), and afterwards cautioned Roosevelt that Bohr be monitored as a security risk: 'It seems to me Bohr ought to be confined,' Churchill said, 'or at any rate made to see that he is very near the edge of mortal crimes.'

Many of Los Alamos' boffins were Jews exiled from Nazi-occupied Europe, and motivated by hard personal experience – chiefly the physicists Bethe, Peierls and Edward Teller, and the astonishingly gifted mathematician John von Neumann, to name a few; Jewish émigrés also had prominent roles at Chicago University and in other parts of the Manhattan Project. These scientists included Szilard and the Nobel laureate James Franck, who had left Germany in 1933. Their job, as they initially saw it, was to build a weapon to defeat Germany. In the pursuit of victory they shared a personal motive: to avenge family members and friends persecuted by the Nazis. Enrico Fermi, who left fascist Italy to protect his Jewish wife, shared their determination; Oppenheimer sympathised; as did Einstein. They scorned 'Aryan scientists' such as Heisenberg who remained in Germany.

•

In 1942–43 the Manhattan Project scientists came under great pressure from Groves to meet his stringent schedule. Events were progressing in several locations, chiefly Chicago, where the scientists' priority was to demonstrate that a self-sustaining nuclear reaction might work. While the theory was sound, and pointed with certainty to this outcome, nobody had yet achieved it. The man responsible for the breakthrough was Enrico Fermi, nicknamed 'the Pope' in deference to his nationality and reputation for omniscience. Fermi, a professor of physics at the Univerity of Rome at the age of 24, received the Nobel Prize in 1938 (aged 37) for his 'demonstrations of the existence of new radioactive elements produced by neutron irradiation, and for his related discovery of nuclear reactions brought about by slow neutrons', according to the citation. That year Fermi and his wife, Laura, immigrated to America. He worked first at Columbia University in New York and from 1940 with Franck, Szilard and Compton at the University of Chicago's 'Metallurgical Lab', where he focused his energy on achieving a nuclear chain reaction.

On 2 December 1942, in a squash court beneath the West Stands of Stagg Field football stadium, near the Metlab, Fermi and his team put the finishing touches on the first 'atomic pile' of graphite blocks embedded with uranium. The dozen or so scientists present included Compton, Szilard and Wigner, who gathered on the spectator stand 2 metres above the court floor, whereupon the young George Weil prepared to remove the last rod that held the reaction in check.

In theory, the reaction was capable of flying out of control and consuming much of Chicago along with the footballers on Stagg Field. In practice, this was highly unlikely. Cadmium rods were inserted in the pile to absorb overeager neutrons, to stop or slow the bombardment process and to avoid a conflagration: 'The same effect might be achieved,' Fermi later wrote, 'by running a pipe of cold water through a rubbish heap; by keeping the temperature low the pipe would prevent the spontaneous burning.' If the rods failed, a three-man 'liquid control squad' on a platform overhead stood ready to drench the pile with buckets of cadmium salt solution.

Always cool under pressure, Fermi gave the signal to initiate the neutron bombardment. George Weil removed the rod. The team above fastened on their indicators, which measured the neutron count and 'told us how rapidly the disintegration of the uranium atoms … was proceeding'. The first attempt failed – the pile's safety threshold was set too low. 'Let's go to lunch,' Fermi said. After lunch and several adjustments, they resumed their places and made a second attempt. The neutron storm rose 'at a slow but ever increasing rate'; shortly the reaction was self-sustaining. It was unspectacular – 'no fuses burned, no lights flashed,' Fermi wrote. 'But to us it meant that release of atomic energy on a large scale would be only a matter of time.' Compton made a coded phone call to Conant, in Washington:

Compton: The Italian navigator has landed in the New World.
Conant: How were the natives?
Compton: Very friendly.

An elated Eugene Wigner presented Fermi with a bottle of chianti. They drank from paper cups, in silence.

•

Meanwhile, in the valley below the Hill in New Mexico, ordinary Americans went about their business in morlock-like ignorance of the goings-on above them, where great experiments were advancing to meet harsh deadlines.

Oppenheimer arrived at Los Alamos on 16 March 1943 with his wife, Kitty – senior scientists were permitted the company of their wives. Their home was a log-and-stone house built in 1929 at the end of Bathtub Row, named because the homes thereon had one. He started work at 7.30am. Periodically the couple entertained. His 'dark moods', 'savage sarcasm' and knack for the brutal riposte were more restrained in the company of his peers. He charmed his staff and endowed his colleagues with 'rare qualities and facets they did not know they possessed'.

Groves and Oppenheimer worked like two cogs in a machine. Oppenheimer met the engineer's demands; Groves respected the scientist's brains. Oppenheimer accepted the militarisation of his bomb-making laboratory; Groves held back at insisting that scientists wear military uniforms. They shared an overweening ambition for the Project, and 'saw in each other the skills and intelligence necessary' to fulfil it. Oppenheimer looked the part – groomed, business-like and efficient; Groves responded with rare empathy, treating the scientist, as the historian Norris observed, 'delicately, like a fine instrument that needed to be played just right'.

On arrival, senior staff – who had 'Secret Limited' level security clearance – received a copy of *The Los Alamos Primer*, a digest of a series of lectures, 'On How to Build an Atomic Bomb', given by the physicist Robert Serber, one of Oppenheimer's colleagues at Berkeley. *The Primer*, edited by Edward Condon, the lab's associate director, was the first, detailed explanation of why they were there. It was issued only to the most senior scientists and lab technicians, many of whom reacted with euphoria at this final confirmation of their mission.

The Primer was an exposition not a cookbook: 'The object of the Project,' it informed the new arrival, 'is to produce a *practical military weapon* in the form of a bomb [Condon used the word 'gadget'] in which the energy is released by fast neutron chain reaction ...' (his emphasis). Serber ran through the process and risks, sprinkling his text with equations of interest to the specialist.

Serber left little to the physicists' imagination in his section on 'Damage': intense radiation ('a very large number of neutrons') would saturate everything to 'a radius of 1000 yards [900 metres]' from the explosion – 'great enough to produce severe pathological effects'; a million curies of radiation would remain 'even after 10 days'. The blast wave's destructive radius would extend to 'about two miles [three kilometres]'. He concluded with a discussion of the detonation methods.

•

Far bigger and more dangerous facilities than Los Alamos were being erected around the country throughout 1943 and 1944; they were neither dark nor Satanic, but clean to a curlicue and hermetically sealed. Among the most important were Oak Ridge and Hanford, whose employees were primarily engineers and technicians, rather than scientists.

Oak Ridge was the administrative heart of the Manhattan Project; and the site of the world's largest factory, Clinton Engineering Works, built on a 24,000-hectare government reservation on the Tennessee Valley floor, 30 kilometres northwest of Knoxville. General Electric, Westinghouse and Allis-Chalmers supplied the main equipment; Stone & Webster designed and constructed it; and Tennessee Eastman, a subsidiary of Eastman Kodak, handled the operations. Oak Ridge's job was essentially to separate the highly fissile U235 uranium isotope from the ordinary U238, and manufacture enough U235 for use in a bomb; Hanford's, in Washington state, was to produce an atomic bomb using the alternative weapons-grade material plutonium, a highly fissile, extremely toxic new element, the discovery of which Lawrence had reported in May 1941. Both projects ran in tandem as an insurance policy should the other fail. Both involved huge amounts of electricity, water, equipment, space and manpower.

Oak Ridge grew into a secret city employing, at the height of construction, almost 70,000 people for whom, at one point, 1000 homes were being built per month (in peacetime the town held 3000 inhabitants). All the social requirements of an inland American city accompanied them: schools, a hospital, a dental clinic, cinemas and sports facilities. By June 1945 Oak Ridge had one high school, eight elementary schools and a grammar school built or under construction, with 317 teachers and 11,000 pupils; nine drugstores; 13 supermarkets and seven cinemas. Seventeen different religious bodies catered for the workers' spiritual needs.

The site at Hanford, which at its peak would employ 45,000 workers, was designed and built by the DuPont Corporation. At first, the task seemed impossible and DuPont demurred. Groves' argument – 'if we succeed in time, we'll shorten the war and save tens of thousands of American lives' – persuaded the company's president, Walter Carpenter. He attached two conditions: that DuPont made 'no profit whatever from the work it did'; and that 'no patent rights growing out of DuPont's work on the project should go to DuPont'. 'Our feeling,' Carpenter told employees (after the war), 'was that the importance to the nation of the work on releasing atomic energy was so great that control, including patent rights, should rest with the Government.'

Construction of Hanford involved the use of 8500 major pieces of construction equipment, and the building of 550 kilometres of permanent roads and 1200 kilometres of railroad; 36,000 tonnes of steel, 38,000 cubic metres of lumber, and 11,000 poles for electric power were also required. Reports that salmon were dying in the nearby Columbia River drew Groves' concern, and he asked his medical chief Dr Stafford Warren to commission a study on the effects of radiation on aquatic organisms. The study, by Dr Lauren Donaldson, concluded that, as far as could be ascertained, radiation emanating from the factory would not harm salmon. Groves was acutely aware of the dangers of radiation to humans, too – 'a serious and extremely insidious hazard', he recalled in his memoirs. He insisted on enclosing the huge reactors in heavy metal walls and concrete tanks to protect workers, and set the safe tolerance dose at one-hundredth of a roentgen per day (that is, 0.12 rems; the current annual US occupational limit in adults is 0.005 rems, according to the MIT Radiation Protection Office).

●

The greatest challenge facing Oak Ridge was how to extract the highly fissile but extremely rare isotope U235 from common uranium. Of several methods, Groves directed his empire to pursue two: electromagnetism and thermal/gaseous diffusion.

The electromagnetic method utilised Canadian scientist Arthur Dempster's 1918 invention of the mass spectograph in conjunction with Ernest Lawrence's cyclotron (nicknamed the slingshot). The process is based on the principle that electrically charged atoms (ions) describe a curved path when accelerated through a magnetic field. With the speed and magnetic field at constant levels, atoms of different mass (for example, the uranium isotopes U235 and U238) will follow different 'flight' paths: heavier atoms have longer radii than lighter atoms. In this way, two kinds of uranium could be separated, neutralised and collected in specially designed containers. It sounds simple; in practice it involved the construction of the world's largest magnet in a facility covering 200 hectares. The electrical conductors required 86,000 tons of silver since the war's demand for copper exceeded national supply. The silver was borrowed from the US Treasury, whose tight-lipped reply to Colonel Ken Nichols' request was, 'Colonel, in the Treasury we do not speak of tons of silver; our unit is the Troy ounce.'

The thermal diffusion method works on the principle that two gases will diffuse through a porous membrane at rates inversely proportional to the square root of their molecular weights. For example, if gas X has a molecular weight of 4, and gas Y has a molecular weight of 9, when they pass through a membrane, the lighter gas diffuses at a rate of three volumes to the heavier gas's diffusion rate of two volumes. To separate the uranium isotopes this way proved exceptionally difficult, however, as their molecular weights differed only slightly. The gaseous ore uranium hexafluoride had to be forced through a cascading network of 'barriers' – or atomic 'sieves'. Each pass yielded a gaseous residue containing a greater proportion of U235 than the previous stage. To visualise this in operation, one could imagine a gigantic tube snaking around a football-stadium-sized facility punctuated along its path by thousands of huge membranes. Everything inside had to be kept spotlessly clean to avoid polluting the enriched gas at any stage of the journey. The challenge involved building membranes with gaps small enough to faciliate the passage of U235 gas molecules: the holes in the screens, or sieves, eventually delivered were one-hundredth of a micron, or four ten-millionths of an inch, in diameter. 'They won't believe it when the time comes when this can be told,' James Conant said of the Oak Ridge site. 'It is more fantastic than Jules Verne.'

•

Security was a paramount priority in the Manhattan Project. Important arrivals at Los Alamos set mouths aflutter. Groves insisted that celebrity scientists use aliases when visiting. Niels Bohr became Mr Baker; the Fermis – Mr and Mrs Farmer; Harold Urey – Hiram Upton; and Arthur Compton – Mr Comstock. Codenames permeated the whole enterprise. The main industrial plants, the plutonium bomb-making factory at Hanford and the uranium enrichment plan at Oak Ridge were Sites W and X, and Los Alamos Site Y. Various key words were forbidden, chiefly 'bomb' and 'uranium'. Wives were kept in the dark: 'I can tell you nothing about it,' husbands would say, before their departure for Los Alamos. 'We're going away, that's all.' Only Groves was permitted a diary, which said little and poorly. A sign in Groves' offices, the mysterious Rooms 5120 and 5121 of the War Department

Building in Foggy Bottom, Washington, said, 'O Lord! Help me to keep my big mouth shut!' The loquacious found themselves on long-term assignments in the Pacific.

Ordinary employees – engineers, factory hands, clerical staff – were carefully screened and worked in cordoned-off 'compartments', or silos. Workers in one silo had no idea what their counterparts were up to in another: 'Compartmentalisation of knowledge, to me, was the very heart of security,' Groves later said. 'My rule was simple ... each man should know everything he needed to know to do his job and nothing else.' When addressing employees, Groves said nothing of the bomb, or the fact that it might 'win the war', only that their work was of 'extreme importance to the war effort'. This charter throbbed in the minds of hundreds of thousands of carefully processed employees, none of whom knew what he or she was making; nor did they see any result for their effort: 'They would see huge quantities of material going into the plants, but nothing coming out.' The rows of young women at Oak Ridge sitting at their calutrons – mass spectrometers used for separating uranium isotopes – had 'no idea what they were doing' or what the machines did; 'merely that they were making some sort of catalyst that would be very important in the war', observed Theodore Rockwell, an engineer at Oak Ridge. Unions were banned: 'We simply could not allow [them],' Groves wrote, 'to gain the over-all, detailed knowledge that a union representative would necessarily gain ...' Black workers were segregated, paid little and housed in poor conditions; racial discrimination against African-Americans and Hispanics did not pause for the war effort.

Notwithstanding the tightest security net, several spies infiltrated the Project, the most damaging of whom was Klaus Fuchs, a German émigré who began spying for the Soviet Union from Britain in August 1941 (two other, relatively insignificant spooks were the technicians David Greenglass and Ted Hall). Having established contacts in Moscow, and insinuated himself into the British scientific community, Fuchs came highly recommended to Los Alamos, where he worked in the most sensitive department dealing with detonation: specifically, the extremely complex implosion system of the plutonium bomb, sketches of which he sent to his Soviet-run handler, Harry Gold. Fuchs was also deeply involved in research on the hydrogen bomb. Unusually hardworking, he drew no attention to himself and made a strong impression on Hans Bethe and Robert Oppenheimer. Washington would learn the extent of his damage after the war: the Russian bomb project drew extensively on Fuchs' information, which had 'great value', according to Soviet physicist Igor Kurchatov, who headed the Soviet atomic program. Indeed, Fuchs informed Moscow of 'essentially everything we were doing at Los Alamos' from August 1944, the physicist Edward Teller wrote in his memoirs. It enabled the Soviet Union to define the dimensions of an atomic bomb as early as April 1945, three months before the test of the weapon codenamed Trinity at Alamogordo in New Mexico.*

Fears of Soviet penetration of the atomic secret coalesced in Washington, where Truman quickly grasped that he had inherited the reins of a clandestine race – initially against Germany: 'They may be ahead of us by as much as a year,' James Conant, who headed the National Defense Research Committee, had warned in 1942. The Allies were then unaware of the woeful state of the Nazi atomic industry, which US intelligence (Project Alsos) would reveal as a miserable failure after the German surrender in May 1945. Albert Speer would later complain of Hitler's inability to grasp the 'revolutionary nature' of 'Jewish physics' (an exception concerned the brilliant Austrian Lise Meitner, whom Hitler had been prepared to exempt from the anti-Jewish laws, an offer she refused). Nor did Japan's atomic research pose any threat, as US intelligence would accurately conclude in early 1945. In fact, Japan had abandoned any hope of making a bomb in March 1943, when a colloquium of scientists concluded that it would take

* After the war, Fuchs returned to the UK where he worked at Harwell, the British nuclear facility. In 1949, using decrypted Soviet intelligence cables – the Venona Program – the FBI, along with British intelligence, began questioning Fuchs. He confessed, was convicted of espionage in a two-day trial and spent 14 years in Wormwood Scrubs prison. After his release, he moved to East Germany.

them ten years. The mysterious Soviet nuclear program under Igor Kurchatov was considered further advanced than Germany's or Japan's but well short of the United States'.

'This country has a temporary advantage, which may disappear or even reverse if there is a secret arms race,' Bush and Conant warned War Secretary Stimson in a letter of 30 September 1944. It was 'the height of folly' for the US and Britain to believe they could sustain their supremacy in nuclear technology, they added – and apprised the War Secretary of the consequences for the world were Russia and America to build secret nuclear arsenals. If Russia beat the US to the development of a 'super-super bomb' – the thermonuclear hydrogen bomb – 'we should be in a terrifying situation if hostilities should occur'. The 'expanding art' of nuclear war might, within a year, involve weapons equivalent to 10,000 tons of high explosive, enabling one nuclear-armed B-29 to inflict damage equivalent to that of 1000 conventionally armed B-29s. In place of the current policy of blanket secrecy, Bush and Conant recommended a 'free interchange' of all nuclear information, under the auspices of an international body, when the war ended. The two scientists' proposals were ignored. The Russians were not to be trusted and strictest secrecy would prevail – a policy that hardened after Soviet deceit at Yalta. America thus entered the last year of the war as the only country to possess the theory, brains and resources to produce atomic bombs, the first test of which was scheduled for the coming July.

CHAPTER 7

SPRING 1945

Today, a B-29 [bombed] the Shirakami Shrine area. Although the resulting damage was very minor, we must not underestimate the enemy simply because there was only one aircraft.

Yoko Moriwaki, Hiroshima schoolgirl, aged 12 (April 1945)

Supply the nation with pine roots – our untapped war power!
Pine oil has the power to crush the Americans and the British!
Lack of oil is our great enemy.

Official Japanese brochure, signed by the Departments of War, Navy and
Paddy Farming, distributed in Hiroshima in 1945

BUCKET RELAY TEAMS, A SIGHT familiar throughout Japanese cities, stood ready to douse the firestorm should it come to Hiroshima and Nagasaki, both of which stood eerily quiet as summer approached. By April 1945, the only other large cities that had been spared America's massed air raids were Kyoto, Niigata and Kokura. Their pristine state seemed a cruel affront to millions of homeless Japanese whose cities smouldered in ruin, and who sheltered in makeshift homes, schools and temples (dubbed the 'playgrounds of the poor') in the countryside. The people in the undamaged suburbs of Hiroshima and Nagasaki were amazed at their good fortune: perhaps we will be spared, they hoped. The mood of some lightened in the warmer weather. It was spring, and the longer spells of daylight reassured the worshippers of the sun.

Others felt a strange dread at the anomaly of their survival, as if their lives were numbered according to each day of peaceful sky, each night of calm. In March, news of the burning of Tokyo had reached them; by May, Chiba, Yokohama, Osaka, Nagoya and dozens of other cities had been firebombed; surely we are next, the people wondered. 'There was just one small air raid once,' recalled Kohji Hosokawa, a Hiroshima resident in 1945. 'Many people were quite worried about why they hadn't been bombed.' A dim sense of great expectations in ruins troubled the daily slog. Workers too weak, or frightened, stayed at home; national absenteeism rose to 49 per cent in July 1945, from 20 per cent in September 1944.

•

Some Japanese clung to the hope that national 'spirit' would win the war. The media's victory bleat grew shriller with the spreading signs of doom: it was Japan's divine destiny, state propaganda insisted, to rule over the 'eight corners of the world' aided, in the mortal realm, by the willing sacrifice of the 'One hundred million' (a wishful exaggeration; in 1945 there were 70 million Japanese), whose hearts and minds – *yamato-damashii* – would conquer the Anglo-American enemy.

Such were the shreds of hope to which old Zenchiki Hiraki clung, as he moved restlessly about his large, old farmhouse of cedar beams, straw thatch and paper screens. His younger son, Chiyoka, was then fighting somewhere in the Pacific. Chiyoka would return a *gunshin*, literally a 'god of war'; of that, Zenchiki had no doubt. His three middle-aged daughters and their children were then hard at work in Hiroshima. Only a small, sickly young woman and her three children shared his house; the woman he persisted in calling *yome* – 'daughter-in-law' – a reminder of her subordinate status. He never called Shizue by her name. '*Yome!*' he would shout, as though summoning a slave –

'*Yome*! You're working too slowly! *Yome*! Work harder!' – careless of the fact that this tearful young woman, bent double in his presence, was his son's wife.

Zenchiki had urged his son to marry because they needed 'extra hands' on the farm. The old man had hoped for an award-winning 'Children's Battalion', of 10 or more grandchildren. Eight or nine were not unusual in villages near Hiroshima. Alas, Shizue had managed only three; and they were too young to work. So Shizue planted, hoed, cooked, ploughed, fetched water, gathered thatch, chopped firewood, laundered and cleaned. Her 10-year-old son, Hisao, and elder daughter, Mitsue, five, cared for their baby sister, Harue, while their mother worked the fields. They rarely went to school in Hiroshima, and only visited the city, with their mother, to run errands for their grandfather.

Unlike Zenchiki, who refused to admit defeat, many other citizens were losing faith in official promises. They did their duty, turned up for work, but a creeping scepticism entered the privacy of their homes. National slogans such as 'We Want Not a Thing Until Victory' did not sit well on an empty stomach. Orders issued in the Emperor's name lost gravitas: 'Most people didn't really believe that orders were coming from the Emperor,' said Iwao Nakanishi, then a Hiroshima student, 'but everything was said to come from the Emperor. It was just a phrase used to make people follow orders. And we knew what would happen if we didn't follow orders.'

•

Most civilians were too exhausted or hungry to care, and mutely obeyed. Their main concern was their stomachs: the ordinary Japanese were gradually running out of food. The chronic shortage demoralised them more than the threat of attack or aerial bombardment: 'Our minds are too occupied with the problems of food,' was a typical reply to a private survey by the Domei News Agency, commissioned by the government in April 1945. 'As far as the war is concerned, let someone else do it. We are not too interested.' Such remarks were 'barometers', Domei reported, of the daily scramble for food, the scarcity of which they feared 'much more than foreign enemies'.

In 1945 the rice harvest was the worst since 1909; rice imports were 11 per cent of their pre-war average; and rice rations tended to be odious admixtures of rice and seaweed or other unidentifiable ingredients. Fruit and vegetable production plunged; the latter down 81 per cent on the previous year. Sugar consumption averaged 3 pounds (1.4 kilograms) per capita compared with 30 pounds (14 kilograms) per capita before the war. Offshore fishery landings, at 348,000 tons (315,700 tonnes), were half their 1944 level. Tuberculosis and beriberi were common; malnutrition endemic.

Potatoes were the staple: potato tempura, potato porridge, rice with potatoes, miso soup with potatoes, dried potato: 'Everything was made out of potato,' one Nagasaki resident complained. Sardines sometimes graced the family table; there was very little pepper, salt or sugar. Larger families could not survive on these rations, and parents or elder children often sneaked out to steal food at the risk of arrest.

They derived no hope from imports or the sale of raw materials for food; the US naval blockade had virtually sealed off the country. By the spring of 1945 oil, metal and other commodities were virtually exhausted. Japan had 12,346,000 barrels of oil in reserve in the first quarter of 1942, the wartime peak; by April 1945, it stored less than 200,000. The blockade ensured that Japan imported no oil in 1945. Turpentine, vegetable oil and charcoal were used as domestic fuel, and alcohol in aviation. The war machine was literally running out of gas.

Metals were so scarce that factories resorted to wood wherever possible: in aircraft construction, for example, wing tips, tail surfaces and fuselages were made of wood. Wood replaced plastics in cockpit control wheels, knobs and handles. Coal production halved between 1944 and 1945, with the latter yield of 22.6 million tonnes the lowest in 20 years (in 1925, Japan had produced 31.4 million tonnes of coal).

Nor did the manpower exist to extract or process it: that spring the cities and their surrounding villages were virtually empty of young men – away at war, wounded or dead. 'We rarely saw any fathers in the town,' one Nagasaki woman recalled, 'there were a lot of grandmothers, mothers and children. I remember seeing one father-like person in my town, but I think he was ill.'

•

These dire circumstances compelled Tokyo to intensify the civilian mobilisation orders and to extend them to schoolchildren. Under successive legislation – the National Mobilisation Law, Decisive Battle Educational Guidelines and Volunteer Enlistment Law – dating back to 1938, the regime had created a homeguard and civilian workforce, the supposed bedrock of the 'hundred million' who would defend the nation in extremis. In reality the mobilised civilians were a ramshackle amalgam of dads' armies, women's brigades and youth fighters: weekend warriors and weekday indentured labour. Their jobs were to make weapons, demolish homes, clear firebreaks, forage for food, collect pine roots, build air shelters and practise bayonet drills. Awls, bamboo spears, axes and picks were their weekend weapons. Men too old or medically unfit to fight were sent to war factories and forbidden light occupations 'so as to concentrate their efforts on the production of munitions', under the Scheme for Strengthening Domestic Preparedness (*Kokunai Taisei Kyoka Hosaku*). All except the very sick, very old and very young were mobilised.

Women – mostly young housewives like Shizue – were the chief targets of the labour drive, because there were so few men. The Women's Emergency Labour Corps (*Joshi Kinro Teishin Tai*) pressed into *monpe* some 12 million 'unoccupied women', including girls aged 15 and over. Seventeen kinds of jobs were reserved exclusively for women and girls, including clerical assistant, cashier, janitor, telephonist, bus conductor, cook and elevator operator. Since late 1944, however, most female recruits were employed in munitions factories.

In April 1945 Hiroshima extended the National Mobilisation Law to all 12- to 15-year-olds and evacuated younger ones to the countryside; infants were to stay with their parents in the cities. A few brave teachers opposed the law when first promulgated: the children working in the cities would have no protection, no means of swift evacuation, they pleaded. In Zakobacho, a Hiroshima suburb, school principals demonstrated outside the prefectural administration building, in reply to which the army threatened to beat them for defying the orders of the Emperor: 'Senior military officers,' recalled Kohji Hosokawa, a Hiroshima resident, 'came to different schools, gathered the teachers and told them of the mobilisation plan. When the teachers opposed it, the military officials withdrew their long swords and hit them on the ground as a threat ...'

Schools were converted into munitions factories and commissioned to make soldiers' uniforms, small arms and aircraft parts; the playgrounds often served as farm plots. The students' most important job was clearing debris to create firebreaks. As in other major cities, great sections of Hiroshima and Nagasaki were demolished to create empty spaces that were expected to check the spread of firestorms (today's wide avenues in Hiroshima are a legacy of this). The children would pass tiles and bricks hand-to-hand down long lines, often while singing anti-American songs. Or they collected metal: the Metal Collection Order of 1941 saw them salvaging any scrap that remained – gateposts, noticeboards, railings and household items, which could be used for melting down into weapon parts and bullets.

And they gathered pine oil, which was supposed to supplement the oil shortage. Teams of women and children were sent into the forests to dig up pine roots. In 1944, Hiroshima Prefecture officials claimed to have harvested 7500 tonnes of pine roots. In 1945 the regime lauded pine oil as the saviour of Japan, as local pamphlets brayed:

Supply the nation with pine roots - our untapped war power!
Pine oil has the power to crush the Americans and the British!
Lack of oil is our great enemy.
Without a constant supply of oil, we cannot win the war. Aeroplanes
and warships that have run dry of oil are nothing but ornaments ...
- The Department of War
- The Department of the Navy
- The Department of Paddy Farming

•

Girls living in Hiroshima were in the frontline of the mobilised labour forces, partly because there were so many girls' schools in the heart of the city. Yoko Moriwaki, a 12-year-old schoolgirl from the island of Miyajima, Hiroshima, was among the youngest of the city's mobilised children. She started a diary that year, her first at Hiroshima First Prefectural Girls' School (Daiichikenjo), one of the city's oldest and most prestigious.

April 6 (Fri)
The 1945 school entrance ceremony was held today. At last, I am
going to be one of those girls whom I have for so long admired - a
Daiichikenjo student! I am going to work hard every day and do all
that I can to not shame the name of Japanese schoolgirls.

Delighted to win a place at the selective school, Yoko pleaded with her mother for something to wear home after class. They unpicked the material of an old kimono and sewed it into a dress.

April 7 (Sat)
Today was the first day of school. In the morning, I sprang out of
bed, bursting with energy. If Daddy were here, he would have been
overjoyed! ... I must do my best for him, as well. Daddy, please be
happy today, because I was made deputy class captain of Class 6!

On April 12 (Thursday) her class performed air-raid siren evacuation drills. 'The first time it took six minutes for us to evacuate,' Yoko wrote, 'but the second time, it took only four. When you evacuate, the most important thing is to be swift and silent.' In her gym class she 'practised walking and dodging things in pitch darkness'. And in the afternoon, 'I was playing in the playground when the air-raid siren sounded again, and so I came straight home.'

April 21 (Sat)
Today, a new student called Asako Fujita joined our class. She is an evacuee from Osaka. She was still in Osaka when 90 of those B-29s attacked the city and will be coming to school every day from Jigozen. I am going to be her friend.

April 30 (Mon)
Today, a B-29 [bombed] the Shirakami Shrine area. Although the resulting damage was very minor, we must not underestimate the enemy simply because there was only one aircraft.

On 17 May her school life changed. First-year students were ordered to join the demolition teams and remove debris to make firebreaks. The area allocated for the girls from Yoko's school was Dobashi, near the centre of town:

Labour service began today, at last. Our job is to clear away 70 buildings, starting from the local courthouse. Most of the rubble has already been cleared away, but I am going to work hard and do the best job I can anyway.

After two weeks of clearing debris, 'for some reason, my body felt very weak'. But her feelings were 'nothing at all' when she remembered her father 'fighting on the battlefield'. She frequently returned to an empty house and had to prepare the family meal: 'After I had finished preparing dinner, I went to meet the 5:30pm boat … When the boat arrived and mother was on it, I felt inexplicably happy.'

•

A little boy, Shoso Kawamoto, was one of 23,500 Hiroshima children under the age of 12 who were evacuated to the countryside that year; they included 460 of his fellow sixth graders. Most of his family stayed in the city: his parents, two elder sisters, and two younger siblings. His eldest sister, Tokie, had a job in the maintenance section of Hiroshima train station; his sister Michiko was in her second year of high school; and the two babies stayed in the care of their parents in the city, as decreed by the new law. An elder brother, aged 17, worked for an electrical appliance maker in Manchuria.

Shoso arrived at Kamisugi village, his evacuation point, about 50 kilometres northeast of Hiroshima on 15 April 1945. He and his classmates slept on the floor of Zentokuji, the local Buddhist temple: 'We hung mosquito nets and tied them to the *tatami* mats … we only had one rice bowl every morning and evening.' Older children like Shoso put on a brave face for the younger kids: 'Though we all wanted to cry, we didn't,' he recalled. The temple had no lights outside and the littlest children, too frightened to go to the toilet at night, 'wet their futons and we would have to clean it up'. Shoso's friends traded pencils and notebooks for rice balls with the local villagers; 'but the third and fourth grade students were too shy to do that, so they were always hungry and crying'. The child evacuees soon abandoned their studies and joined the village children gathering wood, digging pine roots, swimming in the river and stealing fruit. In time, they formed gangs, and made slingshots and spears and taunted the bed-wetters, who lay in terror through the long, black nights pricked by starlight and comet, or perhaps the tail lights of American bombers flying home. Once a month his mother and father visited him: 'We would spend just a few minutes together,' said Shoso, tearful still after

55 years at the memory of those brief reunions. 'They never spoke of the war.' They always left with a prayer that he would return home to Hiroshima safely.

·

Nagasaki refused to evacuate as many children as Hiroshima. Primary-school enrolments in the city fell from 32,905 in April 1944 to 11,315 in July 1945, and while some of the 21,590 not at school did find refuge in the rural temples and villages, most were kept at home. The reasons were several: the authorities were reluctant to alarm the people by packing off thousands of children to the countryside; the railways on Kyushu were clogged with heavy troop movement south; and severe shortages of petrol limited civilian transportation to the countryside. Rural shelter, in any case, was limited: families whose city homes had been demolished were already cramming the rural temples and villages.

The children over 12 who remained in Nagasaki, like those in other cities, were put to work in the factories and demolition sites. Nagasaki imposed a uniqely grim responsibility on its teenage labour: assembling torpedoes in Mitsubishi's underground plant in the hills near Urakami. Tsuruji Matsuzoe, the 15-year-old who hoped to become a teacher, worked here in a team of 10. As the fear of air raids rose, 12-hour night shifts replaced the day shifts. To get to work on time Matsuzoe moved from his downtown billet into a dormitory, a three-storey concrete building painted black, near the torpedo factory.

His daily routine started at 5pm, with dinner – rice and vegetables – after which he walked to the factory through a rice paddy. He would reach the tunnel at about 7pm and work all night. At dawn he returned to the dorm, threw off his oil-spattered uniform, washed, changed into a *yukata* (pyjamas), lined up for roll call, ate a breakfast of barley, soy beans and rice served in a shared bowl ('The food was never shared evenly so sometimes we had more to eat, sometimes less,' he recalled), and fell exhausted onto a futon. He slept until noon, rose for lunch – a rice ball – and resumed sleeping until 5pm, when the next shift began. The poor diet and exhausting work broke the children's health; they rapidly lost weight. Many died from malnutrition, beriberi and tuberculosis. Matsuzoe initially shared his room – the size of a small classroom – with 12 boys; by mid-1945 there were 'just five or six of us'.

Girls as young as 12 worked in the torpedo tunnel too. Fourteen-year-old Kiyoko Mori was ordered to present herself at the factory in May. At the time she and her class were 'too hungry to concentrate' on lessons, her teacher had said. Instead, he told the girls: 'Let me show you what I have inside my lunchbox.' He opened it and revealed a slice of sweet potato. He then asked the class to list 'all the things we could think of that we wanted to eat', Kiyoko recalled. 'Caramel', 'cake' and 'chocolate' headed the list, then cooked food like sukiyaki and fried eggs, then fruit and vegetables. 'Finally, when it came to my turn, I couldn't think of any more food. So I said, "watermelon". I remember that class vividly.'

Her girlfriends used lathes to make screws for torpedoes. Too small to operate the machine, Kiyoko served as a runner between different ends of the tunnel: 'I had no idea what we were making,' she said later. 'I was just running errands … Then, one time, I saw two huge torpedoes … I realised that such things existed but I was really surprised to find out that we were making them. When I returned to my area I asked [my supervisor], "Are we making torpedoes?" I whispered so that nobody would hear me. He realized that I must have seen one and said, "Yes, we are making parts for missiles here." At that time, nobody saw anything, nobody heard anything, nobody knew anything. That was the kind of world it was.'

•

In Tokyo, the military regime ran the war effort from an underground location within the heavily fortified, well-stocked bomb shelters near the Imperial Palace. The government had no direct contact with ordinary people other than through the slogans and lies disseminated by the state-controlled press, and the feared *Kempeitai*, the military police. It exerted its executive power through prefectural and district officials, the armed forces, police and neighbourhood groups.

Civilians – chiefly women, schoolchildren and the elderly – were expected to defend themselves, and pay for it: of the 70,000 yen (Y16.8 million in 2009) Hiroshima Prefecture received from Tokyo that year, just 4000 yen was allotted to the city's defence: all of that went to the police and fire-fighting units. The people were expected to donate – in cash, materials and time – the rest needed to fight fires, demolish buildings and dig shelters.

Voluntary neighbourhood associations (*tonarigumi* or *rimpohan*) had been set up in 1940, under the Imperial Rule Assistance laws, to oversee these tasks and gather donations. The neighbourhood associations were responsible for self-defence in the event of an air raid – until larger government forces arrived to help. They were 'voluntary' in name only; in practice, social pressure and the *Kempeitai* ensured everyone participated, though many did so eagerly.*

The neighbourhood groups were comprised of smaller working parties called 'block associations' – *chokai* – run by families responsible for their city block. The *chokai* typically represented 10 to 20 families and were similar in structure to Germany's wartime self-defence units, the *Selbstutchz*. Lowest on the rung of the prefectural bureaucracy, the *chokai* shouldered the toughest jobs – jobs the elderly and women were least able, physically and financially, to perform. In short, these families were at the bottom of the safety chain, whose links could be traced from Tokyo's government bunkers to the prefectural governor of Hiroshima – nominally responsible for civilian defence – to the Chief of the Prefectural Police, down to Hiroshima's local police chief. He assigned duties to the city's 28 police stations and two fire departments,

---

* The same element of compulsion operated in the Women's Emergency Labour Corps (*Joshi Kinro Teishin Tai*), which applied a form of 'soft' conscription through the tacit warning that if not enough women volunteered 'the Government would be forced to resort to actual conscription'.

which deployed leaders across a range of functions, or 'units': fire-fighting, ration distribution, demolition, gas defence, rescue, medical aid, corpse removal etc. The trouble was the unit leaders had no units; they lacked resources and staff. They relied on the *chokai* block teams, ordinary families, to donate cash and do all the work, unpaid. Here, then, was the practical application of Japanese wartime 'spirit'.

•

As applied everywhere in Japan, Hiroshima's and Nagasaki's fire-fighting facilities were in a pathetic state. The regime failed or refused to provide basic equipment. Households were expected to use their own mops, firehoods, sandbags and hoses, and to pay for the installation of water pumps scattered through town.

Two fire stations, in west and east Hiroshima, employing 440 men equipped with 45 pumper trucks and a single ladder truck (with a 20-metre ladder, the only one in the prefecture) were expected to defend a city of 300,000. Some 3600 volunteers (*keibodan*) had joined the firemen, but only a few hundred turned up for duty in spring 1945, most preferring to stick to their friendly neighbourhood associations. The fire-fighting chiefs reckoned they needed 130 water tanks to combat a firestorm; only 80 tanks were then available in Hiroshima. On the west side of the city the water mains were badly cracked and likely to rupture.

Nagasaki fared no better, with just 287 firemen. Authorities there relied largely on the help of 5000 neighbourhood associations purporting to represent some 140,000 'able-bodied' adults and children – most of whom were malnourished or sick – as well as *chokai* and welfare and sanitation groups. In April 1945 the city police and fire services had about 25 fire trucks, 65 hand-drawn pumps and two fire-fighting patrol boats. The city had an ample water supply – with 761 fire hydrants and 10,000 wells – but the pumping and drawing mechanisms were abysmally maintained. In any event, the tiny streets in the congested city centres were inaccessible to fire trucks. Here 'bucket brigades' of civilians, handing water buckets down long lines, were the first line of defence against a firestorm.

'The crudest methods of fighting fire had to be utilized,' concluded the US Strategic Bombing Survey, in its 1946 study of the Pacific air war. 'The neglect on the part of the government may possibly be [an example of] the proverbially fatalistic attitude of the Japanese, as well as to the small value they place on human life.' This reflected the crudest prejudice of the times, conflating military propaganda with parental love: while the militarists regarded the death of a soldier as, according to *Bushido*, 'lighter than a feather', Japanese parents shared any fathers' or mother's concern for their children, and despaired of their families' exposure to air raids.

A pathetic symbol of the population's desperation were the fist-sized sand balls held together by paste, which the Americans later uncovered in a warehouse in Ujina in Hiroshima and identified as 'intended to be thrown at a fire from a distance'. If they had a million sandballs the people were unlikely to get the chance to throw them because the air-raid warning systems were incapable of warning anyone in time. In Hiroshima the sirens – installed at strategic points around the city – were woefully organised and maintained. With no master switch to turn them on at once, a telephonist had to activate each siren post individually (Hiroshima planned to introduce a simultaneous siren system in September 1945). Telephonists made mistakes; in any case, the authorities preferred not to alarm the cities without good cause. A single plane scarcely warranted a full air-raid warning.

Nagasaki operated a more efficient siren system; the city was one of the few in Japan with an air-raid control centre located inside a bomb-proof concrete shelter on the side of a hill. Japan's Western Army headquarters in Fukuoka, northeast of Nagasaki, and the largest city on Kyushu, picked up any air-raid sightings in the region and sent alerts to the Nagasaki garrison, which had the authority to sound the alarm in the prefecture once it detected approaching planes.

Bomb shelters, if they existed, were in a dismal state in Hiroshima. People dug in a desultory, resigned fashion, survivors recall: the scratching of a trench in the yard, a shallow hole in the kitchen floor, shoddy attempts at group shelters. Few existed in the inner city areas, which were too densely populated to fit them.

A better prospect existed in Nagasaki, thanks to its topography. Tunnel shelters built into the hills could hold 75,000 people, or 30 per cent of the city's population of 240,000. Some of these had survived 500-pound bombs. The city seemed well equipped to withstand severe conventional bombing. The problem was getting to the shelters before the bombs fell. Inner city residents had no time to 'head for the hills'; so they relied on shallow trenches in their living-room earthen floors. These would be death traps under incendiary air raids, as the only exits tended to be inside their (destroyed) homes. In addition Nagasaki had 1160 'covered trenches' (covered by timber and earth) and more 'uncovered trenches' for use as emergency public shelters. Both proved worthless in conventional air raids.

Japanese rescue services were completely ineffectual. This was a nationwide problem, and the same dire pattern repeated itself night after night during incendiary attacks: tens of thousands perished due to the absence of the most rudimentary rescue equipment. Hiroshima and Nagasaki were no better served. The rescue teams were selected from the local police and very poorly trained. Incompetent instructors concentrated on curbing panic and boosting morale rather than actually rescuing anyone. In any case, the rescue workers had no mobile earthmoving equipment to remove rubble; no power cranes or steam shovels – just crude hand tools; and no listening devices to detect the sounds of those alive under rubble. 'It was impossible,' concluded the US Strategic Bombing Survey, 'for this service to do much more than go through the motions, and, at times, even the motions were pointless.'

Darkness did not help. Blackouts were useless against B-29s electronically guided to the target. In any case, area bombing targeted whole cities. Whether the pilots saw speckles of light or an expanse of darkness mattered little during indiscriminate firebombing. Aircrews knew a city existed down there, somewhere. Camouflage was useless. In this, the Japanese were inconsistent and apathetic, leaving the curve of dams and the shape of the Imperial gardens exposed while attempting to disguise buildings with elaborate patterns, bamboo lattice-work, fish nets and shrubbery that offered little if any protection or deception.

In sum, the full weight of the cities' defence against air raids fell upon elderly men, the women and children. One statistic sums up the hopelessness of the situation: by June, 3,400,000 Japanese schoolchildren had been mobilised to work in demolition teams or factories. The majority – like 12-year-old Yoko and 15-year-old Tsuruji – were put to work at the start of the American incendiary campaign. In so doing, the old men in Tokyo knowingly exposed the nation's youth to death by firestorm.

•

As the threat to the homeland rose, Tokyo acted to reinforce the southern islands of Shikoku and Kyushu and the cities around the Inland Sea. Neighbourhood associations dug defences and bomb shelters on the beaches of Shikoku and the southern shores of Kyushu. On 18 April 1945, the Second General Army established its headquarters near Hiroshima Castle, under the command of Field Marshal Shunroku Hata, to prepare for the anticipated American invasion of Kyushu; soon after, the Chugoku Military District made its headquarters in the castle grounds, incorporating an underground communications centre staffed by schoolgirls as young as 12. All were primed to defend the city to the death.

CHAPTER 8

THE TARGET COMMITTEE

*The most desirable target would be a vital war plant employing a
large number of workers and closely surrounded by workers' houses.*

James Conant, member of the Interim Committee on nuclear power, 31 May 1945

ON 10 MAY 1945, TWO days after Germany's surrender, a committee
met in Robert Oppenheimer's office in Los Alamos. It comprised a
carefully selected group of scientists and military personnel, loosely
known as the Target Committee. Its more prominent members were
Brigadier General Thomas Farrell, Groves' second in command of the
bombing mission; Captain William 'Deak' Parsons, associate director
of Los Alamos' Ordnance Division; John von Neumann, the great
Hungarian-American mathematician; and the physicist William
Penney, a leading member of the small British team at Los Alamos,
which included Peierls and the spy Fuchs. The scientists present
(chiefly Penney and von Neumann) were then unaware that the
decision to use the bomb on Japan had long predated Germany's
surrender. Indeed, Washington's powerful Military Policy Committee,
consisting of Bush, Conant, Major General Styer and other top
advisers, had confirmed Japan as the target on 5 May 1943. None of
the scientists then working on the Manhattan Project was informed,
least of all Leo Szilard and the émigré Jewish physicists, who had

tirelessly pursued the creation of the weapon on the supposition that if it were used, it would be used on Germany. In fact, Germany had ceased to be the target precisely two years before it surrendered.

With Germany out of the war, the top minds within the Manhattan Project focused on the choices of local targets within Japan. The question essentially was this: which of the preserved Japanese cities would best demonstrate the destructive power of the atomic bomb? Groves had been ruminating on targets since late 1944, and convened a preliminary discussion of the subject on 27 April 1945, at which he laid down his criteria. The target should:

- possess sentimental value to the Japanese so its destruction would 'adversely affect' the will of the people to continue the war;
- have some military significance – munitions factories, troop concentrations etc;
- be mostly intact, to demonstrate the awesome destructive power of an atomic bomb;
- be big enough for a weapon of the atomic bomb's magnitude.

Groves asked the scientists and military personnel to debate the details: they analysed weather conditions, timing, use of radar or visual sights, and priority cities. Hiroshima, they noted, was 'the largest untouched target' and remained off LeMay's list of cities open to incendiary attack. 'It should be given consideration,' they concluded. Tokyo, Yawata and Yokohama were thought unsuitable – Tokyo was 'all bombed and burned out' and 'practically rubble, with only the Palace grounds still standing'.

A fortnight later, at the formal 10 May target meeting, Oppenheimer ran through the agenda: 'height of detonation', 'gadget [bomb] jettisoning and landing', 'status of targets', 'psychological factors in target selection', 'radiological effects', and so on. Dr Joyce C. Stearns, a scientist representing the air force, named the four shortlisted targets in order of preference: Kyoto, Hiroshima, Yokohama and Kokura. They were all 'large urban areas of more than three miles [five kilometres] in diameter'; 'capable of being effectively damaged by the blast'; and 'likely to be unattacked by next August'. Someone raised the possibility of bombing the Emperor's Palace in Tokyo – a spectacular idea, they agreed, but militarily impractical. In any case, Tokyo had been struck from the list because it was already 'rubble', the minutes noted.

Kyoto, a large urban industrial city with a population of one million, met most of the committee's criteria. Thousands of Japanese people and industries had moved there to escape destruction elsewhere; furthermore, stated Dr Stearns, Kyoto's psychological advantage as a cultural and 'intellectual centre' made the residents 'more likely to appreciate the significance of such a weapon as the gadget'.

Hiroshima, a city of 318,000, held similar appeal. It was 'an important army depot and port of embarkation', said Stearns, situated in the middle of an urban area 'of such a size that a large part of the city could be extensively damaged'. Hiroshima, the biggest of the 'unattacked' targets, was surrounded by hills that were 'likely to produce a focusing effect which would considerably increase the blast damage'. On top of this, the Ota River made it 'not a good' incendiary target, raising the likelihood of its preservation for the atomic bomb.

The meeting barely touched on the two cities' military attributes, if any. Kyoto, Japan's ancient capital, had no significant military installations; however, its beautiful wooden shrines and temples recommended it, Groves had earlier said (he was not at the 10 May meeting), as both sentimental and highly combustible. Hiroshima's port, main industrial and military districts were located outside the urban regions, to the southeast of the city.

Oppenheimer next assessed the radiation risk: aircraft should not fly within 4 kilometres of the detonation point, he advised, to 'avoid the cloud of radio-active materials'. The radiation risk to Japanese civilians was not discussed.

Someone raised the question of whether incendiary bombers should attack the city after the nuclear strike. 'This has the great advantage,' the minutes record, 'that the enemies' fire-fighting ability will probably be paralysed by the gadget so that a very serious conflagration [will start].' The ensuing firestorm, however, might confuse photo-reconnaissance of the atomic damage and subject aircrews to radioactive contamination. For that reason, the meeting decided against following up atomic bombing with an incendiary raid.

In summary, the gentlemen unanimously agreed that the bomb should be dropped on a large urban centre, the psychological impact of which should be 'spectacular', to ensure 'international recognition' of the new weapon. A full report of the proceedings was sent to Groves on the 12th.

•

The Target Committee met again the next day in Oppenheimer's office to discuss technical aspects of the mission. Before this meeting Oppenheimer sent Farrell a longer description of the likely effects of radiation. The uranium bomb, Oppenheimer warned, would release about 10^9 times as much toxic material as would inflict a single lethal dose; and radiation emissions would be lethal within a 1-mile (1.6-kilometre) radius. Within a second of the blast, gamma radiation equivalent to about 10^{12} curies would coat a large section of the targeted city; falling, within a day, to 'about 10 million curies'. 'If the bomb is delivered during rain,' Oppenheimer added, 'most of the active material will be brought down … in the vicinity of the target area.' He warned again that the delivery aircraft and follow-up planes should maintain a minimum distance of 2.5 miles (4 kilometres) from the detonation point to avoid radioactive contamination. Monitoring of ground radiation in the vicinity 'will be necessary' for some weeks, after which it should be 'quite safe to enter'.

The Target Committee regrouped at the Pentagon on 28 May (Oppenheimer sent a representative). The members concentrated on the aiming points within the targeted cities. The plane carrying the atomic bomb 'should avoid trying to pinpoint' military or industrial installations because they were 'small, spread on fringes of city and quite dispersed'. Instead, aircrews should 'endeavor to place ... [the] gadget in [the] center of selected city'. They were quite explicit about this: the plane should target the heart of a major city. One reason was that the aircraft had to release the bomb from a great height – some 30,000 feet (9000 metres) – to escape the shock wave and avoid the radioactive cloud; that limited the target to large urban areas easily visible from the air.

Captain Deak Parsons gave another reason to drop the bomb on a city centre: 'The human and material destruction would be obvious.' An intact urban area would show off the bomb to great effect. Whether the bomb hit soldiers, ordnance and munitions factories, while desirable from a publicity point of view, was incidental to this line of thinking – and did not influence the final decision. The target must be a city centre, they concluded. 'No-one on the Target Committee ever recommended any other kind of target,' McGeorge Bundy, a Washington insider who later became John F. Kennedy's national security adviser, later wrote, 'and while every city proposed had quite traditional military objectives inside it, the true object of attack was the city itself.'

The Target Committee dismissed talk of giving a prior warning or demonstration of the bomb to Japan. Parsons had persistently rejected suggestions of a noncombat demonstration: 'The reaction of observers to a desert shot would be one of intense disappointment,' he had warned in September 1944. Even the crater would be 'unimpressive,' he said. Groves shared his contempt for 'tender souls' who advocated a noncombat demonstration. Oppenheimer, too, later wrote that he agreed completely with Parsons about 'the fallacy of regarding a controlled test as the culmination of the work of this laboratory'. When the meeting ended, the committee had no doubt about where the first atomic bomb would fall: on the heads of hundreds of thousands of civilians.

During June the Target Committee narrowed the choice. On the 15th a memo enlarged upon Kyoto's attributes. It was a 'typical Jap city' with a 'very high proportion of wood in the heavily built-up residential districts'. There were few fire-resistant structures. It contained universities, colleges and 'areas of culture', as well as factories and war plants, which were in fact small and scattered, and in 1945 of neglible use. Nevertheless, the committee placed Kyoto higher on the updated 'reserved' list of targets (that is, those preserved from LeMay's firebombing).

Kokura, too, made the reserved list. Kokura, on Kyushu, west of the Kanmon tunnel, was the most obvious military target. It possessed one of Japan's biggest arsenals, replete with military vehicles, ordnance, heavy naval guns and, as had been reported, poison gas. There were coal and ore docks, steelworks, extensive railway yards and an electric power plant, covering almost 3 kilometres along the shore and 2.4 kilometres inland. A less appealing target, Niigata, had 'fire resistive' industrial plants and houses made of 'heavier plaster' to protect against harsh winters, hence less combustible.

In June Henry Stimson ordered Kyoto's immediate removal from the target list. The War Secretary had discovered the city's presence on it almost by chance. That month he had asked Groves – then in his office on a different matter – whether the target list had been decided. Groves said it had, but refused to name the targets, pending Army Chief of Staff General Marshall's approval. Stimson insisted. It disturbed him to see Kyoto, the ancient capital whose temples and shrines he had visited with his wife in 1926, at the top of the list of cities set aside for the atomic bomb. He ordered it struck off. Groves fudged. 'Hap' Arnold, commanding general of the US Army Air Forces, supported Groves, and favoured keeping Kyoto on the list. Stimson was adamant: 'This is one time I'm going to be the final deciding authority. Nobody's going to tell me what to do on this. In this matter I'm the kingpin,' Stimson told Groves.

Groves was not so easily deterred, and dragged out the argument. They irked him, these meddlesome politicians: the destruction of Kyoto was his to decide; he felt a sense of proprietorial control over how the bomb should be used. The city 'was large enough an area for us to gain complete knowledge of the effects of the atomic bomb. Hiroshima was not nearly so satisfactory in this respect.' For weeks, Groves continued to refer to Kyoto as a target despite Stimson's clear instructions to the contrary. Then, on 30 June, Groves very reluctantly informed the Chiefs of Staff that Kyoto had been eliminated as a possible target for the atomic fission bomb and all bombing, by direction of the Secretary of War. At the same time, he provocatively left the city on the list of 'four places' to be preserved from conventional attack; within weeks, Generals Arnold and MacArthur, as commander of US land forces in the Pacific, along with Admirals Nimitz and King, who together controlled the air and naval attacks on Japan, received the following message: 'Kyoto, Hiroshima, Kokura and Niigata will not be attacked [by conventional forces] ... unless further directions are issued by the Joint Chiefs of Staff.'

•

Another high-powered group ran in parallel with the Target Committee: the Interim Committee of top officials convened by War Secretary Stimson to advise the President on the future of nuclear power for military and civilian use. (The committee was 'interim' because the members anticipated that a permanent body would, in time, control atomic energy.)

On paper, the Interim Committee looked omnipotent. Its permanent members were Stimson; James Byrnes, the President's 'personal representative', pending his appointment as Secretary of State; Vannevar Bush; James Conant; the physicist Dr Karl Compton, Arthur's older brother and president of MIT; Ralph Bard, Under Secretary of the Navy; William Clayton, Assistant Secretary of State;

and George Harrison, Special Consultant to the War Secretary. The scientists Oppenheimer, Arthur Compton, Lawrence and Fermi sat on the committee's Scientific Panel. Generals George Marshall and Leslie Groves received open invitations to attend meetings.

In practice, the committee's influence ebbed away. The problem was Stimson. The War Secretary anchored his authority to the committee's success and personally invited the members. Some turned up as a courtesy, but attendance levels swiftly declined. Groves attended once. The immediate demands of the atomic mission preoccupied him; he had little time for Stimson's visionary talk about the future of atomic power. There was a war to be won.

Stimson soon lost the attention of the President. On the night before the first meeting, on 30 May, he forwarded Marshall and the President a memo that drew attention to the War Secretary's rather quixotic disposition: 'My dear Marshall,' he wrote, 'Here is a letter which I have just received this afternoon, and which I should like you to read before tomorrow's meeting ... I think it is the letter of an honest man ... I think it a remarkable document, and for that reason wish you to have the impress of its logic before the meeting ...'

The author was Oswald C. Brewster, of 23 East 11th Street, New York, an engineer in the Manhattan Project's Oak Ridge plant. Plain terror, he wrote, coupled with a naive sense of civic duty had prompted him to write to the President, Stimson, Groves and Byrnes and share his feelings about the bomb. He was 'appalled by the conviction' that the atomic bomb would lead to the 'destruction of our present-day civilisation'. This was not hysteria or 'crack-pot raving', he insisted. He earnestly hoped that 'the thing' might not work, now that Germany had surrendered. If America became the world's first nuclear power '... all the world would ... conspire and intrigue against us ... We would be the most hated and feared nation on earth ...' Brewster feared a 'corrupt and venal demagogue', who, in possession of atomic weapons, 'could turn on us and the world and conquer it for his own insane satisfaction'.

In his overwrought state Brewster urged the Pentagon to close a factory at Oak Ridge, drawing accusations of treason from co-workers; but they misunderstood him, he complained: the plant should close *after* it had made sufficient fissile material to demonstrate the bomb. The dogged engineer suggested ways of curbing a nuclear arms race: through transparency, weapon demonstrations, international inspectors and the control of the supply of uranium.

Brewster concluded with a heresy that did not reflect well on the judgment of his champion in the War Office: 'I do not of course wish to propose anything to jeopardise the war with Japan, but, horrible as it may seem, I know that it would be better to take greater casualties now in conquering Japan than to bring upon the world the tragedy of unrestrained [nuclear] competitive production.' Truman, for whom saving American lives was a political and moral imperative, ignored the letter. Stimson unwisely insisted on pressing the engineer's argument, and therein lay the letter's historical significance.

After a fitful night Stimson rose early on the 31st, determined to make a good impression on his new committee. At 10am the members filed into the dark-panelled conference room of the War Department. The air was heavy with the presence of three Nobel laureates and Oppenheimer. Stimson opened the proceedings on a portentous note: 'We do not regard it as a new weapon merely,' he said, 'but as a revolutionary change in the relations of man to the universe.' The atomic bomb might mean the 'doom of civilisation', or a 'Frankenstein' that might 'eat us up'; or it might secure world peace. The bomb's implications 'went far beyond the needs of the present war', Stimson said. It must be controlled and nurtured in the service of peace.

Oppenheimer was invited to review the explosive potential of the bombs: two were being developed – the plutonium bomb and the fissile uranium bomb. They used different detonation methods and processes, yet both were expected to deliver payloads ranging from 2000 to 20,000 tons (1800 to 18,000 tonnes) of TNT. Nobody yet knew their precise power. More advanced weapons might measure up to 100,000 tons; and superbombs – thermonuclear weapons – 10 million to 100 million tons, Oppenheimer said. The scientists nodded impassively; they were inured to such fantastic figures.

The numbers, however, and the destruction they implied, 'thoroughly frightened' incoming Secretary of State Byrnes, as he later admitted. He was human, after all; but beyond his horror of the statistics he silently ruminated on the wisdom, or madness, of any talk of sharing the secret with Moscow. As such, Byrnes the politician resolved to pursue his 'go it alone' policy for America that would pointedly exclude the Russians and indeed the rest of the world from the atomic secret: the bomb's power would be the future source of American power. Discussion flared on the question of whether to share the secret with Russia (by which point Stimson had left for another meeting). Oppenheimer advocated divulging the secret 'in the most general terms'. Moscow had 'always been very friendly to science', he rather lamely observed; he felt strongly, however, that 'we should not prejudge the Russian attitude'. General Marshall wondered, too, whether a combination of likeminded powers might control nuclear power; the general even suggested that Russian scientists be invited to witness the bomb test at Alamogordo, scheduled for July.

Such talk alarmed Byrnes, who had observed the Russians at close quarters at Yalta, and Groves, who was violently opposed to sharing with Moscow a secret he had spent almost four years trying to keep. Byrnes swooped: if 'we' gave information to the Russians 'even in general terms', he argued, Stalin would demand a partnership role and a stake in the technology. Indeed, not even the British possessed blueprints of America's atomic factories, chipped in Vannevar Bush.

Byrnes then wrapped up the argument: America should 'push ahead as fast as possible in [nuclear] production and research to make certain that we stay ahead and at the same time make every effort to better our political relations with Russia'. All agreed. If anyone noticed this first official recognition of the start of a nuclear arms race – not with Germany or Japan, but with Russia – he did not say so. Nor were any of the members tactless enough to point out the inconsistency of keeping secrets from those with whom Byrnes hoped to build better political relations.

By accident the talks turned to the use of the weapons on Japan. That morning Ernest Lawrence had suggested staging a demonstration of the bomb, to show off its power and intimidate the Japanese, a move the Target Committee had already rejected. Byrnes took just ten minutes over lunch in the Pentagon (where Stimson rejoined them) to kill that idea: the bomb might be a dud, he warned; the Japanese might shoot down the delivery plane; American POWs might be put in the target zone. A demonstration would not be sufficiently spectacular to persuade Tokyo to surrender, Oppenheimer added. Stimson agreed. 'Nothing,' he later wrote, 'would have been more damaging to our effort to obtain surrender than a warning or a demonstration followed by a dud …'

After lunch the meeting (minus Marshall) examined the next point on the agenda: 'the effect of the bombing of the Japanese and their will to fight'. Would the nuclear impact differ much from an incendiary raid, one of the committee wondered. That rather missed the point, objected Oppenheimer, stung by the suggestion that mere firebombs were in any way comparable: 'The visual effect of the atomic bomb would be tremendous. It would be accompanied by a brilliant luminescence which would rise to a height of 10,000 to 20,000 feet [3000 to 6000 metres]. The neutron effect of the explosion would be dangerous to life for a radius of at least two-thirds of a mile.' The same could not be said of LeMay's jellied petroleum raids. 'Twenty thousand people', Oppenheimer estimated, would probably die in the attack.

At this point, Stimson revived his personal mission to save Kyoto: Japan was not just a place on a map, or a nation that must be defeated, he insisted. The objective, surely, was military damage, not civilian lives. In Stimson's mind the bomb should 'be used as a weapon of war in the manner prescribed by the laws of war' and 'dropped on a military target'. Stimson argued that Kyoto 'must not be bombed. It lies in the form of a cup and thus would be exceptionally vulnerable ... It is exclusively a place of homes and art and shrines.'

With the exception of Stimson on Kyoto – which was essentially an aesthetic objection – not one of the committee men raised the ethical, moral or religious case against the use of an atomic bomb without warning on an undefended city. The businesslike tone, the strict adherence to form, the cool pragmatism, did not admit humanitarian arguments however vibrantly they lived in the minds and diaries of several of the men present.

Total war had debased everyone involved. While older men, such as Marshall and Stimson, shared a fading nostalgia for a bygone age of moral clarity, when soldiers fought soldiers in open combat and spared civilians, they now faced 'a newer [morality] that stressed virtually total war', observed the historian Barton J. Bernstein. In truth, the American Civil War and the Great War gave the lie to that 'older morality', as both men knew. Marshall recommended, for example, on 29 May, in discussion with Assistant War Secretary John McCloy, the use of gas to destroy Japanese units on outlying Pacific islands: 'Just drench them and sicken them so that the fight would be taken out of them – saturate an area, possibly with mustard, and just stand off.' He meant to limit American casualties with whatever means available.

If he drew on outdated civilised values, Stimson was also a far-sighted *éminence grise*, who grasped the moral implications of nuclear war. The idea of the bomb tormented him – so much that he sought comfort in the notion of recruiting a religious evangelist to 'appeal to the souls of mankind and bring about a spiritual revival of Christian principles'. America, he believed, was losings its moral compass just as it might be about to claim military supremacy over the world. The dawn of the atomic era called for a deeper human response, he believed, energised by a spirit of co-operation and compassion. He did not act on his compulsion, but dwelt long on the atomic question – and the question in Stimson's troubled mind was not 'will this weapon kill civilians?' but rather, if we continue on this course 'will any civilians remain?' He poured much of his anxiety into his diary.

Officially Stimson seemed contradictory and muddled. In the meetings he summarised his position on the bomb, thus: (1) 'we could not give the Japanese any warning'; (2) 'we could not concentrate on a civilian area'; (3) 'we should seek to make a profound psychological impression on as many of the inhabitants as possible'. He meant to use the bomb to shock the enemy – 'to make a profound impression' – with a display of devastation so horrible that Tokyo would be forced to surrender. However, he insisted it must be a military target. His statement's inherent contradiction – how could the bomb shock Tokyo without concentrating on a civilian area? – either eluded Stimson, or he lacked the intellectual honesty to confront it – provoked no comment in the Interim Committee meeting, and eased the task of Conant: 'The most desirable target,' then, Conant said, 'would be a vital war plant employing a large number of workers and closely surrounded by workers' houses.'

Stimson persuaded himself that this meant a military target. The physicists on the committee's Scientific Panel agreed; Groves ticked off another victory; and the War Secretary's self-deception was complete. A slightly surreal atmosphere lingered, as the men reflected on what they had done. The meeting that had opened with Stimson's declaration of mankind's 'new relationship with the universe' ended with his approval of the first atomic attack, on the centre of a city, to which he consented moments after he had rejected the bombing of civilians.

The committee unanimously agreed that the atomic bombs should be used: (1) as soon as possible; (2) without warning; and (3) on war plants surrounded by workers' homes or other buildings susceptible to damage, in order to make a spectacular impression 'on as many inhabitants as possible'.

As the meeting drew to a close, a suggestion was made to drop atomic bombs on several cities at the same time. That may indeed be 'feasible', Oppenheimer replied. But Groves objected: 'We would lose the advantage of gaining additional knowledge concerning the weapon at each successive bombing', thus underlining the experimental element of the attack. Nor would the 'effect' be 'sufficiently distinct' from the incendiary campaign of the air force. In any case, which cities should they choose? LeMay was expected to exhaust his targets by October. Dropping an atomic bomb on Tokyo would merely make the rubble bounce, to apply Churchill's future description of the effect of a nuclear war. Indeed, over most Japanese cities the weapon 'would not have a fair background to show its strength', as Stimson had told Truman in a different context – in a bizarre departure from his usual show of concern which had made the President laugh.

There was one last piece of business before the meeting adjourned: the vexing matter of a group of 'undesirable scientists' who had recently expressed their opposition to the use of the bomb on Japan. Most were émigré European physicists who saw their fight with Germany, not Japan; their moving spirit was Leo Szilard, a difficult man whom Washington now regarded as a perennial irritant. How might these meddling boffins be subdued? The Interim Committee's Scientific Panel seemed best equipped to soothe the dissent in their ranks, and Oppenheimer, Arthur Compton, Lawrence and Fermi were asked to prepare a report on whether 'we could devise any kind of demonstration [of the atomic bomb] that would ... bring the war to an end without using the bomb against a live target'. The committee anticipated an answer in the negative.

•

The next day – 1 June 1945 – Truman rose early to prepare a statement for Congress. It was a bright summer's day, and he chose one of his three new seersucker suits – the gift of a New Orleans cotton company. The President felt refreshed after hosting the Prince Regent of Iraq at a state dinner a few nights earlier. He had spent Memorial

Day, 31 May, on the Presidential yacht, cruising the Potomac, playing poker and approving his speech for the San Francisco Conference on the creation of the United Nations, then in session. Yesterday he had resolved the problem of succession in the State Department by finally approving the appointment of James Byrnes to replace Edward Stettinius as Secretary of State.

That 1 June morning Truman received Byrnes' summary of the previous day's marathon Interim Committee meeting. Byrnes had skilfully exploited his position as the President's special representative, laying stress where he saw fit, emphasising the consensus on the weapon's use and, in effect, relegating Stimson to the sidelines. Byrnes' upbeat assessment fortified the President for his important speech:

'There can be no peace in the world,' Truman told a rapt house, 'until the military power of Japan is destroyed ... If the Japanese insist on continuing resistance beyond the point of reason, their country will suffer the same destruction as Germany ...'

On the day of Truman's speech, four of America's most powerful industrialists – presidents George Bucher of Westinghouse, Walter Carpenter of DuPont, James Rafferty of Union Carbide, and James White of Tennessee Eastman – attended the second sitting of the Interim Committee, where Byrnes reiterated, in Stimson's absence, their intention to use the bomb as soon as available without warning on an urban area. All agreed. The talk then turned to the more amenable subject of the forthcoming test of the plutonium bomb in the New Mexican desert.

•

On 11 June, Oppenheimer, Compton, Fermi and Lawrence met in Los Alamos to deliberate on the question of whether a demonstration of the bomb would persuade Tokyo to surrender. The question was redundant: the Target Committee had already answered in the negative. The scientists took three days to reach the same conclusion – they made a point of debating the issues. They recast all the scenarios previously rehearsed: 'If only this [a noncombat demonstration] could be done!' Compton pleaded years later, when he was able to do so. At the time, the scientists decided it could not. On 16 June the four physicists – among whom Oppenheimer exerted the strongest influence – reported to Washington: 'Our hearts were heavy,' Compton would later write: 'What a tragedy,' he insisted, 'that this power ... must first be used for human destruction.' They recommended 'immediate use' of the bomb against a Japanese city in the hope of ending the war and saving American lives: 'We can propose no technical demonstration likely to bring an end to the war,' they concluded. 'We see no acceptable alternative to direct military use.'

In his memoirs, Truman ascribed to Oppenheimer, Compton, Lawrence and Fermi a critical role in influencing his decision of how and where to use the bomb. 'It was their recommendation,' Truman wrote, 'that the bomb be used against the enemy as soon as it could be done [and] that it should be used without specific warning ... against a target that would clearly show its devastating strength. I had realised of course that an atomic bomb explosion would inflict damage and casualties beyond imagination ... It was their conclusion that no technical demonstration they might propose, such as over a deserted island, would be likely to bring the war to an end. It had to be used against an enemy target.' Truman's adroit sharing of responsibility failed to mention the fact that the Target Committee had already made the decision – over which the scientists had had little or no influence. The scientists had served a political role, no more, of thwarting their 'undesirable' colleagues and reinforcing a decision that had been taken.

•

The 'undesirables' would not be thwarted. The dissenting Chicago Group of scientists, then working in the MetLab at Chicago University and led by the eminent chemist James Franck, abhorred a direct atomic strike on a city. In a letter to Oppenheimer's panel, the Franck Committee – whose moving spirit was Leo Szilard – strongly opposed the use of the weapon, 'so close to completion', against 'any enemy country at this time'. To do so would 'sacrifice our whole moral position' and irretrievably weaken America's leadership in 'enforcing any system of international control' designed to make nuclear power a force for world peace rather than 'an uncontrollable weapon of war'.

On 11 June – the first day the Scientific Panel had met at Los Alamos – the irrepressible Szilard laid the Franck Report, the dissenting physicists' petition, at Oppenheimer's door. Its words echoed poor Brewster's:

'The military advantages and the saving of American lives achieved by the sudden use of atomic bombs against Japan may be outweighed by the ensuing loss of confidence and by a wave of horror and revulsion sweeping over the rest of the world … *a demonstration of the new weapon might best be made before the eyes of representatives of all the United Nations, on the desert or a barren island* [Franck's emphasis].

'America could then say to the world, "You see what sort of a weapon we had but did not use?" If the Japanese persisted with their refusal to surrender, then the weapon might be used against them with UN approval and sufficient warning,' Franck recommended. 'To sum up, we urge that the use of nuclear bombs in this war be considered as a problem of long range national policy rather than of military expediency, and that this policy be directed primarily to the achievement of an agreement permitting an effective international control of the means of nuclear warfare.'

It was signed: James Franck, D.J. Hughes, J.J. Nickson, E. Rabinowitch, Glenn Seaborg, J.C. Stearns and L. Szilard.

Groves buried his copy of the Franck Report and had the authors shadowed.

It left a mark, however. Unease spread among the physicists. A whiff of self-exculpation arose in those who, having officially approved a nuclear strike on Japan, felt the gnawing of self-doubt. Pilate-like,

Oppenheimer, Compton, Lawrence and Fermi washed their hands of responsibility: they had approved the bomb's use but ultimately had had no influence, they claimed (in their 16 June report), over what was essentially a military decision: 'We, as scientific men, have no proprietary rights [over the use of atomic energy] … we have no claim to special competence in solving the political, social and military problems which are presented by the advent of atomic power.'

In short, the scientists who were creating the bomb, and who had just recommended its use, sought absolution in advance by claiming they were no better equipped than ordinary people to judge whether atomic power should be used. One wonders why, if they were as incompetent in these matters as they professed, Washington had asked for their expert opinion. Oppenheimer kept his head. He had a huge job to do, and his determination to succeed overwhelmed any ethical concerns about the bomb. And in a sense the scientists were correct: they possessed no power over *how* the bomb should be used. Their job was to build it.

The dissenting scientists infuriated Groves, who fixed on Szilard as the prime agitator. For months the Hungarian had waged a personal crusade against the bomb. Zealous, obstreperous, unconcerned with diplomatic niceties or his raffish appearance, Szilard hardly endeared himself to the powers he hoped to persuade. His meeting with James Byrnes in Spartanburg, South Carolina, on 28 May 1945, only a fortnight ago, had proved an unmitigated disaster. To Szilard's insistence that the bomb be kept a secret, the South Carolinian former judge replied, 'How would you get Congress to appropriate money for atomic energy research if you don't show results for the money which has already been spent?' The suggestion that taxpayers' money validated the use of the bomb flabbergasted Szilard but it little prepared him for Byrnes' next statement. The man soon to be sworn in as Secretary of State said that he believed the weapon would help America contain Russia: 'Rattling the bomb might make Russia more manageable,' Byrnes told the astonished Szilard. Byrnes already saw the role of the bomb in diplomatic, not military, terms, as a political weapon against the Soviet Union. 'I was concerned,' Szilard later wrote, 'that an atomic arms race between America and Russia … might end with the destruction of both countries. I was not disposed … to worry about what would happen to Hungary.'

CHAPTER 9

JAPAN DEFEATED

The view that Japan's defeat is inevitable has achieved number one prominence.

Message to Tokyo from the Greater East Asia Ministry office in Hsingking, 17 June 1945, intercepted by American codebreakers

His Majesty the Emperor ... desires from his heart that [the war] may be quickly terminated. But so long as England and the United States insist upon unconditional surrender the Japanese Empire has no alternative but to fight on.

Message from Tokyo to Moscow, 11 July 1945, intercepted by American codebreakers

IN LATE 1945 A KAMIKAZE airman sat down to explain why his mission had failed and he remained alive:

On April 11, 1945, I, Takehiko Ena, the reconnaissance squadron commander (rank: naval ensign), pilot Mitsuru Umemoto (rank: petty officer, 2nd class) and radio operator Nagaaki Maeda (rank: petty officer, 2nd class) of the Seiki Squadron Special Attack Corps aboard Aircraft Carrier No. 5 set out in a Type 97 Carrier Attack Bomber from the Kushira Naval Air Base in Kagoshima Prefecture on a suicide mission bound for Okinawa.

The previous night Takehiko Ena thought he had eaten his last meal. The food was 'slightly better than usual', he recalled – chicken stew; there was no beef. Fifty kamikaze pilots sat eating with him, all students with a few hours' training. They gnawed the bones and said little. Ena rose: 'I didn't really want to stay there. I went back to my quarters as early as I could and packed my things.'

He packed his flying gear and a few mementoes and lay down for his last night. He couldn't sleep: 'I was thinking of the pain of separating from my parents who had raised me ... I was just past 20.' Ena had five brothers and sisters.

He rose before dawn and joined his crew: 'The most difficult thing was the fear of death inside myself. Of course I was afraid. I couldn't do anything against that fear.' They stood at attention on the tarmac, centred the red sun on their white headbands, bowed low towards the Imperial Palace, threw back their glasses of *sake* (*beppai*) in a formal toast to the Emperor, and recited a farewell poem to the *Kokutai*, 'Overwhelming Gratitude':

Why should I be reluctant to give this youthfulness away for You
When my life achieves its true purpose
By being scattered like a young cherry blossom in the wind ...

They boarded an old trainer aircraft – Japan's few remaining combat planes were reserved for regular missions – packed with 1700 pounds (800 kilograms) of explosives; their orders were to fly it onto the deck of a US ship moored off Okinawa. Unlike most kamikazes, Ena had not volunteered to join the 'Divine Wind', the literal English meaning of the word. He did not want to die. He thought Japan's situation hopeless. Yet as an economics student at Waseda University, and a member of the elite, he had felt a sense of *noblesse oblige* to set an example to less privileged Japanese. By 1945, Tokyo had realised the folly of wasting trained pilots on suicide runs. Students like Ena were expendable.

Ena's crew, he recalled, looked 'cramped, tense and desperate' as they boarded their flying coffin. It was 5am. They said little, just technical talk. As crew leader, Ena outlined their course to Okinawa: 'My mind was fixed only on the mission: to find US ships and fly into them.' His flying skills were meagre; in all, a few hours' training. Regular pilots were then expected to have 100 hours under their belt; and Japanese navy pilots typically completed 650 hours prior to combat missions – about the same as American fighter pilots.

Ena's plane was heavy and unreliable, and their 'Divine Wind' died within moments of take-off: engine failure forced him to ditch in the sea near the island of Kuroshima, 100 kilometres southeast of the Satsuma Peninsula. They swam ashore, staying for 80 days courtesy of the few hundred locals, until a Japanese submarine took them to the mainland. On 30 July they landed at Kuchinozu in Nagasaki Prefecture, and caught a train to the Aircraft Carrier No. 5 Headquarters of the Oita Flying Corps: 'There, we were told that Japan was making its last stand on the mainland and ordered to form a Special Attack Unit at the Ibaraki Prefecture Hyakurihara Naval Air Base, the home unit of Seiki Squadron. With bombs raining down on us, we made straight for the Hyakurihara Flying Corps.' They would pass through Hiroshima on about 6 August.

•

In the flames of the kamikazes and *kaiten* (human torpedoes) lay the last act of a people who had chosen death over life; who would fight to the last before they surrendered their homeland. This was not suicide in the Western sense of despair; most were 'determined to die' in a carefully planned act of honourable self-immolation.

The Americans had drawn those conclusions about the enemy well before their attack on Okinawa. But there they would witness its most sustained and destructive demonstration. The kamikazes' targets were the string of American ships positioned around the main island of the

Ryukyus, a curved sprinkling of atolls which stretched hundreds of kilometres from Kyushu to within 120 kilometres of Formosa (Taiwan). Home to 756,000 people in 1945, the Ryukyus are mostly flat, of coral not volcanic origin, with gently rolling hills, sugarcane plantations and pockets of rainforest, in a sub-tropical climate. The people 'are more hairy than the Japanese', observed a US State Department Bulletin: 'They have higher foreheads, sharper noses, eyes less deep-set and heavy arched eyebrows.'

Okinawans were not deeply religious, with few shrines and temples, and yet they had great respect for the dead. They buried corpses for three years, unearthed and washed the bones, put them in urns in the Chinese fashion, and placed the urns in horseshoe-shaped tombs dug into the hillsides. Some tombs overlooked the beaches where the Americans were expected to land, and might have proved ideal defensive posts were it not for local respect for the dead: only a single enemy machine gun was later found among the Okinawan tombs. A second curious feature was the Okinawans' testy relations with mainland Japanese, who treated them as inferior. This was thought to produce a conciliatory feeling towards the Americans – before the invasion – according to a State Department analysis at the time.

On the morning of 1 April 1945 – codenamed 'Love Day' – some 183,000 American troops and marines came ashore at Kadena Beach in Okinawa. Two days later the Joint Chiefs instructed General Douglas MacArthur, commander of US land forces in the Pacific, to plan the invasion of Kyushu – not to launch it. Many riders were attached to that decision, chiefly the size of the body count in the ensuing invasion of Okinawa; and whether the atomic bomb worked. The marines who waded up Kadena Beach expected immediate resistance; none came. Through a moist dawn they trudged over fields of wild raspberries and sugarcane, through brief forests and rice paddies. The main body of men headed south; the earlier euphoria yielding to apprehension and soon, in some, to a paralysing fear. Veterans of Iwo Jima, an island US forces captured in the long battle from 19 February to 26 March, during which 6812 US servicemen died, knew what lay in store. Soon, concentrated Japanese fire stalled the American advance at the edge of an outer ring of concentric fortifications, trenches, wire and weapon pits, 10 kilometres deep. This was the northern extremity of the 'Shuri Line', the defensive perimeter that the Japanese erected around Naha, the island's capital. Tens of thousands of Japanese troops lay in wait along these rings of fire. From here they would fight to the death.

The Americans charged the Japanese guns with little effect. Thousands fell. Under a hail of shrapnel, in the pouring rain, they inched forward past the inert figures of broken comrades pleading to be returned to the sea; over their own and enemy dead amid the plaintive cries of the terror-stricken; and, as the days passed, through slaughtered civilians, and living women and girls, some raped by advancing Americans or withdrawing Japanese. Over every living thing rose the stench of human detritus – flesh, bones, organs. The Americans and Japanese endured nearly three months of this carnage, until creeping US artillery and frontal assaults forced Okinawa's surrender in late June. Pockets of resistance and civilians in hiding were extricated with flames, smoke and grenades.

The land and sea battles at Okinawa killed 12,500 US sailors, GIs and marines, and wounded 44,000. It was America's most costly naval campaign of the Pacific War: 4907 sailors died on the Okinawan littoral between 1 April and 22 June. Most were victims of the Divine Wind. Ena and his men were supposed to ditch here: perhaps on the deck of the *Howorth* (disabled), the *Abele* (split in half and sunk), or the flagship carriers the *Bunker Hill* and *Enterprise* (severely damaged). Kamikazes sank 27 US ships and damaged 164, at a cost of 1465 aircraft and their crews. Their strike rate at Okinawa was higher than in other battles – perhaps a third of pilots hit a ship, according to some accounts.

The Japanese armed forces lost more than 100,000 dead and wounded; while as many as 75,000 civilians perished.* The figures vary,

---

* Estimates range from 30,000 to 160,000 civilian dead.

according to the source. A great many Okinawan civilians longed to surrender but the Japanese army prevented them, warning that the Americans would rape, torture and murder survivors. An unknown number perished under military coercion, in mute hatred of the business of war that had forced this choice upon them: to obey their unloved countrymen or expire under the rolling American war machine. They were made to see that it was more honourable to commit suicide.

•

The fall of Okinawa sent a powerful message to Washington: an American land invasion of Japan proper would encounter far worse. The likely casualties weighed heavily on Truman, and his private misgivings about the invasion of the Japanese mainland which General MacArthur was planning and expected to lead deepened. Indeed, the President was loath to authorise the land invasion, given the human and political cost, and cast around for alternatives. Several senior members of Truman's cabinet – Chief of Staff Admiral Leahy and War Secretary Stimson – did not believe a land invasion necessary to win the war, with or without the atomic bomb. Truman himself approached this conclusion in June, describing it in his diary on the 17th as his 'hardest decision to date': 'Shall we invade Japan proper or shall we bomb and blockade?' he wondered. In time the President would seek any alternative to an invasion, a position Leahy strongly supported and urged on the Joint Chiefs in June 1945.

The serious reservations in the White House resonated in Congress: the likely casualty rate of a land invasion, shockingly demonstrated at Okinawa, was politically unacceptable; moreover, all senior military commanders and many politicians – of all stripes – knew that by early 1945 the Japanese were an utterly defeated nation. These were the bald facts then known to Washington:

Japan had lost the air war. Since 1944, Japanese air losses had been catastrophic. The shortage of bombers and fighters had forced Tokyo to draw on the one resource the nation still possessed: large numbers of young men willing to die. Of the 2550 kamikaze missions between October 1944 and June 1945, 475 or 18.6 per cent actually struck their targets (a higher strike rate applied at Okinawa). The suicide attacks crippled or sank 45 US vessels, mostly destroyers; they killed 5600 American servicemen, of whom 3389 died at Okinawa. In June 1945, Japan possessed 9000 aircraft – mostly trainers or old planes designated for kamikaze raids, less than half of which had been properly fitted (that is, gutted and packed with explosives). By July America's complete air supremacy, the dire state of Japanese pilots, and the diversion of 2000 B-29s from incendiary missions to raids on the main kamikaze airfields on Kyushu, ensured that few suicide pilots were likely to reach their targets. After Okinawa, in the estimate of Rear Admiral D.C. Ramsey, Chief of Staff of the 5th Fleet, Japan had 4000 aircraft available to defend the homeland. About 100 underground aircraft factories were nearing completion by the end of the war, but the shortage of raw materials severely hampered their likely production rate (only 10 aircraft were eventually built underground). In the last months, the Imperial Army's only operational nightfighter against American bombing raids was the Kawasaki Ki-45 Toryu (Dragon Killer), employed in a task for which it had not been designed; like the more prolific Nakajima fighters, it lacked the armament and high-altitude performance to take on the B-29s; LeMay later boasted that he did not lose a single B-29 to Japanese fighters.

Japan had lost the sea war. American ships, aircraft and submarines had sunk or disabled all 27 Japanese aircraft carriers; both of the apparently unsinkable Yamato-class battleships, the world's largest, each displacing 64,000 tons; and 549 warships (1,744,000 tons) of all categories out of a total 1197 vessels (2,319,000 tons). After the immense sea battles off the Philippines in October 1944, Japan 'no longer had a fleet that could mount an offensive'. With the fall of Okinawa, the few remaining warships were decommissioned or covered in camouflage and used as floating anti-aircraft platforms. Apart from a few unskilled kamikaze squadrons, wretchedly ineffective human torpedoes (the *kaiten*, in which a sailor lay inside and directed the torpedo into an enemy ship) and a few suicide speedboats, the Japanese navy had ceased to exist.

Japanese ground forces were defeated throughout the Pacific and Burma. The Australians had driven the Japanese forces out of Papua and the surrounding islands; the British had prevailed in Burma after a bitter struggle; and MacArthur's forces, in the bloodiest of battles, had defeated them in the Solomon Islands, the Philippines, and on a string of atolls. In China and Manchukuo (Japanese-occupied Manchuria), enemy units were cut off and demoralised: 'The view that Japan's defeat is inevitable has achieved number one prominence,' the Greater East Asia Ministry office in Hsingking warned Tokyo on 17 June in a cable intercepted by Ultra, the American codebreaking system. The occupying units feared that 'contact between Manchukuo and Japan will be broken off' after the fall of Okinawa. Their links with the homeland – and the hope of reinforcements – were indeed completely severed in July, when Japanese commanders in China resorted to conscripting boys aged fourteen and over.

The American naval blockade had choked Japan's capacity to make war. This great arc of US carriers, destroyers and cruisers, aircraft, mines and submarines sank, shot down or bombed the slightest tremor of enemy movement on the water, under the water or overhead. The blockade sealed off the country. Japan's entire island economy relied on trade with China and Korea and transport between Honshu, Kyushu and Hokkaido. When US mines closed the Shimonoseki Straits, separating the main islands of Honshu and Kyushu, in May 1945, 18 of 21 naval repair yards situated on the Inland Sea were placed beyond the reach of Japanese ships, forcing them to use vulnerable, inadequate ports on the Sea of Japan. Hiroshima's port ceased to operate. By the end of July, US interdiction of Japanese imports – even without direct air attack on her cities and industries – had halved the nation's 1944 war production rate. The blockade cut off not only food supplies and vital commodities but also much-needed military reinforcements from China, where millions of stranded Japanese troops ached to get home to defend their country. While as many as 20 divisions did manage to get home by mid-year, most Pacific units remained cut off, defeated and outnumbered: 'We cannot hope to maintain planned communication with the Asian mainland after the end of this month,' wrote Tokyo's military strategists in June 1945. Next month Allied codebreakers heard of the 'severance of communication' between the mainland, Manchuria and China. The Yawata and Hirohata plants of the Japan Iron Manufacturing Company – respectively the country's largest and most modern – relied on shipments of iron ore from Manchuria. Those now ceased. In short, the blockade was 'the most critical single contribution to the American defeat of Japan', concluded the British historian Max Hastings. By the end of July 1945 America had defeated Japan thrice over – by sea, air and strangulation.

Japan had lost its entire merchant shipping fleet, the vital economic lifeline used to deliver food, resources, men and ammunition between Japan and its far-flung armies. By July 1945 American naval forces – chiefly submarines and mines – had sunk or disabled about 8,900,000 tons of merchant shipping, out of a total 10,100,000 tons. The technical supremacy of US submarines played havoc with the Japanese merchant navy, unshielded by any credible anti-submarine force. In 1944, US submarines sank 600 Japanese ships, slashing imports by 40 per cent. In total, American submarines and aircraft killed or wounded 116,000 of 122,000 Japanese merchant seamen, many thousands of whom were strafed in the sea as their vessels disappeared: 'The swine!' said one enraged Japanese survivor. 'Not satisfied with sinking the ship, they must kill those swimming in the sea! Was this being done by human beings?' 'The war against shipping,' concluded the US Strategic Bombing Survey, 'was the most decisive single factor in the collapse of the Japanese economy.'

Japan was defeated economically. In the first half of 1945, imports of crude oil, coal, iron ore and rubber ceased. In March, imports of coal – which powered most Japanese industry – were cut off. In July the oil refineries were nearly out of oil; the alumina plants, out of bauxite; the steel mills short of ore and coke; and the munitions plants shutting down for lack of steel and aluminium. Japan, which had launched the war to acquire resources, virtually had none. Domestic oil and coal reserves were exhausted. Less than 1.5 million barrels of aviation gasoline existed in mid-1945 – forcing a drastic cut in pilot training and combat missions and hastening the last desperate acts of the kamikaze, many of whom flew planes built partly of wood and powered by pine oil extracted from roots dug up by schoolgirls like Yoko Moriwaki in Hiroshima. After Okinawa, Japanese aircraft manufacturers 'may have to resort to more wooden construction', observed Rear Admiral D.C. Ramsey.

•

Meanwhile in the Pacific, America, Britain, the Soviet Union and their allies were amassing the greatest concentration of troops, ships and planes the world has known: millions of men were transferring from Europe to the Pacific theatre, as the largest ever seaborne invasion force converged there. 'Soon we shall have nearly 6 million men in all branches of the services in the actual theater of combat,' noted the director of American War Mobilization.

US bombers and fighters attacked Japanese cities at will, in broad daylight, virtually unopposed. Bombers leaving Okinawa and Iwo Jima were within easy reach of the mainland and resumed precision raids – which had been widely suspended during General LeMay's civilian offensive – on railroads, factories and airfields, over which they laid a 'big blue blanket' – that is, standing patrols to thwart enemy takeoffs. By mid-1945, US bomber losses had fallen to 0.3 per cent per mission. In the second half of 1945, LeMay's crews drew on 100,000 tons (90,000 tonnes) of ordnance per month as they systematically burned Japan to a cinder. So thorough were the firebombing raids on Japan's six largest cities – Tokyo, Osaka, Nagoya, Yokohama, Kobe and Kawasaki – that by 15 June 1945 most of their inhabitants were dead, wounded or forced into the country: 126,762 killed, 315,922 wounded and 1,439,115 homes destroyed. Nationwide, firebombing had wiped out 66 cities, killed 300,000 civilians, wounded about a million, destroyed more than 3.5 million homes and driven some eight million people into the rural areas. The national figures are rough but the scale and proportions are accurate.

The US Navy had multiplied several times within six years: 46,130 ships sailed under the US naval ensign on 1 July 1945, more than the world's entire merchant fleet of 1939. On 1 July, Admiral William Halsey's Task Force 38 of eight huge Essex-class carriers and six Independence-class light carriers, bearing about 1000 planes and accompanied by an escort of battleships, destroyers and cruisers, came 'boiling out of Leyte' to destroy the Japanese homeland. The armada reached the waters off Tokyo on 10 July, and its carrier planes struck deep inside enemy territory. Halsey's precision raids demonstrated the strategic advantage of striking military targets over LeMay's mass napalming of civilians: on 14 July, for example, his carrier aircraft sank eight of 12 huge rail ferries that transported coal from the Hokkaido mines to war factories on Honshu. The American historian Richard Frank described the raid as 'the most devastating single strategic-bombing success of all the campaigns against Japan', crippling coal supplies to the main island.

At the same time, the Soviet armies were crossing Siberia and forming up on the Manchurian border. Here was the fullest expression

– in troops and guns – of Russia's abrogation of the Neutrality Pact it still formally held with Japan. Stalin's convoys extended hundreds of kilometres across the Russian steppe. In May alone, Japanese couriers at a Trans-Siberian Railway shunting yard counted 195 east-bound military trains (28 per day, on average) carrying an estimated 64,000 troops, 2800 trucks, 500 fighter planes, 120 medium tanks, 320 anti-aircraft guns, 131 field guns, 300 collapsible boats and pontoons and ten carloads of bridge girders. In the yards at Irkutsk and Omsk, in central and southwestern Siberia, they counted about 500 tank cars. Members of the Manchukuan consulate at Chita, Manchuria, corroborated the estimates and witnessed 100 Soviet military trains passing during a 12-hour period. On 24 May the Japanese ambassador in Moscow predicted that Japanese–Russian relations 'would reach a crisis in July or August'. America's codebreakers picked up every word.

·

In the face of this challenge, the old samurais scarcely blinked. They would rely on their millions of determined, if poorly equipped, troops, volunteers and civilians, many of them inadequately trained ephebes and 'child soldiers' (15 years and over) with little combat experience; and 5000 effective aircraft of which 4300 had been converted to kamikaze planes, whose past strike rate suggested they would score about 400 direct hits. In any case, Japan possessed enough aviation fuel to put a few thousand planes in the sky, mainly kamikaze missions – according to a study by the US–British Combined Intelligence Committee which pooled the top intelligence in both countries – the effects of which would 'decline rapidly' within days. Japan possessed a cache of weapons and ammunition hidden underground and a few anti-aircraft guns or artillery pieces, but no useful bombers and no effective navy. After July, no reinforcements were forthcoming from Manchuria and China – notwithstanding the repatriation of 16 to 20 divisions in early 1945, which certainly posed a formidable threat, at least in the opening stages of an American landing. Indeed, throughout 1945 Ultra codebreakers regularly updated MacArthur on the solid enemy build-up on Kyushu, as historian Edward Drea has recounted. The same could not be said of the once robust Kwantung Army, an elite unit of the Imperial forces, now semi-trained, demoralised and isolated, almost 600,000 of whom remained cut off in Japanese-occupied Manchukuo.

Despite their relentless invocations to the 'hundred million', the Tokyo leadership knew the nation was militarily defeated – in the sense that it had lost the capacity to wage an offensive war. As early as June 1944, the Japanese Army High Command had privately acknowledged that Japan had no hope of winning the war. The hardliners pledged to fight on, however, to plunge the nation into a bloody defensive campaign in the hope of prising better 'peace' terms from America. The regime persisted in the belief that 'Japanese spirit' would galvanise its residual forces, suicide squads and bamboo-wielding home guard and inflict so many American casualties that the invaders would lose stomach and agree to negotiate. The Imperial Army's Field Manual for the Decisive Battle for the Homeland urged all units to fight to the last man, with their bare hands if necessary; civilians should form a shield against the invaders; any who fled would be shot. The point was to die 'honourably'. The wounded were to be abandoned.

•

With this purpose in mind, on 6 June the Imperial Army called a meeting of the Supreme Council for the Direction of War to confirm the 'Fundamental Policy Henceforth in the Conduct of the War'. The Council was composed of the Big Six: the three hardliners – War Minister Korechika Anami, Army Chief of Staff General Yoshijiro Umezu and Navy Chief of Staff Suemu Toyoda, who backed the army's demand for a clear statement further committing the nation to the battle for the homeland despite recent losses; and the three moderates – Prime Minister Kantaro Suzuki, Foreign Minister Shigenori Togo, and Navy Minister Admiral Mitsumasa Yonai, who wavered and obfuscated: on one hand secretly pursuing peace, on the other, openly supporting war. The hardliners' rousing psychological offensive overwhelmed the moderates: 'Japan was at a fork in the road! The life or death of the nation was at stake! Japan must embrace a fighting spirit bred of conviction in certain victory! She must leave no stone unturned to seize the divine opportunity to triumph!' were typical catchcries of the three hardliners. All day the militants regaled the meeting with arguments in support of a glorious last stand that would bend the American invader to the Japanese will. At the end, the hardliners prevailed, persuading Suzuki to accept the plan, and the Council adopted the decision to commit Japan to national *gyokusai*. Yonai sought refuge in silence; Togo dared oppose the army's daft idea that the deeper the Americans penetrated the homeland, the better the outcome for Japan. He was overruled. The people were ignored; 'the people simply did not count'.

'Here were drowning men,' observed the historian Robert Butow, 'grasping at the proverbial straw, bamboo warriors bending beneath the weight of uncontrollable events, yet never toppling to the ground.'

Two days later, on advice from Prime Minister Suzuki, Emperor Hirohito convened an Imperial conference to discuss the War Council's decision. On this occasion, Suzuki would play the puppet to the militarists' line. To Togo's disgust, the elderly premier, a wavering moderate who feared assassination – army fanatics had wounded him in the 1936 uprising – defended the hardliners. Unconditional surrender, the Prime Minister said, would result in the 'ruin' of the Japanese race and 'destruction' of the state – by which he meant the Imperial system. Japan must 'fight to the very end'. His views were 'accepted' in the sense that nobody opposed him (though afterwards Togo expressed severe reservations). The conference recommitted Japan – on the 'highest authority' – to fight to the last Japanese, 'without reservation, compromise or quarter'.

Hirohito said nothing; nor was he expected to. Shortly afterwards, the Emperor revealed his misgivings about the policy in a private talk with Marquis Kido, the Lord Keeper of the Privy Seal. Since Japan's defeat at the battle of Luzon on 11 February, with 205,535 Japanese dead, Hirohito had lost faith in the military's chances of victory: American air supremacy ensured the annihilation of the Motherland. Somehow a way had to be found to end the war, His Majesty privately intimated to Kido, in a rare breach of custom: the Emperor was expected to act on advice, not give it. Great historical changes hinge on private whispers between great men: ever so quietly the Emperor had dared to intervene, thus setting the Imperial Household on a collision course with the army.

Kido acted immediately. He drafted a counter plan designed to thwart the military clique and end the national death wish. Indeed, it accepted the termination of the war on terms 'only very slightly removed from unconditional surrender', as Butow noted: the laying down of arms; a universal withdrawal from occupied territory; and Manchurian neutrality. Kido, however, refused to countenance the presence of foreign troops on Japanese soil or the destruction of the Imperial system. The success of his plan depended on an uncommonly used measure, invoked in crises (the last such occasion was during the military uprising in 1936): the Emperor's willing intervention. Only Hirohito had the authority to persuade the Imperial forces to sheathe their swords.

As a first step Kido suggested that the Emperor urge the Big Six to set aside their 'death to the last man' decision of 6 June. The Emperor obliged: on 22 June Hirohito summoned the Supreme Council for the Direction of War to the Sacred Presence. After the usual rituals and deep bows, His Majesty spoke: were there any methods of ending the war other than a fight to the death? Anami, Toyoda and Umezu fastened on the policy of resistance; Togo pressed for peace through negotiation, using the Soviet Union as a go-between. All dutifully agreed to search for other avenues to end the war other than national suicide; until such time that policy remained. The Imperial Will had begun to work its silken charm.

In the meantime, it was agreed that Togo would appeal to Moscow to act as a possible mediator in future peace talks with America. Japan believed it held several bargaining chips – chiefly the spoils of the Russo–Japanese war secured by the Treaty of Portsmouth in 1905. And there was the battered Neutrality Pact with Japan, which remained in place, at least nominally, until 13 April 1946, despite Moscow's contempt for the agreement. Indeed, on 5 April 1945 the Kremlin had officially denounced the pact, and given notice to Japan that it would not renew the terms beyond the expiry date. Japan acted in the shadow of this implied threat when the War Council decided, with Hirohito's approval, to send a special envoy to Moscow to negotiate the proposed Russian intervention.

The special envoy chosen was Prince Fumimaro Konoe, a three-time former premier of aristocratic lineage, who privately loathed the military regime, which he blamed for Japan's peril. He had made those views clear to the Emperor in his 'Memorial to the Throne' report of February 1945, in which he told Hirohito that Japan had 'already lost the war'. The hardline militarists, he said, were 'the greatest obstacle to a termination of the conflict'. While they had 'lost confidence in their ability to [win the war], they are likely to continue fighting to the very end merely to save face'. Konoe asked for and received 'carte blanche' to negotiate with the Russians as he saw fit. His mission to Moscow had three principal aims: (1) to prevent the Soviet Union from entering the war in the Pacific; (2) to 'entice' the Kremlin into an attitude of friendship towards Japan; and (3) to persuade the Soviets to intervene (the word 'mediate' was considered defeatist and not used) with the Allies on terms favourable to Japan. It was hoped he would see Stalin before the Soviet leader left for the Potsdam Conference with the US and Britain in July.

•

Washington was attuned to all of these developments. Signals intelligence analysts working in starched shirts thousands of kilometres from the battlefields had cracked the Japanese war codes. In the closed world of Magic and Ultra (diplomatic and military codebreaking systems), American and Commonwealth cryptographers had thoroughly apprised themselves of the 'grand design' of the Imperial war strategy, which they immediately sent to their political masters. Since 1942, when signals intelligence broke Japanese naval codes, the Americans and British had known a great deal about the enemy's ship movements, merchant marine positions, diplomatic communications and resupply activities. Later in the war, signals intelligence broke the more complex army codes and listened into information about troop deployments, garrison strength and even the rations and other stores available. 'We know,' General Marshall wrote in 1944, 'sailing dates and routes of the convoys, and can notify our submarines to lie in wait at the proper points.' The Japanese were utterly unaware of the degree to which the Allies had penetrated their secrets.

Intercepts of diplomatic cables, called Magic, kept the White House fully abreast of Tokyo's repeated appeals to Moscow to intervene. Magic laid bare the regime's delusion that Russia could be persuaded to act on Japan's behalf in 'peace talks' with America. In pursuit of this fantasy, Tokyo sent a flurry of cables to Moscow between May and July 1945. The image of Togo's desperate appeal for Russian help – even as the Red Army assembled on the Manchurian border – encapsulated the final pathos of the Japanese Empire.

Washington interpreted Tokyo's 'peace feelers' sent to Russia and several other unorthodox channels as psychological warfare, conditional, or plain lies, and hence unacceptable. Some cables were without Tokyo's permission, sent by maverick Japanese diplomats appealing for peace to assorted 'middlemen' – Swiss 'third parties', Formosan 'police officials', Western industrialists and European crowned heads. In April, for example, a Japanese counsellor named Inoue told Kurt Sell, a correspondent for Germany's DNB wire service, that a 'negotiated peace' was possible so long as America removed the demand for 'unconditional surrender'. Not even the destruction of 'all the wood-paper houses in Japan' would make the country accept unconditional surrender, Inoue said. On 17 May, Major General Makoto Ono, military attaché to Sweden, whom Tokyo had accused of 'engaging in peace manoeuvres' with 'foreigners', contacted a Swedish oil executive, Eric Erickson, about initiating peace talks. Erickson passed the word on to Prince Carl, younger brother of the Swedish King. Concurrently, and unknown to the Swedish Royal Family, the Swedish Minister to Japan, Widar Bagge, engaged in high-level discussions with Togo about Japanese surrender; these stalled when Togo heard of Ono's unauthorised, private talks. Meanwhile, Allen Dulles, the US Office of Strategic Services representative in Berne, had proposed a 'discussion' between Japan and America in Switzerland, and that a Japanese Admiral be flown in from Tokyo 'in absolute secrecy'. The Japanese took the offer seriously and even discussed the charter of a Swiss mail aircraft to carry the secret admiral. On 22 June, Swiss sources reported – and Magic intercepted – Japanese expectations: 'Japan does not expect to win, but is still hoping to escape [defeat] by prolonging the war long enough to exhaust [her] enemies. Many eagerly desire the landing of the Americans in Japan proper, since they think it would be the last chance to inflict upon the Americans a defeat serious enough to make them come to terms.' These unofficial overtures came to nothing; all were intercepted by Allied codebreakers.

●

While nobody in Washington took these offers seriously, a stream of extraordinary cables between Naotake Sato, Japan's ambassador to the Soviet Union, and Foreign Minister Shigenori Togo, captivated US codebreakers and intelligence officers. The Moscow–Tokyo cables, which did represent the official line, portrayed a defeated regime desperate to find a way of surrendering on its terms.

'What is the meaning of the phrase "unconditional surrender"?' Sato Naotake asked his Tokyo masters in June 1945. The question went to the heart of the matter. A rare voice of sanity, Sato struggled heroically to convey the dark reality that eluded his superiors. A daring and often brutal candour imbued his dispatches to Tokyo. After the loss of Okinawa he predicted that Japan would be 'reduced to ashes' and 'in the same position that Germany was before her defeat'. If Russia invaded Japan, '… we would be in a completely hopeless situation … We would have no choice [but] … to eat dirt and put up with all sorts of sacrifices, fly into her arms in order to save our national structure.'

Sato's warnings went unheeded. Togo, to whom Sato answered in Tokyo, treated him merely as a facilitator, not an adviser, and enjoined the diplomat to 'make a desperate effort to obtain more favourable relations than mere neutrality' with Russia – though 'needless to say we are prepared to make considerable sacrifices in this connection'. Sato, a veteran Russian observer, enacted his masters' orders while scorning their strategy.

On 8 June, while the Big Six were meeting the Emperor to press the militarist line, Sato cabled Tokyo that it would be 'useless' to appeal for warmer Russo–Japanese relations: 'Now that Germany has been annihilated, Russia will hardly be willing to seek closer ties with Japan at the expense of Russo–American relations.' The fact that Sato felt obliged to say this at all was revealing of Tokyo's siege mentality. 'Molotov,' Sato rather brazenly added, 'would undoubtedly be surprised at such excessive naivete.'

On 28 June Togo alerted Sato to the fruitless exchanges between former premier Koki Hirota, and the Soviet ambassador Yakov Malik, in Japan. Through these intercepts, Allied codebreakers discovered the lengths to which the Japanese were prepared to go to appease Russia after the abrogation of the Neutrality Pact. Hirota and Malik had four conversations, between 3 and 14 June. During their third encounter, Hirota effectively surrendered Manchuria and offered to supply Russia with commodities from Japanese-held territory. In return, he asked, would Russia 'consent to reach some agreement with Japan more favourable than the Neutrality Pact'? Malik replied icily that 'until the expiration of the Neutrality Pact, we shall continue to play the role we have been playing ...'. At their last meeting on 14 June, Hirota appealed to Russia for oil, and invoked a future Russo–Japanese military union, in which the Japanese navy (by then non-existent) and the Russian army 'would make a force unequalled in the world'. Under this fantastic arrangement, Russia would supply oil to Japan in return for rubber, tin, lead and other commodities from Japanese-occupied Asia. But Russia had 'no oil to spare', Malik answered. To this Hirota mumbled about their shared hopes for an early peace – which Malik mischievously interpreted as implying the two countries were at war. Yet, since Russia was not a belligerent in the East, 'His Excellency Mr Hirota must be aware that peace there did not depend on Russia'. Malik's menacing dismissal of Hirota did little to dampen Tokyo's hopes of Soviet friendship.

Undeterred, on 5 July Togo ordered Sato to see Molotov before the Soviet Foreign Minister left for the Three Power Conference in Potsdam, 'and do everything in your power to lead the Russians along the lines we desire'. Sato interpreted these instructions as inconsequential, perhaps farcical, but did as he was told, and prepared a samovar of Japanese treats for the gluttonous Soviet bear: 'firm and lasting relations of friendship between Japan and Russia'; 'a treaty ... based on the principle of non-aggression'; 'Manchukuo's neutralization' and the withdrawal of Japanese troops; the renunciation of Japan's fishing rights ... if 'Russia agrees to supply us with oil'; and a willingness to discuss 'any matter the Russians would like to bring up'.

Surely there were no more abject admissions of defeat than this: Japan serving up the shreds of its empire to Moscow in return for Soviet co-operation in putative peace talks that Sato knew would never eventuate. The gesture, however, had the deeper intent of securing and perhaps extending the non-aggression pact, and removing the possibility of a Russian invasion, Japan's gravest fear. In this sense, Togo reasoned, if the Russian gamble was a losing bet, it was one worth taking.

A kind of pattern set in: Sato received his orders, then challenged or ridiculed the instruction as he enacted it; he thus left a trail of blistering critiques of official policy. Had the Japanese government not noticed, he dared to suggest to Togo, 'the general world-wide trend' that considered Japan 'the one obstacle to the restoration of world peace'? 'It seems extremely unlikely,' he added, 'that Russia would flout the Anglo-Americans and the opinion of the entire world by supporting Japan's war effort with either moral or material means. If one looks at the matter objectively, one must see that this cannot be.' With that, he abruptly dismissed any hope of forging another non-aggression pact with Russia.

Further damaging Japanese hopes, 'T.V.' Soong, the Chinese Foreign Minister, was then in talks with Moscow. Sato feared China and Russia might denounce Japan as an 'aggressor' and even sign 'a treaty of friendship'. In such circumstances, 'it would be 'utterly meaningless', he told Tokyo, to presume to sway Molotov with an offer of Manchukuo's neutrality. Apart from being 'entirely out of line with the general trend of events', in Sato's diplomatic idiom – in other words, fantastic in the extreme – 'the enemy would learn of it and this would undoubtedly stiffen his determination'.

Yet on 10 July Togo again overruled him: 'Your opinions notwithstanding, please carry out my orders.'

The next day Tokyo cabled the first official 'peace feeler' to Sato in Moscow, marked 'extremely urgent': 'We are now secretly giving consideration to the termination of the war, because of the pressing situation which confronts Japan both at home and abroad.' Would Sato oblige the new policy by sounding out Molotov 'on the extent to which it is possible to make use of Russia in ending the war'? Furthermore, the Imperial court was 'tremendously interested' in making peace – the first official acknowledgement of Hirohito's involvement. Tokyo offered to divest the empire '... as a proposal for ending the war', adding that it had 'absolutely no idea of annexing or holding the territories which she occupied during the war'.

Yes, Sato would oblige; yes, the Imperial interest was heartening; yet, no, the offer to relinquish territory, as a negotiating tactic, seemed utterly fanciful. In his reply, Sato seemed to read Allied minds: 'The fact is that we have already lost Burma and the Philippines and even Okinawa ... How much of an effect do you expect our statement regarding the non-annexation of territories which we have already lost or about to lose will have on the Soviet authorities? ... We certainly will not convince them with pretty little phrases devoid of all connection with reality.' The poor man had reached the limits of his patience: 'If the Japanese empire is really faced with the necessity of terminating the war, we must first of all make up our minds to do so ... Is there any sense in continuing the war no matter how many millions of our urban populations are sacrificed?'

Togo angrily ordered his errant diplomat to obtain Molotov's answer at once. But Molotov was unavailable, so Sato secured an audience with Vice Commissar Lozovsky, who dimly replied that he had no idea 'what my government's reply will be'.

On 12 July, Togo sent another 'very urgent message': while it may 'smack a little of attacking without sufficient reconnaissance', he ordered Sato to go 'a step further' and arrange a meeting between Special Envoy Prince Konoe and Stalin. This was the first Sato had heard of the special envoy's mission; astonished, he read that Konoe intended personally to deliver the following statement from the Emperor, no less:

His Majesty the Emperor, mindful of the fact that the present war daily brings greater evil and sacrifice upon the peoples of all belligerent powers, desires from his heart that it may be quickly terminated. But so long as England and the United States insist upon unconditional surrender the Japanese Empire has no alternative but to fight on with all its strength for the honor and the existence of the Motherland ...

Here at least was an expression of Japan's desire to end the war, the relieved ambassador felt – which clearly identified the one obstacle, unconditional surrender, preventing it. Sato was asked to prepare the ground for Konoe's arrival, the date of which depended on Stalin's availability. At 5pm on 13 July, Sato asked Lozovsky to transmit 'His Majesty's private intentions' to Stalin – by telephone, if need be. But Stalin and Molotov were leaving for Potsdam the next day and gave no reply. In transmitting this bad news to Tokyo, Sato tacked onto his cable the diplomatic equivalent of a steel-capped boot: 'I believe that in the long run Japan has indeed no choice but to accept unconditional surrender or terms closely equivalent.'

'In spite of your views,' Togo furiously replied, 'you must carry out instructions. Endeavour to obtain the good offices of the Soviet Union in ending the war *short of unconditional surrender* [my emphasis].'

•

The Emperor's apparent intervention astonished American intelligence officers, who immediately advised their Washington superiors. The initiative 'recast the picture in significant ways', conceded the historian Richard Frank (who in general has dismissed Japan's peace manoeuvres as insincere or unworkable). Truman and Byrnes received the news in Potsdam. They ignored it: 'Telegram from Jap Emperor asking for peace,' Truman noted, after talks with Churchill. In the eyes of the President and Byrnes, the Potsdam meeting and the forthcoming Trinity test had eclipsed a putative, conditional offer to share a peace pipe with the Emperor.

Elsewhere, highly placed Washington officials took a closer interest in the Emperor's intervention. Army intelligence chief Brigadier General John Weckerling offered three interpretations: that Hirohito had 'brought his will to bear in favour of peace in spite of military opposition'; that 'groups close to the throne' had 'triumphed over militaristic elements who favor prolonged, desperate resistance'; and that the regime believed it could 'stave off defeat' by buying Russian intervention and proposing peace terms to war-weary America. His interpretations were over-optimistic or incorrect. Officials close to the throne had clearly not triumphed over the military; nor was Japan trying to stave off defeat – the leadership knew it had been defeated. On the contrary, Tokyo was seeking Soviet help to mediate a surrender on terms favourable to Japan – and 'stave off' a Russian invasion.

CHAPTER 10

UNCONDITIONAL SURRENDER

The President would be 'crucified' if he accepted anything less than unconditional surrender.

James Byrnes, Secretary of State, mid-1945

Surrender by Japan would be highly unlikely regardless of military defeat, in the absence of a public undertaking by the President that unconditional surrender would not mean the elimination of the present dynasty ...

Joseph Grew, Under Secretary of State, former American ambassador to Japan, April 1945

THE JAPANESE WERE DEFEATED BUT would they surrender? The question perplexed the White House, War and State Departments. Truman's reiteration of the phrase 'unconditional surrender' had become a populist slogan and unsettled several prominent figures within his administration, who shared Churchill's view that a softening of the terms might end the war sooner. The whole question of surrender hinged on whether or not to grant the Japanese their single, abiding request: the retention of the Emperor. Meeting that condition was utterly unacceptable to hardliners in the State Department and the American people. The moderates, however, advised sending a clear statement to Tokyo to the effect that Japan must surrender all her arms and territory, submit to American occupation and a war crimes trial; but could keep their Emperor as a powerless figurehead.

The moderates' motives were honourable: to impose terms that were close to 'unconditional', in order to secure Japan's capitulation, end the war and limit American casualties; uppermost in their minds were the deaths likely to accompany a ground invasion, if it went ahead. Chief among those calling for softer terms was Joseph Grew, the Under Secretary of State, who had been US ambassador to Japan in the decade before Pearl Harbor. Grew understood the Emperor's place in the Japanese psyche as few in Washington did. A carefully phrased ultimatum that spared the Emperor's destruction would, he believed, compel the Japanese to surrender at little cost to American honour, with a concomitant saving of many lives: 'Surrender by Japan,' he warned on 14 April, 'would be highly unlikely regardless of military defeat, in the absence of a public undertaking by the President that unconditional surrender would not mean the elimination of the present dynasty ...' At various times, War Secretary Henry Stimson, Chief of Staff William Leahy, Assistant Secretary of War John McCloy, Navy chief James Forrestal and their colleagues similarly pressed Truman to ease the terms, to accommodate Hirohito. Leahy went further: he saw 'no justification for an invasion of an already thoroughly defeated Japan', and hoped instead that 'a surrender could be arranged with terms acceptable to Japan'.

Surely this was naive, argued the opponents of granting Japan a 'conditional surrender'. There was no guarantee that Japan would surrender, even with the gift of the Emperor: Tokyo would interpret any lenience as weakness and fight on. In any case, Truman had a political motive to insist on the harshest peace: most Americans agreed with him and felt no compunction to ease the terms of Japan's defeat and humiliation after four years of some of the bloodiest battles the world had seen. From New York to Texas, they longed to exact the most terrible revenge on the country that had inflicted Pearl Harbor and Bataan. Polls consistently showed a large majority in favour of unconditional surrender. A third wished to see Hirohito hang; most supported his imprisonment as a war criminal. Nine times as many Americans wanted [the servicemen] to fight on – 'until we have completely beaten her on the Japanese homeland' – rather than accept any Japanese peace offer, according to a poll on 1 June 1945. Their governing motive was vengeance: so many husbands, sons and brothers were dead, wounded or captive. As often in war, the civilians in the rear were more zealous for blood than the soldiers at the frontline.

Yet the same emotional impulse – to save America's sons – drove many Americans to seek ways of ending the war through what they saw as a harmless compromise: the *Washington Post*, for example, challenged the insistence on 'unconditional surrender' in a powerful editorial on 11 June 1945:

[The two words] remain ... the perpetual trump card of the Japanese die-hards for their game of national suicide. Let us amend them; let us give Japan conditions, harsh conditions certainly, and conditions that will render her diplomatically and militarily impotent for generations. But let us somehow assure those Japanese who are ready to plead for peace that, even on our own terms, life and peace will be better than war and annihilation.

Support for more conciliatory terms came from an unlikely quarter. The Joint Chiefs of Staff – whom none dared call defeatist – circulated a fresh interpretation of unconditional surrender: 'If ... the Japanese people, as well as their leaders, were persuaded both that absolute defeat was inevitable and that unconditional surrender did not imply national annihilation, surrender might follow fairly quickly.' The Joint Chiefs had a sound military reason for retaining the Emperor: as a tool to subdue the armed forces (at Potsdam they would insist, from a purely military viewpoint, the Emperor should remain in office to subdue fanatical elements of the Imperial Army outside Japan).

Truman listened and initially agreed with these arguments. He was ready to consider any alternative to hasten the surrender and avoid the massive losses of a land attack. Abandoning Roosevelt's casually invoked ultimatum of unconditional surrender would not, however, appease a vengeful public or firebrand congressmen – such as Senator Richard Russell of Georgia – who were conspicuously *not* at the frontline. Any amendment had to be sold politically.

•

In June and early July the plan to invade Japan, codenamed Operation Downfall, occupied Washington's top military minds. If it went ahead, history's largest seaborne invasion would realise MacArthur's conception of two successive thrusts: first, the amphibious assault on Kyushu, dubbed Operation Olympic, scheduled for 1 November 1945; then the massed attack on the Tokyo Plain – Operation Coronet – set for March 1946.

On Monday 18 June, four days after Hirohito's official intervention and the day after Truman noted in his diary – 'shall we invade or bomb and blockade?' in the wake of the carnage of Okinawa – the President convened a critical meeting of the Joint Chiefs of Staff in an effort to find an answer. This was crunch time for the invasion plan. The decision of whether to proceed rested, of course, with Truman, not with the Joint Chiefs, the Pentagon or MacArthur (who expected to command Operation Downfall). Truman had little regard for 'Prima Donna, Brass Hat, Five-Star MacArthur', as he had told friends during a sail down the Potomac the previous day. 'It is a great pity we have stuffed shirts like that in key positions.' Shortening the war and saving American lives preoccupied Truman, not soothing MacArthur's considerable ego.

At 3.30pm the masters of America's military strategy filed into the White House: Admiral Ernest King – clever, arrogant and 'perhaps the most disliked Allied leader of World War II' – who saw invasion as a contingency if the naval blockade failed; General George Marshall – honourable, self-disciplined, incorruptible – who advocated a massive, concentrated land invasion while exploring with War Secretary Stimson a workable surrender formula; Admiral William Leahy, Truman's Chief of Staff, who thought strategic bombing of civilians was 'barbarism not worthy of Christian man' and that the naval blockade alone would defeat Japan – in the latter view, he had the support of Admiral Chester Nimitz, Commander in Chief of the Pacific Fleet.* Lieutenant General Ira Eaker represented General Hap Arnold, the gruff, hard-driven chief of US Army Air Forces who shared LeMay's absolute faith in strategic bombing – despite its failure in Germany – as an alternative to invasion. In attendance too were department chiefs Henry Stimson (War), James Forrestal (Navy) and John McCloy, the Assistant Secretary of War.

All were aware of S-1; all knew of the atomic test planned for 16 July; all were attuned to the hope that, if successful, the bomb – or the threat of it – might hasten the end of the war and remove America's reliance on Russia. None entered the meeting disposed to mention this on the record; the elephant in the room remained a state secret officially aired in Target and Interim Committee meetings. The bomb's absence from the minutes, however, did not mean it was not discussed.

Truman called on Marshall, as the senior soldier, to begin. The general outlined the forces and strategies being prepared for the invasion. The plan earmarked 1 November 1945 for the Kyushu landing (as MacArthur had proposed). The circumstances, he said, were similar to those that applied before D-day. By November, Marshall added, American sea and/or air power will have:

- 'cut or choked off entirely Japanese shipping south of Korea';
- 'smashed practically every industrial target worth hitting' and 'huge areas in Jap cities';

---

* Nimitz told King on 25 May that continued blockade and conventional bombardment were enough to defeat Japan.

- rendered the Japanese Navy, 'if any still exists', completely powerless;
- 'cut Jap reinforcement capabilities from the mainland to negligible proportions'.

The weather and the helplessness of the enemy's homeland defences further recommended a November invasion, Marshall said. 'The decisive blow', however, may well be 'the entry or threat of entry of Russia into the war' – Russia's invasion of Japanese-occupied Manchuria, the 'decisive action leveraging [Japan] into capitulation'.

Marshall turned to the likely losses, which aroused intense discussion – most of it inconclusive and hypothetical. The Pentagon estimated that American casualties – dead, missing and wounded – during the first 30 days of an invasion 'should not exceed the price we have paid for Luzon', where 31,000 were killed, wounded or missing (compared with 42,000 American casualties within a month of the Normandy landings). Several caveats qualified this relatively low body count: the invasion of Kyushu would take longer than 90 days, and the figures did not include naval losses, which had been extremely heavy at Okinawa. In any case, Marshall insisted 'it was wrong to give any estimate in number'. The meeting fixed on 31,000 – a far cry from Marshall's later estimate of 500,000 battle casualties, which Truman claims the general gave him after the war, and which has bedevilled debate ever since (see Chapter 23).

Marshall and King concurred that invasion was the 'only course' available: only ground troops could finish off the Japanese Empire and force an unconditional surrender. There must be no delay, King said; winter would not wait. 'We should do Kyushu now,' he urged (his sudden enthusiasm for the attack on Japan marked a departure from his earlier proposal to invade Japanese-occupied China). 'Once started, however,' King remarked, with words Truman dearly wanted to hear, '[the operation] can always be stopped, if desired.'

A dissenting voice was Leahy, who, at Truman's invitation, questioned the surprisingly small casualty estimates, citing America's 35 per cent casualty rate in Okinawa. In what numbers were we likely to invade Japan, he asked; '766,700' US troops were projected, Marshall replied. They would face about eight Japanese divisions or, at most, 350,000 troops and, of course, a deeply hostile people. The dreadful mental arithmetic rattled the room: that left 270,000 Americans dead or wounded. King protested, however, that Kyushu was very different from Okinawa, and raised the likely casualties to 'somewhere between Luzon ... and Okinawa' – or about 36,000 dead, wounded or missing. In this instance, King's arithmetic was almost as dubious as his geography – Kyushu is a mountainous land riven with caves and hilly redoubts, rather like Okinawa.

So the invasion would be 'another Okinawa closer to Japan?' Truman grimly asked. The chiefs nodded. And the Kyushu landing – was it 'the best solution under the circumstances?' the President wondered. 'It is,' the Chiefs replied.

Unpersuaded, Truman asked for Stimson's view. Would not the invasion of Japan by white men have the effect of uniting the Japanese people, he asked, interrupting the War Secretary, who had been regaling the meeting with dubious ideas about a 'large submerged class' of Japanese insurgents. Stimson agreed: yes, the Japanese would 'fight and fight' if 'white men' invaded their country.

His opposition to an invasion deepening, the President examined another card in his hand: the forthcoming Potsdam Conference, and how to get from Russia 'all the assistance in the war that was possible'. This jolted the Joint Chiefs, who were forced to confront the military reality of 'unconditional surrender' – hitherto a political and diplomatic notion: it would mean a war in which the Soviets shared operations and, of course, the spoils. Were the Russians needed at all, several wondered. Silence. King spoke: the Soviets were 'not indispensable' and 'we should not beg them to come in'. His view echoed the feelings in the room.

Leahy then broke ranks and directly challenged the 'unconditional surrender' formula: it would make the Japanese fight harder, he insisted. He did not think its imposition 'at all necessary'. Truman appeared to agree, at least in part, suggesting that the definition of 'surrender' had not yet been fixed.*

Clearly, for Truman, the invasion plan was fading rapidly from the list of possible alternatives. He authorised the continued planning of the operation, but did not, and would never, approve its execution. The collapse of the Japanese economy, the total sea blockade and ongoing air raids had 'already created the conditions in which invasion would probably be unnecessary'. Indeed, Truman had convened the meeting precisely because he hoped to prevent 'an Okinawa from one end of Japan to another'. If the invasion of Kyushu and later Honshu was the 'best solution' of 'all possible alternative plans', demons of doubt lingered between the lines of the President's reluctant imprimatur.

In the days following, estimates of dramatically higher casualties further doomed the invasion plan. Nimitz, King and MacArthur all warned of a greater number of dead and missing than presented at the 18 June meeting. Even MacArthur ratcheted up his modest estimate, to 50,800 casualties in the first 30 days. No one could provide accurate projections, of course, and Truman never received a clear or unanimous calculation of likely losses, as King later said. Since the war, estimates of 500,000 to one million casualties have been crudely cited to justify the use of the atomic bomb – a classic case of justifying past actions using later information which was not applied at the time. At the time, nobody in a position of influence officially projected such astronomical numbers. The bomb, in any case, would not 'save' these hypothetical lists of dead and wounded: in late June and early July Operation

---

* 'It was with that thought in mind that I have left the door open for Congress to take appropriate action [on unconditional surrender],' Truman later claimed. This appears to be a misprint in the minutes – perhaps he meant 'before Congress', as Truman had never consulted Congress on the matter. But he did suggest that the definition of 'surrender' was still open to discussion. Public opinion, however, exerted a powerful hold on the President and he reminded the Joint Chiefs that he could not act 'at this time' to counter huge public support for the exaction of total victory over Japan.

Downfall lost the support of Truman and the Joint Chiefs not because the atomic bomb offered an alternative, but because the invasion plan was seen as too costly and, given Japan's military and economic defeat, ultimately unnecessary – regardless of the success or failure of the atomic test.*

The meeting drew to a close. But as the Joint Chiefs gathered up their papers, McCloy, thus far a quiet observer of the proceedings, spoke. A clever, thoughtful man, the Assistant Secretary of War was not afraid to express himself firmly. Only the day before he had urged Truman to drop the phrase, 'unconditional surrender'. For months McCloy had shadowed the issue as the 'leading oarsman' in Washington opposing the policy: 'I feel,' he noted in late May, 'that Japan is struggling to find a way out of the horrible mess she has got herself into … I wonder whether we can't accomplish everything we want to accomplish without the use of that term.'

He now found himself sitting among 'Joint Chiefs of Staff and security and Presidents [sic] and Secretaries of War', contemplating the weapon nobody dared name. As they prepared to leave, Truman turned to McCloy and said, 'Nobody leaves this room until he's been heard from.' McCloy glanced at Stimson, who nodded. McCloy's words do not appear in the official minutes, but he reprised the discussion in his memoir, and others present later verified his account: The bomb offered a 'political solution', McCloy said, that would avoid the need for invasion.

A hush ensued. McCloy continued: 'We should tell the Japanese that we have the bomb and we would drop it unless they surrendered.' Naming S-1 'even in that select circle … was sort of a shock,' he would recall. 'You didn't mention the bomb out loud; it was like … mentioning Skull and Bones [an undergraduate secret society] in polite society at Yale; it just wasn't done. Well, there was a sort of gasp at that.'

---

* There is solid consensus for this view, from various sides of the debate.

McCloy persevered: 'I think our moral position would be stronger if we gave them a specific warning of the bomb.'

The President seemed interested. He urged McCloy to take up the matter with Byrnes, who would soon be sworn in as Secretary of State. McCloy did so and Byrnes swiftly killed the idea. Byrnes, as Truman knew, firmly opposed any 'deals' with Japan that might be considered 'a weakness on our part', McCloy later wrote. (For the rest of his life, McCloy would regret the 'missed opportunity' of 18 June, insisting that the Japanese would have surrendered had America made clear that they could retain the Emperor and warned them of the bomb. Instead, the President had 'succumbed' – McCloy wrote, at the age of 89, in a letter to presidential adviser Clark Clifford – 'to the so-called hardliners' at the State Department.)

•

The land invasion plans were dealt a terminal blow in early July. Further reports, based on Ultra intercepts, of mounting Japanese strength in Kyushu, turned a blowtorch on the case for Downfall. The horrific example of Okinawa focused American minds on the growing presence of Japanese troops, and armed civilians, in Kyushu. On 8 July, the Combined Intelligence Committee released an 'Estimate of the Enemy Situation' – sourced to Ultra, military appraisals and interrogation of prisoners. Prepared for the Joint Chiefs of Staff, it stands as one of the most authoritative assessments of Japan's military capability in the dying days of the war. By July 1945, the report states, Japan expected to be able to field 35 active divisions and 14 depot divisions – a total of two million men (many of them worn-out or poorly trained conscripts, or civilians pressed into uniform) – in defence of Kyushu and Honshu. There were, however, qualifications. Most of these men had not been deployed as of 21 July, due to service elsewhere and transport delays, leaving 196,000 Japanese troops and perhaps 300,000 male civilians fit for military service stationed in southern Kyushu, according to US Sixth Army estimates. However, Ultra updated these estimates throughout July, with evidence of further homeland divisions moving to Kyushu. General MacArthur, ever anxious to lead the invasion, dismissed the figures as misinformation, or simply ignored them. Meanwhile, the Olympic Medical Plan (published 31 July) estimated 30,700 American casualties within 15 days of the invasion of Kyushu (requiring 11,670 pints/5520 litres of blood); 71,000 casualties after 30 days (27,000 pints/12,770 litres); and 395,000 casualties after 160 days (150,000 pints/71,000 litres). In each case about a third of the projected casualties were listed as battlefield dead and wounded; the rest would be general illness and non-battle injuries.

Regardless of the quality of the enemy troops – and the evidence suggests they were badly equipped, relying more on spirit than any tangible factors (like adequate air cover and artillery) – their huge numbers unsettled and ultimately helped to shelve the US invasion plans. That was not because America feared it would lose the encounter; rather, hurling American lives at a defeated nation, at a people intent on their own destruction, made little sense: why expend American lives playing to the samurai dream of a 'noble sacrifice', a national *gyokusai*? Why assume the role of executioner to a regime determined to inflict martyrdom on its people? And at what cost? The unrelenting roll call of the American dead was politically intolerable at a time when the sea blockade and air war – precision and incendiary – were grinding the enemy under. And there was the wild card of the Soviet Union, whose entry into the conflict Truman continued publicly to encourage, and privately to question. Washington could not overlook the gift of Soviet arms assistance, which, the intelligence chiefs concluded, would 'convince the Japanese of the inevitability of complete defeat'.

The atomic bomb, if it worked, was not seen as a direct alternative to the invasion: the invasion and the bomb were never mutually exclusive; nobody presented the case in terms of 'if the bomb works, the invasion is off'. These events advanced in tandem, in a complex interplay between threat and counter-threat, setback and opportunity. Indeed, some in the Pentagon believed that the bomb, if it worked, made the invasion *more* likely – as a supporting weapon: 'In the original plans for the invasion,' General Marshall later wrote, 'we wanted nine atomic bombs for three attacks' – on three fronts. The risk of irradiating the advancing army did not recommend the strategy.

By early July 1945, regardless of whether the bomb worked or not, Japan's pathetic state, the likely casualties of Tokyo's death wish, and Truman's political sensitivity made it almost inconceivable that MacArthur's invasion plan would proceed. Ultra confirmed Washington's fears – and those of the Joint Chiefs of Staff – that Japan's leaders had not only correctly identified where the proposed invasion would start; they had made the defence of the southern half of Kyushu their 'highest priority'. These developments led to the decision to set aside, if not yet completely cancel, Olympic – MacArthur's cherished invasion plan – a week before the momentous developments in the New Mexican desert.

•

Stimson moved further into the cold during June and July; his influence waned as he made his moral reservations clearer. His fall from grace symbolised the excision of conventional morality from the political heights of Washington. On 10 May the War Secretary had talked privately with Marshall – his closest companion in age and outlook – 'on rather deep matters'. Stimson hoped to hold off the invasion of Japan 'until after we had tried out S-1 ... probably we could get the trial before the locking of arms and much bloodshed'.

Stimson privately paled at the thought of dropping the bomb on a city. And yet he had recently approved the world's first nuclear strike, on 'workers' homes'. At first glance, it is difficult to see how he reconciled these contrary positions. The answer is that Stimson was above all a politician and military strategist, not an ethicist or man of God. His public image worried him more than the dictates of his conscience: he feared that his approval of the atomic attack would damage his public reputation as 'a Christian gentleman', as he later wrote.

The carrot of the Emperor would force Japan to surrender, he maintained. On 6 June, in a private chat with the President, he raised the possibility of achieving 'all our strategic objectives' without the insistence on unconditional terms. Implicit here was the gift of the Emperor. Allow this and the 'liberal men' in Tokyo would have a potent political weapon against their fanatical colleagues; or so Stimson hoped. Surely a class existed within Japan 'with whom we can make proper terms', he repeated in his diary on 18 June, the night of the meeting with the Joint Chiefs; surely the Japanese can be made to respond peacefully to a 'last chance' warning, he wrote, on the 19th. Hitherto, these had been his private musings; henceforth the embattled War Secretary intended to make a more public stand – in line with Grew's moderation.

That day, in talks with Grew and Forrestal, Stimson expressed his abhorrence of the (at that time) anticipated cave-by-cave attack on the Japanese homeland. Were there not reasonable elements within the Japanese regime, he wondered, who resisted Tokyo's death wish? Grew agreed: 'All the blustering the Japanese were now doing about fighting to the last meant nothing; there might be important things going on in the minds of the leaders of Japan at the moment of a quite contrary character ...'

America should clarify what it meant by 'unconditional surrender', Grew advised. For him, like Stimson, it meant letting the Japanese determine their post-war political structure – including, if they desired, the Imperial line – so long as it enshrined freedom of thought and speech, and human rights, and contained no militaristic element. It meant allowing Japan to retain the Emperor as a figurehead. Presented with those terms, he argued, Japan's rulers would 'desist from further hostilities'. The preservation of the throne and the 'non-molestation' of Hirohito, Grew later advised Truman, 'were likely to be irreducible Japanese terms'. The intelligence community lent weight to these deliberations: in early July the Combined Intelligence Committee warned that Japan equated 'unconditional surrender' with the loss of the Emperor and 'virtual extinction'. In this light, it suggested, a promise to retain the Emperor might compel the Japanese to disarm and relinquish all territory.

Stimson and Grew were not the only high officials in Truman's administration willing to abandon the unconditional surrender formula to secure victory over Japan. Some, like McCloy, had even advised offering the Japanese a warning of the atomic bomb. Ralph Bard, Assistant Secretary of the Navy (as well as McCloy and others at different times) urged Truman to make a show of the weapon's power before any military use. In a memo to Stimson on 27 June, Bard favoured an explicit warning to Japan two or three days before dropping the bomb – to demonstrate that America was 'a great humanitarian nation' with a strong sense of 'fair play'. He believed that Tokyo was sincere in its efforts to find a medium of surrender; he even supported peace negotiations. This was, of course, going too far, and few agreed with Bard.

•

On 2 July, the President received Stimson in the Oval Office. The War Secretary looked tired and pale. They discussed the draft of a proposed Presidential statement on the Japanese surrender. With time running out and people fretting at the door, Stimson asked Truman why he had not been invited to join the Presidential party at the Potsdam Conference, which began that month. Had the President declined to invite him 'on account of the fear that I could not take the trip?' Stimson asked, casually referring to his health.

'Yes, that was just it,' replied Truman laughing.

But the Surgeon General has endorsed my condition, Stimson protested. And 'practically every item on the German agenda' – at the Berlin conference – 'was a matter handled by the War Department'. The President said he would think it over and discuss it tomorrow. In such homely slights are powerful men brought low: the official in charge of the war would not be invited to the meeting convened to end it.

Seeing his star wane, Stimson sensed he had nothing to lose by added candour. Later that day he wrote to the President, setting forth a nightmare vision of fanatical resistance and terrible American losses, far greater than at Okinawa, which would leave Japan 'even more thoroughly destroyed than was the case with Germany'. Was this necessary, he wondered – not fully realising the extent to which the President agreed with him about the redundancy of the invasion plan. Surely the Japanese were on the brink of defeat? Japan had no allies, virtually no navy, and was prey to a surface and submarine blockade that deprived her people of food and supplies. Her cities were 'terribly vulnerable' to air attack. Against her marched not only the Anglo-American forces but also 'the ominous threat of Russia'. America enjoyed 'great moral superiority' as the victim of [Japan's] first sneak attack. The difficulty, he conceded, was to impress upon the Japanese warlords the futility of resistance.

To this statement Stimson appended a new draft of what would become known, with important amendments and deletions, as the Potsdam Declaration (officially, the Potsdam Proclamation): a warning to the Japanese leadership to surrender or face annihilation. His words resonated with those of an earlier draft by Joseph Grew (which the President had considered 'sound' at the time). Both drafts allowed Japan to retain Hirohito as a powerless head of state; and promised not to enslave or 'extirpate' the Japanese as a race 'or destroy them as a nation' – but to remove all vestiges of the military regime so that Tokyo could not mount another war. The Japanese, it concluded, should be permitted 'a constitutional monarchy under the present dynasty' if it be shown to the complete satisfaction of the world that such a government 'will never again aspire to aggression'. Crucially, the draft listed the Soviet Union as one of the four signatories, with America, Britain and China.

•

It was to no avail. A new, hardline force had entered the Truman administration. On his swearing in as Secretary of State, on 3 July, Byrnes swiftly assumed greater powers than his position entailed. He acted in some ways as a de facto president – and moved at once to stifle the air of compromise. In coming weeks, Truman sat back to watch Byrnes tear apart these dovish tendencies, stifle any softening of the surrender terms and thwart Stimson's expectation of an invitation to Potsdam (the War Secretary would invite himself and attend under his own steam). Byrnes ensured that Grew, McCloy and Bard (hitherto a member of the Interim Committee) were excluded from critical meetings and their views largely ignored.

Under the new Secretary, the State Department pointedly refused to entertain ideas about retaining the Emperor. The President would be 'crucified' if he accepted anything less than unconditional surrender, Byrnes, with an eye on public feeling, confided to his secretary. Curiously, *official* US foreign policy (on unconditional surrender) made no direct reference to the Emperor – stating only that Japan must disarm and dismantle its military system – a state of ambiguity that left Hirohito's fate the subject of raging debate and confusion in Washington and Tokyo. Nowhere was the debate more intense than in the State Department under Byrnes, which affirmed that the 'only terms' on which America would deal with Japan were those listed under 'unconditional surrender' – as announced by Roosevelt at Casablanca in 1943 – which prescribed the elimination of the military system, implicitly including Hirohito as supreme commander.

The State Department duly fell in step with Byrnes' hardline view. The new Secretary had influential backers: Assistant State Secretary Dean Acheson, Director of the War Department's Office of Facts and

Figures Archie MacLeish and their supporters reacted violently to any suggestion of retention of the Emperor: it would be seen as exonerating a war criminal and allowing an abhorrent enemy to set the terms of surrender; the Emperor stood at the pinnacle of an odious military system, and his continuation, even as a powerless figurehead, risked the resurgence of that system. In any case, the perpetrators of Pearl Harbor, Bataan and innumerable atrocities against prisoners and civilians were in no position to impose conditions on America. The State Department hammered out these views at a staff meeting on 7 July, over which Grew awkwardly presided as Acting Secretary (Byrnes being away). Nor were there any 'liberal-minded Japanese', the hardliners argued: Ultra's intercepts had revealed Tokyo's continuing, bitter determination to fight to the last.

Byrnes' obsession with privacy has obscured many of his words and deeds, leading some to infer what a man of his character might have done, rather than what he did, during the coming events. The Protestant convert (he grew up a Catholic) from South Carolina has been variously described as deceitful, pathologically secretive, a master of the dark arts of political arm-twisting and openly racist. Some of these criticisms are unfair. For instance, while he opposed the principle of racial integration, the central tenet behind Roosevelt's civil liberties program, he refused to join the Ku Klux Klan at a time when it was politically expedient to do so. He shared the Klan's basic ideas but baulked at their methods; the lynching of black men was not the politician's way. His restraint was thought courageous at the time because, as an ex-Catholic, he had much to prove to the hooded Protestants who tended to persecute papists when blacks were scarce.

Whatever Byrnes' flaws or strengths, his actions must be seen in the light of his record. He was a skilled judge and administrator, and a highly experienced politician of the kind that excelled behind the scenes on committees. His work as head of the Office of War Mobilization was exemplary at a time of national emergency. His deep knowledge of Washington and his thwarted ambition – he had hoped to succeed Roosevelt as president – quickly established him as Truman's 'big brother' in political terms. As Truman's personal 'coach' on sensitive areas of foreign policy, Byrnes enjoyed great influence over the President well before his elevation to Secretary of State. It was Byrnes who, handing Truman a leather-bound transcript of his Yalta notes, urged the inexperienced new leader to adopt a much tougher line on Russia. Byrnes also served as Truman's eyes and ears on the Interim Committee, at whose 21 June meeting he overruled Stimson and drove the decision to revoke Clause Two of the Quebec Agreement with Britain and Canada, signed by Churchill and Roosevelt on 19 August 1943, which folded British atomic research into the Manhattan Project and bound the signatories not to use the atomic bomb against a third country without mutual consent. Washington had lost faith in the agreement in 1944, when it emerged that Britain had shared secret details with France in exchange for post-war patents on nuclear reactors. At Byrnes' urging, America had thus freed herself to use the weapon unilaterally without any need to consult her allies.

•

The question of Japanese 'peace feelers' exercised Byrnes on taking office: en route to Potsdam, he received a cable from Grew outlining the latest 'peace offer' – this time, from the Japanese military attaché in Stockholm. Of itself, it did not warrant Byrnes' close attention – merely one more in a flurry of Japanese 'peace' proposals, of dubious provenance, sent to an assortment of intermediaries during the last months of the war. Few were sent through normal diplomatic channels, and none officially reached Washington. If their credibility varied, their messages were consistent: the Japanese sought a negotiated peace on the precondition that America agreed to ensure the survival of the Emperor. That was unthinkable, of course; yet Byrnes' chief concern, tweaked by the enemy's latest initiative, lay in the growing media and congressional interest in Japanese 'peace feelers', and the effect this would have on public opinion. What did 'unconditional surrender' actually mean? What, precisely, were America's demands? Powerful voices in the media, such as the *Washington Post*, *Time*, *Newsweek*, the *New York Times* and the influential broadcaster Raymond Swing, wanted clarification. Politicians, too, joined the chorus: Senator Homer Capehart, the Indiana Republican, demanded a published definition that would set the minimum terms that America would accept so that 'those who hereafter must die will know exactly what is to be accomplished by their sacrifice'.

Byrnes asked his Under Secretary, Joseph Grew, to put a stop to the growing media speculation about 'whether the Japanese government had or had not made a *bona fide* peace offer'. 'We have received no peace offer from the Japanese government,' Grew dutifully announced in a press statement on 10 July, 'either through official or unofficial channels' – which was technically true (Hirohito's first peace offer – sent to Moscow – came the next day). 'The alleged "peace feelers",' he wrote, 'have invariably been inquiries as to our position.' They were merely a form of psychological warfare intended to divide the Allies and should be ignored, he said.

In the same statement, Grew wrote a trenchant defence of 'unconditional surrender' that seemed to fly in the face of his previous opposition to the policy: 'I wish ... to drill home into the consciousness of our people, namely, that we must not, under any circumstances, accept a compromise peace with Japan ...' The Under Secretary had travelled far in a few days under his new boss. Soon, all of Washington would march to Byrnes' tune. 'The policy of this government,' Grew continued, 'had been, is, and will continue to be unconditional surrender.' So just what did it mean, under the new broom?

'It does not mean the destruction or enslavement of the Japanese people. It means the end of the war. It means the termination of the influence of the military leaders who have brought Japan to the present brink of disaster. It means provision for the return of soldiers and sailors to their families, their farms, their jobs. It means not prolonging the present agony and suffering of the Japanese in the vain hope of victory.'

Grew, with Byrnes looking over his shoulder, had refused to clarify the crucial question, the fate of Hirohito, on which all depended, as he knew. In the Japanese mind the loss of the Emperor *did* mean the destruction of the state and the Japanese people, as Grew had constantly advised. Only 10 days earlier, on 29 June, Grew and others had drafted an agenda for the Potsdam meeting. It recommended that any ultimatum to Japan should 'eliminate the most serious single obstacle to Japanese unconditional surrender, namely concern over the fate of the throne'. Byrnes saw no reason to ease Japan's concerns in this regard, and as the Presidential party headed for Potsdam, the split in Washington between those who supported Byrnes' hardline policy and those who, like Grew and Stimson, privately opposed it, deepened. One thing was clear to Byrnes as he sailed east: any clarification of the fate of the Emperor would be pointedly removed from the script.

CHAPTER 11

TRINITY

My God, we're going to drop that on a city?

Chemist Henry Linschitz, after witnessing Trinity

AT DAWN ON 7 JULY 1945 President Truman and Secretary of State James Byrnes boarded the USS *Augusta*, at Newport News, Virginia, bound for the Three Power (America, Britain and USSR) peace conference in Potsdam. The voyage took eight days through peaceful waters; ships were no longer darkened at night or preceded by mine sweepers. The 9050-ton vessel, known affectionately as *Augie*, had a fittingly illustrious past: on 5 June 1944 she had joined the invasion fleet on D-day, bearing General Omar Bradley, commander of the US land forces, to his observation point 3 kilometres off the beachhead.

Truman had asked to postpone the conference until 15 July, the deadline set for the atomic test. The delay dismayed Churchill who, gravely concerned at the presence of the Red Army in the heart of Europe, had pressed for 3 or 4 July. Churchill saw the containment of Soviet designs on Europe as the main priority of the conference. But Truman had other priorities. He insisted on the later date because the outcome of the Potsdam negotiations rested in part on the atomic test result. If the bomb worked, America conceivably had the power to force Japan to surrender without Russian help, as Byrnes had quietly argued. Byrnes already perceived a wider, diplomatic role for the weapon – to curb Russian aggression in Europe – and he expanded on this theme to the President as they sailed across 'the pond'.* 'Had a long talk with my able and conniving Secretary of State,' Truman later wrote. 'My but he has a keen mind! And he is an honest man.'

Truman's stated priority for going to Potsdam was 'to bring Russia into the Pacific War' (he continued this line on his return: 'Truman's Aim at Berlin: Get Reds into War' ran the headlines on 9 August). Russian support would hasten victory and save hundreds of thousands of Americans from injury or death, a White House staff meeting concluded on 4 July. But the President's private agenda was less clear – and hinged on the success of the atomic bomb. Truman privately hoped to finish off the Japanese without the Russians; the bomb was his 'master card'. In any event, the mission to secure Russian help was somewhat superfluous – a point lost on the press – given that Stalin had already pledged at Yalta to enter the Pacific War. The Soviet abrogation of the Neutrality Pact with Japan and the vast troop build-ups on the Manchurian border, of which Ultra had full knowledge, were clear signs that Stalin meant to fulfil his commitment.

Over the journey's course, Byrnes immersed himself in the job of negotiating a path through this confluence of events that would see America emerge as the ultimate victor over Japan – *without* the Soviet Union. He boarded the *Augusta* in possession of the latest draft of the Potsdam Declaration – a synthesis of the work of Stimson, Grew and McCloy – which the War Secretary had handed him on 2 July. It warned of 'prompt and utter destruction' if Japan refused to surrender unconditionally. Byrnes fastened on to two elements of the draft: first, the authors had left open the possibility of 'a constitutional monarchy' under the present dynasty; second, it included the Soviet Union as a signatory (along with the United States, Britain and China). Stimson had inserted Russia's name as an 'additional sanction to our warning'. In short, the draft offered the Japanese a continuation of the Imperial

---

* Stimson shared Byrnes' faith in the bomb as a diplomatic weapon against Russia. A fitting moment to settle 'the Polish, Rumanian, Yugoslavian and Manchurian problems', he wrote in his diary, would arise after the first bomb fell on Japan.

system, and named their historic and most feared enemy, Russia, among the punitive forces if they refused.

Byrnes loathed the document. Any deal that retained Hirohito would outrage American public opinion and prove politically explosive – as his old friend Cordell Hull, Secretary of State under Roosevelt, had warned during a telephone conversation on 6 July. And the sight of the Soviet Union listed as a co-signatory repelled him. Moscow had not participated in the Pacific War; yet Stalin's signature on the ultimatum gave the dictator a seat at the negotiating table – with the dreadful prospect of a re-run of Russia's Eastern European land grab in Asia. Byrnes resolved once and for all to remove the gift of the Emperor and dampen Moscow's hopes of being 'in at the kill'.

In fact, the Secretary of State had more ambitious plans: he intended to win the war *without* Russian help or *any* concessions to the Japanese. His 'winner take all' gambit appealed to the presidential pride – and poker player – in Truman.

'I must frankly admit,' Byrnes remarked later in his memoir, 'in view of what we knew of Soviet actions in eastern Germany and the violations of the Yalta agreements in Poland, Rumania and Bulgaria, I would have been satisfied had the Russians determined not to enter the war. Notwithstanding Japan's persistent refusal to surrender, I believed the atomic bomb would be successful and would force the Japanese to accept surrender on our terms. I feared what would happen when the Red Army entered Manchuria ...'

And so, as Grew and Stimson feared, their precious draft was indeed 'ditched' during the voyage 'by people who accompany the President'. Byrnes had set his ideas in motion in the weeks before his departure. With the President's backing he sought to delay Soviet intervention in the Pacific by urging Peking's Kuomintang government to return its Foreign Minister, T.V. Soong, to Moscow to prolong Sino–Russian negotiations over the spoils of Japan's defeat. Stalin saw an accord with China over the carving-up of Japanese-occupied Manchuria as a prerequisite to any declaration of war against Japan, because it guaranteed the spoils in advance. In the meantime, Byrnes and Truman 'had, of course, begun to hope that a Japanese surrender might be imminent and we did not want to urge the Russians to enter the war'. Their ideal scenario was that the Sino–Russian negotiations stall Moscow long enough for American planes to drop the bomb – and thus force Japan's surrender – to America.

•

Accompanying Truman and Byrnes to Potsdam were the President's White House staff, led by his personal adviser, Admiral Leahy, and press chief, Charlie Ross, as well as other close confidants, bourbon drinkers and poker lovers. Byrnes took his loyal aides Ben Cohen, H. Freeman Matthews and Charles Bohlen. Conspicuous by their absence were the War Department's Stimson and McCloy, whom the ascendant State Department had neatly sidelined. Stimson, however, determined to be heard, travelled at his own expense to Berlin. Though excluded from the conference, he would serve a valuable role, as the recipient of dispatches from his assistant George Harrison, who had been deputed to relay news of the atomic test then being prepared in New Mexico. As Acting Secretary of State, Grew was obliged to stay at home. Those excluded had all, at one point or another, urged Truman to moderate the surrender terms. Byrnes, their most strident opponent, now monopolised the President's attention on the matter.

Truman spent his days at sea in energetic form: he rose at dawn, exercised on the deck, breakfasted in his cabin, and, from 9.30 until noon, met with Byrnes and their advisers. On Sunday 8 July, the President attended a Protestant church service, held below deck due to bad weather. He usually dined in the Presidential cabin and ate lunch in a different mess each day, joining the 'chow line' with the sailors, aluminium tray in hand.

In the evenings before dinner a symphonette performed Elgar, Mozart, Strauss and Brahms, as well as modern tunes such as 'Over the Rainbow'. After dinner the official party repaired to Byrnes' cabin to watch the Pathe news service and a feature film, perhaps a Bob Hope comedy or a Walt Disney animation. The *Augusta's* library contained a wide range of books selected to satisfy Truman's eclectic taste: under 'G' were his Under Secretary of State's *Ten Years in Japan* and L. Goodman's *Fireside Book of Dog Stories*. If he lacked time for these, Truman certainly read the morning and afternoon *Augusta Press* – typed sheets of president-friendly news of the world. On 7 July, for example, Truman read with presumable satisfaction that 600 Superfortresses, a record, had the day before dropped 4000 tons (3600 tonnes) of incendiaries on five Japanese cities, losing no planes, facing no enemy aircraft and 'meagre' ground fire; Japan's Home Ministry, he further read, had transferred new powers to regional authorities 'in preparation for the decisive battle to be fought on our own soil'; the *Red Star*, the Soviet army journal, called for the 'gang of Polish Emigré Provovateurs [sic] and Warmongers' (that is, the exiled Polish government in London) to be 'rendered harmless'; and the Duke and Duchess of Windsor (Edward and the former Mrs Simpson) were to visit England, discreetly – to spare the Royal family embarrassment.

The next day, Truman read in the *Augusta Press* of the 'tightening' of the Allied blockade around Japan as US warships freely prowled the East China Sea – 'Tokyo's ... confounded war lords will order resistance to the death ... even the propaganda-drugged Japanese people must eventually discover that there is another and honorable way out.' Between 8 and 12 July he read of the Fifth US Air Force's attack on Kyushu for the fourth straight day; of the 423,000 Japanese dead tallied by MacArthur's forces in the Philippines; and, to Truman's great satisfaction, of Admiral 'Bull' Halsey's 3rd Fleet 'knocking the hell out of the Japs' after a 1000-plane raid over half a dozen Japanese cities, with not a single enemy aircraft in sight.

Byrnes and Truman spent much time in private discussion. They trimmed their official business at Potsdam into four manageable

American objectives: to lay the foundations for post-war negotiations; to establish a council of foreign ministers that would negotiate the detail; to adopt a fresh approach to German reparations in view of ongoing disagreements; and to persuade the Soviet Union to implement the Yalta agreements on the future of Eastern Europe. Unofficially, they talked of the atomic bomb and their distrust of Russia; but nowhere was 'getting Russia into the Pacific' treated as a priority.

•

At dawn on 14 July the *Augusta* neared Portsmouth. The rising sun burned a hole through the morning fog, as a British light cruiser and six destroyers escorted her up the mine-swept English Channel. Bugles sounded and a British navy band struck up 'The Star-Spangled Banner'. The President received the salute in a grey tweed cap and olive overcoat, standing in the 40-millimetre anti-aircraft gun on the bridge deck. As each ship withdrew their escort, their crews shouted, 'Three Cheers for Mr Truman, President of the United States!'

The next day the *Augie* sailed up the Scheldt estuary towards Antwerp where the Acting Burgomaster of Flushing sent a message of welcome: 'May your arrival in Europe contribute to the building … of a spirit of peace and friendship between the peoples of the earth.' Hundreds of 'wildly enthusiastic' Dutch and Belgians lined the southern shore; a sullen horde of German prisoners held behind barbed wire dimmed the applause in passing. At Antwerp 47 automobiles collected the presidential party and drove through the grey countryside, cheered on by thousands of recently liberated Belgians. From Brussels, the C-54 'Sacred Cow' flew the presidential entourage to Berlin, where Stimson and McCloy were waiting on the tarmac at the head of the welcoming party.

The President's motorcade wound through Soviet-controlled East Berlin, past green-capped Russian frontier guardsmen lining the roadsides, towards Potsdam and its wealthy suburb of Babelsburg on

Lake Griebnitz, former home to several prominent Nazi film producers and directors, all now imprisoned, dead or in exile. A few weeks earlier drunken invading Soviet troops had broken into a film studio, dressed themselves in the pride of the costume department – Spanish doublets, white ruffs and Napoleonic uniforms – and danced in the streets to accordions, as they fired their weapons into the night sky and the war raged around.

The American headquarters, dubbed the 'Little White House', were situated at No. 2 Kaiser Strasse. It was a three-storey stucco mansion of austere ugliness set on a pretty lawn, bounded by groves of trees which rolled to the lake shore. 'It is a dirty yellow color,' the President wrote, '... stripped of everything by the Russians – not even a tin spoon left.' Their Soviet hosts had thoughtfully supplied German furniture plundered from surrounding castles: 'Oppressive and awesome in its gloom,' noted the *New York Times*, the 'nightmare of a house' was filled with depressing still lifes and hideous lamps. At least the food was American: Truman's own cooks, loyal Filipinos, were brought over; and bottled water sent from France. A private map room and communications centre were installed, with a direct wire service to Frankfurt and Washington. Nonetheless, all phone users were advised not to discuss confidential matters 'as telephone facilities are not secure'. The Soviets had in fact installed hidden microphones in both American and British residencies in advance of the occupants' arrival.

Truman would read of the brutal treatment of the former owners, the Müller-Grotes, a prominent publishing family, in a letter from one of their sons, years later. It described how Russians soldiers sacked the house as the family hid in the cellar. 'Ten weeks before you entered this house,' Müller-Grote told the President, 'its tenants were living in constant fright ... By day and night plundering Russian soldiers went in and out, raping my sisters before their own parents and children, beating up my old parents.' Furniture and books were dumped in bomb craters and the family's collection of Dutch and German paintings stolen (some later emerged in American army possession). A hint of the crime lay outside the back door, where White House staff noticed a mound of earth, the fresh grave of the *hausfrau*, shot by a Red Army guard when she returned to retrieve her possessions.

Two blocks away, at 23 Ringstrasse, Winston Churchill, Foreign Secretary Anthony Eden and the British entourage made themselves comfortable in a slightly more sumptuous home. Generalissimo Josef Stalin and his Soviet delegates were shortly to move into a far better appointed estate a kilometre or so down the road, en route to the Cecilienhof Palace – where the conference would be held. A sickness, rumoured to be a mild heart problem, had apparently delayed the Soviet leader, who feared flying and travelled from Moscow in a heavily armoured train.

Churchill called on Truman on 15 July at 11am. 'A most charming and very clever person,' the President wrote of the British leader. 'He gave me a lot of hooey about how great my country is and how he loved Roosevelt and intended to love me etc etc.' They would get along if Churchill 'doesn't give me too much soft soap'. Churchill praised Truman's 'precise, sparkling manner and obvious power of decision'. The pair drank to liberty with Scotch, to Truman's distaste.

That afternoon Truman, Byrnes and Leahy toured Berlin in an open sedan, topped and tailed by security vehicles. They paused on the autobahn to review the world's largest armoured division – 'Hell on Wheels' – the US 2nd, whose massed Sherman tanks lined one side of the road, and then drove towards the ruined city. A never-ending procession of sad old men, ragged women and dirty children, 'from tots to teens', picked over the rubble and along the buckled and cratered streets, dragging or pushing their belongings in little carts nowhere without hope.

The eyes of Berlin were a wounded animal's, shadowed, dying, helpless. The sounds were the clatter of carts, the hiss of steam, piercing screams, distant gunfire and the eternal whimper in unseen places of the aftermath of war. Tens of thousands of Berliners were dead; many more wounded, starving, sick and homeless. Drunk Russian soldiers swarmed over the slain beast, sniffing out souvenirs, alcohol, women. Mongolians who had never seen electricity unscrewed and pocketed light bulbs, meaning to show the miracle of light to their home villagers. The mass rape of German women, the slaughter of children got up as Hitler Youth, and the destruction of works of art, medicines and food supplies marked out the trail of the barbarian. German men, who had inflicted no less on Russia, were nowhere to be seen: 'Wholesale raping and looting by Russian soldiers for 10 days after Berlin fell. Of 15–50 age group, no-one missed,' observed Walter Brown, Byrnes' assistant. 'Most able-bodied men taken to mines in Russia. ½ million gallons of milk spoiled for lack of cooling apparatus.'

On his return to the Little White House, the President stood alone on the back porch with the breeze coming in off the lake. Tomorrow they would test the atomic bomb. The buglers were playing 'Colours' at the base of the American flag. Profoundly moved, he returned indoors and sank into his papers. Among them were transcripts of the Togo–Sato 'peace feelers' to the Soviet Union, which the President had received a day or two earlier. He paid them scant attention – mere offers to negotiate, and not genuine acts of surrender. Here were Togo's orders to Sato to present Japan's 'peace proposal' to Stalin before the dictator left for Potsdam ('We immediately grasped its significance,' one cryptographer had noted; 'The Japanese were seriously suing for peace.') And here were the svelte words of the Emperor himself, 'desiring from his heart that the war be quickly terminated' – as always, on Japanese terms.

•

On Sunday 15 July, at around 5pm New Mexico time (1am, Monday 16, Berlin time), a black Buick, three buses, other automobiles and a truck left Santa Fe and snaked through the New Mexican desert toward Alamogordo. The Buick contained Groves and his top advisers; Manhattan Project scientists and military observers occupied the other vehicles. The physicists Sir James Chadwick and Ernest Lawrence joined the convoy at Albuquerque. They wound down Highway 85, past the old Spanish-American outposts of Los Lunas, Socorro and San Antonio. Their destination was 'Trinity', codename for the base camp of the atomic test – a huddle of military huts, protective earthworks and trenches on a disused reservoir at Alamogordo Air Base, 340 kilometres south of Santa Fe. Brigadier General Thomas Farrell, Robert Oppenheimer, Enrico Fermi, Edward Teller and other prominent scientists and engineers were already there, hard at work on the final assembly of the plutonium bomb. Harvard physicist Professor Ken Bainbridge, the field commander, ran the operations: issuing instructions, checking equipment, delegating tasks.

Oppenheimer had chosen the name 'Trinity' after the 'three person'd God' – the Father, the Son and the Holy Ghost – of John Donne's 'Holy Sonnet' 14. The poem marries the self-flagellatory torment of the Old Testament with the devotional self-sacrifice of the New, and holds meaning for Christian and Jew:

> Batter my heart, three person'd God; for you
> As yet but knocke, breathe, shine, and seeke to mend;
> That I may rise, and stand, o'erthrow me, and bend
> Your force to break, blowe, burn and make me new ...

Donne intended the poem as a plea for redemption from a tripartite God. Oppenheimer hauled the poem into the 20th century as an appeal to the god of atomic energy, who would 'break, blowe, burn and make me new' – that is, purify me by threatening to destroy me. The theme of redemption through destruction possibly appealed to the scientist's troubled conscience; and furnished his hope that the bomb could yet redeem mankind and end war forever, a dream Bohr and other scientists shared. The heathen Japanese, broken, blown and burned, were, implicitly, not to participate in the peaceful rebirth of a post-nuclear Judeo-Christian world.

The convoy arrived at around 8pm that night. The military compound stood on a patch of desert in the Jornada del Muerto – the 'Day of the Dead' – in the roughest section of the Valley of the Camino Real, 32 kilometres from the detonation point. In the distance were the Sierra Oscuro mountains, the home of golden eagles, mountain lions, bighorn sheep and burrowing owls. On the desert floor rattlesnakes, jack rabbits and kangaroo rats lived among the yucca plants. Every morning technicians, who had been on site for weeks, checked their boots for centipedes and scorpions.

The newcomers went through the bomb drill, read aloud by torchlight. At the short sound of the siren – 'minus five minutes to Zero' – all observers were to prepare 'a suitable place to lie down on'; at the long siren – 'minus two minutes to Zero' – all personnel were 'to lie prone on the ground … the face and eyes directed towards the ground and the head away from Zero' – to avoid flying rocks, glass and other debris 'between the source of the blast and the individual'. Open all car windows, the instructors advised those, like Ernest Lawrence, who chose to stay in their vehicles.

Long trousers and long-sleeved shirts were recommended, to 'overcome ultraviolet light injuries to the skin'. Dr Teller gave a short lecture on sunburn: 'Someone produced sunburn lotion and passed it around,' reported William Laurence, the nervous *New York Times* correspondent (and Groves' personally appointed PR tool). The 'eerie' sight of famous scientists and military men daubing sunburn cream on their noses in pitch darkness spooked Laurence, a short, pugnacious man with a keen eye for 'local colour'.

Welders' goggles and special sunglasses were issued. 'Do not watch for the flash directly,' they were told, 'but turn over after it has occurred and watch the cloud. Stay on the ground until the blast wave has passed.' The probable brilliance of the explosion would be 'so bright it would blind us looking directly at it for sometime', noted one scientist.

A Plymouth sedan delivered the plutonium core from Hanford to the George MacDonald Ranch house, 3 kilometres southeast of the detonation point, where a team assembled the plutonium bomb in a dust-proofed bedroom. The plutonium weapon relied on a completely different detonation system from the uranium 'gun-fired' bomb – which would not be tested, partly because the scientists were convinced it would work. Plutonium was too impure, too unreliable, to chain-react under a gun blast, as the physicist Emilio Segrè had shown. The eccentric genius Seth Neddermeyer had solved the detonation problem while working at Los Alamos that year: he had suggested surrounding the plutonium core with a sphere packed with high explosives that would crush, or implode, the ball of plutonium to a supercritical state. If the reaction failed, however, the scientists feared it could blow rare and highly toxic bits of plutonium 'all over the countryside'.

The mood oscillated between gloom and hope. Data on the weather, the technical apparatus and the schedule flowed in from Los Alamos. The weather reports worsened. The inclement conditions risked blowing a radioactive mist across populated areas – Amarillo, 480 kilometres away seemed most exposed, Groves thought. And rain, 'would bring down excessive fallout over a small area'; the general was well aware of the risks of radiation borne on water droplets. Lightning lit the eastern sky and distant thunder growled across the desert camp. The rain continued. 'What the hell is wrong with the weather?' Groves yelled at the meteorologists, as if it were in their power to improve it. Conscious of the President's deadline, the general refused to postpone the test. 'We were under incredible pressure' to complete it before the Potsdam Conference, said Oppenheimer later. They shivered in the cold by-now morning air and sporadic drizzle. Cars were positioned to monitor the movement of the radioactive cloud; troops on hand to evacuate local people if a wind change blew it towards their homes. Two lead-lined Sherman tanks prepared to drive into the crater after the test and collect soil samples.

Groves grew stern, calm, as imperturbable as the Sierra Oscuro range; Oppenheimer looked fretful, chain-smoking, intense, according to witnesses. The general walked up and down with his chief scientist, trying to relieve Oppenheimer's tension. Unable to sleep earlier that night, the physicist had haunted the mess hall; Groves now took him in hand, and the pair drove to the forward bunker, the nearest observation point, at 9 kilometres from ground zero. In the distance was the tower, 30 metres high, on top of which the plutonium bomb hung in the sky. At about 4am the rain eased; the damp, overcast sky admitted a little starlight. 'Conditions holding for the next two hours,' said the weather report. The test would go ahead, they decided, at 5.30am.

Groves left Oppenheimer with other scientists in the forward bunker and returned to the main observation point, a control tower 6 kilometres further back. At 5.10 Sam Allison began the countdown, briefly interrupted by strains of 'The Star Spangled Banner' – interference from the Voice of America morning radio show in California.

The mood wavered between faith and doubt: 'Lord I believe; help thou mine Unbelief,' Brigadier General Farrell would describe the feeling; he was not the only hardened soldier to seek comfort in prayer. Groves thought of how he would react if the count reached zero and nothing happened. 'I was spared that embarrassment,' he later wrote. Nervous scientists prayed their own input would not be responsible for a dud. A hundred of them had earlier placed dollar bets on the force of the blast: Teller wagered it would be the equivalent of 45,000 tons of TNT; Hans Bethe, 8000 tons; the physicist Isidor Rabi, about 18,000 tons; explosives expert George Kistiakowsky, 1400 tons; and Oppenheimer, 300 tons. Most of the rest agreed with Oppenheimer, except Conant, who reckoned 4400 tons but did not bet. Groves found the gambling distasteful; Fermi, in black comic mood, angered the general by taking bets on whether or not the bomb would ignite the atmosphere and destroy New Mexico, or the world. In fact, Groves had warned the state governor that he might have to declare martial law if a disaster were to occur: that is, if the blast set off an uncontrolled nuclear reaction in the atmosphere's nitrogen. As the deadline approached everyone – 'Christian, Jew and Atheist' – prayed 'harder than they had ever prayed before'.

The young physicist Donald Hornig was the last to leave the bomb tower; his job was to kill the test if anything went wrong. The observers put on their goggles and lay in silence with their feet pointing towards the detonation point. Conant lay between Bush and Groves, with his eyes open. The countdown continued, shadowed by further radio interference, Tchaikovsky's 'Serenade for Strings'. At minus 45 seconds, the physicist Joe McKibben initiated the firing mechanism. Oppenheimer held his breath. A gong from the control tower signalled 10 seconds to detonation – the longest 10 seconds Ernest Lawrence had known, he later wrote.

•

The first man-made nuclear explosion detonated at 5.29 and 45 seconds. Within a millionth of a second the 32 detonation points on the outer sphere fired; the conventional explosives burst; the shells collapsed under the implosive power, triggering, through a complex series of 'synapses', a chain reaction inside the plutonium core. Radiation waves fled the bomb casing at the speed of light. Billions of neutrons liberated billions more in conditions that 'briefly resembled the state of the universe moments after its first primordial explosion', wrote Bethe. The flash was 'like a gigantic magnesium flare'. Conant witnessed the hills 'bathed in a brilliant light, as if somebody had turned the sun on with a switch'. A bell-shaped fireball rose from the earth, whose 'warm brilliant yellow light' enveloped Ernest Lawrence as he stepped from his car. It was 'as brilliant as the sun ... boiling and swirling into the heavens' – about a kilometre and a half in diameter at its base, turning from orange to purple as it gained height, illuminating the ordinary clouds.

The nuclear dawn was visible in Santa Fe, 400 kilometres away; a blind woman later claimed to have seen the light. The shock wave broke with a sharp crack like the near report of artillery fire. 'What was that!?' shouted Laurence, the 'terribly afraid' journalist. A sustained roar ensued, belching ash, debris and vegetation across the desert then sucking the mess back as it withdrew. Every sign of life within a 3-kilometre radius ceased to exist. The wave knocked down men standing at 16 kilometres. From the centre of the fireball a column of hot gases and radioactive dust shot into the sky and swelled outwards in the shape of the head of a jellyfish, its underbelly scarred by 'yellow flashes, scarlet and green', a sight hitherto unseen. The head reached 12 kilometres (40,000 feet) and lingered; the purple afterglow represented 'the enormous radioactivity of the gases', wrote Ernest Lawrence.

Mutual congratulations and 'restrained applause' – a few men indulged in a triumphant jig – greeted the success. Hushed murmurs 'bordering on reverence' followed. The sight produced 'solemnity in everyone', noticed Ernest Lawrence. As they found their voices, allusions from the banal to the biblical tumbled forth: 'Like the end of the world,' said Conant; 'the greatest single event in the history of mankind,' said Dr Charles Thomas of Monsanto; 'the nearest thing to Doomsday … the last man will see what we saw,' claimed Professor Kistiakowsky; '… as if God himself had appeared among us' and 'a vision from the Book of Revelation', observed Chadwick. Some felt personally menaced: 'It seemed to come toward one,' feared Isidor Rabi. Others felt touched by the hand of the divine: 'I was privileged to witness the Birth of the World,' wrote the journalist Laurence, who felt 'present at the moment of Creation when the Lord said: Let There be Light!'

Some laughed, some cried, then the mood grew sombre, stupefied. Conant wept with 'relief, hope, fear and gnawing responsibility': he had a premonition of the end of the world, and the poor man found it difficult to carry out a coherent conversation. The only off note came from the operation's field commander, Ken Bainbridge, who snorted at this 'foul and awesome display', which had made 'us all sons of bitches'. Oppenheimer outdid them all, in his self-referential awe, quoting a line from the Bhagavad Gita – describing the moment when Krishna's avatar, Vishnu, demonstrated his power in multi-armed form: 'I am become death – the destroyer of worlds' – at which the Project leader adopted a kind of strut, 'like High Noon', as though he had just created the fastest weapon in the west. Oppie's quote was at least accurate: man had indeed acquired the power to destroy worlds, and himself: 'A new thing had just been born; a new control; a new understanding, which man had acquired over nature,' mused Rabi, failing to notice that nature had just exhibited a new power over him. Others felt they had defied God's creation: 'We puny things,' wrote Brigadier General Farrell, 'were blasphemous to dare tamper with the forces heretofore reserved for the Almighty.'

Fermi dared. Impatient for data, the Italian scientist busied himself with an ad hoc experiment. He tried to calculate the yield of the 'gadget' at his 16-kilometre observation point, by dropping small pieces of paper before, during, and after the passage of the blast wave. 'Since, at the time, there was no wind, I could observe very distinctly and actually measure the displacement of the pieces of paper that were in the process of falling while the blast wave was passing.' The blast shifted Fermi's confetti about 2.5 metres, which corresponded to the energy produced by 'ten thousand tons of TNT'. This understated the figure: the first atomic explosion released energy equivalent to about 18,600 tons (17,000 tonnes) of TNT and Rabi won the bet.

The superlatives and biblical references failed to capture the consequences of what they had witnessed and what they intended to do with this power. Understandably, words were inadequate and the moral conundrum too great: the chemist Henry Linschitz was reduced to asking himself, 'My God, we're going to drop that on a city?' The scientists sought refuge in a litany of statistics and data that quantified the magnitude, the intensity, the destructive force: milliseconds after the blast the core temperature was 10,000 times that of the surface of the sun; the earth groaned beneath 100 billion atmospheres of pressure; the radioactive fallout was a million times stronger than the world's radium supply; and so on. It remained for the generals to decide how the new power should be used, and they were mercifully glib:

'The war is over,' Farrell told Groves soon after the test.

'Yes,' Groves replied, 'after we drop two bombs on Japan.'

The Sherman tanks entered the crater and found the explosion had vaporised the bomb tower, leaving a few struts poking out of the crystallised sand. It had torn from its concrete foundations, more than half a kilometre from ground zero, a massive test cylinder made of 40 tons of steel, 'twisted it, ripped it apart and left it flat on the ground'. Groves had not expected any damage to this: 'I no longer consider the Pentagon a safe shelter from such a bomb,' he declared, with uncharacteristic modesty: he had built the Pentagon. Radioactive material covered a wide area, with some concentrations found 190 kilometres away. 'Dust outfall was potentially a very dangerous hazard over a band almost 30 miles [50 kilometres] wide extending almost 90 miles [145 kilometres] northeast of the site,' Colonel Stafford Warren, an American radiologist (and later inventor of the mammogram), told Groves.

Corrugated-iron strips and boxes filled with wood shavings – set up by Oppenheimer's brother, Frank, to resemble flimsy Japanese homes – were charred at 900 metres. All exposed surfaces heated instantly to 390 degrees Celsius at 1.4 kilometres.

Press releases in the name of the commanding officer of the Alamogordo Army Air Base were dispatched to quell local media interest: 'Several inquiries have been received,' it stated, 'concerning a heavy explosion which occurred [here] this morning. A remotely located ammunition magazine containing a considerable amount of high explosives and pyrotechnics exploded. There was no loss of life or injury to anyone and the property damage ... negligible. Weather conditions affecting the content of gas shells exploded by the blast may make it desirable for the Army to evacuate temporarily a few civilians from their homes.' (Various drafts had been prepared to cover all eventualities. One announced the 'deaths of several persons', including 'some of the scientists engaged in the test'.) Thoroughly deceived, the New Mexican media relegated the incident to a routine news item.

•

Groves telephoned the words 'New York Yankees' ('success beyond imagination') to his office ('Brooklyn Dodgers' had meant 'as expected'; and 'Cincinnati Reds', 'utter failure'). His secretary handed the message to George Harrison at the Pentagon, who cabled Stimson in Potsdam. Groves then composed a fuller report – 'not a concise, formal military report,' but a straight attempt to describe the test, in his words and Farrell's: 'For the first time in history there was a nuclear explosion. And what an explosion!' he began. The test had been 'successful beyond the most optimistic expectations of anyone … All seemed to feel that they had been present at the birth of a new age – The Age of Atomic Energy.' His team had discovered something 'immeasurably more important than the discovery of electricity or any of the other great discoveries that have so affected our existence'. The searing light had a beauty and clarity that 'the great poets dream about but describe most poorly'. The report confirmed 'huge concentrations of highly radioactive materials' in the mushroom cloud.

Groves then flew back to Washington; his colleagues' excited chatter about the bomb irritated him: 'My thoughts were now completely wrapped up with the preparations for the coming climax in Japan.'

•

Meanwhile, the Chicago dissenters, none of whom had been invited to or were then aware of the test, had been very busy. Three days before Trinity, the Franck Committee and its tireless campaigner, Leo Szilard, who no longer had involvement in the daily operations of the

Manhattan Project, got an inkling of the event when new rules prohibited telephone calls to Los Alamos. At the time, Szilard had conducted a secret straw poll of scientists and engineers involved in the Manhattan Project: 69 (46 per cent) believed the weapon should be demonstrated and Japan given a chance to surrender before use; 39 (26 per cent) believed the weapon should be demonstrated in America, with Japanese representatives present, followed by full use if Japan refused to surrender; 23 (15 per cent) believed the military should use the weapon as they saw fit; and three said the weapon should not be used and kept a secret. The results fortified the Franck Committee's decision to launch a petition, whose signatories registered their 'opposition [to the bomb] on moral grounds'. The 'gadget' should be used only if America's livelihood were endangered and the power of the new weapon 'made known to the peoples of the world'. It should be 'described and demonstrated' before use on the Japanese, who should be given a chance to surrender. Szilard tried to have the draft circulated, but Oppenheimer banned it: 'Scientists had no right to use their prestige to influence political decisions,' he insisted.

A minority of scientists opposed the Franck Committee, and their counter-petitions packed an emotional punch: 'Are not the men of the fighting forces a part of the nation?' one asked. 'Are not they, who are risking their lives for the nation, entitled to the weapons which have been designed? ... Are we to go on shedding American blood when we have available a means to a speedy victory? No! If we can save even a handful of American lives, then let us use this weapon – now!' The signatories somewhat damaged their case by adding: 'Furthermore, we fail to see the use of a moral argument when we are considering such an immoral situation as war.' Were there no limits, then; no rules governing the behaviour of nations at war? Were Japanese methods – death marches, torture and the massacre of prisoners and civilians – similarly excusable? If so, highly intelligent Americans had thus acquiesced in the descent of the United States to the barbaric level of the regimes against which they were fighting.

The day after Trinity, Szilard, unaware it had gone ahead, issued a final version of his petition, duly watered down to escape Oppenheimer's ban. Signed by 68 Manhattan Project scientists, it warned: 'A nation which sets the precedent of using these newly liberated forces of nature for purposes of destruction may have to bear the responsibility of opening the door to an era of devastation on an unimaginable scale.' It directly appealed to the President, 'the commander in chief', to rule that the United States 'shall not resort to the use of atomic bombs in this war unless the terms which will be imposed upon Japan have been made public in detail and Japan knowing these terms has refused to surrender …'

Oppenheimer permitted its transmission to the President, on the condition that it went through normal diplomatic channels. On 24 July, after sitting on it for six days, Arthur Compton, the leader of the Chicago Metlab, passed a copy to Colonel Nichols, chief engineer of the Manhattan Project, who sat on it for a further five days before ordering its dispatch by military courier to Groves. The general did nothing until 1 August, when he received assurances that both the uranium and plutonium bombs had reached Tinian Island and were ready for departure to Japan. The petition never reached Truman, then in Potsdam; Stimson did not see it until late August. Either way, the President's mind was decided, and the dissenting scientists little more than an unwelcome distraction, their petition one of an 'endless succession of memoranda to be read if time permitted'.

On the same day – 1 August – the management of the Hotel Quadrangle evicted Szilard. The staff had complained of his repeated refusal to drain the bathtub and flush the toilet, which he deemed to be 'maid's work'. If darker forces had engineered Szilard's reduction to this pathetic state, evicted and unemployed, destiny could not have chosen a less presentable 'whistle-blower' five days before an atomic bomb was dropped, without warning, on Hiroshima.

CHAPTER 12

POTSDAM

We call upon the Government of Japan to proclaim now the
unconditional surrender of all the Japanese armed forces ...
The alternative for Japan is prompt and utter destruction.

The Potsdam Proclamation, 26 July 1945

AS SZILARD DESCENDED INTO PATHOS, the delegates in Potsdam began the tortuous negotiations that would design the post-war world. The night before the first day of the talks, a cable from Trinity arrived that would transform the mood of the American delegation from gravity to elation. After 7.30pm on 16 July, Berlin time, Truman and Byrnes met Stimson, their roving War Secretary, at the Little White House in Babelsburg. Stimson dutifully carried the news from Alamogordo:

> 16 JULY 1945
> EYES ONLY FROM HARRISON FOR STIMSON
> Operated on this morning. Diagnosis not yet complete but results
> seem satisfactory and already exceed expectations ... Dr Groves
> pleased ...

The President and Secretary of State were immensely relieved. Stimson shortly withdrew and retired to the comfort of his diary:

'... Mr Harrison's first message arrived ... President and Byrnes ... were delighted with it.'

The next day, Harrison sent further news:

17 JULY 1945

TOP SECRET SECRETARY OF WAR FROM HARRISON

Doctor has just returned most enthusiastic and confident that the little boy is as husky as his big brother. The light in his eyes discernible from here to Highhold and I could have heard his screams from here to my farm.

Decoders might have wondered at the virility of the 77-year-old Stimson in producing a baby boy. 'Doctor' referred to Groves; the 'big brother' was actually the plutonium bomb being tested, and the 'little boy' the uranium bomb they were confident about, already on its way by ship to Tinian Island; Highhold, Stimson's farm on Long Island, 400 kilometres away; and 'my farm', Harrison's, 60 kilometres from the Pentagon.

'I send my warmest congratulations to the Doctor and the consultant,' Stimson replied.

Stimson delivered Harrison's second cable to Truman the next morning. The President looked 'greatly reinforced', and Churchill similarly delighted at the 'earth-shaking news'. The British Prime Minister swiftly drew two conclusions: the invasion of the Japanese homeland would not proceed (Churchill was unaware of the extent to which the invasion plan was already redundant) and, more significantly, the Allies believed they no longer relied on Russia in the Pacific War, as Churchill informed Eden: 'It is quite clear that the United States does not at the present time desire Russian participation in the war against Japan.' The President later confided in his journal, 'We have discovered the most terrible bomb in the history of the world. It may be the fire destruction prophesied in the Euphrates Valley Era, after Noah and his fabulous Ark.'

Truman first met Stalin at the Little White House at noon on 17 July, the day after Trinity. After handshakes and pleasantries, he told the Generalissimo: 'I am no diplomat, but usually say yes or no to questions after hearing all the argument.' Pleased to hear it, Stalin said he had more questions to add to the conference agenda. 'Fire away,' Truman replied. Stalin's questions were 'dynamite', Truman noted in his diary, 'but I have some dynamite too which I am not exploding now'. Stalin casually made clear that he would enter 'the Jap War on August 15th'. 'Fini Japs when that comes about ...' Truman later noted in his secret Potsdam journal. It was a tantalising message to posterity, suggesting that the President believed Russia's intervention would not only end the war with Japan, but completely obviate the American invasion plan.

'I had gotten what I came for,' the President wrote to his wife, Bess, the next day: 'Stalin goes to war August 15 with no strings on it ... we'll end the war a year sooner now, and think of the kids who won't be killed.' With no disrespect to their conjugal relationship, Truman misrepresented his position – a point lost on those who continue to read his correspondence with his wife as 'evidence' that Truman's main priority at Potsdam was to get Stalin into the Pacific. 'I want the Jap War won and I want 'em both [Britain and Russia] in it,' he added. In truth, by then, the President, with Byrnes at his ear, was contemplating how best to keep the Russians 'out of it': America had borne the brunt of the Pacific War and had effectively defeated the enemy; now Trinity had transformed the stakes in America's favour. Publicly, Truman continued to welcome Soviet sabre-rattling, as an insurance policy in a widening mix of options. His private feelings on the subject were contingent upon the availability of the atomic bomb, highly receptive to Byrnes' nimble-minded persuasion, and deeply qualified by his distrust of Moscow. Henceforth the American delegation worked on the assumption that they did not need, nor would they seek, Russian intervention in the Pacific. Of course, he could not divulge that to his wife: Bess was one of millions of Americans necessarily in the dark.

•

The Potsdam Conference began 21 hours and 30 minutes after Trinity – a day late due to Stalin's illness – on 17 July at the Cecilienhof Palace, a mock-Tudor estate built by Kaiser Wilhelm II for his eldest son, Crown Prince Wilhelm, and daughter-in-law, Duchess Cecilie zu Mecklenburg. Completed in 1917, this Hohenzollern family country manor served as a hospital during the war; by its conclusion the royal owners had exiled themselves, and the Soviet conquerors had commandeered and stripped the palace.

The Union Jack, Star-Spangled Banner and Hammer and Sickle fluttered over the palace's motley pile of Elizabethan, Victorian and Gothic, sharing the rooftops with a crazy assortment of chimneys and turrets, as if designed 'by a mad illustrator of children's books'. Heavy ivy clung to the four wings that enclosed a central courtyard where Russian advance units had planted their signature: a 7-metre Red Star fashioned out of red geraniums, pink roses and blue hydrangeas. Inside the Soviets had hastily refitted the desecrated halls with their plundered arrangements of garish furniture, paintings and sculptures.

The Soviet delegation occupied the Crown Princess's study, dubbed the 'Red Salon', wallpapered in deep red, with mahogany bookcases and, around the fireplace, 18th-century Delft tiles. In the opposite wing, across the Great Hall, the Crown Prince's Smoking Room, panelled in dark oak and pine, contained the American delegation. Here Truman sat at an elegant mahogany desk beneath a painting of the Mönchgut Peninsula. Next door, the British occupied the Prince's former library, refurnished in the neo-Gothic style, a Russian attempt to please British taste. His Soviet hosts had considerately hung a painting of the head of a Saint Bernard in recognition of Churchill's affection for dogs.

The Great Hall that divided the Soviet from the Anglo-American rooms rose through two storeys, lit by a great bay window overlooking the lake. In the centre of the room stood a heavy circular table, 3 metres in diameter and purpose-built in Moscow that year, surrounded by two concentric circles of red-upholstered chairs: the inner circle, consisting of three large armchairs for the leaders, flanked by smaller chairs for their foreign ministers; and, in the outer circle, smaller chairs for their advisers. The Big Three would enter through separate doors heavily guarded by Soviet troops.

The meeting proceeded in the strictest secrecy. Over the next 10 days the world would hear nothing of the debate over the future boundaries of post-war Europe and the fate of the Soviet satellites. Three men and their advisers would decide the destiny of the continent, inside a sealed hunting lodge, ringed by bayonets. To their chagrin, some 200 reporters then in Potsdam were refused entry; there would be no press conferences until the final day. Correspondents were reduced to filing gossip about 'Who Had Lunch with Whom' and 'All Comforts of Home Set for Big Three'. Potsdam, reported the *Stars and Stripes*, had become a 'dream community of clipped lawns and super service', surrounded by ruin and starvation. The Allies made an ostentatious display of victory amid the squalor: fresh strawberries – 'big, juicy' ones, insisted a US mess officer – melons, berries, tomatoes and lettuce hearts were flown in. Old-world silver and Bavarian china graced the dining tables. The PX sold luxury cigars, the latest cameras and self-winding shockproof wristwatches; French perfume and Parisian handbags were presented to the attendant wives (Bess Truman was not among them). The maintenance of comfort at the Cecilienhof required 1000 white orderly coats, 500 mosquito bars, 200 fly swatters, 250 shoe brushers and 250 corkscrews.

At the opening of the first meeting Stalin nominated Truman as chairman. The President, dressed in a polka-dot bow tie, dark double-breasted suit and two-tone summer shoes, in his usual jaunty style, accepted, but doubted whether he could fill the shoes of his great predecessor Roosevelt, on whom Churchill lavished praise. Truman then ran through the agenda, after which Churchill insisted they debate the Polish question; Stalin wanted to negotiate the division of the German navy and merchant fleet; Truman had pet ideas about freeing up Europe's waterways; and so on. For 10 days they argued over the division of Germany, Poland and Eastern Europe. They made little progress; Stalin was abrupt and belligerent; Churchill at his cavalier worst; and Truman's big guns, which demanded Soviet compromises, were repeatedly plugged. Most issues of substance were referred to the Council of Foreign Ministers, and a future 'peace conference', to be thrashed out later.

They rarely mentioned Japan. The Pacific War was not on the official agenda – despite Truman highlighting it as his main priority in coming. In fact, Washington had removed the subject from the official agenda several weeks earlier. The Soviet commitment to the fight against Japan – which Stalin had conditionally agreed to at Yalta – scarcely raised an official murmur. The Americans pointedly avoided the subject.

Conspicuous by its absence, the subject of Japan pricked Stalin's keen antenna. In one of his disarming tangents, during a discussion of the division of the German navy, Stalin suddenly raised the question: 'Are not the Russians to wage war against Japan?'

'When Russia was ready to fight Japan,' Truman replied, 'she would be taken in the shipping pool the same as the others.'

Stalin, however, was 'interested in the question of principle' of entry into the Pacific War – a question that Churchill felicitously deflected on a point of detail. Stalin persisted: did his allies want Russia in the Pacific or not? The dictator detected in their obfuscation the whirr of furious backpedalling.

Instead, most discussion of the Pacific War, and Russia's role in it, tended to flow over informal exchanges at morning tea, dinner and cocktail parties. These conversations could be startlingly candid: in one meeting on the morning of the 18th, Stalin revealed to Truman and Byrnes – as noted by Byrnes' assistant – 'that Japan had asked to send mission to Moscow to talk peace. Said Emperor did not want to continue bloodshed but no way out under unconditional surrender terms.' To which Byrnes inquired whether Russian policy on unconditional surrender had changed at all? 'No change,' Stalin replied. In the absence of any fresh 'suggestions' – that is, softer terms – from the United States or Britain, Stalin said he would continue to reject Japanese 'peace offers' and 'be ready to move against Japan' on 15 August'.

Their agreement with Russia at Yalta obliged the American delegates to welcome this gesture; in the privacy of their rooms, however, feelings were decidedly cooling. Just as the Americans were trying to disentangle themselves from Stalin's embrace, the Soviet leader was showing himself more than willing to join his comrades in arms in the Pacific War. By now, however, Byrnes had lost all enthusiasm for the idea: he 'no longer desired Russia's declaration of war against Japan', observed Walter Brown, his loyal aide. '[Byrnes] thinks United States and United Kingdom will have to issue joint statement giving Japs two weeks to surrender or face destruction. (Secret weapon will be ready by that time).'

•

It is unclear exactly when Byrnes put a line through the Soviet Union's name on the draft copy of the Potsdam Declaration – probably in the days before the conference, or aboard the *Augusta*. He initialled and wrote 'Destroy' beside his amendment (a copy of which survives). At a stroke it removed the name of Japan's most feared enemy – and Stimson's 'additional sanction' – from the surrender ultimatum. The Russians were not informed of this; they presumed they would be co-signatories to the declaration and were busy drafting a suggested wording of their own. Byrnes defended his editing to Truman and colleagues on the grounds that Russia had no stake in a nation they had not fought. The act gave the lie to Truman's publicly stated intention to get the Reds into the Pacific. Privately Truman swiftly gravitated to this new 'unofficial' policy, which Byrnes had engineered: to force the Japanese to surrender solely to America and deny Stalin what he so dearly sought – to be 'in on the kill'.

For his part, Stalin was determined to seize – as agreed at Yalta – what he saw as rightfully his: control of Dairen, Port Arthur and the Manchurian railroads, among other assets. Bolshevism demanded a foothold in Asia. Byrnes perceived this danger and quietly considered the possibility, in his talks with Truman, that atomic power would serve a twin role: to end the war with Japan and serve as a diplomatic stick against further Soviet incursions in Asia and Europe. Byrnes' memoirs make clear his position: that he saw a diplomatic role for the bomb, as drawing a line in the sand to the footscraping of the Soviet Union. Truman relished the prospect of a double victory.

Truman and Churchill lunched alone on 18 July at the British residence in Babelsberg. The President, unable to contain his excitement at the 'world-shaking news' of the bomb, showed the Prime Minister the Trinity cables. Awed, Churchill wondered how, and when, to tell Stalin of the discovery – if indeed he should be told? The news might jolt the Russians into the war in a bid to claim their share of territory: 'The President and I,' Churchill later wrote, 'no longer felt we needed [Soviet] aid to conquer Japan.' They agreed, however, that failing to inform the Russians of the bomb would deepen their nominal ally's distrust – little realising that the spy Fuchs had kept Stalin abreast of the developments in Los Alamos. The British and American leaders decided to inform Stalin, but not until the bomb was almost ready; then, at an appropriate time, Truman would casually mention to the Soviet leader that America possessed an 'entirely novel form of bomb ... which we think will have decisive effects upon the Japanese will to continue the war'.

Churchill then raised the vexed subject of unconditional surrender, warning of the 'tremendous cost in American and to a lesser extent in

British life' if they enforced it. Were there not words that ensured victory and gave 'some assurance' of Japan's military honour and national existence'? To which Truman sharply interjected that Japan had no military honour left after Pearl Harbor. 'At any rate,' Churchill responded, 'they had something for which they were ready to face certain death in very large numbers.' Truman would hear no more talk of compromise, given the terrible resonsibilities upon him in regard to the 'unlimited effusion of American blood', Churchill later noted. The terms of 'unconditional surrender' would remain.

That afternoon Truman visited Stalin. The Soviet leader handed the President a copy of Hirohito's message to Moscow which outlined the Konoe peace mission – the contents of which the President was aware, via his Magic diplomatic summaries. Stalin suggested three responses to the cable: 'to lull the Japanese to sleep' by asking them to clarify the 'exact character' of the message; 'ignore it completely'; or 'send back a definite refusal'. Truman preferred the first suggestion: it bought time and was 'factual', as Soviet Foreign Minister Molotov agreed, because nobody in truth understood exactly what the Japanese had proposed.

Truman knew Soviet intervention in the Pacific was inevitable – he could not unmake Yalta. But a creeping awareness of Russian designs on Asia heightened his anxiety that Japan should be made to surrender exclusively to America on American terms. The bomb was the ace in his pack, as he confided in his Potsdam diary: 'Believe Japs will fold up before Russia comes in. I am sure they will when Manhattan appears over their homeland. I shall inform Stalin about it at an opportune time.'

•

Henry Stimson cut an isolated, shuffling figure in Potsdam. Excluded from the conference, he dropped in for unofficial chats with Truman, Byrnes and Churchill when they deigned to see him. His great age and experience, however, lent him gravitas and his candid advice kept Truman's choices dimly alive at a time when Byrnes was monopolising the President's attention. On the 17th, Stimson met Byrnes and recommended two last-ditch actions: to warn the Japanese of the bomb before use; and to assure them of the continuation of the Emperor. Byrnes rejected both: 'Byrnes was opposed to a prompt and early warning ...' Stimson wrote. 'He outlined a timetable on the subject ... which apparently had been agreed to by the President, so I pressed it no further.' No statement more poignantly illustrated the War Secretary's diminishing influence, but he doggedly stuck to his mission.

Indeed, Stimson took to his role as roaming political minstrel; he performed one job that put him in high demand: as the eyes and ears of the events in New Mexico. His dispatches from Trinity opened doors and were a source of great relief to the American party. At 3.30pm on 21 July Stimson arrived at the Little White House brandishing Groves' 'immensely powerful' account of the atomic test, which revealed 'far greater destructive power than we expected'. Stimson read it aloud: '... A massive cloud ... reaching the sub-stratosphere ... huge concentrations of highly radioactive materials ...' etc. When he finished, Truman and Byrnes looked 'immensely pleased', Stimson wrote. It gave the President 'an entirely new feeling of confidence'. Here was a crystalline moment in the blur of events, the confirmation that the bomb had worked from the very achitect of the Project. It prompted Truman to call in his political and military advisers – Byrnes, Marshall, King, Arnold and Leahy, all of whom were present in Potsdam – to review the military strategy in light of this 'revolutionary development'.

Later that day George Harrison sent more news from the Pentagon: 'Patient progressing rapidly, and will be ready for final operation first good break in August ...' – which Stimson relayed to Truman. The President had all the information he needed and arrived at the Cecilienhof on 21 July 'tremendously pepped up' and determined to stare down the Soviet steamroller in Eastern Europe. The bomb was the 'master card' in his hand, noted Stimson. The Big Three duly launched on a long and complex debate over the location of Poland's western border and precisely where the Soviet zone of occupation began and ended. Stalin, in his usual obstructive manner, resisted Truman's demand for a clear definition of the Russian zone of occupation, which was obscured by the Polish presence in East Germany: 'We have withdrawn our troops to our zones,' Truman said, 'but it now appears that another government [Poland's] has been given a zone of occupation and that has been done without consulting us ...'

Truman was stern, uncompromising, and several times forced Stalin on the defensive. He refused 'in a most emphatic and decisive manner' Russian demands, Stimson later noted. In reply, Stalin 'squirmed' and 'whined', according to one account. Truman later told Bess how he had 'reared up on his hind legs and told 'em [the Russians] where to get off ...'. (He neglected to say that the news from Alamogordo had produced this burst of self-confidence.)

The bomb thus performed its first official role, as a tacit diplomatic weapon – and presidential confidence-booster – in negotiations over Eastern Europe. It failed. The President drew no concessions from Stalin; the Soviets would not be told 'where to get off': the Lublin Poles would stay where they were, answerable to Moscow, as far as Stalin was concerned. The talks ended in a mute standoff, and the early frost of the Cold War continued its silent ministry.

All was temporarily forgiven that night – the Generalissimo's turn to host dinner. 'Was it a dandy,' Truman wrote to his daughter: caviar and vodka and mare's milk butter, followed by smoked herring, white fish, venison, duck and chicken; with toasts 'every five minutes' to 'somebody or something'. A quartet of Russian musicians played Chopin and Tchaikovsky. Churchill, who loved words – preferably his own – hated these long musical interludes and meant to get his revenge at the British banquet the next night.

·

Stimson was in a foul mood over dinner. That afternoon he had received a cable from Washington, which requested, to his chagrin, that his 'pet city' be returned to the atomic target list; Groves wanted it ranked a 'first choice' target, as Harrison wrote: 'All your local military advisors engaged in preparation definitely favour your pet city and would like to feel free to use it as first choice if those on the ride select it out of 4 possible spots in light of local conditions at the time.'

Stimson sent a blunt reply: no new factors had arisen that made Kyoto a target. The bitterness wrung by 'such a wanton act might make it impossible during the long post-war period to reconcile the Japanese to us'. The destruction of Japan's oldest shrines might prevent a Japan 'sympathetic' to the United States 'in case there should be any aggression by Russia in Manchuria'. However strange his desire for Japanese 'sympathy', an eye on post-war political gain partly motivated the saviour of Kyoto; clearly, he had also grown adept at dressing his personal crusade in political gloss.

Stimson received the final target list, via a cable from General Arnold, on the 22nd. Kyoto was not on it. The four chosen cities were all 'believed to contain large numbers of key Japanese industrialists and political figures who have sought refuge from major destroyed cities' – adding to their suitability. The strikes would be 'visual' (not radar-guided), 'to ensure accuracy'. The bombardiers would require clear skies, and if weather favoured one city over another, the crews would have to divert in mid-attack to the more visible target. 'Two tested type [plutonium] bombs are expected to be available in August, one about the 6th and another the 24th.' There were more bombs in the pipeline, with news of 'future availabilities' forthcoming in a few days.

Stimson rose early on the 22nd after a fitful sleep. En route to a meeting with Churchill, he stopped at the Little White House for a chat with Truman. The President mentioned in passing that he did not think the Russians were needed in the Pacific War (as Stimson noted in his diary). The British leader wholeheartedly agreed: having read the Groves report in full, Churchill said he better understood the President's feisty performance the day before. Truman was a 'changed man' who 'bossed the whole meeting', he said. Churchill similarly felt the bomb should be used 'in our favour in the negotiations'. At some point, however, the Soviets should be told that 'we intended to use it'.

Churchill then leaned forward and, with a flourish of his cigar, declared, 'Stimson, what was gunpowder? Trivial. What was electricity? Meaningless. This atomic bomb is the Second Coming in wrath.'* Stimson, a devout Christian, never doubted that Christ was on 'our' side; he later wondered, however, whether the Son of God would have condoned the use of weapons of mass destruction on a civilian population centre.

In conference that afternoon the Russians were on the warpath again. Having jettisoned their post-war designation as 'only a continental Power', they now sought to branch 'in all directions'. They were not only vigorously working to extend their influence in Poland, Austria, Rumania and Bulgaria, but also desiring bases in Turkey, Italian colonies in the Mediterranean, a firm footing in Asia, and an 'immediate trusteeship' over Korea. Most of these demands were bluffs, Truman concluded, and stoutly resisted them.

The nightly entertainment was a welcome respite from these fraught daily encounters, and Churchill took his revenge for Stalin's music that evening, subjecting the delegates to loud, interminable renditions of 'Carry Me Back to Green Pastures', 'Serenade Espagnole' and Irish reels, courtesy of the Royal Air Force Band. Stalin requested quieter songs. Throughout this terrific din, the delegates proposed raucous toasts and exuberantly signed each other's menu cards. In the

---

* Churchill had worked himself into a great euphoria over the bomb. He poured out glorious visions to his generals, in which atomic power would 'redress the balance with the Russians'. Unmoved, Sir Alan Brooke (later Field Marshal Lord Alanbrooke) stepped in 'to crush his overoptimism'; the Prime Minister 'painted a wonderful picture of himself as the sole possessor of these bombs ... capable of dumping them where he wished ... capable of dictating to Stalin!' Others were similarly appalled. Lord Moran, Churchill's doctor, felt 'deeply shocked' when he heard of the 'ruthless decision to use the bomb on Japan': 'There had been no moment in the whole war,' he wrote in his diary, 'when things looked to me so black and desperate, and the future so hopeless. I knew enough of science to grasp that this was only the beginning ... I thought of my boys.'

midst of the revelry Stalin pointedly rose and drank to the armies of the Big Three, 'joining forces against Japan'. Truman and Churchill smilingly raised their glasses.

•

The ever-loyal Stimson continued his peregrination as Truman's unofficial messenger. On the 23rd the President asked him to sound out General Marshall on the military role of the bomb, and what to do about the Russians. The general was expansive: America, Marshall told the War Secretary, probably 'did not need the assistance of the Russians to conquer Japan'. But he warned that Russia would invade Manchuria regardless, and that America should prepare for this. Marshall was not persuaded that the bomb alone would end the war (in fact, he had earlier suggested that several 'tactical' atomic bombs should accompany an invasion). Stimson returned to the Little White House to find another telegram from Harrison stating the 'exact dates as far as possible when they expected to have S-1 ready'. This news, and Stimson's upbeat version of his encounter with Marshall, greatly cheered the President on his lonely path.

Stimson chose the moment to appeal to the President: the next day, in an act of extraordinary persistence, he made one last tilt at retaining the original wording of the Potsdam Declaration, now moving towards its final draft. Byrnes had cut Stimson's critical sentence that offered the Japanese people 'a constitutional monarchy under the present dynasty ...'. (In so doing, Byrnes had acted with the support of the Joint Chiefs of Staff, who recommended on 17 July that the offending phrase be struck out, lest 'radical elements in Japan construed [it] as a commitment to continue the institution of the Emperor and Emperor worship'. It remains a mystery why the Chiefs did this, as they had previously stressed the vital role of the Emperor in quelling those very 'radical elements' at the surrender. The act smacked of political intervention.)

And so, on 24 July, as Truman awaited Chinese leader Chiang Kai-shek's approval of the text, Stimson requested that the sentence be reinstated: 'The insertion … might be just the thing that would make or mar their acceptance.' The President firmly rebuffed him: Truman's and Byrnes' minds were made up; the text could not be changed. The timely arrival of a Magic intercept of 21 July helped to clinch the decision: the cable, sent in the Emperor's name, declared that the Japanese would fight to the last man unless America modified the surrender terms. Truman would not be dictated to by the nation that destroyed Pearl Harbor. As a last resort, Stimson urged Truman to reassure the Japanese 'through diplomatic channels, if it was found that they were hanging fire on that one point [retention of the Emperor]'. The President glibly replied that he 'had it in mind, and that he would take care of it'.

In any case, events had moved well beyond Stimson's weary purview. That day, 24 July, the President formally approved the use of the bomb in a meeting with Churchill and each of their military advisers. Groves had sent a two-page cable seeking approval 'of our plan of operation'. Truman rubber-stamped the plan; nothing was recorded, no minutes taken, according to witnesses. Later Washington cabled news of a 'good chance' that the 'patient' (bomb) would be ready on 4 or 5 August. This 'highly delighted' the President, who told Stimson: 'It's just what I want and gives me the cue for my warning [to Japan].'

Truman jotted down his reflections of these events in his 'Potsdam diary', often written with a self-justifying eye on his place in history: 'Even if the Japs are savages, ruthless, merciless and fanatic,' Truman wrote at about this time, 'we as the leader of the world for the common welfare cannot drop this terrible bomb on the old capital [Kyoto] or the new [Tokyo] … The target will be a purely military one and we will issue a warning statement asking the Japs to surrender and save lives. I'm sure they will not do that, but we will have given them the chance.' This was plainly self-serving and false in spirit and fact: Truman's humanitarian concern for Tokyo rang hollow given that he knew the city lay in ruins; the bomb would be dropped without warning on a city centre, as he also well knew; nor could he be in any doubt that it would erase from the face of the earth a population centre.

Back at the official conference, the delegates struggled to carve out a future world. That same 24 July day the Polish delegation, 'helpless victims of the visions and designs of others', naively presented their case for a socialist state free of Soviet control. Unbeknown to these 'dreadful people', as Churchill dismissed them, the Poles' miserable fate had been decided before they arrived; they faded into the corridors and wallpaper, largely ignored. Potsdam stifled the last gasp of a democratic Poland, whose people would not taste freedom for another 44 years. Stalin's position was as immovable as his crippled left arm. The day's session ended with near-breakdown in the negotiations over issues whose legal complexity, at one point, reduced the delegates to helpless mirth and prompted Churchill to wonder – in one of his typically melodramatic versions of events – whether each side had in fact declared war on the other.

In this atmosphere of anxiety and distrust the American and British leaders decided to reveal the gist of S-1 to Stalin. They chose not to mention that it was an atomic bomb, fearing the Soviet leader would press for technical details and even a nuclear partnership. As the delegates stood about in groups awaiting their chauffeurs, Truman walked around the conference table and nonchalantly told Stalin that America possessed 'a new weapon of unusual destructive force'. Stalin showed no special interest. He was glad to hear it and hoped that his allies 'would make good use of it against the Japanese'.

Churchill and Byrnes, standing nearby, closely watched Stalin's expression during this exchange. 'He seemed to be delighted!' Churchill recorded. 'A new bomb! Of extraordinary power! Probably decisive on the whole Japanese war! What a bit of luck! I was sure that he had no idea of the significance of what he was being told.'

'He didn't realise what I was talking about,' Truman later claimed.

Stalin played a more subtle game. He knew exactly what they were talking about: on 2 June the spy Klaus Fuchs had informed his Soviet contact of the forthcoming Trinity test. Stalin received that intelligence in the middle of June. Hence the Soviet leader's casual reaction: the first poker-faced gambit of the nuclear age. But the timing of the news surprised him: Stalin had not realised the pace of American progress on the bomb. He returned grimly to his rooms where, according to Soviet sources, he ordered Lavrentiy Beria, the NKVD – Soviet secret police – chief, to 'speed up the work' on the Russian bomb; another account has Stalin ordering Molotov to 'talk it over with Kurchatov', the head of Russia's nuclear program. His instructions had the urgency of a race.

•

By the end of July the political atmosphere had seriously degenerated. The official conference reached a stalemate. Allies in name, smiling comrades over the canapés, the Soviet and Anglo-American delegations fundamentally disagreed on all major points: the terms of the German settlement, the Polish question, the end of the Pacific War. A darker truth loomed; none dared speak its name. The democratic powers had lost Eastern Europe to Soviet communism; they were determined not to lose Asia. Byrnes was most anxious to 'get the Japanese affair over with' before the Russians got into Manchuria and claimed Port Arthur. Once in, he wrote, 'it would not be easy to get them out'.

To break the impasse, Japan must be made to surrender swiftly. Truman and Churchill acted: at the bracing hour of 7am on 26 July 1945, outside the Cecilienhof Palace, the American and British delegations issued the Potsdam Proclamation [Declaration] – to a ravenous press. The final version enshrined Byrnes' amendments: it excised any reference to the Emperor or constitutional monarchy; removed the Soviet Union's name; and made no mention of the atomic bomb.

The time had come, it declared, for Japan, 'to decide whether she will continue to be controlled by those self-willed militaristic advisers whose unintelligent calculations have brought the Empire of Japan to the threshold of annihilation, or whether she will follow the path of reason. Following are our terms. We will not deviate from them. There are no alternatives. We shall brook no delay ...'

The Declaration called for the destruction of Japan's war-making powers; the elimination 'for all time' of the authority and influence of those who had misled the Japanese people 'into embarking on world conquest'; and the complete disarmament of the Japanese forces. Her vanquished armies would be allowed to return home and the Japanese people permitted to pursue peaceful industries, and enjoy freedom of speech, religion and assembly. War criminals would meet stern justice. The occupying forces would be withdrawn only after Japan had established, 'in accordance with the freely expressed will of the people', a peaceful, democratic government.

The Declaration's final words conveyed its lethal intent – without specifying how America might complete the enemy's annihilation: 'We call upon the government of Japan to proclaim now the unconditional surrender of all the Japanese armed forces ... The alternative for Japan is prompt and utter destruction' (see Appendix 3 for full text).

The document was not unreasonable, the terms more lenient than those imposed on Germany. A probing Japanese eye would surely read in the gently nuanced 'freely expressed will of the people' the opportunity to retain the Emperor as a constitutional monarch. And 'unconditional surrender' referred explicitly to the 'armed forces', not the people. Hirohito's role, however, remained ambiguous, as he was titular head of the armed forces; it was this very ambiguity that left the Declaration open to a slurry of interpretations in Tokyo.

The Russians, when they heard, were furious; for once Stalin had been comprehensively outmanoeuvred – and thoroughly deceived. The night before, Molotov had sent a message to Byrnes requesting the Declaration be delayed two or three days; Byrnes claimed the message arrived two hours too late. The furious commissar pressed the Americans: Why had the ultimatum gone out without their joint consent? Why had the Americans ignored the Soviet request to delay it? Why had the Americans ignored the Soviet draft (which even referred to Japan's 'treacherous' attack on Pearl Harbor, 'the same perfidious surprise attack by which it had attacked Russia forty years ago')?

What particularly incensed the Soviet leadership was not that the Potsdam Declaration offered 'proof' that America hoped to secure Japan's surrender without Soviet help (Stalin presumed as much); rather that it ostentatiously, *publicly*, declared the exclusion – and therefore, in Stalin's eyes – the humiliation of the Soviet Union. It amounted to the craven deceit of an ally. While the Russians had difficulty swallowing a dose of their own medicine, Byrnes made several attempts to mollify his humiliated Soviet counterpart. But Molotov sulked, refusing Byrnes' three invitations to lunch. Truman was absent, inspecting American troops in Frankfurt.

Byrnes tried to explain the situation to Molotov in conference that day. The Americans had not received Molotov's request in time, to which the Russian responded that 'we were not informed until *after* the press release went out'. Byrnes tried another tack: 'We did not consult the Soviet government since the latter was not at war with Japan and we did not wish to embarrass them.' Molotov fell silent; he was 'not authorised' to discuss the matter further; Stalin, he ominously implied, would attend to it. Byrnes promptly changed the subject but the note of menace remained.

Later Byrnes persisted. The Declaration had to be sent before Churchill's resignation, he explained, as it bore the British leader's signature. Indeed, in a sensational electoral upset, Labour's Clement Attlee had defeated Churchill at a general election the day before, and the great wartime leader would not be returning to Berlin. This hardly satisfied Molotov, and the episode rankled. For their part, Byrnes and Truman had no faith in Russian fair dealing, not least Molotov's cynical gesture three days after Truman and Churchill issued the

Declaration, calling on America, Britain and their Pacific allies to issue 'a formal request to the Soviet Government for its entry into the War'.* It was an awkward invitation which Truman dispatched with an empty reference to international treaties that obliged Russia to assist in 'preserving world peace' etc. Stalin had resolved to enter the war as soon as possible; the bomb now brought forward his invasion plans – to stake a claim on Asia before Japan succumbed to a nuclear-armed America. A race was on for the spoils.

•

It was Sunday, and the Christian members of the American and British delegations, including Truman, attended Morning Worship at an improvised chapel. They read Psalm 106 and sang 'Holy Holy Holy', 'How Firm a Foundation' and 'Fairest Lord Jesus'. Chaplain Northen led them in prayer:

> ... Draw me nearer, nearer, blessed Lord
> To the cross where Thou hast died;
> Draw me nearer, nearer, nearer, blessed Lord
> To Thy precious bleeding side.

The conference ended on 2 August. The Big Three – with the new Prime Minister Clement Attlee in Churchill's place – beamed out of the official photograph with an air of accomplished goodwill. Very little had been achieved. The official communiqué was a travesty of the truth. 'Important agreements and decisions were reached,' it stated; 'views were exchanged on a number of other questions'. The discussions had 'strengthened the ties between the three governments and extended the scope of their collaboration and understanding'. President Truman, Generalissimo Stalin and Prime Minister Attlee departed

---

* Truman sidestepped the awkward proposal in a letter to Stalin suggesting that the Russians make the case for their declaration of war on Japan according to the Moscow Declaration of 1943 and the UN Charter.

'with renewed confidence, that their governments and peoples, together with the other United Nations, will ensure the creation of a just and enduring peace'.

Beyond this official feather dusting, the Potsdam Conference had agreed, in effect, that Russia would swallow Poland; Poland digest a slice of Germany; and Germany be carved up, into four zones of occupation. Far from strengthening ties, Potsdam drove America and the Soviet Union into chronic psychological conflict. Had they known what transpired here, the peoples of the world would have gone forth not with renewed confidence but with feelings of dread and despair at the victors' failure to ease the misery of, or salvage a strong and abiding peace from, the worst clash of arms in history: 50 million people lay dead; their memory deserved at least this. Potsdam split Europe along bitter lines and cast the die for another global conflict. Stalin's aggression and indifference to the right of self-determination of nations caught in Soviet-occupied territory forced a tougher line from Truman; the bomb raised the President's volume, little more. The delegates left the conference with premonitions of a future war more terrible than any hitherto imaginable; perhaps understandably, they sought refuge in warm communiqués and one stern ultimatum to their common foe. The Americans sailed home in anticipation of a prompt Japanese reply.

CHAPTER 13

MOKUSATSU

*I remember seeing the flyers coming down through the sky, and how
big they were, coming down beautifully like snow in the sky ...
We all wanted to go and pick the flyers up and see what was on
them, but we were afraid of being called spies. We were so scared.
We were scared of the* Kempeitai, *the military police, coming
and getting us.*

Miyoko Watanabe, on seeing pamphlets outlining the Potsdam Declaration
dropped over Hiroshima

THE GOVERNMENT INTENDS TO IGNORE IT.

The Japanese newspaper *Asahi Shimbun* in reply to the Potsdam ultimatum

ALLIED CODEBREAKERS CONTINUED intercepting the secret dialogue
between Sato, Japan's ambassador to Moscow, and Togo, his Foreign
Minister in Tokyo, and sending the Magic summaries to Truman and his
staff in Potsdam. The cables offered American observers a ringside seat on
a relationship that portrayed a regime in its death throes, tearing itself
apart over how to extricate itself from the war with 'honour'. Yet nobody
outside Japan believed the country had a scrap of honour worth fighting
for.

A distant check on his leaders' fantasies, Sato perceived the futility
of these deliberations and prevailed upon Tokyo to surrender –

unconditionally if need be – to avoid the destruction of Japan. Foreign Minister Togo – representing the views of the Supreme War Council – replied that Japan would never submit to unconditional surrender if it risked the Emperor's life. Instead, he instructed Sato to continue to appeal to Russia to initiate 'peace negotiations' with America.

Togo was reduced to desperate measures. On 16 July, the day of Trinity and the day before the Potsdam Conference began, he told Sato: 'We have … decided to recognise Russia's [territorial claims] on a broad scale in exchange for Russia's good offices in concluding the war.' The Supreme Council believed they could buy Moscow's goodwill regardless of the fact that Russian troops were poised to grab them anyway. Another futile toe in the mire, then, Sato thought; in reply, he warned that Japan must propose concrete peace terms or suffer annihilation. Tokyo refused to listen. The depth of the Council's denial found expression elsewhere – in the blame they heaped on 'Anglo-Americans' for failing to end the war, a war that Japan had launched: 'If,' Togo continued to his ambassador, 'the Anglo Americans were to have regard for Japan's honour and existence, then they could save humanity by bringing the war to an end. If, however, they insist unrelentingly upon unconditional surrender, the Japanese are unanimous in their resolve to wage a thorough-going war.'

Togo repeatedly impressed upon Sato the importance of the visit to Moscow by Prince Konoe, the Emperor's special envoy. Tokyo persisted in the delusion that Russia would receive this eminent person, and agree to negotiate with the Allies, a notion as removed from reality as *Horai*, the mystical realm where flowers never die and rice bowls replenish themselves until 'the eater desires no more'.

Moscow's dismissive reply, sent in Stalin's name, arrived via Sato on 19 July. The Supreme Council met in Suzuki's office at 6pm the next day to examine it; Anami was absent. Togo read it aloud: 'The message from the Japanese Emperor … does not contain anything concrete. It is also unclear to us for what purpose Prince Konoe is to be sent … Therefore we won't be able to reply to you … at this moment.'

Here is how Moscow chose to 'lull the Japanese to sleep', as Stalin and Truman had agreed at Potsdam: they would draw out the Japanese; they would buy time. Both the Russian and American leaders needed time: the Soviets, to build their forces in the east; and the Americans, to ship and prepare the atomic bomb.

The next day Sato sent a long impassioned appeal to Tokyo, expressed 'without reserve': Japan had little chance of getting Soviet co-operation, he warned; the Japanese leaders were 'out of touch with the atmosphere prevailing here'. Tokyo should instead 'be ready for an invasion' by America. He salted these grim tidings with a cheerless note on Allied strategy: America, he claimed, planned to destroy the autumn rice crop by firebombing the drained paddies before the October harvest: 'If we lose this autumn's harvest, we will be confronted with absolute famine and will be unable to continue the war. Furthermore, the empire, stripped of its air power, will be able to do nothing in the face of the situation and will be at the enemy's mercy.'

This decidedly undiplomatic diplomat then questioned the fighting quality of the Imperial forces: 'All our officers, soldiers and civilians – who have already lost their fighting strength ... – cannot save the Imperial House by dying a glorious death on the battlefield.' Sato had dared impugn the troops' ability to defend the Emperor. He then went further, presuming to read the Imperial mind, a blasphemous act: 70 million people were 'withering away', he warned, and that must 'disturb' the Emperor's thoughts. 'There is nothing else for us to do,' Sato despaired, 'but ... to make peace as quickly as possible and suffer curtailment.'

Japan must insist on one condition, however, Sato wrote: 'the safeguarding of our national structure [that is, the *Kokutai*, or the Imperial House].' He proposed an ingenious path forward: that the future of the Imperial dynasty be treated as a domestic, not national, concern – and thus struck from any surrender terms, freeing Japan to 'surrender unconditionally'. His scheme defined the *Kokutai* narrowly

– as the Imperial line, and not the nation as a whole. His overriding motive was to remove the Imperial system from the negotiating table. He hoped the Allies would see the point, in return for which Japan would surrender 'unconditionally'. It was a shrewd distinction and an inspired offer, which appeared to give each side what it wanted – and save the Japanese people: 'It is meaningless to prove one's devotion by wrecking the nation,' he warned. Japan had reached the point where 'we have no assured production', and 'even Honshu will be trampled under foot'; Japan had 'completely lost control of the air and of the sea'; she 'cannot repulse the raids carried on day after day … Now that we are being scorched with fire, I think it becomes necessary to act with all the more speed' – and surrender 'to preserve the lives of hundreds of thousands of people who are about to go to their death needlessly …'. The American cryptographers passed on this intercepted proposal to Washington; the White House ignored it.

In any event, the Supreme Council dismissed Sato's extraordinary idea and turned its attention to the more pressing concerns of the Imperial mission. Russia's uninterest in seeing Prince Konoe, the Emperor's voice, hurt Japanese pride; the Imperial way, it seemed, cut little ice beyond Japanese shores. After some discussion, the Council decided to fortify Konoe's role. The special envoy's 'concrete' purpose would henceforth be to urge Stalin to 'mediate an end to the Greater East Asian War' (the word 'mediate' used here for the first time); and strongly impress upon Moscow that His Majesty the Emperor supported the idea.

Then they dealt with Sato. The Council were accustomed to their ambassador's unusual candour, but his latest outburst presumed to challenge his Highness's wisdom. In the clearest expression yet of Japan's refusal to surrender, Togo told his errant diplomat, 'We are unable to consent to [unconditional surrender] under any circumstances whatever … the whole country as one man will pit itself against the enemy in accordance with the Imperial Will so long as the enemy demands unconditional surrender. It is in order to avoid such a state of affairs that we are seeking a peace, which is not so-called unconditional surrender, through the good office of Russia …'.

•

Such language understandably fortified Truman's dismissal of 'peace feelers'. In Potsdam and Washington, the President, the Secretary of State, their senior staff and the Joint Chiefs rejected the Japanese dialogue as futile banter, unworthy of interest. Togo's most recent cable underlined Japan's fantastic refusal to yield: Tokyo's 'final judgement and decision', concluded James Forrestal, the US Navy Secretary, was to fight on 'with all the vigor and bitterness with which the nation was capable so long as the only alternative was the unconditional surrender'.

Tokyo anxiously awaited an answer from Moscow; the hours anticipated the freight of days, but no answer came. The Council of Six, believing their message had reached Stalin, wondered: were the British and Americans at Potsdam privy to the news of the Imperial mission? A 'friendly atmosphere' prevailed in Potsdam, Sato grimly reported, with 'frequent private meetings' among the Big Three. Togo interpreted this as an opportunity not a threat. On 25 July – the day after Truman had signed off on the use of the atomic bomb – Togo instructed his ambassador to seek another meeting with Molotov, to stress the grave importance of Konoe's mission. Japan, Togo declared, was ready to 'restore peace' according to the terms of the Atlantic Charter: 'It is impossible for us to accept unconditional surrender … but there is no objection to the restoration of peace on the basis of the Atlantic Charter.' That Charter, signed on 14 August 1941, and supported by 26 nations, respected 'the right of all peoples to choose the form of government under which they will live'. Tokyo now decided that its people should receive that right, which had been utterly denied them under more than a decade of totalitarian rule.

Molotov was unavailable; Vice Commissar for Foreign Affairs Solomon Lozovsky agreed to pass on Sato's fresh approach. 'Thank you for your kindness,' the Japanese ambassador said. 'I personally hope that your reply will be expedited.' The pathos of Japan's enduring faith in Russia's 'friendly offices', as Sato described them, faintly moved Lozovsky, who knew the truth.

Another answerless day passed. The old samurai, in frock coats and winged collars, sitting at attention at the conference table in the government's well-stocked Tokyo shelter, continued to observe, *in extremis*, the ancient forms of deference and decorum of the warrior class; they lived in the shadow of that antique past, in darkened codes of 'honour' and 'sacrifice' in whose interests they were willing to destroy their nation and race. Throughout they acted in the thrall of the armed forces, who were deaf to the agonies of firestorm, hunger and homelessness. They heard only the dull, slow chime of the Imperial Will.

•

In Washington the State Department felt isolated and ignored. Acting Secretary Grew and his staff – including the hardliners Dean Acheson and Archie MacLeish – had heard little from Potsdam. They were not consulted over the final wording of the Declaration, and sounded alarmingly ignorant when the media called. Byrnes had left Grew comprehensively out of the picture.

And so it disturbed the State Department, at this critical juncture, to hear of a series of strange radio broadcasts to Japan by one Ellis Zacharias, a naval captain and fluent Japanese speaker who claimed to be an 'official spokesman' for the US government. The State Department knew little of Zacharias. Yet the officer was not an obvious maverick; he served with the Navy's psychological war team and the Office of War Information, and was known to Truman. He later claimed, with no reason to lie, that Forrestal had appointed him and approved his broadcasts (nor would his errant activities damage his career: Zacharias later rose to rear admiral).

In one extraordinary broadcast, on 21 July, Zacharias announced that if Japan surrendered immediately she would be entitled to choose her own government under the terms of the Atlantic Charter – exactly what Tokyo had asked for. Zacharias seemed a credible source – but was he a designated spokesman for American policy, Tokyo wondered. His broadcasts were released to the American press, which gave them an official imprimatur. Yet no recognised Washington authority followed up the broadcasts, which were soon taken off air.

In fact, Zacharias had not received clearance to speak on behalf of the US government; the State Department had neither approved, nor in fact seen, the scripts. Zacharias had made himself a one-man propaganda unit, with potentially lethal consequences. The 21 July address directly contradicted official policy: Washington had expressly *not* offered Japan the rights stated under the Atlantic Charter. Tokyo and the US media were similarly confused during the last weeks of July: who in fact spoke for the American government? Which American message was *the* American message? Did the broadcasts reflect a softening of America's position or not, asked the *New York Times*.

'Secret contacts' purporting to act as peace mediators deepened the confusion: the BBC, an 'Australian Radio' and a 'Montevideo broadcast' claimed to know of a 'Japanese peace offer' that Stalin had brought to Potsdam; meanwhile, an exotic lineup of 'intermediaries' continued to issue a variety of 'peace feelers'. They received little recognition in, or encouragement from, Tokyo or Washington.* Under Secretary of State Joseph Grew attempted to dismiss these 'peace feelers' as 'the usual moves in the conduct of psychological warfare by a defeated enemy', but they persisted and sounded shriller as the war ground on.

---

* The most persistent were Toshikazu Kase, the Japanese Minister in Berne, who 'bombarded' Tokyo with 'all kinds of material' designed to persuade his leaders to end the war; and General Onodera, the Japanese military attaché in Stockholm, who embarked on unsanctioned peace talks in June prompting a terrific denunciation from the Supreme Council, which redoubled its determination to prosecute the war.

In late July word of these vexed affairs reached the American public. On 26 July, the day of the Potsdam Declaration (which Japan did not see until the following day), Tokyo's moderates defied the hardliners by daring to issue an unusually explicit public offer to surrender, on condition the Emperor be allowed to stay on the throne. The offer made front-page headlines in America: Japan 'pleads for an easing of unconditional surrender', reported the *International Herald Tribune* in a 'clear-cut peace bid in the face of devastating American and aerial attacks'; 'Tokyo Radio,' reported the *Stars and Stripes*, 'in a startlingly frank broadcast beamed to the US ... said today [26 July] that Japan is ready to call off the war if the US will modify its peace terms.' Tokyo further warned that the 'world's future depends much on Stalin's action'. The media presented the offer as a genuine attempt by the Japanese to lay down their arms and surrender all occupied territory in exchange for the preservation of the Imperial House. There were too many unanswered questions, however; too many strings attached. Had Japan agreed to foreign occupation? To abandon the machinery of totalitarian government? To renounce militarism in perpetuity? In any case, who were the Japanese to impose conditions on America? Those were the issues occupying the minds of the US leadership, who were in no mood to negotiate them; the White House dismissed the 'peace offer' as one more in a long stream, with unacceptable conditions attached.

Nonetheless, the media coverage alarmed the State Department. According to the official Washington line, Japan had refused to surrender on any terms; suddenly terms appeared to be in play. Grew received no enlightenment on the matter from Potsdam, and in a meeting on the 26th he bemoaned the fact that America's ambassador to Spain knew more of the goings-on there than his department did. 'A great deal of harm had been done,' he said, 'by the [media] speculation on Japanese surrender terms.' His department agreed to complain to Berlin – with Byrnes in their sights – about this gross failure of communication. Japan's offer was ignored.

•

'I remember seeing the flyers coming down through the sky, and how big they were, coming down beautifully like snow in the sky.' The leaflets Miyoko Watanabe saw were A5 sheets of white paper with something printed on them. They landed all around her, on the streets of Hiroshima, but she never read them.

'We never looked at them. We all wanted to go and pick the flyers up and see what was on them, but we were afraid of being called spies. We were so scared. We were scared of the *Kempeitai*, the military police, coming and getting us. I don't know anyone who actually did look at them directly, unless they did it in secret and didn't tell anyone.'

Copies of the Potsdam Declaration lay untouched on the streets, parks and public places of Japan's major cities. By night, out of sight of the *Kempeitai*, brave citizens secreted copies in their homes.

'We knew the words Potsdam Declaration, but not through the news,' said Miyoko. 'They tried to hide this but somehow the words made it through.'

Tokyo picked up shortwave radio announcements of the Declaration, from San Francisco, early on 26 July; and Tokyo Radio reported the terms of the ultimatum at 6am on 27 July. That day, the men with the power to end the war gathered to examine the Declaration. The three 'moderates' on the Supreme Council, conscious of the disaster facing Japan, were anxious to secure an 'honourable' peace. Two of the most senior, Togo and Prime Minister Suzuki, were deeply disliked by the Council's three hardliners and the broader military: Togo, 'an arrogant man of sixty-two', observed a Japanese history, 'is inclined to be contemptuous of other people's opinions; he was far more outspoken that his Premier [Prime Minister], Kantaro Suzuki, who was then over seventy-seven, deaf and drowsy, saying one thing today and its opposite tomorrow, willing to let other men hug the limelight while he dozed his way through meetings that never seemed to come to an end or a conclusion'. The third moderate, Yonai – perhaps the most realistic, and sympathetic to the suffering of the people – lacked the strength of personality to impose his opinions and was constantly overruled, chiefly by War Minister Anami, who loathed him.

Two points on the Declaration met with subdued approval: the Soviet Union was not a signatory, which suggested Moscow remained neutral and encouraged Togo to continue his efforts to pursue Soviet mediation; and the Japanese people were offered, of their 'freely expressed will', the opportunity to establish the government of their choice – implying the possible retention of the Emperor. (The leaders assumed the people would never abolish the Imperial system of their own volition.)

The document's overbearing tone, ominous references to 'war criminals' and 'occupying forces' and, most importantly, its ambiguity in relation to the Emperor, were less palatable; indeed, they formed the topics of intense discussion with Hirohito that morning – in two separate meetings with Togo and Kido: did the Allies mean to preserve or punish him? And who were the self-willed militaristic advisers whose 'unintelligent calculations' had brought the Empire of Japan to the threshold of annihilation? While the leadership sought clarity on these and other questions, they agreed to continue the Soviet peace feelers; for now, the Potsdam Declaration would be set aside.

The three hardliners, however, drew the darkest interpretation of the document: to them, the conspicuous absence of any reassurance of the Emperor's safety pointedly implied his punishment and probable execution as a war criminal – tantamount to the destruction of the soul of Nippon. The hateful document must therefore be rejected, they concluded; posterity would never forgive any other action. Those responsible for surrendering the national godhead would be condemned for eternity as 'the most reviled figures in all Japanese history'. The militarists played on this fear; none was willing to sign a paper they interpreted as the Emperor's death warrant.

•

So the Japanese rulers chose to ignore the Potsdam Declaration – specifically, they would 'kill it with silence' – *mokusatsu* – a Japanese negotiating tactic that treated offence with silent contempt. The Kenkyusha Dictionary variously defines *mokusatsu* as 'take no notice of; treat [anything] with silent contempt; ignore [with silence];' or 'remain in wise and masterly inactivity'. Togo held the latter, more dignified definition in mind when he read, to his dismay, the headline 'LAUGHABLE MATTER' in the *Yomiuri Hochi* newspaper, alongside an edited version of the Potsdam Declaration. (The *Asahi Shimbun*, the official organ, more mildly reported, 'THE GOVERNMENT INTENDS TO MOKUSATSU [ignore it]'.) The report, leaked by the army, quoted sources purporting to represent Togo, but who were almost certainly military placemen: 'Since the joint declaration of America, Britain and Chungking [China] is a thing of no great value, it will only serve to re-enhance the government's resolve to carry the war forward unfalteringly to a successful conclusion!' Indeed, the government exhorted factory supervisors, like Toyofumi Ogura, to impress upon their employees the need to prepare for the decisive battle of the homeland, and 'an honorable death for a hundred million people'.

Togo later claimed in his memoirs that the paper misrepresented him. Yet at the time he did nothing to counter the impression conveyed. Presumably his reason was fear of public denunciation: the inner terror of 'seeming' wrong, of showing weakness, of offending the Imperial forces and, by extension, His Royal Highness, with the concomitant risk of assassination and ignominy. Togo could not express his feelings so publicly in case he lost face.

'Face' and 'feeling' were divergent and often contradictory in the Japanese culture of 1945. What one said bore little relation to how one felt, according to the code of *haragei*, meaning, literally, 'stomach art': 'Our mouths could not speak what our "stomachs" felt', was how Chief Cabinet Secretary Sakomizu explained the Japanese practice of 'communicating almost wordlessly by inference and indirection and getting a "sense of the meeting" by gut feeling and tacit understanding'. It was the 'skill of building, or changing, a consensus by unspoken communication'. The experienced stomach artist felt the truth in his belly, though he dared not utter it; one never laid one's soul on the table (as Sato had brazenly done). To some, Prime Minister Suzuki seemed a mute, nodding old fool; a weathervane politician, who spun with the wishes of the armed forces. To his comrades he was a master of *haragei*.

In this sense, the six councillors understood each other perfectly; their apparent indifference to Potsdam encoded a genuine understanding of the document and its import, which they were loath to articulate. Their silence was neither a case of indifference nor an attempt to call America's 'bluff'. It served several purposes: to appear 'wise and masterly' – that is, in unhurried control; to express contempt for the offensive passages of the ultimatum; and to create an atmosphere in which they might grope towards a consensual response.

The American leadership naturally interpreted this delay as an emphatic rejection of Potsdam, as Truman had foreseen. Washington could hardly be expected patiently to work through the nuances of Japanese stomach art – a point lost on Grew and others who reckoned Washington ought to have done more to comprehend *mokusatsu* – at a time of war. To suggest the President had the time or inclination to inquire into the cultural aspects of Japanese diplomacy was absurd.

Nor were Tokyo's hardliners satisfied with the press leaks of the *mokusatsu* policy. Media denunciations of the document were not enough: they demanded a direct, official refutation of the hated Potsdam ultimatum. Its dictatorial tone – 'There are no alternatives … We shall brook no delay' – infuriated the proud military mind. The pamphlet – dropped all over Japan – dared to call them 'unintelligent'. In high dudgeon the hardline faction met on 28 July and compelled

Suzuki officially to *mokusatsu* the Potsdam Declaration (Togo was absent from this meeting and Yonai overruled) with a firm statement committing the nation to the war. Suzuki wearily obliged, and fronted the Japanese media: 'The government does not think that [the Potsdam statement] has serious value. We can only ignore [*mokusatsu*] it. We will do our utmost to fight the war to the bitter end.' Various translations of this statement are extant; for example the Potsdam Declaration 'is not a thing of any great value … We will press forward resolutely to carry the war to a successful conclusion.'

The foreign press leaped on Tokyo's statement: JAPAN OFFICIALLY TURNS DOWN ALLIED SURRENDER ULTIMATUM, the *New York Times* duly led on 30 July. Western newspapers and their governments had forgivably, or wilfully, mistranslated the word 'ignore' to mean 'reject', 'turned down' and so on. But their cultural, or linguistic, misunderstanding hardly exonerated the Japanese regime from the charge of failing to respond clearly and promptly to a demand that impinged upon the livelihood of millions. If Suzuki and Togo had sincerely meant 'wait and see' or 'comment withheld' or 'we need more time', they might have said so.

Truman seized on the salient point: 'Radio Tokyo had reaffirmed the Japanese government's determination to fight,' the President said on 28 July. 'Our proclamation had been referred to as "unworthy of consideration", "absurd", and "presumptuous".' There was nothing, then, in the Japanese reply that restrained Washington from its course: by this stage, only the most abject surrender in line with all points on the Declaration would have persuaded Washington to consider it. Japanese pride would not tolerate such humiliation. Tokyo's stated resolve to continue fighting was a depressing example of the triumph of hope over experience.

The President and Byrnes had expected as much: the wording of the Declaration, they well knew, had elicited precisely the response they had predicted. It did not follow, however, that the punishment for Tokyo rejecting, or ignoring, Potsdam, should be the atomic destruction of a population centre. That ethical consideration held no sway, as the atomic bombs travelled across the Pacific, in the hold of an aircraft and the belly of a ship, on the swell of an unstoppable momentum that had reached the point of no return. And so Tokyo, without a friend in the world, waited silently to learn the full meaning of the words 'prompt and utter destruction'.

•

Sato injected some reality, if little else, into these dire events. In the last week of July the Japanese stoic peppered Tokyo with a series of gritty dispatches. The Potsdam ultimatum, he warned, 'seems to have been intended as a threatening blast against us and as a prelude to a Three Power offensive' (that is, America, Britain and China); the Konoe mission – which had not even departed – had little hope of success unless a 'concrete' peace offer accompanied it; the chances of Soviet mediation were 'extremely doubtful'. According to the BBC, he added, Stalin had discussed 'for the first time' the war in the Far East with his Anglo-American counterparts. Here was the true face of the Soviet Union, Sato warned: a co-belligerent in an Allied invasion of Japan.

Tokyo ignored him and dared to hope that Russia remained neutral, if not exactly friendly. The Council of Six had noted with relief the absence of Stalin's signature on the Declaration. However, that left a hole. What in fact was Russia's position? Did Stalin support the ultimatum? That was 'a question of supreme importance' in determining Japan's future, Togo cabled Sato on 29 July. Would the ambassador kindly set aside his personal views and 'attempt to sound out the Russian attitude' to the Three Power Proclamation? In the meantime the Big Six would return to their 'policy of careful study' of Potsdam while continuing to kill it with silence.

Sato, in an act of extraordinary insubordination in wartime Japan, chose to disobey these fantastic instructions. He would seek an interview with the Russians only on condition that the Imperial government produced 'a concrete and definite plan for concluding the war'. The exercise was useless without one. Sato's fellow diplomats grew bolder too. From Switzerland Toshikazu Kase joined the attempt to persuade Tokyo to see reason: the Potsdam ultimatum imposed 'unconditional surrender' on the Japanese armed forces, not the people, Kase noted; it did not seek to enslave the nation; in fact, though it did not mention the Atlantic Charter, it did offer the people the democratic right to choose their government (the Imperial system, if they so wished). Germany had fared far worse, Kase stressed: America and Britain meant to occupy, not to dismantle, Japan. He ended this hopeful assessment on an ominous note: the Soviet Union 'probably did not raise any objection' to the ultimatum.

On 30 July, Sato's patience snapped. The exhausted man had had enough. He warned that Russia could not be persuaded to receive Konoe, the voice of the Emperor. There was 'no alternative to immediate unconditional surrender if we are to try to make America and England moderate and to prevent [Russia's] participation in the war'. He had stated the unthinkable – a clear warning that the Russians, their most feared adversary, were likely to participate in the invasion of Japan. If he intended to shock, the attempt fell dead: the warning failed to penetrate the Tokyo fog. Like monks cloistered with their myths, the Council of Six seemed to inhabit a different realm.

·

There were rare moments of clarity. On 2 August, for example, to Sato's relief Togo conceded that the Three Power Proclamation might, after all, form a basis on which to decide 'concrete peace conditions'. After all, the enemy may land on the Japanese mainland at any time, Togo warned. But he soon slipped back into his familiar mantra: the Council had unanimously decided to seek 'the good offices of Russia' in ending the war 'in accordance with the Imperial Will. Would Sato please resume his efforts to seek an interview with Molotov and 'do your best to … furnish us with an immediate reply … if we should let one day slip by, it might have [consequences] lasting for thousands of years'.

It was 'extremely auspicious', Sato replied, with a rally of wishful thinking, that 'you are disposed at least' to make the Potsdam terms the basis for negotiation. And it pleased him to read that the militarists were, for the first time, willing to consider peace. But surely this was too little, too late? He warned that if the government and the military 'dilly dally … all Japan will be reduced to ashes'.

Sato made a final, 'unreserved' plea: the absence of Russia's name on the ultimatum led him to hope that, in some way – he was clutching at straws – the Soviet Union may be 'favorable to us', or at least neutral. Konoe's mission was, however, doomed to fail: 'It is absolutely unthinkable that Russia would ignore the Three Power Proclamation and then engage in conversations with our special envoy.' Unless Japan surrendered immediately, the Motherland was finished. Read this cable, he pleaded, not only to the Supreme Council but to the Emperor himself: 'I implore you,' he begged, 'to report this to the Throne with all the energy at your command.' The men in Tokyo were aghast: read a field diplomat's report to His Majesty? The poor man had clearly lost his bearings and should probably be recalled. Sato sent it on 5 August 1945, the day before an atomic bomb was scheduled to fall on Hiroshima.

CHAPTER 14

SUMMER 1945

A rifle in your hand, a hammer in mine –
But the road into battle is one, and no more.
To die for our country's a mission divine
For the boys and the girls of the volunteer corps.

Mobilised children marching home after demolition duties, Hiroshima, August 1945

A WARM FRONT CAME OVER the Inland Sea and crept into the saucer
of Hiroshima. The mountains kept it there and the people sweltered.
Paper umbrellas lent a hint of elegance to the kimono-less women.
Some wore face masks and elbow-length gloves to protect them from
the sun as they went to work.

The nation lay in ruins. By June–July 1945, American air raids had
firebombed 66 cities, destroyed 2,510,000 Japanese homes and rendered
30 per cent of the urban population homeless. Estimates of the number
killed and wounded vary: General LeMay, Commander of America's
Pacific air offensive, was apt to boast of more than a million dead; others
have placed the figure at several hundred thousand. 'In its climactic five
months of jellied air attacks,' the US Army Air Force official history
states, LeMay's bombers 'killed outright 310,000 Japanese, injured
412,000, and rendered 9,200,000 homeless ... Never in the history of
war has such colossal devastation been visited on an enemy at so slight
a cost to the conqueror.' In fact, the cost to the US Air Force (including

the Army Air Force) in the Pacific was 88,119 airmen (killed and missing), compared with 160,000 Allied airmen killed over Europe. By the end of July 1945, 43 per cent of Japan's largest cities were destroyed and half of Tokyo – two million people – had fled the city. The Japanese military regime looked askance; civilian losses would never force it to surrender, despite the claims of the architects of terror-bombing, Generals Giulio Douhet and William Mitchell.

The people did not defy, far less rise against, the regime (as they were supposed to, according to the theory of terror-bombing). Yet nor were they as responsive to the defiant political slogans and die-hard messages on posters and radio broadcasts, which Tokyo disseminated throughout the country: the *ichioku*, the magical 'hundred million' hearts beating as one, would 'never surrender'; the 'hundred million as a shattered jewel' (*ichioku gyokusai*) would fight to the bitter end (*uchiteshi yamamu*). Ancient samurai poems, reissued in leaflets as martial songs, girded the nation for glory:

Across the ocean, corpses soaking in the water;
Across the mountains, corpses covered by the grass.
We shall die by the side of our lord.
We shall never look back.

This induced little cheer in a people anxious for their next rice ball. Ordinary Japanese struggled to reconcile the regime's rampant triumphalism with the devastation, hunger and the signs, everywhere, of imminent defeat. The rice ration, distributed through local women's groups, was virtually exhausted. In response, official brochures offered suggestions: 'Food Substitution: how to eat things [that] people wouldn't normally eat'. One government pamphlet, 'The Diet for Winning the Greater East Asian War', advised the peasants around Hiroshima to eat locusts and chrysalises. In such circumstances, many dared to *think* against the law and hope for peace; some even defied the *Kempeitai* and secreted American air-dropped leaflets warning of their defeat. 'We did know; we heard that Japan was gradually being defeated,' said Miyoko Matsubara, who lived in Hiroshima. Her father and uncle, she recalled, 'were just pretending to be confident. Most of the information was being hidden by the government'.

The frequency of military postmen bearing little boxes containing a bone or relic of another dead soldier symbolised what could not be said: that Japan had lost the war. The dead soldier's mother, sister or wife would proudly place the remains by the lock of hair he had left before departure. Even with the signs of defeat crowding in on them, the citizenry at home rallied to the national spirit when it came to their own. Thank heavens he wasn't captured, the neighbours would say, relieved that the young man had not allowed himself to become a prisoner, thus disgracing his family and the community. Until 1945, his widow might have received a one-off payment of 1000 yen in War Death Insurance, if her husband had maintained his annual 10-yen premium before he died. In the summer of that year, the government revoked the payments; casualties had risen to an intolerable level.

Japanese families knew very little of their husbands, fathers and sons, then fighting overseas. Soldiers' letters home, if they arrived at all, were heavily censored. Much more lenient censorship had operated in 1937, the year of the Rape of Nanking, when a Hiroshima schoolboy was allowed to receive a photo of a Japanese officer about to chop off the head of a bound Chinese man. By 1945, such openness was unheard of.

Shizue Hiraki, who tended her father-in-law, Zenchiki, along with her young children, had heard only that her husband, Chiyoka, was doing his duty for Japan and the Emperor; she had no idea where. Aged 32, he had trained in Kure, near Hiroshima, and was posted to Burma. Her daughter Mitsue later recalled her father 'strongly hugging us as he said goodbye and wiping my little sister's runny nose'. That was at the Hall of Triumphal Return, in Ujina, a suburb of Hiroshima. Most nights, like millions of Japanese housewives, Shizue sat down to write her husband a 'comfort letter', strictly in the prescribed manner of blissful optimism. Chiyoka wrote one or two letters to his wife and children, then silence. In July 1945, Shizue did not know whether her husband was dead or alive.

It was a familiar experience all over Japan. Kikuyo Nakamura, in 1945 a 21-year-old Nagasaki housewife with a newborn baby, had received two letters from her husband during his entire two-year absence. He had been posted to southern Malaya. 'In his final letter, everything had been censored and blacked out except the words, "We are heading south to an undisclosed location. I hope you are all well."'

Kikuyo and Shizue continued their volunteer duties conscious of an impending catastrophe. On weekends Shizue wrapped her head in the *hachimaki* (bandana), tied the sash of the National Defence Women's Association across her chest and joined the local villagers brandishing bamboo spears in self-defence classes. She even tried to raise the family's donation to the village war fund, promoted throughout Japan as, 'Savings for Annihilating the Americans and the British!' But she did it in mute obedience, with little thought or relish. The days of 'walking with her chest puffed out', as the official propaganda instructed, were past.

•

As fears of air raids increased and rumours of losses on distant battlefields spread, the mobilised children of Hiroshima and Nagasaki were pressed to work harder. The schoolgirl Yoko Moriwaki, like thousands of others, worked from dawn to dusk. She went everywhere on foot to fulfil her duties, as she recorded in her diary. During June and July she walked 'to Hara village to help with farm work' (round trip 16 kilometres); 'to Yoshijima airfield to plant sweet potatoes and soybeans' (round trip 7 kilometres); and 'to Yagi Hall to put books in safe storage. On the way there, I felt like throwing them away! I did my best, though,' (round trip 25 kilometres). The child had turned into a marathon walker.

In early June Yoko visited the Gokoku Shrine in Hiroshima, in memory of the war dead. The girl paused at the great gates, exhaled and entered. She bowed low and 'worshipped with all my heart'. It was a cloudy summer's day. 'I prayed that we would be ultimately victorious ... that Daddy would have lasting good fortune in battle and that I would do a great job as class captain.'

On 5 June, 'right at this very moment, a fierce battle is being waged in Okinawa', she wrote. 'I am sure that the British and American schoolgirls are working hard doing all sorts of things to win the war. We must not be outdone by those schoolgirls, we simply mustn't!'

That day, American aircraft bombed the Kobe/Osaka region: 'Schoolgirls like me were hit by enemy fire and some people might even have been killed.' It occurred to her later that 'If enemy planes attack the mainland ... the mainland is a battlefield.'

The child's awareness of, and obligation to, the people around her filled her every thought and action. 'God, please protect my brother Hiroshi,' she wrote on 12 June, the day he was sent away to Kyushu.

•

In Nagasaki that summer, Dr Takashi Nagai's faith in a Christian God survived a tragedy that may have disabused less ardent converts. In early 1945 two of his four children, Ikuko and Sasano, died of illnesses related to malnutrition. The radiologist prayed daily for their souls; he found little consolation in any other reality. He worked longer hours to blot out the grief, conducting riskier experiments with X-rays that exposed him to large doses of radiation. In June he self-diagnosed chronic myeloid leukaemia and gave himself two to three years to live.

His wife, Midori, heard the news in silence, without apparent emotion. Nagai feared she would blame him for flinging himself recklessly at his work, knowing the risks. 'I lowered my head towards her in reverence,' he wrote. She accepted his fate as God's will: 'You have given everything you had for work that was very, very important. It was for His glory.'

Religious consolations were applied in other contexts as part of the war effort. Frequently, Catholic and Buddhist religious leaders were expected to 'console' the workers and mobilised students. The priests' and monks' blessings were intended not to ease the children's hardship but to encourage them to work harder in the name of victory. That did not mean Japanese clergy mucked in with the kids. Like their Buddhist counterparts, most priests had petitioned for, and received, exemption from mobilised labour. Theirs was the spiritual, not the manual, realm; their state-decreed role to imbue the war effort with intimations of the divine.

The number of Christian priests and Buddhist monks who defied this instruction as a grotesque perversion of the spirit of the Gospels and Mantras is not recorded. Nagasaki's Catholic clergy, however, seem to have taken a dim view of the regime: in July army officials 'savagely berated' several lay Catholics, including Nagai, as potential fifth columnists and ordered them to report to police headquarters if Americans invaded. Urakami's parish priest was arrested and accused of 'praying for Japan's defeat'. He responded that he prayed only for peace, to which the police chief barked, 'You can continue that prayer in your church only if you put the Emperor in place of your Almighty God.' The priest bravely reminded the police chief that the Meiji Emperor's rescript to soldiers began, 'I, in obedience to the Grace of Heaven ...'. 'That Heaven, which even he obeys, sir, is what we call Almighty God.' The priest was told to go home.

•

Dr Nagai had other concerns: the care of his patients in the event of an air raid. Most Japanese cities' medical facilities were in a dire state. Nagasaki had about 600 nurses, 25 'first aid stations' located in doctors' homes or schools, and several 'mobile' medical teams – mobile in name only, as they lacked vehicles. There were few ambulances. The first-aid units possessed bandages, disinfectant, burn medications, some opiates, and heart and respiratory stimulants; but only a small and insufficient amount of anti-tetanus serum. The armed forces had commandeered most of the supplies of blood plasma. Surgical instruments were sterilised by boiling.

Urakami presented a more reassuring picture than downtown. Here, the Nagasaki University Hospital, where Dr Nagai worked, provided more than three-quarters of the city's hospital beds and most of its medical supplies. The reinforced concrete structure with a wooden interior was built on a hillside above the cathedral. Nearby stood the Nagasaki Medical College, home to 800 medical students and constructed mostly of wood.

The mortuary services were rudimentary and relied on public volunteers for whom brawn, resilience and a distinct lack of squeamishness were more highly valued than the technical skills of a mortician. The mortuary squads were answerable to the district policemen. In Nagasaki some 230 men, organised in three platoons, were responsible for bearing away the dead to central assembly points – field clinics in schools and temples – for identification. The people's nametags were sewn into their *monpe*, otherwise morticians relied on local familiarity; Japan did not fingerprint citizens. The system was utterly ill equipped to handle the body count of a massed incendiary raid.

Local officials knew the likely casualties of an air raid; the terrible experiences of Osaka and other nearby cities were well broadcast. Incendiary and high explosives had hitherto destroyed 969 Japanese hospitals with a loss of 51,935 beds, 20 per cent of the country's total (the exact number of patient casualties is unknown). In Tokyo, medical practitioners fled the city after the firestorm: St Luke's Hospital lost 100 of 150 nurses in the firebombing, and 15 of the 30 physicians then in the city (another 30 had been seconded to the armed forces). Notable exceptions were the student nurses, almost all of whom stayed at their posts. Many died bravely trying to rescue their patients.

Hiroshima was in a worse state than Nagasaki. In July 1945, just 60 of that city's 288 doctors remained. The rest had gone to help the victims of firebombing in nearby cities. Those who stayed lectured neighbourhood groups on first aid – chiefly how to treat burns and bandage wounds. 'School patriotic groups' trained in first aid supplemented the lack of medical personnel. Girls were in the vanguard, chiefly as devoted student nurses. The doctors gave no guidance on shock treatment and few had any but the most rudimentary knowledge of radiation; not that they had any reason, then, to study its effects.

Hiroshima's 40 civilian hospitals contained a total of 1000 beds. Five larger hospitals, with 5000 beds, served the exclusive needs of the army. The city had no emergency wards or first-aid stations, and just 10 official ambulances. In the event of an air raid, trucks, buses, carts and streetcars were expected to carry the dead and wounded to 'emergency areas' – schools and temples – or just cleared spaces. Neighbourhood associations were to provide 'at least one litter' each.

Drugs and medical supplies were extremely scarce because the armed forces had requisitioned most medicines 'for their exclusive use'. Morphine and other painkillers were virtually non-existent; bandages, cotton and disinfectant were in short supply. Zinc oxide and oils were available to treat burns. Six medical supply dumps were situated outside Hiroshima, in caves and concrete shelters – beyond immediate reach in a crisis. Families were obliged to buy primitive first-aid kits and given instructions on how to treat themselves. In sum, Hiroshima's medical preparations for aerial attack were 'totally inadequate for any type of emergency', concluded the US Strategic Bombing Survey. It was not that Japanese medical services were backward; rather, that the civilian areas simply lacked the resources and skills, as the military had sequestered them.

Red Cross hospitals were the bright spots in this grim portrait. They were better equipped and well staffed, thanks to international assistance. The armed forces had commandeered many to train nurses, but Hiroshima's continued to function as a hospital. In 1945 it stood as a solid concrete structure near Hijiyama hill, built to withstand incendiary or conventional bombardment.

If the B-29s did not directly target medical facilities, hospitals were impossible to avoid in the mass burning of cities. Generally situated in congested, highly combustible urban areas made of wood, Japanese hospitals did not enjoy the spacious settings typical of American university hospitals – so observed the US Strategic Bombing Survey, with a trace of reproof. Japanese town planners, however, could hardly be blamed for failing to anticipate the possibility that their hospitals would be subject to a conflagration powerful enough to melt the metal shutters and incinerate the wards.

•

The clearest evidence of Japan's hopelessness was the drive to recruit child soldiers and aircrews. The shortage of men on the home front pressed into uniform boys aged 15 and up, usually with their parents' consent. In Hiroshima, Nagasaki and elsewhere, advertisements for 'child soldiers' appeared in the newspapers, encouraging parents to enlist their sons. Posters exhorted children to worship and imitate the death squads and kamikazes. Captions such as, 'Mother! Father! Send me into the skies too!' accompanied dreamy pictures of boys gazing at the sun in aircraft goggles, against a backdrop of Zeroes crashing into US ships. The Intelligence and Aviation Bureaus and the Great Japan Aeronautic Association were responsible for these desperate appeals.

The army was not to be outdone: since the first *gyokusai* on the Aleutian Islands in May 1943, where 2576 Japanese soldiers gave their lives – many in a last-ditch suicidal frenzy – the military regime had continued to promote the idea of the 'honourable death' and 'determined to die' spirit, and instil in the young a sense of suicidal revenge:

Take revenge on the enemy of Atto Island!

More than 2000 soldiers of God were killed!

Let us continue the fight of those noble heroes!

Give all that you have to increase our fighting power!

Thus stated a poster distributed by the Hiroshima division of the Imperial Rule Assistance Association (*Taisei Yokusankai*).

The minimum draft age was 20, but 15-year-olds could enlist if they passed the recruitment test, an aptitude and medical exam. 'My mother was not against me becoming a soldier, because I was doing it for our country. Everyone was like that,' said Yukio Katayama, then 17, of Onomichi city in Hiroshima. He enlisted but failed the exam. Some were drawn to the armed forces not necessarily to fight, but in the hope of being fed. Hungry children sought jobs in the Hiroshima barracks and the castle grounds as junior batmen or quartermasters' assistants. Iwao Nakanishi, 15, of Aosaki, just east of Hiroshima Station, reckoned that if he enlisted, 'at least I'd be able to eat'. Like most boys, he was mesmerised by the images of kamikaze heroics: 'I wanted to become a kamikaze pilot and sink American battleships.' He passed the army's enlistment exam, and in the first week of August found himself assembling soldiers' departure kits in Hiroshima Castle: 'I'd organise their shoes, hats, clothing and other gear. I remember being very hungry and feeling faint as I worked.' In early August American planes dropped millions of leaflets on Hiroshima. At the risk of arrest, Iwao secretly picked one up and read an American ultimatum to the Japanese people – it did not specify Hiroshima – to surrender or face annihilation. It was an extract from the Potsdam Proclamation.

Meanwhile, families in the wrecked cities were ordered to prepare for the battle for the homeland – *Ketsu-Go*. Imperial General Headquarters had adopted a strategy in March whereby every man, woman and child was expected to give their lives for the *Kokutai*; those who retreated would be disgraced and shot. The Field Manual for the

Defence of the Homeland and Hiroshima's Mass Evacuation Implementation Procedures forbade the evacuation of any able-bodied person over 12: 'In the event of a disaster, each person must at all costs protect the battle station he/she is allocated ... one must devote oneself to fighting fires right to the end.'

The time rapidly approached. News of the fall of the Ryukyus and the Marianas reached the Japanese people that summer, and the American forces were expected to land at southern Kyushu at any time. The young were particularly susceptible to the battle cry. If their parents had lost or were losing faith, the young were easy prey to the propaganda of their samurai rulers, whose program for national self-immolation would spare no one. A strange, unfailing optimism resided with the students in their white headbands, the young women in self-defence sashes, the volunteer teenage nurses, the boys brandishing bamboo sticks. Their glazed-eyed belief in victory delighted the old men in Tokyo, who knew better. On the night of 5 August, the students left their factories and demolition sites and marched home, singing:

A rifle in your hand, a hammer in mine –
But the road into battle is one, and no more.
To die for our country's a mission divine
For the boys and the girls of the volunteer corps.

CHAPTER 15

TINIAN ISLAND

Results clear cut, successful in all respects. Visual effects greater than Trinity test. Target Hiroshima. Conditions normal in airplane following delivery. Proceeding to regular base.

Captain Deak Parsons, in-flight message, 6 August 1945

Its effects may be attributed by the Japanese to a huge meteor.

General Leslie Groves, memo to the Pentagon, 6 August 1945

IN THE FIRST WEEK OF August, then, Tokyo's Supreme Council remained divided between the war faction, which was determined to fight to the death, and the peace faction, which played a worthless hand with Moscow. The Potsdam Declaration lay between them, unanswered, on the table, thoroughly dead; and the Emperor's special envoy's mission forlorn. The Japanese leaders were still waiting, hoping, for a Soviet reply to their 'peace feelers' when the plane bearing the first atomic bomb departed Tinian airfield.

FROM: TERMINAL [POTSDAM]
TO: WAR DEPARTMENT [WASHINGTON DC]
FROM STIMSON ... TOP SECRET FOR HARRISON'S EYES ONLY
23rd JULY 1945
[Cable: Victory 238]

Please wire when the weapon of the kind recently tested will be ready for use and give approximate time when each additional weapon of this kind will be ready. Matter of greatest importance here request answer as soon as possible. End.

FOR EYES ONLY SECRETARY OF WAR REFERENCE VICTORY 238
[FROM HARRISON]
23rd July 1945
First one of tested type should be ready at Pacific base about 6 August. Second one ready about 24 August. Additional ones ready at accelerating rate from possibly three in September to we hope seven or more in December. The increased rate above three per month entails changes in design which Groves believes thoroughly sound. Groves sees OPPIE in Chicago tomorrow Tuesday for discussion as to future plans with respect to this ...

FROM: TERMINAL
TO: WAR DEPARTMENT
FROM STIMSON ... TOP SECRET FOR HARRISON'S EYES ONLY
23rd July 1945
[Cable: Victory 218]
We are greatly pleased with apparent improvement in timing of patient's progress. We assume operation may be any time after the 1st August. Whenever ... a more definite date [is available], please immediately advise us ... Also, give name of place or alternate places, always excluding the particular place [Kyoto] against which I have decided ...

SECRETARY OF WAR EYES ONLY FROM HARRISON
23rd July 1945
Operation may be possible any time from August 1 depending on state of preparation of patient and condition of atmosphere ... some

chance August 1 to 3, good chance August 4 to 5 and barring
unexpected relapse almost certain before August 10.

SECRETARY OF WAR EYES ONLY FROM HARRISON
23rd July 1945
Reference ... your Victory 218 Hiroshima, Kokura, Niigata in order of
choice here.

On 25 July, Nagasaki's name reappeared on the target list – on the
final directive authorising the use of 'atomic weapons' against Japan.
Chief of Staff General Marshall formally approved the order – the
only written direction to use the weapon – in response to requests by
Army Air Force commanders who did not wish to take personal
responsibility for the first nuclear attack on an urban target.

That day, as the Allied leaders continued their negotiations at
Potsdam, General Carl Spaatz received the directive via General
Thomas Handy, Deputy Chief of Staff, on behalf of Marshall, then in
Potsdam. Spaatz had recently been promoted to command the US
Strategic Air Forces in the Pacific – with authority over the atomic
bombing and the conventional air raids on Japan. Groves in fact
drafted and prepared the order, which arrived at Spaatz's HQ in
Guam, the forward operations base for the US Army Air Force, the
day before Truman issued the Potsdam Declaration.

Truman had had no hand in the order's creation; at no time did he
issue an official instruction to drop the bomb. He 'let the military
proceed without his interference'. Later, he would claim, 'I made the
decision', and told Stimson that the 'order would stand' unless 'the
Japanese reply to our ultimatum was acceptable'. (The archives contain
no such instruction; nor did Stimson, a thorough recorder of events,
mention it in his diary.) But let us take Truman at his word: in the
President's account, he issued the Potsdam Declaration *after* he gave
official authority to drop the bomb. By his own reckoning, then,
Truman released the ultimatum warning the Japanese of their doom

the day after he authorised the nuclear attack on a Japanese city. According to this sequence the bomb was pre-ordained and inevitable, the Potsdam ultimatum a perfunctory piece of political grandstanding, and the Japanese response superfluous. For it begged the question: how on earth would Truman have *stopped* the process he claimed to have started had Tokyo miraculously fallen on its sword in reply to Potsdam and offered an unconditional surrender?

The ultimatum's careful editing offered an insurance policy against that outcome: the removal of the Soviet Union and the uncertainty over the Emperor's future precluded the likelihood of Japan's unconditional surrender; for that the President had Byrnes to thank. In any case, regardless of whether Truman 'decided' to authorise the dropping of the bomb before or after his Potsdam ultimatum, the result was the same: Little Boy was being prepared for use; the complex processes involved in delivering the weapon were set in motion. The action had received the approval stamp of Marshall and Groves, whose directive stated:

To: General Spaatz, Commanding General, United States Army Strategic Air Forces. 24 July 1945:

(1) The 509 Composite Group, 20th Air Force will deliver its first special bomb as soon as weather will permit visual bombing after about 3 August 1945 on one of the targets: Hiroshima, Kokura, Niigata and Nagasaki. To carry military and civilian scientific personnel from the War Department to observe and record the effects of the explosion of the bomb, additional aircraft will accompany the airplane carrying the bomb. The observing planes will stay several miles distant from point of impact of the bomb.

(2) Additional bombs will be delivered on the above targets as soon as made ready by the project staff. Further instructions will be issued concerning targets other than those listed above ...

Meanwhile, Groves examined the order of targets. Hiroshima should be given first priority among the four cities, he advised Spaatz; however, three 'much less suitable targets' – Osaka, Amagasaki and Omuta – should be bombed if the priority cities were unreachable due to the weather or unforeseen factors. A grave question arose during these final deliberations: were any American POWs held in the targeted cities, and should this alter the target list? The Japanese were well known for moving POWs to vulnerable industrial areas. Every large Japanese city kept prisoners of war as slave labour, noted American intelligence. Mitsubishi Shipyards' 'Fukuoka 14' prison camp in Nagasaki held several hundred prisoners – mostly Dutch, British and Australian – and an 'Allied prisoner of war camp' existed 'one mile north of the centre of the city', revealed Japanese prisoners under interrogation. POW camps existed in Hiroshima Prefecture but none was located in the city; however, at least a dozen American airmen was reportedly imprisoned in Hiroshima Castle – a fact unconfirmed by Washington. While perplexing, considerations about prisoners of war could not be allowed to interfere with the atomic mission.

Groves stuck to his plan to drop at least two and 'as many as three [plutonium bombs]' in conformance with 'planned strategical operations'. In a message to Oppenheimer on 19 July, the general had again suggested that nuclear weapons might accompany a land invasion were Japan to refuse to yield. Marshall had made a similar suggestion. Yet at this stage the Joint Chiefs did not believe that an invasion was militarily necessary – to Truman's relief and MacArthur's chagrin.

On 30 July Groves sent General Marshall fresh detail on the destructive power of the bomb and the risks, if any, of radiation to the aircrews and ground troops. He drew this from analyses of the New Mexico test. The blast, he said, should be lethal to a radius of at least 1000 feet (300 metres) from the point on the ground directly below the explosion; heat and flame should be fatal to about 2000 feet (600 metres). The light 'for a few thousandths of a second' will be 'as bright as a thousand suns; at the end of a second, as bright as one or possibly two suns'. The effect would blind an onlooker half a mile away; and inflict temporary sight impairment on those at 10 miles (16 kilometres) or beyond. All structures within 1 to 2 square miles (2.5 to 5 kilometres) would be completely demolished. Groves was oddly sanguine about the risks of radiation: 'No damaging effects are anticipated on the ground from radioactive materials' (because, unlike the ground blast of Fat Man at Alamogordo, Little Boy would be detonated in the air). Ground troops, he claimed, would be able to move through the area immediately after the blast, 'preferably by motor but on foot if desired'. He drew Marshall's attention to the faster bomb production rate: In September, 'we should have three or four bombs'; another three to four in October; five by the end of November; and seven in December. Groves believed they would be used, if the war continued, possibly as an accompaniment to a land invasion. His memo was sent on to Tinian.

·

The Manhattan Project devolved on Tinian, a flat atoll in the Marianas, east of the Philippines and north of New Guinea, and within range of the Japanese homeland, whence the Little Boy (the uranium bomb) and Fat Man (the plutonium bomb) were sent by sea and air respectively in conditions that entailed the strictest secrecy, 'maximum reliability' and 'maximum speed of delivery'.

The USS *Indianapolis*, a Portland-class cruiser, departed San Francisco on 16 July, en route to Tinian through submarine-infested seas. In its hold were one 300-pound (136-kilogram) box containing 'projectile assembly of active material for the gun-type bomb', one 300-pound box full of 'special tools and scientific instruments', and one 10,000-pound (4500-kilogram) box containing 'the inert parts for a complete gun-type bomb'. The ship arrived on 28 July, on the same day as three C-54 planes carrying the plutonium sphere, detonator and other requirements for the plutonium bomb, as well as the uranium inserts. Brigadier General Thomas Farrell accompanied the bomb parts.

Farrell was responsible for supervising the delivery of the 'vial of wrath' (as he called it, in a post-war press release) to Japan. He was Groves' 'handyman', he would say. Slight in stature, fit and energetic, this fast-talking officer possessed an unusual ability to manage a multitude of problems at once. A veteran of five major World War I operations, he had earned the Distinguished Service Cross, Silver Star (with two oak leaves), Croix de Guerre (with palm), Legion of Merit (with oak-leaf cluster) and a Purple Heart.

As technicians began to assemble the bombs, the carefully selected aircrews of the 509th Composite Group rehearsed the delivery process and refitted their planes. Vengeance fired their determination when news arrived on 4 August that four days earlier the Japanese had torpedoed the *Indianapolis*, which had just departed Tinian; the ship sank in 12 minutes, taking 300 crewmen to the seabed; of the remaining 880 floating in the water, most died of dehydration, shark attack and exposure. Only 316 men survived the US Navy's greatest single loss-of-life mishap at sea.

General Hap Arnold, chief of Army Air Forces, had activated the oddly named 509th Composite in December 1944 with the specific purpose of delivering 'certain special bombs'. The group's origins were unforgettable: in September that year, its future commander, Lieutenant Colonel Paul Tibbets, stood on a soapbox outside a hangar at Wendover Air Base in Utah and told them, 'Your mission would end the war'. They were not to ask questions; simply, that they had been chosen 'for something very special and very secret'.

Over the ensuing 10 months the 509th airmen flew as many as 30 visual drops a week over the barren salt flats surrounding Wendover. They dropped orange projectiles, nicknamed 'pumpkins' or 'blockbusters', modelled on the shape and size of the plutonium bomb. Each contained about 5500 pounds (2500 kilograms) of heavy explosive. The men had not experienced training like this: bombing the centre of a series of concentric circles; flying sudden steep high-altitude turns; dropping out of the sky at terrific speed on their wing tips; and negotiating low-level navigation exercises along the Cuban coast and around the Caribbean.

By May 1945, the 15 starting aircrews (reduced from 21 initially selected) had had four months' training in visual and radar bombing. Each bombardier had released 80 to 100 'pumpkins' at Wendover, Tibbets told the Target Committee on 28 May. In June and early July 1945, each crew member gained at least four weeks' operational experience over Japanese target areas – which explained the single B-29s frequently seen over Hiroshima and Nagasaki that summer. As these cities were 'reserved for destruction', the aircraft dropped their pumpkins on the surrounding countryside. The local people grew accustomed to the appearance of a few solitary planes and did not think to evacuate or seek shelter; in Hiroshima Yoko Moriwaki, after she saw a B-29 overhead while walking home from school, noted in her diary: 'We must not underestimate the enemy simply because there was only one aircraft.'

•

These 'dress rehearsals' took off from a flat coral-bound island of sugarcane 2300 kilometres southeast of Tokyo. American forces had fought for seven days in July 1944 for control of Tinian, a pinprick 16 kilometres long and 5 kilometres wide. Near the end of the battle, Japanese troops and civilians who refused to surrender jumped to their deaths at a place the Americans named Suicide Cliff.

The US occupying forces converted Tinian into a natural aircraft carrier. On 1 March, the island came under the authority of Brigadier General Frederick Kimble and LeMay's XXI Bomber Command. By then the SeaBees, 13 Construction Battalion of the 6th CB Brigade, had built airfields, roads, sewage works, warehouses, barracks, fuel depots and docks. The Pacific outpost would soon become, for a brief time, the world's biggest airbase: six huge runways of crushed coral 2.4 kilometres long crisscrossed the island; four latticed the North Field where as many as 265 B-29s were assembled, nose-to-tail, on any day. Their orders were to fulfil LeMay's area-bombing missions – in his words, to 'scorch, boil and bake to death' the Japanese cities. From Tinian flew the Superfortresses that had firebombed Tokyo on 9 March; and from here, over the ensuing months, hundreds of missions slogged up the 'Hirohito Highway', night after night, to attack dozens of residential areas.

In early June, a mysterious collection of buildings appeared on the island: five warehouses, 10 magazines, several bombsite shops, tent sheds, generator buildings and a compressor hut organised around a central administration building fitted with the only air-conditioned offices in the Marianas. Electricity generators were sent up from Guam until a fresh supply could arrive in July. These were the headquarters of Project Alberta, the codename for the atomic bomb mission.

The 509th Composite Group arrived on Tinian that month, and were instantly set apart from conventional air units on the island. They were tribal, secretive and conscious of their special status – but precisely to what use none yet knew. LeMay's air forces took umbrage at this arrogant secret air unit. Who were these upstarts? Why were civilian scientists attached to an army air unit? The 509th's pilots, navigators and bombardiers had no precise answers to these questions. They knew only their 'cover' story: to deliver a new kind of bomb, a very big bomb, which would make a 'big bang' – so they were told on leaving America – using an improved variety of heavy explosive. Until they completed their mission, the men must not 'divulge what they knew or were briefed'. LeMay and Groves resolved the command issues during a potentially explosive meeting but which passed without a shout.

The 509th's perks included the best available accommodation on the island, to the intense irritation of the conventional air units. Their showers were hot; they ate steak and drank whisky. They enjoyed the use of five refrigerators, washing machines, and a private movie theatre – the Pumpkin Playhouse. Their attitude, however, was 'most unfortunate', Colonel Elmer Kirkpatrick complained to Groves. 'At times they have acted as spoiled children as I sometimes think they are ... I don't believe that any other organisation has had half the deference and consideration they have, and yet their tone is one of not having had enough consideration. To use our language, there is somewhat of a "dumped on" attitude by some.'

•

Colonel Paul Tibbets stood above these petty, intraservice rivalries. Tibbets seemed unperturbed by the strains of his position; he had fortified himself against the anxiety brought on by being among the very few to know the truth about the mission – or perhaps he simply did not feel as other men do. No plane had dropped an atomic bomb; none had flown out of a radioactive cloud; this mission might kill his men; his plane might not return. Tibbets carried on coolly and never mentioned the task ahead; he was forbidden to say the words 'atomic' and 'radioactive'.

Tibbets' technical skill, proven courage and success within the exacting standards of the US Army's Strategic Air Service had drawn the attention of the top brass. A veteran of dozens of combat missions over Europe and North Africa, he was one of the most experienced B-29 test pilots. General Arnold handpicked him for the atomic mission in the northern summer of 1944. During the security test he startled Groves' examiners with his honesty: asked whether he had ever been arrested, Tibbets admitted he had, 10 years earlier, for having sex in the back seat of a car on Florida Beach.

Tibbets actually enjoyed life at Wendover Field, the loathed air base on the Utah–Nevada border, which Bob Hope, on a brief visit, called 'Leftover Field'. He trained his men there until pumpkin runs were their second nature. The isolation, the rigid command structure, the thoroughly scheduled days, all appealed to this ascetic young man, who seemed somehow much older than his 29 years. Orders, methods and results – the stuff of carefully planned action – sustained him. He described his mission to his superiors with the brevity of one for whom words, unless in the service of the task, were a waste of time. It was, he said, 'to wage atomic war'.

Tibbets had unlimited security clearance. He needed only to say the code word 'silverplate' and the young colonel got what he wanted – the power to raise several squadrons, known as Tibbets' Private Air Force. The nucleus of the 509th Composite Group was the 393rd Heavy Bombardment Squadron, chosen for its high reputation. The complete unit of the 393rd numbered 225 officers and 1542 enlisted men, and incorporated a troop carrier squadron and all relevant supporting units. Thirty special agents secretly scrutinised every man and reported the slightest security breach to Tibbets, who learned the details of each member's drinking habits, sex life, family and political orientation. Those who failed were packed off to remote air bases – in North Alaska, for example, where they could talk to 'any polar bear or walrus' willing to listen, Tibbets later wrote.

Those who survived Tibbets' spies and exacting standards formed the kernel of the 509th. Three men convicted of manslaughter and several ex-criminals who had falsified their names to enlist were among the successful. Tibbets, who knew of their criminal records, offered to return their conviction files – with matches to burn them – if the men succeeded. He valued their air skills over their moral rectitude. The group trained all day, every day. Tibbets would 'drill, drill and further drill his crews, until the best of them could hit the ground within just twenty-five feet of the bull's eye'. Tibbets' confidence and notoriety rose in tandem with the distant respect that attached to his name. He dared to correct LeMay: the atomic delivery aircraft must fly *above* 25,000 feet (7600 metres), he told the commander at a meeting in Guam. The special weapon would destroy a plane flying under that, he explained.

'Tibbets was very co-operative with [LeMay],' wrote Kirkpatrick, but '... I have the feeling he is being a bit cocky with lesser staff officers and may be inclined to rub his special situation in a bit. However, he is a capable man and, I think, smart enough to know how far he can go with different people. He plays his cards well. I hope he can continue to do so ...'

•

On Saturday 4 August, Tibbets addressed his restless crews, some 80 men, in the Operations Briefing Room at Tinian. Highly ranked, unknown officers and slightly dishevelled scientists were among the strange new faces present. Two blackboards draped in cloth stood behind him. 'The moment has arrived,' he announced. 'This is what we have all been working towards. Very recently the weapon we are about to deliver was successfully tested in the States. We have received orders to drop it on the enemy.' Two intelligence officers removed the cloth. Maps of three cities – Hiroshima, Kokura and Nagasaki – appeared, one of which would be bombed on the morning of 6 August, weather permitting. Seven B-29s would fly the mission: three reconnaissance aircraft would radio weather conditions over each city – Major John Wilson's *Jabit III*, Major Ralph Taylor's *Full House*, and Captain Claude Eatherly's *Straight Flush*; three would perform special functions – Captain Charles McKnight's *Top Secret* would fly to Iwo Jima as a reserve aircraft, Captain George Marquardt's *Necessary Evil* would photograph the bombing, and Major Charles Sweeney's *The Great Artiste* would carry scientists and bomb-measuring equipment; an as yet unnamed aircraft coded 'Victor 82' would drop the bomb. Each plane carried 10 crewmen and usually a civilian observer.

A tall balding man with a dome-like forehead rose to speak. Navy captain William 'Deak' Parsons, 44, the officer in charge of the 509th's Project Alberta, exuded the inscrutable calm of the complete insider. The airmen had heard all the rumours about this man and the secret project he led; none realised he was also a director of the secret weapons laboratory at Los Alamos. Parsons' involvement went deeper: he had served on the Target Committee and worked closely with Groves and Oppenheimer. He rarely left the side of Little Boy; he accompanied it across the Pacific, rather like an archaeologist returning with priceless contraband from some ancient tomb. All communications about Project Alberta, between Tinian and Washington, went through him or his technical deputy, Dr Norman F. Ramsey, who were answerable to Groves, who in turn liaised with the Pentagon and the White House.

Parsons' absorption in his work gave the impression that he had a closer kinship with machines and systems than with his fellow human beings. A pioneer in the discovery of radar, and the inventor of the proximity fuze, he was a superb technician and ordnance expert – perhaps the finest the navy had produced. He understood the innards of the atomic bomb with the care and respect of a Swiss watchmaker. He had no time for problem-makers, only solutions; he performed with the reflexive precision of a man curiously fated to make the 'gadget' work. Parsons shared with Tibbets a disdain for human fallibility: the compromise that tempted the average man, the path of least resistance that drew the lazy or weak, these were but pathetic entreaties to this arctic soul.

Like most American servicemen, Parsons admitted no feeling for the Japanese people; the idea of a distinction between civilian and combatant held no sway over his mind. The Japanese cities were military centres, names on a map; the ordinary people members of an inhuman race, unworthy of consideration. It was not that the Americans, at command level, bothered openly to hate the Japanese people; rather, that Japanese civilians simply did not figure in any calculation; they were simply the targets of an experiment that might end the war. Parsons' touchstones were Okinawa, Iwo Jima, Bataan and Pearl Harbor. His vengeance had a personal motive, too: days earlier, he had seen the mutilated face of his 19-year-old half-brother, a casualty of Iwo Jima, who lay in a San Diego naval hospital, the victim of a Japanese mortar. The shrapnel had ripped off the young man's jaw and blinded his right eye.

'The bomb you are going to drop,' Parsons began, 'is something new in the history of warfare. It is the most destructive weapon ever produced. We think it will knock out everything within a three-mile area.' The reaction in the room was one of 'shocked disbelief', Tibbets observed. A film of the Trinity test which Parsons had intended to show got chewed up in the projector, so he continued with his own description: 'the loudest, the brightest, the hottest thing … since Creation'; 'ten times more brilliant than the sun'; and '10,000 times hotter than its surface'. The flash will blind anyone looking at it within a radius of five miles, Parsons added (he had rejected, incidentally, a suggestion by two Los Alamos scientists to switch on a 'super-powerful' siren at the same time as the bomb fell, blinding curious Japanese who looked up to seek the sound). 'No one,' he concluded, 'knows what will happen when the bomb is dropped from the air.'

Throughout their presentations neither Parsons nor Tibbets used the words 'nuclear', 'atomic' or 'radioactive'. Nor did they warn the airmen that the mushroom cloud, which Parsons drew on a chalkboard, would contain intense carcinogens. The crewmen did not actually know the nature of what they were about to do.

Later, the 12-man crew of the delivery plane was selected. As well as Parsons (bomb commander) and Tibbets (airplane commander) they were: Captain Robert Lewis, assistant pilot; Major Tom Ferebee, bombardier; Captain Theodore 'Dutch' Van Kirk, navigator; Second Lieutenant Morris Jeppson, bomb electronics test officer; Lieutenant Jacob Beser, radar countermeasures officer; technical sergeant George 'Bob' Caron, tail gunner; technical sergeant Wyatt Duzenbury, engineer; Sergeant Joe Stiborik, radar operator; Sergeant Robert Shumard, assistant flight engineer; and private first class Richard Nelson, radio operator. All were from the 393rd Bomb Squadron except Tibbets, Ferebee and Van Kirk, of 509th Headquarters; Jeppson, of the 1st Ordnance Squadron; and Parsons. Their mission, in brief, was to drop a 5-ton (4.5-tonne) atomic weapon from a height of 5 miles (8 kilometres) to a point less than 2000 feet (600 metres) above the centre of the specified city with a margin of error of less than 250 yards (230 metres).

They were mostly in their 20s or early 30s (Parsons, the eldest, was 44) – tough, resilient airmen, exceptionally skilled in their particular fields. Lewis, 26, confident to the point of recklessness, was often pulling stunts – such as tipping the wings of his B-29 as he screamed past the Wendover Control Tower back in Utah; his skill as a pilot compensated for this. Ferebee, 26, a big mustachioed bloke from Monksville, North Carolina, and superb baseball player, had flown 63 combat missions over Europe and North Africa and seemed 'nerveless' and 'impossible to rattle', said colleagues. Others were quieter: gunner Bob Caron, 26, married for less than a year, flew to Dodge City to see his newborn daughter before undertaking the 6000-mile (9600-kilometre) trip to the Marianas. On the ground they drank and played hard, salvaging a crate of bootleg whisky, a box of condoms and a garter peeled off the leg of a dancer after the celebrations for their departure for Tinian. In the air they were perfectionists.

·

There were delays on Tinian. Parsons' exacting specifications for the bomb-assembly plant on the island postponed its completion until early August. The crews had to comprehend the exceptional demands of the mission, and how it differed from a conventional air raid. A special technical group had drawn up a long checklist of these: 'The principal difference lies in the fact that the potency and small number [of bombs] available tremendously increases the need for absolute reliability of delivery,' noted one cable to LeMay. And the poor weather – severe storms predicted for 2 August – pushed the departure date back, at least until 5 August: 'LeMay now advises at least 24 hours later for Little Boy operations than indicated on account of weather,' read a cable from Tinian to the War Department on 4 August.

On 5 August an unnamed Superfortress – 'No. 82' – stood on the North Field tarmac stripped, rewired and specially modified to accommodate the 5-ton atomic bomb. That night Tibbets scribbled the name of his 57-year-old mother on a scrap of paper, gave it to a signwriter, and ordered the name be painted on the strike ship. His mother had supported Tibbets against his father when he dropped out of medical school to become a pilot; this mission, he felt, was for her. The crew, however, were unimpressed when they saw the giant letters; Bob Lewis furiously wondered what in hell the name was doing on his plane. The boys had expected something daring, funny – sexy: a blonde triumph or an anti-Jap cartoon, and a proper, gutsy name like, *Necessary Evil*, *The Great Artiste*, *Straight Flush* and so on. But the damage was done and no crew member dared paint a large-breasted blonde over the skipper's mother's name: 'ENOLA GAY'.

The crew prepared through the night and went over the brief: the flight to Japan and back, via Iwo Jima, would take 12 hours; the choice of target would depend on the weather reports from the three reconnaissance planes. The crews set their watches and repaired to the Mess Hall (the 'Dogpatch Inn') for a midnight breakfast of eggs, sausages and pancakes. Tibbets spoke briefly, instructing the men to do their job and obey orders, and then invited the unit chaplain, William Downey, to bless the mission: 'Almighty Father, we pray Thee to be with those who brave the heights of Thy heaven, and who carry the battle to our enemies ... May they, as well as we, know Thy strength and power, and armed with Thy might may they bring this war to a rapid end ...' The day before, Charles Sweeney, a Catholic who would command *The Great Artiste* on this mission, attended Mass and received Holy Communion. He decided not to reveal the nature of his job at the confessional; instead, he resolved to 'commune silently with God and tell Him about our mission'. At least one other airman asked for absolution.

On the way to the planes the trucks stopped at the Personal Equipment Supply Hut, and each airman signed for a parachute, flak vest, combat knife, water-drinking kit, survival kit, flotation device, fish hooks, food rations and a .45-calibre automatic pistol with ammunition. The trucks proceeded to North Field through a web of security, floodlights and military police.

The *Enola Gay*'s four engines fired at 2.27am (Tinian time), 6 August, and Lewis taxied slowly towards the Hirohito Highway. The three reconnaissance planes had already rumbled away towards Japan. In recent weeks an alarming number of aircraft, weighed down with ordnance, had crashed before lift-off; on one morning four planes had burst into flame on North Field, with few survivors. Parsons could not risk an accident on take-off, so he had resolved, a day earlier, to arm the weapon mid-flight. He had not attempted the complex 11-step process in the air and spent the days before departure continually rehearsing.

The aircraft took off without mishap, Bob Lewis wrote in his 'pilot's log', which he scribbled at the request of William Laurence, the ubiquitous *New York Times* reporter whom Groves had selected to witness the Trinity test (Tibbets had refused to allow Laurence on the mission). Lewis wrote the log in the style of a letter to 'Mom and Dad', scrawled 'in almost complete darkness' in the cockpit (halfway through the flight he ran out of ink). Despite these setbacks, he managed to express something of the atmosphere on board the *Enola Gay*:

Little Boy Mission #1

First Atomic Bomb

August 6th 1945

By Capt Robert A. Lewis, Pilot Aboard Ship

Briefing at 2300 [Tinian time; subtract an hour for the time at Hiroshima]

Eating at 0030

Dear Mom and Dad,

… we got off the ground at exactly 0245. Everything went well on take-off, nothing unusual was encountered …

At 0320 [Tinian time] … items 1-11 were completed satisf. by Capt. Parsons [that is, Parsons had successfully completed the 11 steps involved in arming the bomb.]

To do this, Parsons had crawled back through to the bomb bay, as the plane flew through dense cloud, where he: (1) checked that the green plugs were installed; (2) removed the rear plate; (3) removed the armour plate; (4) inserted the breech wrench in the breech plug; (5) unscrewed the breech plug and placed it on the rubber pad; (6) inserted the charge, four sections, with red ends to breech; (7) inserted the breech plug and tightened home; (8) connected the firing line; (9) installed the armour plate; (10) installed the rear plate; and (11) removed and secured the catwalk and tools. Tibbets radioed the steps to Tinian, but lost contact at step nine.

Lewis continued his log:

At 0420 Dutch Van Kirk sent me up an ETA of Iwo Jima of 0552 ... The colonel, better known as the 'old bull' [Tibbets], shows signs of a tough day, with all he had to do to help get this mission off he is deserving of a few winks. So I'll have a bite to eat and look after George [the auto-pilot] ...

At 0430 we started to see signs of a late moon in the east. I think everyone will be relieved when we have left our bomb with the Japs and get half way home. Or better still all the way home ...

The first signs of dawn came to us at 0500 and that also is a nice sight after having spent the previous 30 minutes dodging large cumulous clouds. It looks at this time 0515 that we will have clear sailing for a long spell.

Our bombardier Maj Tom Ferebee has been very quiet and one thinks he is mentally back in mid-west part of the old U.S.

By 0552 it is real light outside and we are only a few miles from Iwo Jima. We are finishing a second climb which is to 9,000 ft [2700 metres]. We'll stay here until we are about 1 hr away from the Empire ...

At 0710 ... Outside of a high, thin cirrus and the low stuff it's a very beautiful day. We are now about 2 hrs from Bombs away ...

At 0730 ... it's a funny feeling knowing it's right in back of you. Knock wood.

We started our climb to 30,000 ft [9000 metres] at 0740. Well folks its not long now. I checked with crew at 20,000 feet and all stations report [in] satisfactory.

At 0830 [Tinian time] Claude Eatherly's *Straight Flush*, then flying over Hiroshima, sent weather data. Nelson decoded: 'Bomb primary [target]'. 'It's Hiroshima,' Tibbets said over the intercom.

'We will make a bomb run on Hiroshima,' wrote Lewis. 'Right now we are 25 miles [40 kilometres] from the Empire and everyone has a big hopeful look on his face.' He added: 'There'll be a short intermission while we bomb our target ...'

·

At 7.50am Hiroshima time Lewis and Tibbets took control of the plane from the auto-pilot. The crew put on their flak jackets; Van Kirk checked the radar, ground speed and air speed.

'We are about to start the bomb run,' Tibbets told the crew. 'Put on your goggles and place them on your forehead. When you hear the tone signal, pull them down over your eyes and leave them there until after the flash.'

Their goggles shut out light on the edges and bridge of the nose, and dulled the blue sky that came through a gap in the cloud. Shortly Hiroshima's outline and the ships at the mouth of the Ota's tributaries appeared. Duzenbury steadied the aircraft and Caron watched for enemy interceptors, while Beser monitored Japanese radar frequencies for interference with the bomb's four radar units (nicknamed 'Archies' and invented by a Japanese scientist). Parsons kept his check on the bomb's instruments. The crew buckled their parachutes onto their harnesses (Jeppson attached his oxygen mask to an emergency oxygen bottle because, he later said, 'the blast from this bomb might blow out the windows').

At 8.12 (Hiroshima time) they reached the initial point (IP) of the bombing run. 'It's all yours,' Tibbets told Ferebee. The bombardier set his left eye against the Norden bombsite. The target was the T-shape formed by the Aioi Bridge in the city centre, 25 kilometres north, which the plane approached at a ground speed of 285 miles (460 kilometres) per hour.

The 'T' appeared in Ferebee's sights at 8.14:17; he initiated the 60-second action that would culminate in the blast. The radio tone started; the crew lowered their goggles and at 8:15:17 the bomb-bay doors opened. The projectile fell out and the aircraft jerked violently up with the weight loss. The tail fins on the bomb straightened it into a nose-dive, and the three last steps ending in detonation, at 1850 feet (560 metres) over the city, began.

Lewis took the *Enola Gay* into a sharp 155-degree turn – 'a right-hand diving turn at the limit of the plane's capabilities,' said Caron – and *The Great Artiste* similarly swung away to the left. The crew counted down the 43 seconds expected before detonation – one-thousand one, one-thousand two; Jeppson reached the number too soon and thought they had dropped a dud; Van Kirk focused on his watch; Caron mistook the sun for the flash and closed his eyes; then a rippling front of heat, travelling at 7200 miles (11,590 kilometres) per hour, tossed the plane about like 'a huge hand', which Parsons confused with anti-aircraft fire.

Caron took photographs of the world below: '… everything was burning,' he observed. 'I saw fires springing up in different places, like flames shooting up on a bed of coals. I was asked to count them. I said, "Count them?" Hell, I gave up … it looked like lava or molasses covering the whole city and it seemed to flow outward up into the foothills where the little valleys would come into the plain … pretty soon it was hard to see anything because of the smoke.'

'Fellows,' Tibbets announced over the intercom, 'you have just dropped the first atomic bomb in history.'

Parsons' omniscient calm momentarily deserted him: he looked shocked and awed. 'Jesus Christ,' said Jeppson. 'If people knew what we were doing, we could have sold tickets for $100,000.' Caron continued photographing the cloud; Lewis scribbled, 'Just how many did we kill?' and then, 'My God, what have we done?' – not an expression of collective regret; rather a reflexive burst by a man lost for words ('My God, look at that sonofabitch go!' he is said to have also shouted, according to other crew members).

The mushroom cloud reached 45,000 feet (13,700 metres) and obscured the city; the sight made the crew feel like 'Buck Rogers 25th century warriors,' Lewis thought. 'I honestly feel the Japs will give up before we land at Tinian.'

'Down below all you could see was a black boiling nest,' Tibbets later said. 'I didn't think about what was going on down on the ground ... I didn't order the bomb to be dropped, but I had a mission to do.'

The standard flight report noted: 'Target at Hiroshima attacked visually $1/10$th cloud at 052315Z. No fighters and no flak.' Parsons sent one of 28 pre-coded outcomes to Tinian: '82 V 670. Able, Line 1, Line 2, Line 6, Line 9.' On receipt, Farrell decoded: 'Results clear cut, successful in all respects. Visual effects greater than Trinity test. Target Hiroshima. Conditions normal in airplane following delivery. Proceeding to regular base.'

The message flew directly to the Pentagon and Lewis wrote his last line to his parents: 'Everyone got a few catnaps, Love to all, Bud.'

•

At Tinian news of the mission's success sped around the base. With whoops and back slaps, crowds rushed to receive the planes as they thundered in at about 3pm that afternoon. More than 200 airmen, soldiers, press and scientists surrounded the *Enola Gay* as it taxied to a standstill. Tibbets emerged first, followed by Parsons and the crew. Spaatz strode up and pinned the Distinguished Service Cross on Tibbets' chest (the crew and Parsons later received Silver Stars).

They were taken through a parting crowd to the medical hut and checked for radiation under Geiger counters; their eyes for flash burn. All were healthy and unaffected; nor were harmful levels of radiation found on the *Enola Gay*.

The debrief was an exuberant account of the flight, washed down with bourbon and lemonade. Relief was the prevailing emotion: 'Here was the successful climax to about eleven months of awfully hard and demanding work,' Tibbets later said. Similarly, Van Kirk: 'What a relief. We'd gotten the thing there and we'd done the mission.' Laurence, entranced, had already squirrelled away Lewis' log.

The party ran all afternoon. The mess officer, Charles Perry, ordered in thousands of sandwiches, hot dogs and meat pies (for a pie-eating contest). The official program offered 'free beer from 2pm', an 'all star' softball game, a jitterbug contest, prizes, music, a movie (*It's a Pleasure*, with Sonja Henie and Michael O'Shea) and an 'extra added attraction' – a blonde, vivacious, curvaceous, starlet …

Farrell, meanwhile, attended to the serious responsibilities of informing Groves and the Pentagon; Groves amended and forwarded his report to Marshall, announcing 'Mission Accomplished':

Confirmed neither fighter or flak attack and one-tenth cloud cover with large open hole directly over target. High speed camera reports excellent record …

Flash - not so blinding as New Mexico test because of bright sunlight … Cloud was most turbulent. It went at least to forty thousand feet … observed from combat airplane three hundred and sixty-three nautical miles away … Observation was then limited by haze and not curvature of earth …

Blast - there were two distinct shocks felt in combat airplane similar in intensity of close flak bursts. Entire city except outermost ends of dock areas was covered with a dark grey dust layer ... It was extremely turbulent with flashes of fire visible in the dust. Estimated diameter of this dust layer is at least three miles. One observer stated it looked as though whole town was being torn apart with columns of dust rising out of valleys approaching the town ... visual observation of structural damage could not be made.

Parsons and other observers felt this strike was tremendous ... even in comparison with New Mexico test. Its effects may be attributed by the Japanese to a huge meteor.

Groves personally took the report to Marshall's office in the Pentagon at 6.15am Washington time; General Arnold and George Harrison shortly joined them. Stimson, then at Highhold, his Long Island retreat, was informed over a secure telephone and offered his 'very warm congratulations'. Groves' elation and the shared sense of relief received a slight dampener from Marshall, who warned against excessive gratification over the bomb's success 'because it undoubtedly involved a large number of Japanese casualties'. Groves replied that he had not thought about those casualties; rather, he had in mind 'the men who made the Bataan death march'. In the hall, Arnold slapped him on the back: 'I am glad you said that – it's just the way I feel.'

Parsons' last act that night was to leave his mark on the bomb's receipt issued when the weapon arrived in Tinian: 'I certify that the above material was expended [sic] to the city of Hiroshima, Japan, at 0915, 6 August. Signed W.S. Parsons.' Two days later he wrote to his father: 'Dear Dad, This will be a short note to say that you have lived to see atomic weapons used and that your son was the "weaponeer" and also delivered the first one to the Japanese!'

The rest of the crew of the *Enola Gay* joined the fag ends of the party, got drunk and fell asleep.

CHAPTER 16

AUGUSTA

If they do not now accept our terms they may expect a rain of ruin
from the air, the like of which has never been seen on this earth.

President Harry Truman, after the atomic bombing of Hiroshima, 6 August 1945

EARLY ON 2 AUGUST – four days before the bomb fell – the Presidential party flew to Plymouth, England, landing unexpectedly at Harrowbeer airfield, where the plane put down due to fog. They boarded the *Augusta* at 2pm, relieved to be away from Berlin. 'I am very sure no-one wants to go back to that awful city,' Truman wrote. Nor was he eager to see the Soviet delegation: 'If those SOBs want to see me again, they'll have to come to Washington!'

Before the ship departed, the President dined with King George VI in the Royal cabin of HMS *Renown*, moored nearby. 'Welcome to my country!' the King said, clasping Truman's hand. There were pipers, buglers, an inspection of the guard of honour and the singing of anthems, followed by a brisk luncheon of soup, fish, lamb chops, peas and ice-cream with chocolate sauce. The King was 'a very pleasant and surprising person', Truman wrote. His Highness showed him a sword that Queen Elizabeth I had presented to Sir Francis Drake. They discussed Potsdam with Lord Halifax, Admiral Leahy, Secretary Byrnes and other officials. The King 'was very much interested in ... our new terrific explosive', Truman noted. It was a candid, relaxed discussion, during which Leahy expressed doubt about whether the bomb should be used – and whether it would work. The President and Byrnes chided the admiral; but Leahy insisted that Magic intercepts of enemy cables showed the Japanese were seeking a peaceful solution. The three men cheerfully agreed that 'the Japs' were probably 'looking for peace' but the President thought they would sue for peace through Russia. Truman had clearly seen, or been briefed on, the intercepts during Potsdam; he was aware of Tokyo's 'peace feelers' and the special envoy's proposed mission to Russia (indeed, Stalin had alluded to this in their meetings). But Truman did not believe this communication worthy of a reply: the Japanese 'feelers' had said nothing that changed the President's resolve to press for surrender on American terms.

The Presidential party left the *Renown* and steamed west. There were the usual on-board entertainments, concerts and mess hall enjoyments. Truman rose early and devoted his morning hours to work: preparing an address to the nation; perusing the terms of the Potsdam documents; and holding long discussions with his advisers, chiefly Byrnes and Leahy.

On 4 August the ship's *Press* reported that Japan had suspended 'suicide attacks by the kamikaze after a heavy sacrifice of pilots and planes' – the enemy's first admission that its death squads were almost exhausted. The remaining kamikazes were to be held in reserve for plunges into large ships, Tokyo Radio said. At the same time, Japan had moved vital industries underground or inland in readiness for the expected land invasion.

Truman rose at 5.30am on 5 August and almost completed a full day's work by 9am. He strolled the deck in a seaman's cap, smiling at everyone, and looked in the best of health. The waters were sheet-like, and the skies fair. Further good news arrived: Tokyo Radio reported that 'not a nook or cranny' in all the 'sacred islands' offered protection from American air raiders 'wheeling out from carriers, the Marianas and Okinawa'. In recent days, US aircraft had dropped three million copies of the Potsdam Declaration over Japan and warned 12 mid-sized Japanese cities, ports and industrial centres of 30,000 to 70,000 people – including the 'Pittsburgh of Japan', the giant steel-producing city of Yawata – that they faced imminent destruction. With no big cities left to destroy, LeMay was determined to unleash the air war on smaller towns (he had avoided Yawata, the B-29s' first target, in June 1944, after the decision to switch to civilian bombardment in March 1945). Tokyo defied these warnings, declaring that Japan would meet the American invasion with 'a force several times larger, regardless of where the enemy might choose to land on the mainland'.

•

That night the President attended performances of Brahms, Schubert and Tchaikovsky. He slept well. He awoke on the 6th refreshed and energetic. The *Augusta* sailed south of Newfoundland; the weather clear and warm, the seas calm. At 11.45am the President appeared in the mess hall for an early lunch. Moments before noon, as Truman sat eating with the sailors, Captain Frank Graham brought a cable from the United States Pacific Fleet Headquarters, sent via the War Department. Four hours earlier an atomic bomb had been dropped on Hiroshima, it said: 'Results clear cut. Successful in all respects. Visible effects greater than in any test ...'

Truman firmly shook Graham's hand and exclaimed to the sailors at his table, 'This is the greatest thing in history!' He passed the message to Byrnes, sitting nearby: 'It's time for us to get on home!' he called out.

Within minutes another cable arrived:

Big bomb dropped on Hiroshima 5 August at 7.15 P.M. Washington time. First reports indicate complete success which was even more conspicuous than earlier test. Stimson.

The President then rose, silencing the noisy mess hall; the sailors put down their forks. He had just received news, he said, of 'our first assault on Japan with a terrifically powerful new weapon, which used an explosive 20,000 times as powerful as a ton of TNT'. Cheers greeted the announcement. 'I guess I'll get home sooner now,' said a young serviceman sitting at the President's table. The applause pursued Truman out of the hall; he entered the wardroom, unannounced, and the ship's officers jumped to their feet.

'Keep your seats, gentlemen.' Truman waved for them to sit. 'I have an announcement to make … We have just dropped a new bomb on Japan … It is an atomic bomb. It was an overwhelming success.' A thunderous burst of applause drowned the President's words. 'We won the gamble,' he added, with a broad smile on his face. There was complete elation, the mood infectiously happy. The officers expressed the hope the atomic strike would hasten the end of the war. Later Truman wrote that no announcement had ever made him happier: 'I was greatly moved.'

•

At 10.30am Washington time on 6 August, Eben Ayers, a White House press adviser, found himself fronting the media to announce 'a darned good story'. Ayers, whom a reporter later derided as 'mousey', was determined to impress upon the capital's top reporters that he was in possession of a great news event. It was not in Ayers' nature, however, to overdo the enthusiasm; nor did he think it proper to smile or look triumphant – something Truman would not hesitate to do, he later reflected in his diary.

The reporters gathered for the usual morning conference, a relaxed affair, peppered with jokes and mild heckling. Ayers had rarely faced the pack alone; his possession of a 'story' of this magnitude added to his intimidation. (His boss, Charlie Ross, was then on the *Augusta* with Truman.) He shyly appeared and urged the press to wait as the details had not yet arrived.

A week earlier General Alexander Surles, PR chief at the Pentagon, had warned Ayers to prepare for 'a tremendous story', which, the general hoped, would be ready for release within a few days. The statement had had a difficult genesis: Groves had hoped to announce to the world the complete destruction of Hiroshima. The lingering mushroom cloud, however, obscured the evidence for such a claim. F-13 fighters had flown over the city four hours after the blast but could take only 'oblique pictures', LeMay cabled Groves. In reply Groves asked whether Brigadier General Farrell, then unavailable, had any reason not to release the press statement on the bomb. Farrell 'strongly recommends release', LeMay replied.

And so, at 11am, Ayers called the meeting to order: 'I have got what I think is a darned good story,' he began. 'It's a statement by the President, which starts off this way …' He began reading the first paragraph, 'Sixteen hours ago, an American plane …' He paused. 'Now the statement explains the whole thing. It is an atomic bomb, releasing atomic energy. This is the first time it has been done.'

The press blinked at this fussy man as if he were telling them a great joke. Ayers continued: 'This statement is a little over three pages,' he said, handing out copies as though it were a school quiz. 'The first page is loose. I will give it out to you here. Wait until you get it before you go out so you won't ball it all up. I'm not going to tell you anything about it because I haven't read it all myself; it tells the whole story better than I can. It's a big story.'

On reading it, Joe Fox of the *Washington Star* shouted, 'It's a hell of a story!' The reporters were struck dumb, 'unable to grasp what it was about', Ayers noted. None panicked or bolted for his or her phone. They absorbed the narrative in a state of silent wonder – as though the White House had just verified an alien invasion or asteroid strike; the story did, in fact, change the world. Some reporters had trouble persuading their editors, who refused to believe them. The mass of information quite overwhelmed them. Where did the story begin and end? The world's first atomic bomb; the existence of the secret enterprise; Japan's response; the science of the atom; the end of the war; the future of a nuclear world … all cried out for front-page attention.

'Perhaps I underplayed my hand,' Ayers reflected later. 'My announcement was gross understatement,' he wrote in his diary, 'and I wondered if I had been too matter-of-fact about it. But a good reporter does not need to be told when he has a good story and I did not want to seem to be trying to sell [it].'

•

A few minutes after Truman's revelation the *Augusta* broadcast a pre-recorded radio bulletin of the President's statement, released through the White House. The sailors listened, incredulous, to news of the first 'atomic bomb'.

'Sixteen hours ago,' the President announced, 'an American airplane dropped one bomb on Hiroshima and destroyed its usefulness to the enemy ... It had more than 2,000 times the blast power of the British "Grand Slam", which is the largest bomb ever yet used in the history of warfare.

'It is an atomic bomb,' Truman continued. 'It is a harnessing of the basic power of the universe. The force from which the sun draws its power has been loosed against those who brought war to the Far East.' He spoke of the scientific marvel; the British contribution; the race of the laboratories; the sheer scale and cost.

And he addressed the enemy: 'We are now prepared to obliterate more rapidly and completely every productive enterprise the Japanese have above ground in any city. We shall destroy their docks, their factories and their communications. Make no mistake: we shall completely destroy Japan's power to make war.

'It was to spare the Japanese people from utter destruction that the ultimatum of July 26 was issued at Potsdam. Their leaders promptly rejected that ultimatum. If they do not now accept our terms they may expect a rain of ruin from the air, the like of which has never been seen on this earth.'

He concluded with a promise to recommend to Congress the establishment of a commission to control the use of atomic power; and to explore ways of harnessing it 'towards the maintenance of world peace'. There were no details of the destruction of Hiroshima, or its inhabitants: nobody in Washington yet knew what the bomb had done.

Thousands of troops then stationed throughout the Pacific whooped for joy at this news. Their relief was understandable: the bomb would surely end the war, they believed; they were going home. Truman, in his cabin with Charlie Ross and other staff at the time of the broadcast, was in a contemplative mood. 'The atomic bomb,' he told them, 'will turn out to be the greatest power for the good of mankind ever known because it is in the hands of two peaceful nations.' The secret will be kept 'until it is certain beyond any shadow of a doubt that the peace of the world is secure'.

Later, in Byrnes' cabin over bourbon, the two men reflected on the achievement and the likely Japanese reply. The Secretary of State recalled how he had once doubted whether the bomb would work, and how he feared the huge cost. 'This goes to show,' Truman said, 'how important it is to have men in key placed [sic] who have the respect of the people. Jim, you and I know several people who, if they had been in charge of this project, it would never have succeeded, because Congress would not have had the confidence in them to permit these huge expenditures in secrecy.' Byrnes heartily agreed.

That afternoon they attended a program of entertainment and boxing held on the well deck: the ship's orchestra played and comedians performed. The boxing abruptly ended when the ring posts collapsed, slightly injuring a spectator's head. But the celebratory spirit continued into the night at the Warrant Officers' dinner. Truman drew the line at the movie presentation – a puppetoon, *Hot Lips Jasper*, followed by *Nob Hill* – and went to bed.

·

The President awoke to a different world and an effusive media. The *Augusta Press* reported that 'the most terrible destructive force ever harnessed by man' had been dropped on a 'Japanese Army Base'. Truman's reference to the sun presented a 'dramatic possibility' for propaganda, the paper added: 'They regard their Emperor Hirohito as a direct descendant from the Sun Goddess. Now ... the very power of the sun itself is being turned to their destruction.'

•

Bales of press releases tumbled from the White House and Pentagon printers in a seeming race to get the lion's share of the coverage. Groves launched his own press offensive lest the White House steal the limelight, in response to which the security-conscious Stimson ordered the general to desist – unless cleared by the War Department.

On the morning of 7 August Americans heard for the first time of the vast secret to produce the 'deadliest weapon ever devised'; 'a spectacular new discovery in the field of science'; costing '$1,950,000 ...'

NEW AGE USHERED ... HIROSHIMA IS TARGET ... 'IMPENETRABLE'
CLOUD OF DUST HIDES CITY AFTER SINGLE BOMB STRIKES
By Sydney Shalett
WASHINGTON, Aug 6. The White House and War Department
announced today that an atomic bomb, possessing ... a destructive
force equal to the load of 2,000 B-29s ... had been dropped on
Japan ...

[The bomb struck] an important army centre ... about the time
that citizens on the Eastern seaboard were sitting down to their
Sunday suppers.

... 'an impenetrable cloud of dust and smoke' masked the target
area from reconnaissance planes ...

Secretary Stimson said that this new weapon 'should prove a
tremendous aid in the shortening of the war against Japan' ...

Hiroshima, first city on earth to be the target of the 'Cosmic Bomb', is a city of 318,000 which is – or was – a major quartermaster depot and port of embarkation for the Japanese. In addition to large military supply depots, it manufactured ordnance, mainly large guns and tanks, and machine tools and aircraft ordnance parts ...

'What has been done,' [President Truman] said, 'is the greatest achievement of organised science in history.'*

On page 10 of the *New York Times* veteran correspondent Hanson Baldwin offered a palliative to the march of triumph:

It is almost useless, to talk of the 'rules' of war. Yet when this is said, we have sowed the whirlwind. Much of our bombing throughout this war – like the enemy's – has been directed against cities and hence against civilians. Because our bombing has been more ... devastating, Americans have become a synonym for destruction ... We may yet reap the whirlwind. Certainly with such God-like power under man's imperfect control we face a frightful responsibility. Atomic energy may well lead to a bright new world in which man shares a common brotherhood, or we will become – beneath the bombs and rockets – a world of troglodytes.

As news of the devastation emerged, the headlines grew sombre: 'Atomic Bomb Wiped Out 60% of Hiroshima', ran the *New York Times* on the 8th. 'With a single bomb we were able to destroy in a matter of seconds an area equivalent to one-eighth of Manhattan ...' 'Hiroshima Inferno – 4 Square Miles Obliterated – Huge Death Roll', led the London *Times* on the 9th.

No reporter had actually seen the remains of the city; they were merely regurgitating Pentagon press releases. Nor had any journalist

---

* William Laurence later claimed to have written the President's press releases, a job the *New York Times* reporter described as 'unique in the history of journalism ... no greater honour could have come to any newspaperman, or anyone else for that matter'. His first draft was rejected and a friend of Stimson's, Arthur Page, wrote the final version.

– thus far – attempted to uncover the effect of a nuclear weapon and its radioactive fallout on human beings.

A spoiler to the near-universal elation came from an unexpected quarter: the Catholic Church. The Vatican deplored the atomic attack. 'The use of the atomic bombs in Japan has created an unfavorable impression on the Vatican,' wrote the Vatican City newspaper, the *Observattore Romano*. The Holy See praised the example of Leonardo da Vinci, who destroyed his plans for a submarine for fear mankind might use it to the ruin of civilisation – ominously charging that those who built the atomic bomb 'did not think as did Leonardo'. The Pope concluded with a spray of brimstone: 'Force, and its cult and its exaltation, have their punishment and their nemesis ... Christianity, its charity and its law, which condemns force, expects a prize from these tremendous lessons. Its nemesis is not the goddess of vengeance but of justice.'

The British response, issued in Churchill's name, was solemn and portentous: 'It is now for Japan to realize,' the former Prime Minister declared, in a statement drafted on 31 July, 'in the glare of the first atomic bomb which has smitten her, what the consequences will be of an indefinite continuance of this terrible means of maintaining a rule of law in the world. This revelation of the secrets of nature, long mercifully withheld from man, should arouse the most solemn reflections in the mind and conscience of every human being capable of comprehension. We must indeed pray that these awe-striking agencies will be made to conduce to peace among the nations, and that instead of wreaking measureless havoc upon the entire globe, they may become a perennial fountain of world prosperity.'

·

A crowd of 200 colleagues and supporters met the President at Newport News, where the *Augusta* moored. After brief speeches of congratulations, Truman boarded the train to Washington, arriving back at the White House at 10.45pm, 7 August, after a round trip of

15,000 kilometres. He stepped from his limousine looking fit and tanned, and swept up the stairs with a small group of cabinet members and staff to his study on the residence floor. He played a few bars on the piano and rang Mrs Truman, then at the family home in Independence, Missouri. Drinks were ordered – bourbon – and presents handed out. Truman spoke of the trip, of Stalin and Churchill, and enlightened his staff to the reality of the Soviet Union. The bomb was the harbinger of peace; the Japs must soon surrender; the war would end: these were exhilarating days.

The next morning Truman attended the usual conference with his department heads. A note of concern at the future repercussions of nuclear power slightly overshadowed his pleasure at the weapon's success. He laughed again at Admiral Leahy's fear that, up until the last minute, it wouldn't go off. He frowned at the Pope's attitude, but sought to mollify the Vatican, whose co-operation would be needed in coming days, he said, to calm the Catholic countries of Europe. Somehow he must 'reach the Pope and reassure him'. Archbishop Spellman of New York, and the Bishop of Detroit were suggested as intermediaries between the White House and the Pontiff.

•

At 2pm on 6 August Oppenheimer received a call from Groves:

'I'm very proud of you and all your people,' the general said.

'It went alright?'

'Apparently it went with a tremendous bang …'

'When was this, was it after sundown?'

'No, unfortunately it had to be in the daytime on account of security of the plane …'

'Right. Everybody is feeling reasonably good about it and I extend my heartiest congratulations. It's been a long road.'

That afternoon the Los Alamos public-address system announced the 'successful combat drop' of one of the laboratory's 'units'. The scientists assembled in the Tech Area amphitheatre – their feelings ranged from jubilation to unease. Oppenheimer entered in the unlikely pose of a prize fighter, shaking his clasped fists over his head to loud cheers and the stamping of feet.

In other parts of the Manhattan Project, in Hanford and Oak Ridge, thousands of Manhattan Project employees woke on 7 August to discover what they had been building; those who knew were suddenly free to talk about it. James Hush, an employee at Los Alamos, wrote to his parents that day:

> Dear Folks,
>
> This morning at 9.15 our public address system at work heralded the most startling radio announcement of this war and so completely electrified all the workers here that ... the rest of the morning and all afternoon were spent in bull sessions and general expressions of happiness and excitement over the very great news that at last our work, so veiled in secrecy for the past three years, is now before the eyes of the world. Now we can brag in front of and expect praise from our friends ... From the sound of things the whole world shares our excitement and rejoicing ... The first atomic bomb to be used against the enemy was dropped on Hiroshima this morning and must have caused tremendous damage to that city, which is about the size of Denver, not to mention scarring [sic] the pants off all the Japanese for miles and miles around ...

Security remained paramount, however: Groves kept the lid on technical detail and loose gossip. The commander of the Los Alamos Engineering Corps warned his staff not to disclose 'even to our families and friends ... any information in addition to that contained in official authorised releases'. The same message circulated throughout the Manhattan complex.

The workers celebrated long and hard at Oak Ridge and Hanford and in the smaller factories of the Manhattan Project. Congratulations flowed for their meritorious service, and their unselfish and tireless effort. 'Our first objective has been reached,' Colonel Nichols announced to all military and civilian personnel. 'We must press on lest the Jap catch his breath. Remember … he has not yet quit. Let us produce with unslackening zeal and speed so that final victory can be won at the earliest possible moment.'

The scientists' exuberance quickly faded: that night a dormitory party broke up after an hour. Oppenheimer briefly dropped by, found 'a usually cool-headed young group leader', the physicist Volney Wilson, retching in the bushes, and said, 'the reaction has set in'. If Oppenheimer's earlier triumphalism seemed out of character – reminiscent, perhaps, of his High Noon act after Trinity – it was consistent with the physicist's wardrobe of personas. Oppenheimer would reveal a more authentic one after studying the effects of his laboratory's creation.

•

Far from the elation in Washington and Los Alamos, Leo Szilard sat alone in his new quarters – two rooms at the back of a drab three-storey brick apartment block on South Blackstone Avenue, Chicago, overlooking an alley and some garages. He had enjoyed, until his recent eviction from the Quadrangle Club, an uninterrupted view of the club's neo-Gothic spires. Such was the price of refusing to empty his bathtub. Szilard's suitcase contained a few rumpled clothes and papers, and yet, despite his relative squalor, the tireless Hungarian meant to continue his crusade.

Re-energised by news of the atomic bomb – proof of the chain reaction he had imagined on a London street corner – Szilard resolved to bring his petition against the weapon to the attention of the American public (the copy he had sent to the President lay in Groves' drawer, ignored). Oblivious of the fact that his condemnation was especially unwelcome that day, 6 August, when most Americans were celebrating the marvel of atomic power – the victory weapon that would bring their boys home – Szilard wrote to Arthur Compton requesting security clearance for a press statement; he was anxious not to release anything that could be construed as a 'military secret'. Compton passed Szilard's request to the Intelligence and Security Division of the Manhattan Project, which promptly refused it. Szilard was muzzled. Ten days later army intelligence conditionally agreed to the petition's release, pending the approval of the White House. Before Truman could decide, however, Groves intervened and reclassified the petition 'Secret'. It remained classified until 1957; a complete copy was not published until 1963.

Szilard's was not a lone crusade; an element of the scientific community, chiefly the Chicago group – the 68 signatories to his petition – shared his views. Two more scientists would join the petition that eventually reached the White House, and another 85 signed an Oak Ridge version. In sum 155 Manhattan project scientists registered their moral opposition to dropping the bomb without warning on a Japanese city. These dissenting voices – many of whom worked in the lab that had built the bomb – so irritated the White House that Truman issued a press statement about the merits of the weapon 'because so many fake scientists were telling crazy tales about it'.

CHAPTER 17

HIROSHIMA, 6 August 1945

There was nothing I could do to help in this hell on earth, so I simply clasped both hands together tightly in front of my face and made my way out of Hiroshima ...

Flight Navigator Takehiko Ena, a kamikaze pilot who returned home through the atom-bombed city after his plane ditched at sea

HIROSHIMA ROSE, WEARY OF THE sound of air-raid sirens. The previous night two warnings, at 9.27pm and 12.25am, had sent the people scurrying for the few shelters. Both were false alarms and the all-clear quickly sounded. At 7.09am a 'yellow' siren wailed, alerting the city to another enemy aircraft (this was the atomic mission's weather reconnaissance plane, *Straight Flush*). Most residents shrugged off the alert and carried on. Single planes were commonplace. 'We didn't pay that much attention because it happened all the time,' said Iwao Nakanishi. The warning lifted at 7.31am.

It was going to be a hot August day, with blue skies. Some 12,000 mobilised children rose early and set off for the demolition sites around the city, 12- and 13-year-old girls like Yoko Moriwaki, Seiko Ikeda and Tomiko Matsumoto; and the boys Iwao Nakanishi, who worked at an army supply depot (and who had secretly picked up one of the Potsdam leaflets dropped by American planes), and Takashi

Inokuchi, 14, a labourer at a seafood cannery; others headed for the communications centre beneath the Castle barracks, on the grounds of which some 4000 troops too old or inexperienced for combat duty dressed for the usual parade drill. Elsewhere military policemen, like Takashi Morita, prepared to prowl the city centre; government clerks filed into the Hiroshima Industrial Promotion Hall next to the Aioi Bridge; doctors and nurses arrived at their hospitals; and the great assortment of volunteer corps and neighbourhood associations went about their duties, morning exercises, prayers, donation collections and defence training.

At 7am 12-year-old Yoko Moriwaki joined 220 First Year girls in the First Prefectural Girls' High School (Daiichikenjo) at a demolition site in Dobashi under the supervision of Kenichi Sasaki, their form master. The night before she had pledged in her diary: 'Tomorrow, I am going to clear away some houses that have been demolished. I will work hard and do my best.' 'The End', she wrote, as children everywhere like to do, at the end of this chapter. She and her classmates assembled at Koamicho Tram Station, recited the Imperial Rescript to soldiers and sailors, removed their school uniforms, laid them with their lunchboxes under a grove of trees, and put on their *monpe*. Forming two lines, they began passing tiles and bricks to one another, relay-style.

At 8.15am another B-29 flew high over Hiroshima; its silver underbelly flashing in the morning sun. Some looked at it incuriously; most got on with their work. No air raid sounded; there were no warnings, none of the standard flyers, which the US Air Force usually dropped to warn a targeted city of its imminent destruction and the people to evacuate (the Potsdam Declaration did not specify which cities would experience 'prompt and utter destruction'). Some Hiroshimans believed they saw Little Boy falling from the *Enola Gay*, others saw the testing instruments dropped by parachute from *The Great Artiste*.

At 1900 feet (580 metres) above the heart of the city the bomb's detonation sequence ended: the gun mechanism fired the bullet that slammed the two halves of the uranium core together, creating critical mass. The fast chain reaction emitted a soundless flash, like a huge magnesium flare, that produced, for a millionth of a second, conditions comparable to the surface of the sun; the temperature inside the bell-shaped fireball reached one million degrees Celsius. Neutron and gamma rays – the most penetrable forms – saturated the city centre.

The weapon exploded directly above Shima Hospital, in the centre of Hiroshima, instantly killing all patients, doctors and nurses. The heatwave charred every living thing within a 500-metre radius, and scorched uncovered skin at 2 kilometres. Those who saw the flash within this circle did not live to experience their blindness. The ground temperature ranged briefly from 3000 to 4000 degrees Celsius; iron melts at 1535 degrees Celsius. Water in tanks and ponds boiled. Leaves in distant parks turned crinkly brown, then to ash; tree trunks exploded. Tiles melted within 1100 metres – kilns achieve that effect at 1650 degrees Celsius.

Shock and blast waves rippled over the city, punched the innards out of buildings and homes, and bore the detritus on the nuclear wind. Brick buildings two storeys and higher were completely destroyed within a 1.6-kilometre radius, and concrete buildings severely damaged; all wooden structures collapsed within 2.3 kilometres of the detonation point. Within the immediate vicinity, the blast pressure was 32 tonnes per square metre; and the wind speed 440 kilometres per second; 3 kilometres away, these fell to 1.2 tonnes and 30 metres per second.

Tens of thousands of people within a 2-kilometre radius were burned, decapitated, disembowelled, crushed and irradiated. The sudden drop in air pressure blew their eyes from the sockets and ruptured their eardrums; the shock wave cleaved their bodies apart. They were the lucky ones.

Honkawa National Elementary School – the school nearest the detonation (350 metres to the west) – was completely gutted; the principal, all 10 teachers and 400 children, killed immediately (two, behind a wall, survived). Thermal rays charred most of the victims as they played games in the playground. Fukuromachi National

Elementary School, a few hundred metres south, was similarly devastated. Elsewhere in the centre of the city thousands of government officials, soldiers, shop-owners, student labourers and volunteer workers were burned beyond recognition.

Child telephonists performing morning exercises on the roof of the Hiroshima Telephone Exchange were killed instantly; members of the earlier shift were electrocuted at their desks. A late arrival, Taeko Nakamae, 14, jumped out of a window and landed on melted glass. She ran to a river. The bridge was on fire so she tried to swim. She fainted. Someone dragged her to the bank, where she lay among the dead and wounded.

The troops on the castle's West Parade Ground were so badly scorched, 'it was hard to tell front from back'; only their teeth were visible. People died where they stood or sat: on trams, on park benches, at their desks. One man's body was found melted to his bicycle against a bridge railing. Streetcars packed with commuters were thrown off their tracks and consumed by flames. Far away, window panes shattered and children's ears and noses bled. In the village of Hiraki, six-year-old Mitsue Hiraki left the farmhouse to go to the toilet in the barn: 'I saw the cloud 40 kilometres away,' she told me more than 60 years later. Cadets at the Etajima Naval Academy, 110 kilometres away, claim to have felt the impact.

The heatwave instantly dehydrated those exposed to it, and extreme thirst overrode the pain of their wounds: '*Mizu! Mizu!* [Water, Water]' they cried. The very source of life seemed to have become a form of poison. 'You'll die if you drink water,' someone warned the crowd at Hijiyama Bridge. They drank; they died. But water was not the cause, it was simply inadequate. The victims needed a comprehensive rehydration that would replace the electrolytes and proteins lost. None was available and the people thought water was killing them. A girl screamed, 'The faster I die the better,' and jumped into the river. The rivers, the ponds, the tanks seemed deadly oases. 'Wherever a puddle of water had collected from burst water pipes, people gathered like ants around a honey pot.' They slaked their thirst at the rock pools of Asano Park and died amid the gardens, bamboo groves and maple trees. Hundreds perished in the swimming pool of the First Prefectural Middle School.

One youngster left his parents' side to join other children in shredded school uniforms leaning over a tank. He ran away in tears: they were corpses, draped over the rim like harlequins, their lifeless faces reflected in the putrid liquid below. The people sought water to escape the heat, another lethal delusion. 'I saw fire reservoirs filled to the brim with dead people who looked as though they had been boiled alive,' said Dr Michihiko Hachiya, who suffered severe burns on his way to work at Hiroshima Communications Hospital.

•

That morning Seiko Ikeda, aged 12, travelled to the city exhausted; the night before her family had visited her brother in Yamaguchi, who had recently been conscripted, but air raids had delayed their return home, and she had slept for only two hours. Her father insisted that she go to work – her first day as a war labourer. 'To be mobilised,' he had said, 'is all the more reason that you should go to the city, since it is for the country.' So Seiko and her mother put on their work clothes and set off for the train to Hiroshima; her mother carried a first-aid kit. She dropped Seiko at school and boarded a tram.

The tram pitched into the sky; Seiko's mother suffered severe wounds. Seiko felt her body rise in the air; then she lost consciousness. She woke in darkness, flecked with red and purple light and lit with surges of yellow fire. The atmosphere was primeval, livid, hissing; dust and ash fluttered over the dull, shifting silence: had she been deafened, she wondered. Why were there no sounds? Why is it already night? She looked up: a mountainous reddish-black cloud obscured the sun.

Strange dirty people – urchins – shuffled about the rubble. Their hair stood on end, smouldering; their clothes were partly burned off and their bodies, bloody. Those able to walk carried their arms out in front, palms hanging down in a sort of unco-ordinated prayer. Their skin seemed to fall from their fingertips, 'like gloves turned inside out'. To her horror, Seiko realised they were her school friends and teachers.

She examined her body: burnt, bleeding and covered in rags that fell away with her skin. She held her arms 'above my heart', like the others, to avoid rubbing her scalded flesh: 'It was the least painful position.' The flash had exfoliated her face, the shocking appearance of which she was unaware. She tried to find her mother. The ground crawled with the dying: 'We stepped over people charred totally black. We felt pity for them at first, but the more bodies we saw the more we treated them as objects.' They came to the Kyobashi River. The 8am high tide lapped the banks. 'I jumped into the river, as many did ... Some just sank and didn't come up.' The rivers were so congested with human forms, 'you couldn't even see the water'. Seiko got across the river – 'I just wanted to go home' – and lay briefly among strange, distempered creatures, moaning and crying. Voices cried out from beneath the flattened wooden houses that crammed the bank. She covered her ears, repeating, 'sorry, sorry', as she staggered past them: 'That was all I could do.'

With the crowds on the eastern bank Seiko moved up Hijiyama hill, past hundreds of dugouts, from which the city had hoped to defend itself from invasion, now receptacles for the damned, whose swollen faces peered out like raw pumpkins. She crawled to the summit and looked back over the city. Spot fires were merging into a bigger fire. She fled the scene in tears and stumbled down the other side of Hijiyama, where she came to a highway covered in debris and people. 'There were so many things all over the road you couldn't walk. I was crawling on my hands and knees to make my way.' Thousands thronged the street, seeking escape. A strange woman tore up a curtain and wore it around her, reminding Seiko of her own nakedness. 'I remember being so embarrassed.'

The lines of walking – and crawling – wounded moved along the ash-encrusted road, stepping over or around charred remains and the carcasses of trucks, cars, sidecars, horses, bicycles and handcarts, and the riot of fallen power lines that 'tangled everywhere like giant cobwebs'. Among them was Toyofumi Ogura, a young historian at Hiroshima University. He witnessed this 'swarm of people, all of them burnt or injured … teeming up the long, wide roadway. They looked like fragments or scraps of living organisms, motivated not by any personal desire to seek refuge but by some vast, tenacious "life force" that transcended individual will.' A military truck packed with mobilised male students negotiated the wreckage and stopped. They were heading for Kaitaichi, near Seno, Seiko's home village, and offered her a ride.

•

Iwao Nakanishi, 15, worked at the red-brick army supply depot, situated 2.7 kilometres from the blast. A little before 8am he and his class received an order to go to the warehouse in the centre of town; but the truck used to transport them had engine trouble, saving their lives. The blast threw them in the air. A wall protected Iwao, who was sent into town on a scouting mission.

The boy stopped at Miyuki Bridge – around the time a local photographer, Yoshito Matsushige, captured a crowd of people there, all barefoot as their shoes had stuck to the asphalt. In the picture a policeman gives cooking oil to badly burnt citizens; a half-naked mother cradles her child, either dead or in shock: 'She was running around crying to her child, "Please open your eyes,"' Yoshito recalled. He took four more photos and then tearfully gave up: the sight of the charred corpses on a burnt-out tram 'was so hideous I couldn't do it'.

As Iwao approached the scene he noticed hundreds sitting near the foot of the bridge, nursing their wounds. A little boy who had lost his eyes screamed, 'Soldier-san, please help!' Iwao recoiled: 'I wasn't a soldier but I grabbed his arms and tried to help him stand up. His flesh came off and I let him go. I can never forget that ... When I look back, I regret not carrying him on my back and saving him.'

Iwao returned to the depot. The red-brick walls had partially withstood the shock wave, and its shelter drew the city's wounded, who lay all over the floor. He and other children applied vegetable oil to their burns with light brushes, and helped to roll those unable to move: 'They screamed when the brushes touched the glass pieces lodged in their bodies.' Iwao gave water to those who pleaded for it: '... they died instantly. I was told not to give water, but they begged for it. I was just a kid and did not know what to do.'

·

Tomiko Nakamura, 13, was one of the few survivors of 320 girls from Shintaku High School, 1.6 kilometres directly south of the blast, most of whom perished in the playground. She remembers the day with terrible clarity. The flash 'felt like the sun had fallen out of the sky and landed right in front of us'. Struck unconscious, she came around in complete darkness: 'There were flashes of light coming from everywhere, in all directions, like so many sunrises.'

She examined her body. Glass shards covered her scalp; her skin 'rolled off my legs like stockings'; bone poked through the skin of one knee. Her shirt and trousers were burnt and stuck to her flesh. 'Once I realised the state I was in, I felt very sick, so I sat down on the ground. I sat there for a while, but I could see the flames coming closer.'

Picking over the rubble, she heard voices crying from beneath the timbers, yelling 'Help! Help!' and 'It's so hot!' 'I just kept walking.' She reached Tsurumi Bridge where adults were jumping into the river. She passed 'people with black and red faces ... I couldn't tell whether they were men or women.' Nobody helped her. She tried to climb Hijiyama hill but 'the ground was covered in wounded ... there was literally nowhere to take a step.' Some held their inner organs in their hands, staring at them with appalled curiosity.

A military truck took her to Fuchucho, a village in Aki Province, where she was reunited with five friends and a teacher. They lay down on sheets in the school assembly hall. A friend beside her started talking gibberish, then went quiet: 'When I looked at her, I realised that she had died.' The body was placed on a pile of child corpses in the middle of the assembly hall. There were no doctors or nurses and no medicines. Volunteers ladled 'oil from a bucket' onto burns with a wooden spoon. The next day, one by one, the rest of Tomiko's friends passed away: 'As each one died, I thought that I would be next. I didn't stop crying the whole time.'

•

Brave little groups of teachers and students stayed in the city to help. Yoshiko Kajimoto and her fellow workers – 100 14- to 16-year-olds from Yasuda School – were making parts for aeroplane propellers in a factory in Misasa, 2.3 kilometres from the blast. She and nine other survivors tried to get back into the partly destroyed factory to rescue their friends, who were shouting beneath fallen walls and beams. Nobody could find the first-aid kits, so the children tore off their blouses and headbands and wrapped them around their friends' wounds: 'Those headbands that everybody wore really saved a lot of people!' she would recall.

Their teacher rushed to find help. 'Don't leave this place,' he shouted. 'Stay here!' As the children waited, crowds of the wounded approached: 'They would sit down and just die before us.' The teacher returned to find a pile of corpses in front of his students. He had three stretchers.

The children went back and forth, dragging out their classmates, stepping over the dead: 'You just couldn't avoid stepping on them … It was just horrible. Some didn't have heads … It was just like hell. It is not something that young children should ever see.' They worked for hours until the encroaching fires ignited nearby houses: 'We had to carry them as fast as we could. My leg was terribly painful but of course I couldn't stop ...' They dragged their few surviving friends away from the flames to safety.

The cloud turned dirt-brown and hung low over the city. Moisture condensed on the rising ash and dust and fell as oily, sooty droplets. The 'black rain' pelted the northwest areas. No one understood the dangers of exposure to this greasy, radioactive slime. 'It's gasoline! The Americans are dumping gasoline on us!' cried one witness. 'We thought the Americans were trying to burn us,' said the military policeman Takashi Morita – which was true of any other major Japanese city. Katsuzo Oda tasted the droplets: 'It certainly didn't seem to be gasoline.'

Yoshiko walked home, to the suburb of Koi, 3 kilometres from the centre. On the way she bumped into her father, who had seen a wooden sign saying that the students from Yasuda School had escaped north: 'I was so happy when I saw him coming towards me!' She remembers him saying, 'You've done a great job surviving – you made it!'

Many teachers and students showed similar courage. The telephonist Taeko's teacher – a 22-year-old woman of astonishing resilience – discovered her lying on a river bank and placed her in the care of soldiers of the Akatsuki Corps, a relief unit, who took her to safety. The teacher continued searching the city for the rest of the missing class. Elsewhere, near the Seigan temple, a teacher threw himself over four children in a vain attempt to protect them from the approaching flames. Another emerged from a collapsed building at Takeya School, 'in tatters, looking like a black ghost', yet her first act was to find her class and tell them 'to run', remembers one, Kenji Kitagawa. She then lay down among 30 teachers at Takeya who died trying to protect the children. The Japanese poet Shinoe Shoda later visited the wreckage of the school and found the little bones of children's skeletons clinging to adult ones in the playground.

Most people fled, however, without helping others; there was little they could do. The adult mind saw the hopelessness of a situation that eluded the young. For a day, at least, the bomb defeated the Japanese spirit.

·

The soldiers and military police were as helpless as the population. The bomb killed 3243 troops on the parade ground, out of an estimated 10,000 to 20,000 in and around the city (the total varies depending on the time and source). Many soldiers were outside the city or at the port. Military policeman Takashi Morita, then 21 – whose five older siblings were at the time interned in America – led a small group back into the city despite suffering severe burns to his head. At the Aioi Bridge they came upon a badly wounded member of the Korean Royal Family: 'We could not leave such a person unattended. We saw a small boat. We told all the bomb victims on it to hop off and six of us carried him on board.' They made little progress against the incoming tide and soon transferred the Korean prince to a larger vessel. 'A crowd of mobilised students were crying for help, asking why we couldn't take any more, but we just couldn't … in those days a person of such high rank could not have shared the space with other people. I feel consumed with guilt when I think back on it. Surely we could have taken some of these children on board.' The Korean prince died in Hiroshima Port the next morning.

Two scenes remain scored into Morita's memory. One was the sight of a woman giving birth in the rubble: 'I found a girl who brought hot water and helped deliver the baby – the baby and mother survived.' The other was the image of an American prisoner bound to the base of a burned tree, rocking back and forth. He was one of some 14 American airmen who ditched near Hiroshima in late July, some of whom were interned in the castle. Later that day, according to Morita, Japanese soldiers dumped the pilot's near-dead body at the Aioi

Bridge, where people 'started throwing stones at him, screaming, "You Americans! You did this to us!" Eventually they killed him.' (About a month after the war, Morita went back to work with the military police. One day a jeep full of American servicemen arrived to investigate the alleged stoning to death of a US soldier by Japanese military personnel. 'The man who had been primarily responsible for dragging the American to the bridge turned white with fear when he heard this,' Morita said. 'But he didn't give away a thing, and neither did any of us, so nothing happened to him. The US soldier's death was just recorded as a war casualty.')

•

Takeshi Inokuchi, 13, had shared the morning ferry journey from Miyajima Island, with his girlfriend, Naoko, 12. They had travelled hand in hand. 'She was very pretty; I would call her my first love.' She worked at a factory nearer the city centre. He was taking the roll call in the Toyo seafood cannery in Tenmacho when the bomb exploded 1.3 kilometres to the east. The shock wave destroyed his building and wounded or killed all 300 workers, aged between 12 and 14. Takeshi recovered consciousness, escaped and tumbled down a river bank. Despite his wounds, he thought only of his girlfriend. They had spent the previous evening watching fireflies on Miyajima: 'We wouldn't have kissed,' he recalls, 'but we hugged. Just touching each other was fantastic.'

Takeshi tried to reach her factory but the fires drove him back. In despair, he stumbled towards the movie theatre, the Shuraku, in the Koi district, where they had met: 'It was one of the few places where boys and girls could be together,' he said; they had seen Kenji Miyazawa's *Ame No Hi Mo, Kaze No Hi Mo* (On Rainy Days and on Windy Days Too) there. Today, the theatre was on fire and the city ablaze: Takeshi escaped by sticking to the rail tracks: all the wooden bridges on the Tenma River had burned, or collapsed, but the iron railway bridge had withstood the shock wave.

The diarist Yoko Moriwaki's First Year demolition group bore the full force of the blast. All but a dozen or so were killed instantly. Her teacher, Kenichi, though badly wounded, tried to save the survivors; his last words were: 'I am done for. Everyone head for the Koi evacuation centre!' and then he collapsed into the flames.

Yoko was yet alive; she suffered severe burns to her back, legs and hands. Her face was unhurt, because she had been looking away from the flash. She found the strength to crawl. She crawled for hours, west, in the direction of home across a railway bridge high above a river – something she would never have ordinarily done. A military vehicle picked her up and took her to Kanon village, where the local school served as a relief centre.

There, a housewife called Hatsue Ueda fanned her burns and lay a *yukata*, a summer kimono, over the child's naked body. All the time Yoko pleaded for her mother. With no medical training or medicines, Ueda tried to comfort the girl: 'I did as she asked and held her hand in mine.' Ueda applied oil and dripped green tea from the ends of chopsticks into the child's mouth, which opened and closed like a small bird's. She never let go of Yoko's hand.

A doctor arrived and managed to get a call through to Yoko's home, via the village's central phone. The girl's eyes remained fixed on the clock; she whispered for her mother. 'I don't know how many times she asked me, "Isn't Mother here yet?",' Ueda later wrote to a friend. 'I comforted her by saying, "You'll see her soon. Be strong and stay with me, OK?"'

So severe were Yoko's burns, 'it was obvious she couldn't be saved'. The volunteer and the doctor did everything possible to ease the child's final agonies. They stroked her back, drummed lightly on her chest and fanned her burns. She passed away as the doctor took her pulse, still wondering when her mother would arrive. 'I simply have no words to express how sad it was,' Ueda wrote. 'That poor, poor, poor girl.'

Within a few hours, Yoko's mother reached the Kanon National School. At the sight of Yoko's body the poor woman screamed and collapsed, sobbing by the mat on which lay her daughter's remains. Her grief would render her insensible to those around her: 'My mother never smiled again,' her son would later say.

•

Witnesses remember the walking wounded filing out of the city in long, silent lines, each advancing carefully to avoid brushing their burns. Their arms hung forward, praying-mantis-like; their heads were bowed; their eyes fixed on something ahead. Families staggered along, supporting their worst-off members. When asked from where they had come, they pointed back to the city and said, 'That way;' when asked where they were going, they pointed ahead, and said, 'This way.' 'Not a sound came from them,' one local doctor observed. 'They seemed to have given up. The pity they engendered is quite beyond expression …'

There was no mass panic. The people had had no warning; they were not *prepared* to panic. Their shock turned to stupefaction, then to urgent physical needs: '*Mizu! Mizu!* [Water! Water!'] It hurts! It hurts!' – not loudly or hysterically, rather a soft and insistent plea: 'A hum of voices,' remembered Yoshiko Kajimoto, who had tried to help her friends in the propeller factory.

Behavioural traits and cultural idiosyncrasies of the world that had ceased strangely lingered. So sudden had been their transformation from ordinary workers to hideously disfigured survivors of the world's first nuclear attack, their comprehension of what had happened lagged their new circumstances. Many clung to remnants of normality, the adherence to orders and schedules and small courtesies. Echoes of deference and duty persisted: soldiers continued to observe form and rank; people died politely in queues at relief centres. Cries from the wreckage were pitiably courteous: '*Tasukete kure!* [Help, if you please!']'

The sick and wounded were ashamed: 'embarrassed' and 'humiliated' at being seen in a state of such wretchedness; their smell, their nakedness, their appearance affronted the fastidious social forms that persisted in their minds. Soldiers gave women rags to wrap around their naked bodies, a nod at decorum in a country that frowned on displays of female flesh. Dr Michihiko Hachiya was 'disturbed to realise that modesty had deserted me', and requested a towel from a soldier to hide his nakedness. Tamiki Hara 'shuddered rather than felt pity' at the sight of two monstrous victims of indeterminate sex squatting on stone steps beside the river who pleaded, 'That mattress over by those trees is ours. Would you be *good enough* to bring it over here?'

Hysteria was individual, the manifestation of incomprehensible private grief: the sudden sight of the charred remains of a child, who a moment earlier had been smiling on their backs, or by their side, induced despair verging on madness in mothers, who wandered around in circles holding up their dead offspring to the sky. Or they clung fast to the inert bundles as if the very possession would somehow resurrect the child's life. Some ran shrieking through the rubble, careless of their wounds: 'A mother, her child clasped to her bosom, ran by us, screaming as if she had gone mad. She was stark naked and burnt and swollen all over. Her baby had already died.' Toyofumi Ogura, the young Hiroshiman historian, heard her shrill cries before he saw the woman, barefoot, her hair dishevelled, wearing only her *monpe* trousers. Others quietly, selflessly, wrapped up their deceased children and together passed away. Thousands of the wounded slid Ophelia-like into the tributaries of the Ota and drowned. Those who couldn't walk sat along the river banks. Crowds of children gathered in the sandy spits where they might once have played ball and cried out for their parents who never came. Such scenes recurred all over Hiroshima that day.

•

From the surrounding hills the escapees looked back at the 'jellyfish cloud', billowing out to east and west, emitting a fierce light in 'ever-changing shades' of red, purple, blue and green. Its head loomed over the city 'as though waiting to pounce': 'Hey, you monster of a cloud!' screamed one woman, her arms shooing it away: 'Go away! We're civilians! Do you hear – go away!'

On the ground the scattered fires merged into a firestorm, which, whipped up by the winds, consumed what was left of the city. The remains of wood-and-paper homes were powerful kindling to the conflagration, which devoured the spot fires and leaped across firebreaks – as it had done in other cities – hurling out long, burning ropes of flame that seemed to haul the furnace forward. The firestorm blacked out the sun for the second time that day and burned for about four hours. Thousands still trapped in the rubble expired in the flames.

•

The emergency services were either destroyed or hopelessly overwhelmed. More than 80 per cent of the city's doctors and nurses were killed in the explosion, their hospitals levelled or severely damaged. There were few medicines or painkillers. The shock wave tore through the Red Cross Hospital: ceilings and partitions collapsed; windows blew in, showering everyone with glass; instruments were smashed and scattered; patients ran about screaming. One doctor there – Terufumi Sasaki – survived, with minor wounds; the rest were killed or disabled. The fire and police stations were ruined; the water pumps destroyed or ruptured; the mortuary services, such as they existed, helpless.

The Japanese gradually that day responded with a sense of care and determination that belied their paltry resources. By early afternoon, police and volunteers had established 'relief centres' in tram stations and the ruins of schools and hospitals, or open areas on the city fringes cleared of debris, laid with *tatami* mats. What remained of the Asano library became a makeshift morgue. Thousands of children lay in school playgrounds, classrooms and gymnasiums, covered in rough straw mats that were raised in the glow of lanterns by traumatised parents anxious to know if the unidentifiable remains underneath had once been their son or daughter.

One photograph shows a wounded police officer in a tram station writing out Disaster Victim Certificates to patient queues awaiting relief. Those unable to move were laid in trucks and taken to emergency field clinics. Scenes of stupefied people sitting or lying around were commonplace. A photo of a relief centre at one school shows shocked patients with no nurses or doctors in sight and the only form of medication a bucket of cooking oil.

On his way to the Hiroshima Communications Hospital, about a mile from the blast, Dr Michihiko Hachiya tripped over a human head. 'Excuse me! Excuse me please!' the doctor cried hysterically. He was half-naked and bleeding. His home had collapsed. 'An overpowering thirst seized me,' and he begged his wife, Yaeko, for water. She went to find help. He lay down on a road: 'With my wife gone a feeling of dreadful loneliness overcame me'. He regained his strength and together they reached the hospital, partly destroyed and now filling with wounded. He was too weak to work, and his surviving staff laid him on a stretcher in the garden as they tended to others. Then the fires came and panic ensued: the violent updraughts hurled the zinc roofing into the sky; the windows became squares of hissing flame. They attempted to evacuate the patients. His friends moved him to safety; the walking wounded tried to flee. Many perished in the flames.

A functioning water hose saved part of the hospital, and the fires died in the afternoon. The medical team returned to the wreckage. Unable to walk, Dr Hachiya sat on his stretcher and watched. Fresh crowds arrived, begging for treatment. 'They came as an avalanche and overran the hospital.' One was Kohji Hosokawa, brother of Yoko, the young diarist: 'The doctors and nurses were all injured,' he said, 'the medicines and equipment were destroyed and scattered around. The hospital was not functioning at all.'

•

The military gave priority to its own, clearing bodies and tending to the wounded that littered the West Parade Ground. Within hours public fury forced the army to help ordinary people; the piles of corpses demanded immediate removal to prevent the spread of disease.

One relief worker, Sadaichi Teramura, had been evacuated from eastern New Guinea and now served with a Construction Battalion in Murotsu, 70 kilometres from Hiroshima. Ordered by telegram to provide 'emergency relief' to the city, where a 'previously unknown type of bomb had fallen that morning', he set out by fast boat for Ujina port. The experience, he later wrote, rendered him numb to the memory of numerous battlefields. When he set foot on the jetty that night he heard something in the darkness humming; he stopped and listened: 'The indescribable pathos of moaning assaulted my ears,' he wrote. 'I realized to my horror that all around me were thousands of wounded people lying bereft of hope on the cold concrete.' These were the triage rejects, 'living corpses … with only the peace of death before them'. He stepped over them.

The morning after the bomb Teramura took a boat upstream. His battalion's job was to recover the bodies from the water. When the tide ebbed the current dragged rubbish, ash and corpses towards the Inland Sea. Teramura encountered human remains as far south as Hiroshima Bay, about 10 kilometres from the city. Upstream, hundreds, then thousands, of bodies of indistinguishable sex and age jumbled around the boat's gunwale. He and his men hauled them out with ropes, taking five or six ashore at a time for cremation. The soldiers 'preserved a deep silence in the face of the brutality around them', he recalled. Soon the smoke of funeral pyres 'rose from all points of the devastated land'.

Well-drilled volunteers of the Women's Defence Corps and neighbourhood associations in the suburbs and surrounding rural areas moved fearlessly into the city. Dispensaries were established in shrines and temples. In a typical scene at the Toshogu Shrine, a policeman asked the wounded their names and – despite the fact that their homes no longer existed – their addresses, and issued them with identification slips. They sat in rows in the scorching sun. On the second day, food trucks arrived, distributing sweet potatoes, near Hiroshima Station.

At the train station and field clinics volunteers performed immediate triage; the hopeless cases were set aside; at times the dead and living were indistinguishable. Children sat tearfully by their dead or dying parents. One little girl of five or six refused to leave the side of her dreadfully sick mother and had to be dragged away, wailing, 'But she's alive!' This appalling scene was commonplace. Toyofumi Ogura witnessed a little boy, barely four years old, trot up to one of the 'living corpses', probably his mother, and pour water into the woman's open mouth, then run off, happily enough, to refill his watering can. The woman lay drenched in blood with one arm torn off; she was already dead. 'I didn't want to be there when the boy came back,' Ogura wrote, 'so I averted my gaze and started walking.'

Trucks and trains transported thousands of survivors to the rural villages and temples around the city. Hiromi Hasai, a school student who had been mobilised to work in a factory in the village of Hatsukaichi, about 15 kilometres from Hiroshima, volunteered to help lift the casualties off the trucks: 'They were just silent. But there was nowhere to lay them down, no beds left, so we laid them on the ground.' Tetsuo Miyata, a university student, witnessed the first train load arrive at Kakogawa National Elementary School, where he was an assistant teacher. They disembarked and filed into the relief centre: 'It was an eerily silent procession, a scene of wretched humanity, people who … were trying somehow to carry on living: a man limping along, wounds all over his body and carrying on his back a child whose whole face was so blistered from burns that it could not see … They had no objective; all they could think of was to automatically keep following the person in front … Even if one of those at the front had lost their way, the others would probably just have followed along without a word …'

These were the relatively fortunate. The relief workers shuddered to think of the deranged souls left behind in Hiroshima: a man in rags cycled around and around with what appeared to be a piece of charcoal fastened to his bicycle: it was the remains of his child. 'The man himself seemed crazed.' Or the utterly wretched, who spent the night in the Yasu Shrine in Gion and in the caves and dugouts on Hijiyama hill 'groaning with pain, their bodies covered with maggots, and dying in delirium,' observed a witness. And the tribe of enraged soldiers camped under a railway bridge near Hiroshima Station who pranced madly around with Japanese flags shouting the battle cry, '*Banzai!*' And there were the animals, the fallen birds charred black; the burned dogs, sniffing the debris; and the blinded cavalry horses that galloped, or wandered, or walked lamely about the city.

•

From the hills of Eba, from Hijiyama hill, the survivors looked down on the first night of the nuclear age; the bowl of Hiroshima held the city's fading embers like the crater of an active volcano. In the sky the stem of the mushroom cloud lingered, but the head had diffused, noticed young Iwao Nakanishi who, joyfully reunited with his family, headed into the hills.

The morning was the dawn of another world, witnesses remember; the first day of a planet that had changed utterly. The sun shone murkily red through the dissipating vapour; a sort of white soot caked the roads and debris. Flight navigator Takehiko Ena and his two-man kamikaze crew – who had been forced to ditch their plane off Kyushu – reached Itsukaichi Station near Hiroshima. They were returning to their air base in Ibaraki Prefecture to join another suicide mission: 'We were expecting to defend the invasion of the mainland,' Ena said. Survivors streaming from the wasteland told the airmen that 'a special kind of huge bomb' had destroyed Hiroshima; nobody knew how many were dead.

The kamikazes walked several kilometres to a hill near Koi that offered a complete view of the place where the city had been: 'A vast expanse of debris and burnt fields,' Ena recalled, 'extended as far as the eye could see; there was nothing left; the dome was the only building we could see.' They descended into the devastated city, following the tram tracks to Aioi Bridge, which had buckled but withstood the blast.

Army and navy relief forces were erecting tents around the city. Antiseptics were their only medication; painkillers had not yet arrived. Terrible screams issued from the tents: 'I didn't have the courage to look inside,' said Ena. For a moment he imagined that he had succeeded in his suicide mission, and had woken in some kind of hell: 'The acrid smell of burnt living things wafted all around us,' Ena said. 'Everywhere we went, corpses lay on the ground in heaps … It was a hideous sight. Severe burn victims were roaming the streets in groups, dragging their feet like ghosts. The relief effort had barely begun. As one soldier in transit, there was nothing I could do to help in this hell on earth, so I simply clasped both hands together tightly in front of my face and made my way out of Hiroshima …'

•

People searching for their loved ones were trying to enter the city as the kamikazes left it. Iwao Nakanishi's family returned from the hills to search for his grandmother. They found her in a shelter in Hijiyama hill but a soldier barred their entry. The occupants were suffering from a strange disease, the soldier said, like dysentery. The symptoms were fever and diarrhoea, then death, and he warned of an epidemic. It was something Dr Hachiya at the Communications Hospital had observed – patients had a strange sickness that induced violent nausea and other symptoms. It mystified him and he, too, had initially confused it with dysentery. Iwao's little brother had in fact died of dysentery, so the family heeded the warning and left the city. It saved them from overexposure to residual radiation, which poisoned many families searching among the ruins.

One family then trying to enter the city was Shizue Hiraki and her three children. Shizue's father-in-law, Zenchiki – concerned for his three daughters in Hiroshima – had ordered her to travel from their village to Hiroshima and find them. 'Yome! [daughter-in-law]' he shouted. 'Don't come back without them!' He demanded that she take the children, as he was too old to care for them. 'He told my mother to carry my little sister, Harue, on her back,' daughter Mitsue said, while she and her brother, Hisao, aged 10, would walk.

On 8 August the Keibi line resumed running trains from the surrounding towns to Hiroshima's outskirts. Shizue and her three children walked for an hour to Mukaihara, boarded the 11am train to the station nearest Hiroshima and walked the rest of the way. They reached the threshold of an expanse of rubble and smoke, flames still flaring in the wreckage. People held masks over their faces as they entered.

Shizue dragged her mortified children forward. 'We walked and walked,' Mitsue recalled. The ground heat penetrated the straw of the children's *waraji* (straw) sandals and 'made the soles of our feet very hot', she said. Here and there people were 'digging out the burned bones of loved ones and putting them into urns'.

The family tried to find a familiar community hall; it no longer existed. They walked to the Tenmachi area, past the wreckage of the 400-year-old castle, which had disappeared in seconds. They turned south, passed the ashes of Shima Hospital, above which the bomb had detonated – the 'hypocentre' as it would be called – and crossed the Motoyasu Bridge to Nakamichi, the centre of town. The dome of the

Industrial Hall – the future symbol of Hiroshima – was still smouldering.

'No one yet understood what had happened,' said Mitsue. 'If we had known the horror of the atomic bomb we would have had masks, protective clothing, but we just thought it was a conventional bomb. So we were unprotected. We just moved around as we were.'

Shizue tried to shield her children's eyes; but it was impossible to avoid the survivors peering out of ruins, crawling forward and 'begging us to help them'; or the corpses that crammed the rivers. 'My mother just held me tightly by the hand and dragged me along,' said Mitsue. Her elder brother walked alone.

Mitsue spoke to me with profound tenderness of her mother; occasionally, her voice rose in anger at her grandfather, but her tears were tears of love for her mother: 'My mother didn't scold us for crying; she just told us to hurry and keep walking. I don't ever remember seeing her angry. She always tried to comfort us. But she was just afraid, and desperate to find her sisters-in-law.'

Without signposts or familiar landmarks, Shizue lost her way in the rubble. In despair, she resorted to searching bodies in the vicinity of the family's home: 'My mother went around lifting the faces of dead people to try to see whether or not they were my aunties,' Mitsue said. 'I felt desperately sorry for her. If she didn't find them my grandfather would be furious. So we searched. My mother kept saying that she wanted to find at least one by sunset.'

By evening, the family were hungry and exhausted; Shizue's children's feet, blistered. They had a few rice balls to eat and drank from a burst water main. They abandoned the search, curled up on the ground in the Danbara neighbourhood and slept. 'We slept out in the open, amongst dead people … We clung to my mother,' Mitsue said. 'People were crying out around us.' Before dawn the sound of a Buddhist incantation to the dead rose over eastern Hiroshima. The family returned to the village in a train full of wounded and empty-handed relatives.

Hiraki Zenchiki flew into a rage when they arrived home. 'He hated my mother for this,' Mitsue recalled tearfully. In coming days, rescuers retrieved two of his daughters, who were badly hurt. The rest of the Hiraki family died in the city. From that day, Shizue Hiraki was too weak to work – despite her father-in-law's orders. Nobody understood her medical condition; the people simply called it the 'A-bomb disease'.

CHAPTER 18

INVASION

Is the Kwantung Army that weak?
Then the game is up.

Prime Minister Kantaro Suzuki, on hearing of the Soviet Union's invasion of Manchuria
and crushing of the Imperial Army

THE DAY AFTER THE DESTRUCTION of Hiroshima, American aircraft dropped millions of leaflets on Japanese cities, and the US-operated Radio Saipan broadcast the same message at regular intervals:

'To the Japanese people,' the flyers said:

... We are in possession of the most destructive explosive ever devised by man. A single one of our newly developed atomic bombs is actually the equivalent in explosive power to what 2000 of our giant B-29s can carry on a single mission. This awful fact is one for you to ponder and we solemnly assure you it is grimly accurate.

We have just begun to use this weapon against your homeland. If you still have any doubt make inquiry as to what happened in Hiroshima when just one atomic bomb fell on that city.

Before using this bomb to destroy every resource of the military by which they are prolonging this useless war, we ask that you petition the Emperor to end the war ...

You should take steps now to cease military resistance.
Otherwise we shall resolutely employ this bomb and all our other
superior weapons to promptly and forcefully end the war.
EVACUATE YOUR CITIES

•

Tokyo's leaders refused to believe that America had dropped an atomic
bomb on Hiroshima, and suppressed all media reference to the claim.
Waves of B-29s had struck the city, as well as many other cities, ran the
official Japanese line on the night of 6 August. This squared with the
experience of millions of people: on the day before the bomb, American
leaflets warned 12 mid-size Japanese cities of their imminent destruction
(Hiroshima was not among them); and on 6 August, dozens of B-29s
flew incendiary raids against four of LeMay's designated 'death list'
cities, including the already battered Tokyo, after which waves of
Mustang fighters strafed the civilian survivors.

Little more than terrible rumours issued from Hiroshima.
Communication lines were broken. The local office of the Domei
News Agency was destroyed; the city's communications officials dead.
An employee of the Hiroshima post office returned to the smoking
ruin on the 7th but was unable to transmit telegrams (the service
resumed two days later from an alternative office on the outskirts).

The armed forces, however, were aware of the devastation: 15
minutes after the explosion, the Kure Naval Depot informed Tokyo of
the enormously destructive weapon. From then on, military sources
dispatched 'spot reports' of the bomb's effects, which the military
leaders did not share with the civilian members of cabinet. A garbled
message on 7 August to all Japanese naval commands stated: 'Although
we are [investigating] atomic bomb attacks by the enemy, you are to
make every effort to [minimise] damage, as follows: (1) …You will
have fighters intercept [enemy aircraft delivering further bombs];
(2) …You are to endeavour to [shoot the bomb down].'

Civilian ignorance at the highest level of government persisted until the morning after. At 1.30am, 7 August, Japanese time, a phone call woke the director of Domei News Agency in Tokyo. The caller was an employee with urgent news: American radio had reported that an 'atomic bomb' had destroyed Hiroshima, the caller said. 'Since I didn't know how terrible the atomic bomb was,' the Domei director later wrote, 'I felt I was shaken out of bed for a trifling matter.' He got up and relayed the message to Foreign Minister Shigenori Togo and Chief Cabinet Secretary Hisatsune Sakomizu. 'But neither of them knew anything about the atomic bomb. The military knew it, but suppressed the fact: news of an "atomic bomb" should not reach the public.' The military leaders had earlier sent a propaganda report to Tokyo stating merely that 'a new bomb' had struck Hiroshima, and the public need not worry, it said, so long as they 'covered themselves with white cloth'.

Later in the morning Togo and Prime Minister Suzuki studied Truman's broadcast of 6 August: '... Unless Japan is willing to surrender we will drop bombs in other places,' was how a Japanese interpreter rendered the phrase '... If they do not accept our terms they may expect a rain of ruin from the air ...' Togo's immediate reaction was to issue an official protest against this indiscriminate attack on civilians. First, however, he sought confirmation – that the weapon was 'atomic' but military intelligence denied it: the bomb, they said, was an 'extremely powerful conventional weapon' – probably a 4-ton (3.6-tonne) bomb (they later upgraded this to 100 tons). Again, the military sources made no mention of an 'atomic' bomb. Unconvinced, Togo ordered an urgent investigation; scientists were dispatched to Hiroshima. He and Suzuki then conferred with the Emperor.

Cabinet Secretary Hisatsune Sakomizu, meanwhile, privately seized on the news of an atomic bomb as a 'chance to end the war'. Japan, he well knew, had no choice other than to surrender; his study of Japan's war economy showed that the country would be unable to fight longer than two months: the US naval blockade had crippled 'land and sea communication and essential war production'. In his eyes the atomic bomb offered Japan an honourable deliverance, the perfect opportunity to abandon the war: it was impossible to fight a nuclear-armed enemy, Sakomizu explained to close colleagues. The bomb thus exonerated the military from the blame or responsibility for surrender: 'It was not necessary to blame the military,' he later said, 'or anyone else – just the atomic bomb. It was a good excuse. Someone said the atomic bomb was the kamikaze to save Japan.' In time, Togo and Suzuki came to share this appreciation of the weapon as a face-saving expedient: 'Suzuki tried to find a chance to stop the war and the bomb gave him that chance,' Sakomizu later said.

To this end, Sakomizu asked the Cabinet Information Bureau to disseminate all known facts about the atomic bomb in the newspapers and on radio, 'in order to tell the people just how fearful it was'. The General Staff Information Office (the military censor) refused. Sakomizu struggled all day with the Chief of Military Information, who finally relented on one point: cabinet would be permitted to confirm only that the weapon had, in fact, been an atomic bomb.

The Japanese cabinet met in the underground war rooms in Tokyo that afternoon. Togo had satisfied himself that Truman was telling the truth and argued for a swift surrender in line with the Potsdam Declaration. This met with strong dissent: the war faction, led by War Minister Korechika Anami, insisted they await the results of the investigation into the weapon. The delay 'downplayed the bomb's effects' and ranked Hiroshima's emergency at the level of a conventionally bombed target. Implicit in the military's unhurried attitude was their dismissal of the city's peculiar claim: even if rumours of a nuclear attack were true that did not justify any special treatment over other firebombed cities. Togo, the chief 'dove', persevered: the weapon was atomic, he repeated; the war was hopeless; the invasion would not proceed; Japan must surrender in line with the terms issued at Potsdam.

TOP LEFT: The Big Three at Yalta. Deep division lay between the Anglo-Americans and the Russians. Stalin soon reneged on agreements reached on the future of Poland; while Churchill and Roosevelt kept the development of the atomic bomb a close secret from the Soviet Union, their then ally. (Time Life Pictures/Time & Life Pictures/ Getty Images)

TOP RIGHT: Bodies litter this street of Dresden after the Allied firebombing in February 1945. At least 100,000 people lost their lives. (Keystone/Hulton Archive/ Getty Images)

LEFT: B-29s over Tokyo in May 1945. (Mainichi Photo Bank)

TOP & ABOVE: Emperor Hirohito walking through firebombed Tokyo. The U.S. Air Force judged the first firebombing of Tokyo a great success. It destroyed the homes of 372,108 families and killed close to 100,000 people. More than 1.5 million people fled the city. (Mainichi Photo Bank)

RIGHT: A Chinese boy is beheaded—his crime was being a member of a household suspected by the Japanese of aiding Chinese guerrillas. (Bettmann/Corbis)

Revilement of Japan was rife in Allied countries. By 1945 U.S. authorities and the public knew ⁓e brutal treatment by the Japanese of local populations and POWs, as well as of Japan's ⁓gical warfare plans. The apparently demented fighting by Japanese troops, typified by the ⁓kaze suicide raids, signified to soldiers they were fighting a different kind of enemy, who ⁓fied death, while the attack on Pearl Harbor and news of the Bataan Death March left the U.S. ⁓c extremely vengeful. (Bettmann/Corbis)

⁓OM: Groves and Oppenheimer made a formidable, if unlikely, team on the Manhattan Project. ⁓es, direct, unyielding, and efficient, supported the genius scientist, who just after the ⁓essful Trinity test, resorted to quoting the Bhagavad Gita in reply to the might of the explosion ⁓had just witnessed: "I am become death—the destroyer of worlds." (Corbis)

LEFT: Charles Sweeney, at the controls of The Great Artiste which carried scientists and bomb-measuring equipment, attended Mass the day before the Hiroshima bombing. He decided to "commune silently with God and tell Him about our mission." (Corbis)

BOTTOM: The mushroom cloud that appeared over Nagasaki. For two days after the Hiroshima bombing, because of cloud cover, U.S. fighters were unable to photograph the ground devastation to support the press release that told Americans of the total destruction of the city. (Library of Congress)

A-58450 A.C.

TOP: A Japanese baby sits crying in the rubble left by the explosion in Hiroshima. In a radio broadcast 16 hours after the attack, President Truman said the United States had dropped the bomb "in order to shorten the agony of war, in order to save the lives of thousands and thousands of young Americans." About 80,000 people died instantly in the bombing; virtually every building in Hiroshima was destroyed or damaged. (Bettmann/Corbis)

LEFT: Around 30,000 people were killed outright in Nagasaki on the morning of August 9. The bomb detonated over the Urakami district, home to the city's Christian community, where most of the medical and education facilities were located. (Joe O'Donnell, courtesy Kimiko Sakai)

ABOVE: A Nagasaki boy standing to attention while he waits to lay his little brother on a funeral pyre. (Joe O'Donnell, courtesy Kimiko Sakai)

RIGHT: A woman rather incongruously smiles for the camera amid the devastation in Nakamachi, 2.5 kilometers southeast of the Nagasaki hypocenter. She appears to be in one of the earthen bomb shelters some citizens had dug under their houses. (Yosuke Yamahata Photo, copyright Shogo Yamahata, courtesy: IDG films, Nagasaki, 10 August 1945)

LEFT: The Mitsubishi steelworks near the detonation point was destroyed in the blast, although the torpedo factory underground, where Tsuruji Matsuzoe worked as a 15-year-old, remained relatively undamaged. A dead horse lies outside the steelworks. (Yosuke Yamahata Photo, copyright Shogo Yamahata, courtesy: IDG films, Nagasaki, 10 August 1945)

BOTTOM: Months after the bombings of Hiroshima and Nagasaki the death toll had started to rise again, as people succumbed to their wounds and apparently healthy people developed radiation sickness. Two months after the Hiroshima bomb, a mother attends to her badly burnt child. (Keystone/Hulton Archive/ Getty Images)

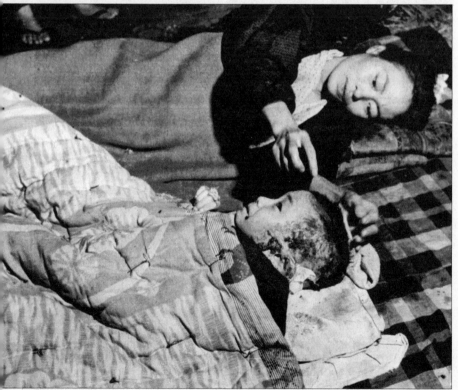

ABOVE: A group of homeless children warm their hands over a fire on the outskirts of Hiroshima after the end of the war. (Alfred Eisenstaedt/ Time & Life Pictures/Getty Images)

RIGHT: A victim in 1951 displays his burn and keloid scars from the attack on Hiroshima six years earlier. (Keystone/Hulton Archive/Getty Images)

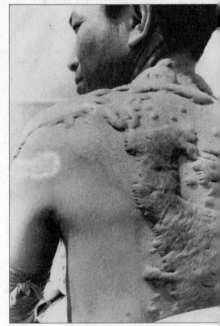

The Foreign Minister failed to persuade the cabinet: far from their being 'shocked into submission', Anami and his fellow hardliners ignored Togo, whose proposed course was not even listed as an agenda item for further discussion. The military chiefs refused to believe America possessed a nuclear arsenal: 'I am convinced that the Americans had only one bomb, after all,' Anami said immediately after the destruction of Hiroshima. They were determined to suspend any decision until they knew the facts.

The elderly, hard-of-hearing Prime Minister Suzuki acquiesced in the militants' course; clearly he had little grip on the machinery of state. The ascendancy of Anami's war faction rose inversely with the hopelessness of Japan's situation. Even now, with a nuclear war hanging over them, the militarists refused to soften their terms. They persisted in the delusion that fighting on would force negotiations – over Japan's claim on Manchuria, a right to conduct their own war crimes trials and other pie-in-the-sky notions that bore no connection with reality. The American people, they believed, would not tolerate the casualties of a land invasion. This scenario played well to the Imperial Army's dreams of battlefield glory but found little concordance with any sane reading of events. America's total air supremacy and devastating naval blockade had all but removed the prospect of a land invasion before the use of the atomic bomb; and Russia's refusal to engage with Japanese diplomacy should have sent a further ominous note to Tokyo. The militarists failed to see that they held no bargaining chips, no cards.

Anami personified this defiant nihilism, at least in public; privately, he was forced to acknowledge the humiliating truth. In a long meeting with Togo that night he accepted the reality that Japan's surrender was inevitable, but he never abandoned his insistence on a negotiated peace. How would he communicate 'surrender' to defiant officers who refused to lay down their weapons? He himself veered between championing the army (and fighting on, thus courting the nuclear annihilation of Japan) and a grudging realisation that Japan's war-making powers were finished. The bomb to some extent catalysed this realisation, but was not its cause. 'On the outside and officially [Anami] pretended that we must continue the war,' observed Sakomizu, 'but inside himself he had made his decision that it must be brought to a stop. He alone could have broken the Suzuki cabinet at any time. It shows his character that he didn't, despite what he knew of our negotiations.' Anami had shown 'one way of being a brave man'. Unable to reconcile these positions, the War Minister contemplated the only course deemed honourable to a Japanese samurai: 'When you have a choice between life and death, always choose death.'

That day (7 August) Domei put out a statement in response to the American reports, only for international release; it was not broadcast domestically – the Japanese people were to be kept utterly in the dark. Reprinted in the *New York Times* and other leading US papers, it blamed the destruction of Hiroshima on 'a new type of bomb' (the Japanese media were not permitted to use the adjective 'atomic' until 11 August) and vowed that Japan would 'cope with it immediately' and 'check the damage'. 'By employing the new weapon designed to massacre innocent civilians,' the statement said (quoting 'informed quarters' – that is, the military), 'the Americans unveiled to the eyes of the entire world their sadistic nature ... What caused the enemy to resort to such bestial tactics, which revealed how this is the veneer of the civilisation the enemy has boasted of, is his impatience at the slow progress of the enemy's much-vaunted invasion of Japan's mainland ...' Thus began the slow insinuation into the public mind of the notion of the people of Hiroshima and Nagasaki as the world's first atomic 'martyrs' – the public sacrifice for Tokyo's attempt to exit the war with face. The survivors of the bomb – they eschewed the word, 'victim' – would reject this characterisation; they were, and should have been treated as, the civilian war wounded.

On the morning of the 8th the Emperor received Togo in the shelter beneath the Imperial Palace. His Majesty and the Foreign Minister shared a deep concern at the course of events; both were mindful of the 'new type of bomb' and the military's refusal to capitulate. Togo gravely advised the Emperor that the country had no option other than to accept the Potsdam Declaration. His Majesty appeared to agree; the war should end 'without delay'. Yet even at this late hour Hirohito wondered whether the Americans would accept a 'figure-head Emperor' – a self-serving tilt that revealed the limit of his duty of care to the Japanese people.

'His Majesty observed that, now with this kind of weapon in use, it had become even more impossible than ever to win the war,' Togo later wrote. The Emperor advised, however, that Japan should not 'completely discard the possibility of negotiating conditions'. Togo informed Prime Minister Suzuki and Kido, the Lord Keeper of the Privy Seal, of the Emperor's position. The militarists were kept out of these secret deliberations, as their fanatical junior officers were bound to interpret Togo's influence as perverting the Imperial Will, which would endanger the lives of the 'peace faction'. For this reason, in part, Prime Minister Suzuki tended to say whatever the hardliners wished to hear. The destruction of Hiroshima had not changed Suzuki's outward refusal to surrender. Regardless of what his 'stomach art' advised him, the atomic bomb, for now, had failed to move the stubborn old man, who knew little of what was happening in the country. Togo similarly paid lip service – in public – to continuing the war, while privately urging the Emperor to intervene. Only the Emperor's word, he knew, could impose surrender *and* control the army's malcontents. In the meantime, he would continue to talk, and scheduled a meeting of the Supreme Council next morning, 9 August.

Here was the last testament of a delinquent regime beyond the reach of reason. The advent of nuclear war had manifestly not achieved the desired outcome; the atomic bomb had not shocked Tokyo into submission, as Washington intended (and later claimed). The nuclear bludgeon failed to deter the militarists, men like Anami, Toyoda and Umezu, from their disastrous course. To them, another city had died in a country that had hitherto suffered the loss of more than 60.

•

A more ominous threat, in the regime's eyes, emanated from the gathering storm on the Manchurian border. The Russians had massively underlined their deadly intent on 28 July – two days after the Potsdam Declaration – when Tokyo received news of a further 381 eastbound Soviet military trains, carrying 170,000 troops, hundreds of guns and tanks, and – vital for an invasion – 300 barges, 83 pontoon bridges and 2900 horses. The Japanese had in fact grossly underestimated Russia's resolve: by night over the past four months, rail carriages had shifted more than a million men and materiel 10,000 kilometres to the Pacific theatre in one of the greatest redeployments in the history of warfare.

Meanwhile, a mood of despair and signs of panic gripped parts of Japanese-occupied Manchuria (Manchukuo), where Japanese forces and civilians were conscious of being caught in a Chinese and Soviet vice, and severely weakened by the repatriation in April of 16 to 20 divisions to defend the homeland. A Chinese communist uprising in southwest Manchukuo sowed dread in Japanese civilians: women were evacuated and all Japanese boys over 14 drafted into the army. In the Canton area Japanese commanders, fearful of Chinese reprisals, ordered these child soldiers to prepare for 'cave warfare' from the dugout encampments built outside the city.

In this light, while the atomic bomb had not deterred Japan's militarists, it had had a grave impact on Soviet thinking. The Kremlin knew all about the impact of the bomb via radio reports and their diplomats in Washington. It depressed Stalin that his allies had so casually excluded him; his paranoia cleaved to the bomb 'as an act of hostility directed against the Soviet Union'. He feared the loss of the prizes agreed to at Yalta were Japan quickly to surrender to the Americans. 'Russia's own self-interests now demand that she actually share in the victory,' warned a Magic summary of late July, 'and it seems certain that she will intervene ... although it is impossible to say when. At the proper time Stalin will say the word ... the far-sightedness and genius of their supreme leader ... has not forgotten the defeats of 1905.' Nor had Stalin's 'far-sighted genius' forgotten what he saw as America's perfidy at Potsdam. At a stroke he brought forward the start of the invasion to midnight, 8 August, a week earlier than the date he had offered Truman in Berlin.

•

Thirty-four hours after the destruction of Hiroshima, Togo cabled Ambassador Sato for any word on Soviet intentions as a possible mediator. No message more grimly demonstrated the pathos of the Japanese leadership: 'The situation is becoming more and more pressing,' Togo wrote, 'and we would like to know at once the explicit attitude of the Russians. So will you put forth still greater efforts to get a reply from them in haste.' Molotov had earlier agreed to see Sato at 5pm, Moscow time, on 8 August.

The credulity of the Supreme War Council would have seemed farcical were its consequences for the Japanese people not so tragic. Sato's diplomatic screeds had made little dent on the mind of his Foreign Minister. Until the last moment, the Japanese leaders failed to perceive a truth of which Sato had constantly warned them. On 7–8 August the Japanese government still held out hope of a breakthrough in Moscow, as Admiral Takagi revealed in his talks with Navy Minister Yonai:

Takagi: I think the real problem is not whether the enemy [that is, the Americans] will invade our mainland, and when it will be ... but rather the diminishing spirit of the people ...

Yonai: I met the Foreign Minister yesterday and he told me that
no telegram [from the Soviet Union] had come ... Perhaps we may
have to be ready for a situation where we won't receive any
response from Russia.

•

At the prescribed time, Molotov received Sato in the Kremlin as
Tokyo slept. If the Japanese ambassador held out the faintest hope of
Soviet transigence – perhaps Stalin would receive the Emperor's
special envoy? – he was soon brutally disabused. Molotov strode in,
waved aside the ambassador's diplomatic pleasantries, gestured Sato to
sit and read aloud the Soviet declaration of war: after Japan rejected
the Potsdam Declaration 'the Allies approached the Soviet Union',
Molotov lied, 'with a proposal to join the war against Japanese
aggression ... Loyal to its duty as an ally, the Soviet Government has
accepted the proposal of the Allies and has joined in the declaration
of the Allied powers of July 26.'

Russia thus imposed its name on the Potsdam ultimatum without
invitation or encouragement from its 'Allies'. Molotov abruptly
concluded: 'In view of the foregoing the Soviet Government declares
that from tomorrow, that is of 9 August, the Soviet Union will consider
itself in a state of war with Japan.' 'Tomorrow' meant almost
immediately: it neared midnight in Eastern Siberia.

The Soviet announcement realised Japan's darkest fears. The chief
Russian goals in invading Manchuria were, according to the Russian
specialist Dr Raymond Garthoff: (1) to acquire a voice in the future of
the Northern Pacific, including Japan; (2) to seize and incorporate
into the Soviet Empire southern Sakhalin and the Kurils; and (3) to
eliminate Japanese and 'pre-emptive Western presence on the North
Asian continent'. The Russian invoice for 1905 and the communist
hunger for a platform in Asia culminated in the Red Army's footfall
across the Manchurian border. Here at last was a mass invasion of

Japanese-occupied territory from the quarter Tokyo least expected and most feared.

Near midnight, on 8 August, Transbaikal time, the advance units of 1.5 million men supported by tens of thousands of armoured vehicles, tanks, artillery pieces, and aircraft, entered Japanese territory. The Soviet invasion extended across a 4400-kilometre front, from the Mongolian wastes to the Sea of Japan. Stalin eagerly described the onslaught to US Ambassador Averell Harriman that afternoon in Moscow: 'Who would have thought that things would have progressed so far by this time?' Stalin asked with a facetious smile. Soviet aircraft had already bombed Changchun and Harbin, he said; shock troops had attacked Grodekovo in the east, where the railroad from Vladivostok crosses the frontier; another column was striking south from the Soviet border towards Hailar; a third was moving east through the mountain pass in the vicinity of Solunshan, a railhead in northern Korea; the cavalry forces were advancing across the Gobi Desert south of Ulan Bator into the Mukden region. And these were just the advance forces, Stalin informed the American ambassador.

Harriman seemed impressed. He reminded the Generalissimo that a year ago Stalin had said that 'things would go fairly fast once Russia entered the [Pacific] War'. Stalin replied that if things 'went fast' now, it was due not only to Russia's entry. Nobody had anticipated the triumph, so soon, of the US Navy, he said.

But what of the atomic bomb, Harriman asked. What effect had or would it have on Japanese resistance? Stalin answered that he thought the Japanese were, at present, looking for a pretext to replace the present government with one 'which would be qualified to undertake a surrender. The atomic bomb might give them this pretext.' The bomb, he felt, would serve as the catalyst for a regime change. He sought to downplay its expected impact; in fact, he feared the bomb would end the war soon and deny Russian claims on Japanese-occupied territory. It was palpably clear to Harriman that Stalin was in a race for the spoils of the Pacific.

The Soviet forces fell on the exhausted Kwantung Army, once the pride of the Japanese Empire, with all the discrimination of a hurricane: more than 1.5 million Red Army troops, 410 million rounds of ammunition, 3.2 million shells, and 100,000 trucks and armoured vehicles tore into the Japanese forces stationed along the border. America donated 500 Sherman tanks and 780,550 tonnes of dry goods to this, the last great military operation of World War II.

What followed was the eclipse of Japan's Imperial adventure: an immense envelopment conceived along three axes as the prelude to the swift and complete destruction of the Japanese army in Manchuria. The Kremlin conceived a fourth offensive with a greater prize: the capture of the Sakhalin and Kuril Islands in advance of the invasion of Hokkaido – the knockout blow Japan feared but refused to accept. Soviet propaganda hailed the invasion as vengeance for 1905: 'The time has come to erase the black stain of history from our homeland,' one colonel told his men. Red Army infantry with pistols at their backs – many fresh from European battlefields – swarmed over river and forest, desert and marshland, deep into Chinese territory. Their only resistance, initially, were the wretched 'smertniks' – the Russian word for Japanese suicide squads – who leaped from the roadside to attack the advancing tanks, with little effect. The Japanese shells were not armour-piercing, and the Soviet T-34s smashed through the enemy positions unmolested. Notwithstanding a few ferocious last stands, the Japanese resistance was quickly overwhelmed.

The Soviet forces 'swept into Manchuria from the desert wastes of Mongolia, bypassed Japanese defensive positions, thrust across the undefended, yet formidable, terrain of the Grand Khingan Mountains, and erupted deep in the Japanese rear,' wrote David Glantz, an expert on the campaign. Detailed planning, total surprise, and a willingness to attack in appalling weather and over the most unyielding terrain defined Russia's 'August Storm'. The breadth and depth of the offensive

– and the awesome salvoes of the katyusha rocket-launchers that preceded each assault – shocked the Imperial forces, who numbered some 713,000 soldiers strung out in the crumbling fortifications of Manchukuo (with another 70,000 in Korea). The Manchukuo name died with their destruction. Though numerically strong, their combat effectiveness was at a nadir; and their equipment woefully outclassed. Russian tanks outnumbered the Japanese five to one (5556 to 1155, noted the historian Richard Frank) and the Japanese 'Air Army' had just 50 first-line aircraft.

The Japanese troops were neither warned of nor equipped to meet the Russian juggernaut, despite Tokyo's knowledge of the Soviet deployment across Siberia; the Kwantung Army were told only of a slight possibility of an attack in August. As usual, the 'evasion of unpalatable reality prevailed over rational analysis', in historian Max Hasting's words. The Russians claimed to have captured 594,000 and killed 80,000 Japanese troops, for 30,000 Russian casualties (dead and wounded).

Like the Japanese civilian, the ordinary Japanese soldier perished in the thrall of a regime to whom they were prepared to give everything – their lives, their children's lives – but from whom they would receive nothing except the certainty of a miserable death.

•

At 10.45am (Washington time), 8 August, Stimson showed Truman the reports of the Strategic Air Forces in Guam, and photographs of the remains of Hiroshima taken by the Army Air Forces and those on Tokyo's wire bulletin, which together described the radius of damage. A sombre Truman reflected that such devastation placed a 'terrible responsibility' on himself and the War Department. Oppressed by that responsibility, Stimson urged the President 'to proceed ... in a way which would produce as quickly as possible [Japan's] surrender'. He proposed 'kindness and tact' rather than the brutal methods used on the Germans: 'When you punish your dog you don't keep souring on him all day after the punishment is over,' Stimson said. 'If you want to keep his affection, punishment takes care of itself. In the same way with Japan. They naturally are a smiling people and we have to get on those terms with them ...' It was an odd remark after four years of carnage, rather as if Stimson had exhausted the American arsenal and found 'kindness' the only weapon left.

That night the White House received news from Averell Harriman of the Russian invasion of Japanese-occupied Manchuria. A little after 3pm the next day, the President called a snap press conference: 'I have only a simple announcement to make ...' he told the reporters. 'This announcement is so important I thought I would call you in. Russia has declared war on Japan. That is all.' He took no questions. When he heard the news, Senator Alexander Wiley of Wisconsin remarked, 'Apparently, the atomic bomb which hit Hiroshima also blew "Joey" off the fence.'

At the same time Secretary of State James Byrnes released a press statement that put a welcoming face on this unwelcome development: 'This action by the Soviet Government should materially shorten the war and save the loss of many lives,' Byrnes declared. The Allied powers would continue their co-operation in the Far East and 'bring peace to the world'. Byrnes thus applauded in public what he had so vehemently opposed in private. As for Japan, further resistance to the Allied nations 'now united in the enforcement of law and justice' was futile. 'There is still time,' Byrnes concluded, 'but little time for the Japanese to save themselves from the destruction which threatens them.' Both Truman and Byrnes were then aware that the plane carrying a second atomic bomb to Japan was airborne. It was too late to abort the mission, even had they wanted to.

That day Truman briefed his press officers Charlie Ross and Eben Ayers on the sequence of events leading to the Russian invasion. It was a fair representation of events – to a point. Moscow, he said, had agreed 'to come in' to the Pacific War three months after Germany's defeat (that is, Stalin had specified 15 August as the day of Soviet entry); Stalin had then been told of 'a powerful new explosive' at Potsdam; Hiroshima's destruction confirmed the weapon's atomic power and spurred Stalin 'to get in [the war] before Japan could fold up'. Truman had insisted on one condition before assenting to Soviet claims in Asia, he told his press men: that Stalin must agree to recognise Chiang Kai-shek's Nationalist government and not the Communist Party's alternative under Mao Tse-Tung's leadership. This Stalin had agreed to do, Truman claimed. In fact, Stalin recognised neither Chiang's nor Mao's regime, though clearly Stalin expected and planned for the Chinese Nationalists to remain in power after the war, even in Manchuria. That suited him – far better a weakened Nationalist government than an empowered Communist one. Stalin wanted to manipulate Mao, not create a rival. In any case, the supposed deal with Truman was not a key part of their Potsdam discussions, and hardly in the spirit of Truman's avowed intent in Berlin, where the President was supposedly urging, not imposing conditions on, Soviet entry in the Pacific. Truman's version presumed that he had been in a position to impose conditions on Soviet entry – he was in no such position at any time. In short, the Soviet invasion was an awkward and unwelcome development, which the White House had had no choice other than to publicly embrace, persisting with the charade that it desired military help from its erstwhile ally.

•

Early on 9 August Tokyo time, Cabinet Secretary Sakomizu received a call from the Domei News Agency, informing him of the Soviet declaration of war. It came as a profound shock: official channels were muzzled; Moscow had refused to allow Sato to cable the announcement, to ensure maximum surprise. At 5am Sakomizu took the text to Prime Minister Suzuki and suggested two paths: (1) the cabinet could resign because their policy of suing for peace through Russia had failed; or (2) the leadership could take 'some step of a positive sort'. To such floundering expressions of helplessness was the empire reduced.

Suzuki replied: 'If we resign it will take two or three days for a new cabinet to be formed. The loss of two or three days is intolerable, since that lapse of time might decide the national destiny. It is necessary for us to take some positive step.' The steps the Prime Minister proposed were to declare war on the Russians and continue fighting until the Japanese people were annihilated; or to accept the Potsdam Declaration.

Suzuki went to see the Emperor at 7am and returned 'an hour or two later' with an answer: His Majesty had agreed to accept the terms issued at Potsdam. Togo and his senior staff had reached the same conclusion that morning, with the condition that 'the acceptance of the Potsdam Proclamation shall not have any influence on the position of the Imperial House'. The Prime Minister scheduled immediate meetings of the Supreme Council and the full cabinet: the Big Six would meet at 10am, an hour before the B-29 *Bockscar* – bearing the plutonium bomb – would reach the vicinity of Nagasaki.

It is palpably clear from these events that the Soviet declaration of war made a deeper impression on Tokyo than the atomic bombing of Hiroshima. When Suzuki heard the news that the Russians had overrun the Imperial Army, he responded: 'Is the Kwantung Army that weak? Then the game is up.' The Japanese leaders had anticipated – many desired – an American land invasion, which would, they believed, ennoble the last sacrifice of the Japanese people. To their shock, it had come from Russia.

The Japanese commanders could *imagine*, from their own battlefield experience, the scale of the confrontation with the Americans, and knew what to expect. The Russians were a very different beast: the Japanese knew they could expect no quarter from the conquerors of Berlin and the losers of 1905; the spectre of a communist Japan haunted the regime. The bomb played a lesser role in this spectrum of the Japanese leaders' anxieties – and without as yet any photographs (and no television) they could not picture the effects of the weapon, the great mushroom cloud. In this sense the samurai leaders experienced the last stand of the Kwantung Army in his guts, not his head; the clash of blood and iron moved them in a way the slaughter of helpless civilians by a single projectile had not: the atomic bomb was a cruel and cowardly weapon, they believed, a vast incendiary dropped from a safe height on innocent people. Hiroshima's was a helpless death, shameful in its pathos: indeed, there was something ignominious – to the Japanese military mind – about collapse without resistance in the blink of an eye. On the other hand, the Soviet invasion of Manchuria was a battle they understood, a clash they could *respect* ... if not win.

•

At 10pm on 9 August, Washington time – 11am, 10 August, Nagasaki time – Truman addressed the nation. The President sought to justify the use of the atomic bomb and explain what had been achieved at Potsdam. In an intriguing mixture of patriotism, vengeance, lies and prayer, Truman began by 'gladly' welcoming the Soviet Union to the Pacific War as 'our gallant and victorious ally'.

He misrepresented the nature of the target: 'The world will note that the first atomic bomb was dropped on Hiroshima, a military base. That was because we wished in this first attack to avoid, insofar as possible, the killing of civilians. But that attack is only a warning of things to come. If Japan does not surrender, bombs will have to be dropped on her war industries and, unfortunately, thousands of civilian lives will be lost. I urge Japanese civilians to leave industrial cities immediately, and save themselves from destruction.

'Having found the bomb we have used it,' the President declared. 'We have used it against those who attacked us without warning at Pearl Harbor, against those who have beaten and starved and executed American prisoners of war, against those who have abandoned all pretense of obeying international laws of warfare. We have used it to shorten the agony of war, in order to save the lives of thousands and thousands of young Americans. We shall continue to use it until we completely destroy Japan's power to make war.'

The secret of atomic power would remain in the hands of America, Britain and Canada because it was 'too dangerous to release in a lawless world'. Nuclear weapons presented an 'awful responsibility' to America, the President concluded; however, 'we thank God that it has come to us instead of to our enemies; and we pray that He may guide us to use it in His ways and for His purposes'.

The President spoke the day after a second atomic bomb had been unleashed on the people of Nagasaki.

CHAPTER 19

NAGASAKI, 9 August 1945

Let us carry the war to them until they beg us to accept the unconditional surrender ... If we do not have available a sufficient number of atomic bombs with which to finish the job immediately let us carry on with TNT and firebombs until we can produce them ... we should continue to strike the Japanese until they are brought grovelling to their knees.

Senator Richard B. Russell of Georgia to Truman, the day before Nagasaki

8 August 1945

Memorandum to General Somervell:

I missed you but informed Lutes [General LeRoy Lutes], who is on his way to dinner. Our second attempt is on. The first Fat Boy is on the way and by morning I hope he has done his job.

Will you please return this in the envelope provided.

[signed]

LR Groves

Major General, USA

Underneath Groves' memo, in pencil, General Somervell had written, 'SWELL!'

The second atomic mission left Tinian Island before dawn on 9 August without Truman's verbal or written consent. That is not to say the President objected, of course; simply, that nobody sought the President's approval. The US Commander in Chief's imprimatur was assumed. On Tinian Captain Deak Parsons, the cool engineer in charge of Project Alberta (codename for the weapons-delivery mission within the Manhattan Project), and his technical specialist, the physicist Norman Ramsey, were anxious to deliver the second bomb as swiftly as possible – to deliver the 'one-two' punches considered essential to force the Japanese to surrender. Groves, who in practice exerted complete control over America's atomic war in the Pacific, concurred. Two days after Hiroshima the general decided to bring forward the second atomic attack, from 11 to 9 August, when clear skies were forecast. The weather was his paramount concern; he paid no attention to whether the Japanese might or might not reply to the destruction of Hiroshima. He set the mission in motion on the evening of the 8th. This new schedule chimed with Groves' strategic plan: the second atomic blow, he believed, should 'follow the first one quickly, so that the Japanese would not have time to recover their balance'. Another criteria for a swift, second atomic strike was the perception it would leave in the minds of the Japanese. 'Everyone,' recalled a member of the bomb assembly team, 'felt that the sooner we could get off another mission, the more likely ... the Japanese would feel we have large quantities of the devices.' And of course, this was a plutonium weapon – with a detonation mechanism completely different from the uranium gun-activated device used on Hiroshima; while it had been tested at Trinity, it had never been dropped from an aircraft. Groves and his senior staff were anxious to see whether the experiment would work.

The questions of whether Tokyo should be given more time to respond to the levelling of Hiroshima, and whether the Russian declaration of war made the second bomb unnecessary, were not examined. Truman received news of the Soviet invasion of Manchuria *after* the second nuclear-armed aircraft had left Tinian for Japan, at 3:47am on 9 August (Tinian time); an order to abort mid-flight meant jettisoning the weapon in the sea. That course was never contemplated.

On the contrary, news of the Russian invasion accelerated the decision. Groves believed the Soviet offensive made a stronger case for the second bomb. For several years the general had seen Russia as America's ultimate enemy and recently perceived, as did many in Washington, that a contest had begun; a race whose prizes were unknown, and whose outcome uncertain. In the immediate term, the Americans were in a race to force Japan's surrender and deny Soviet participation in a completely American victory. In the longer term, Pentagon planners were lining up the Soviet Union, in mock war scenarios, as America's most likely foe in a future nuclear conflict (see Epilogue: Dead Heat).

Groves' preferred target was Kokura, site of Japan's biggest weapons arsenal, and a city of 168,000. The Target Committee had listed Kokura as a 'primary' back in May. The city met the Joint Chiefs' understanding of a military strike better than any in Japan, and LeMay's conventional bombers had dutifully left it unscathed. The choice of target, however, depended on mid-flight weather reports from the reconnaissance planes. The crew of the B-29 that would drop the bomb, *Bockscar* – a pun on boxcar, and after its usual commander, Captain Fred Bock, who commanded *The Great Artiste*, a reconnaissance plane, on the Nagasaki mission – would decide the final target well inside Japanese air space.

·

The people of Nagasaki were vaguely aware of the events in Hiroshima. The *Nippon Times* had warned them under the headline 'A Moral Outrage Against Humanity' of a 'new type of bomb' which 'should not be made light of'. The enemy were 'intent on killing and wounding as many innocent people as possible … to end the war speedily', the paper reported. US leaflets had hitherto announced, in macabre, mock haiku, 'In April Nagasaki was all flowers. In August it will be flame showers.' And flyers were dropped over several Japanese cities the day before Nagasaki's destruction:

ATTENTION JAPANESE PEOPLE – EVACUATE YOUR CITIES
Because your military leaders have rejected the thirteen-part
surrender declaration, two momentous events have occurred in the
last few days. The Soviet Union ... has declared war on your nation.
Thus all powerful countries in the world are now at war against you ...

The second event was Hiroshima's destruction, the prelude to a series of nuclear attacks, the message threatened. Atomic bombs would be dropped 'again and again' unless Japan surrendered.

The flyers lay untouched on the streets of Nagasaki; the authorities alone were permitted to read them. The latest message was treated as another general threat; it made no specific mention of the three shortlisted targets, Kokura, Nagasaki and Niigata.

•

Before dawn on 9 August the unit chaplain led the airmen at prayer; after a hot breakfast they boarded a convoy of trucks, which carried them to the waiting planes on Tinian's North Field. Leading the mission was Commander Frederick Ashworth, weaponeer and officer in charge of the plutonium bomb, codenamed Fat Man; the flight commander was Major Charles Sweeney, 25, who had trained the Manhattan Project crews and flown the observation plane on the Hiroshima mission. The bombardier was Captain Kermit Beahan, recipient of the Distinguished Flying Cross, Air Medal and Purple Heart (among other honours), and veteran of air raids over Germany in 1942. As on the *Enola Gay*, there were 10 other crewmen – co-pilots, gunners, flight engineers, radar and radio operators.

The reconnaissance and observation planes were *The Great Artiste* and *The Big Stink*. Two young Los Alamos physicists, Luis Alvarez and Philip Morrison, accompanied the mission (one replaced Robert Serber, who had forgotten his parachute and missed the flight). Prior to departure the scientists placed letters in the data canisters which were to be parachuted near the target, imploring the Japanese physicist Professor Ryokichi Sagane to persuade the regime's leaders that America possessed atomic weapons: 'As scientists we deplore the use to which this beautiful discovery has been put, but we can assure you that unless Japan surrenders at once, this rain of atomic bombs will increase many fold in fury.' (The Japanese military would suppress the letter; Sagane did not see it until after the war.) There were two British observers: Group Captain Leonard Cheshire, VC, DSO, DFC, Britain's most decorated pilot, whom Churchill had delegated as British witness; and Dr William Penney, Professor of Applied Mathematics at London University, who worked at Los Alamos. The only reporter was, again, the *New York Times'* William Laurence; this time he accompanied the mission as passenger on *The Great Artiste*, along with the nine crewmen.

Before departure Sweeney gathered his crew beside *Bockscar*: 'This is what we have been working for,' he said, 'testing and thinking about for the past year. You were all with me the other day at Hiroshima. It was a perfect mission ... I want our mission to be exactly the same – for Colonel Tibbets. He has chosen us and we owe him and our country the same. Perfectly executed, perfectly flown and dropped on the button. We *will* execute this mission perfectly ... I don't care if I have to dive the airplane into the target, we're going to deliver it.' The target of which he spoke was the Kokura Arsenal; if that attempt failed, the secondary target was the Mitsubishi Shipyards, on the shore of Nagasaki Harbour, opposite the city centre, and the surrounding industrial area.

His hopes of a perfect mission were shortly disappointed. In the first of a series of mishaps, just before starting engines, the crew discovered that the fuel pump to the 600-gallon bomb-bay tank was inoperable, meaning Sweeney had 6400 instead of 7000 gallons (26,500 litres) for the flight, a fault that would have delayed a conventional mission. The mission needed every gallon: the aircraft carried more weight than *Enola Gay* (Fat Man weighed 1300 pounds – 480 kilograms – more than Little Boy) and would have to fly at high altitude to avoid a storm. Tibbets left the decision to Sweeney: 'The hell with it, I want to go,' he said. 'We're going.'

Unlike Little Boy, the plutonium weapon had to be armed before take-off – this could not be done in-flight due to the complex firing mechanism – creating some anxiety as *Bockscar* rumbled down Hirohito Highway. An hour from Tinian the mission flew into an electrical storm. Three hours later a serious malfunction in the circuitry set the red warning light on the bomb's fuse monitor flashing; it meant the firing circuits had closed and some or all of the fuses had been activated. If so, Sweeney faced the immediate prospect of being forced to ditch the bomb. The weaponeer Ashworth's calm inquiry discovered the cause of the malfunction and the crisis passed; the crew, however, had a growing sense of the mission as jinxed. Sweeney, a devout Catholic, thought, 'To have come this far and end in a vaporizing flash. My only response was to whisper, Oh, Lord.'

In *The Great Artiste*, flying nearby, Laurence was taking notes. Empurpled by the grandeur of the occasion, his prose portrayed himself as a charioteer, 'riding the whirlwind through space on a chariot of blue fire' – a reference to St Elmo's fire, the static electricity that gathers around aeroplane propellers.

At dawn the radio operator Sergeant Ralph Curry took off his earphones and said, 'It's good to see the day.'

'It's a long way from Hoopeston, Illinois,' Laurence remarked.

'Yep. Think this bomb will help end the war?' Curry asked.

'There is a very good chance this one may do the trick,' Laurence assured him. 'If not then the next one or two surely will. Its power is such that no nation can stand up against it very long.' It was not his own authority: Laurence conceded that he had heard this opinion expressed 'all around' Tinian a few hours before take-off.

As the only journalist present, Laurence's vengeful anticipation of the coming slaughter is of passing historical interest:

> Somewhere beyond these vast mountains of white clouds ahead of me, there lies Japan, the *land of our enemy* [his emphasis]. In about four hours from now one of its cities, making weapons of war for use against us, will be wiped off the map by the greatest weapon ever made by man. In one tenth of a millionth of a second ... a whirlwind from the skies will pulverize thousands of its buildings and tens of thousands of its inhabitants ... Does one feel any pity or compassion for the poor devils about to die? Not when one thinks of Pearl Harbor or the Death March on Bataan.

Before 8am another crisis struck: one of the two observation planes failed to rendezvous at the correct location on the Japanese coast. Major James Hopkins, commanding *The Big Stink*, had climbed too high and lost contact with *Bockscar*, which flew around for 30 minutes awaiting his arrival. Hopkins, meanwhile, had radioed Tinian, 'Has Sweeney aborted?' Brigadier General Farrell, Groves' second in command, interpreted the garbled message as 'Sweeney aborted' and panic ensued. Had Sweeney ditched the bomb? Had there been an accident? Calm returned with radio contact.

At 9.45am *Bockscar* approached the primary target – Kokura – and the crews strapped on their parachutes. The weather planes reported hazy skies with broken clouds. 'The winds of destiny seemed to favour certain cities,' Laurence observed. The winds had changed direction between the weather reports, and Kokura was now blanketed in heavy smoke from the US bombing of nearby Yawata the night before. The first run on the arsenal failed – 'I can't see it! I can't see it!' yelled Beahan. Sweeney decided on a second attempt – 'something a bomber rarely, if ever, does' – as it risked drawing enemy fire: at the second approach ground fire started 'crawling the flak up towards us'. The run also failed – 'visibility nil' – and Sweeney banked away again and yelled into his intercom: 'Pilot to crew: No drop. Repeat. No drop.' Japanese fighters were scrambling when Sweeney dared a third attempt, from a different angle. But the winds of destiny spared Kokura: 'Made 3 runs on primary but each time target was obscured by haze and smoke,' Ashworth cabled Tinian. After 50 minutes, and critically low on fuel, the crew resorted 'to attack secondary'. Sweeney flew to Nagasaki.

To Sweeney's frustration, cumulous cloud obscured 80 per cent of the second target and so he resolved on a radar-directed attack. Time and fuel were running out: the plane had 300 gallons left with which to get home: 'If we didn't drop we were out of options,' Sweeney later wrote, and 'forced to crash land on the ground in Japan or in the ocean.' The plane began its descent a few seconds after 11am. A gap in the clouds revealed buildings to the north of the city. Beahan released the weapon.

'Bombs away!' he shouted, and corrected himself, 'Bomb away.' Sweeney swung the plane into a deep dive.

The flash was brighter and the bumps greater than after the Hiroshima bomb, Sweeney later wrote. Three shockwaves struck the aircraft; fires were seen to the east and west of the 'luminous' mushroom cloud. The sight induced in Leonard Cheshire 'great feelings of power and relief'. His first conscious thought was, 'It's the end of the war … that's a weapon you cannot fight.'

Laurence witnessed an altogether different sight, rather like a hallucinogenic vision: 'A giant pillar of purple fire,' he wrote, 'like a meteor coming from the earth instead of from outer space, becoming ever more alive as it climbed skyward … It was a living thing, a new species of being, born right before our incredulous eyes.' The cloud was now 'a living totem pole … grimacing at the earth'; next, 'seething and boiling in a white fury of cream foam'; and finally, 'a flower-like form, its giant petals curving downward, creamy white outside, rose-coloured inside'. The diabolical had metamorphosed in the reporter's head into a giant fetish, then a thing of beauty, possibly god-like.

'Well, Bea,' said co-pilot Captain Charles Albury, 'there's a hundred thousand Japs you just killed.' Beahan said nothing.

'Bombed Nagasaki … visually with no fighter opposition and no flak,' *Bockscar* cabled Tinian, to the immense relief of those on the island. 'Results "technically successful", but other factors involved make conference necessary before taking further steps. Visual effects about equal to Hiroshima … Fuel only to get to Okinawa.'

'Column of smoke and mushroom,' the message added, '… soon reached at least 40,000 feet. Dust covered area at least two miles in diameter. Probably fair amount of blast on unprofitable areas [that is, non-military residential areas north of the city]'.

Bockscar barely reached Okinawa. Sweeney issued a mayday to the Yontan control tower as the plane came in to land in a blaze of flares – and abruptly ran out of petrol on the runway. Sirens approached – and questions. A head appeared.

'Where's the dead and wounded?' asked the ground crew.

'Back there,' Sweeney said, pointing north.

On the ground Fred Ashworth began to have misgivings about the efficacy of the mission: 'Gasoline consumption at high altitude,' he later told Brigadier General Farrell, 'failure to rendezvous, and time over primary target forced decision to drop rather than attempt questionable chance of reaching Okinawa with unit.' The delays and severe fuel shortage – and not a fix on the targeted part of the city, the Mitsubishi Shipyards – had compelled them to release the bomb; the alternative was dropping the 10,000-pound projectile at sea to enable them to get home.

Fat Man detonated somewhere over Nagasaki – but where, precisely, Ashworth wondered. Was the target of any military significance? The stony-faced General Jimmy Doolittle, one of America's most famous pilots, who led the first retaliatory air strike against Japan in 1942 and who now commanded the Eighth Air Force, asked Sweeney a similar question:

'What was the extent of the damage?'

'I can't be sure, General. Smoke obscured the target.'

'But you hit the target?'

'Yes. Definitely, sir.'

Back at Tinian, anxious to quell his 'small doubts', Ashworth interviewed the crews of the weather and reconnaissance planes. The bombed area appeared to contain a steel or weapons factory: 'Preliminary conference ... places impact approximately on Mitsubishi Steel and Arms Works, target number 546.' The bombardier, Beahan, reckoned the weapon landed about '500 feet [150 metres] south of the end of the Mitsubishi Steel Works'. These conclusions correlated, and Ashworth felt confident enough to say 'that the bomb was satisfactorily placed and that it did its job well'. He thus divined the point of detonation, as a military target, after the event. It was a salve to a difficult mission, no doubt, and a strong line for the press, but in August 1945 the city had little, if any, practical military significance: the shipyards no longer made ships; the steelworks had few resources to make anything of use to the war effort. Both bombs' primary purpose, as the Target Committee had made amply clear, was to demonstrate their power to destroy cities and shock the regime into surrender.

Farrell cabled Groves with news of the success. The men had carried out 'a supremely tough job ... with determination, sound judgment and great skill ... Ashworth and the pilot Sweeney were men of Stamina [sic] and sound heart. Weaker men could not have done this job.' The rest of Sweeney's crew returned to Tinian that night to a gloomy reception: no lights, cameras or crowds, as had attended the *Enola Gay*'s return. It was late; the men were asleep. Only Tibbets and an admiral standing in the darkness welcomed them home. The beer had run out, so they celebrated with medicinal whisky.

·

A clearer picture of the 'unprofitable areas' to which Ashworth had referred soon emerged: the plutonium bomb detonated 1640 feet (500 metres) above Matsuyama, several kilometres north of the city centre, over the densely populated Urakami district – Japan's largest Christian community and the city's medical and educational district, then crammed with additional, mostly Buddhist, citizens who had been evacuated there. Fat Man exploded a few hundred metres from Urakami Cathedral, the spiritual heart of the area, at precisely 11.02am local time. Subject to a force equal to 22,000 tons (20,000 tonnes) of TNT – almost twice as powerful as the Hiroshima bomb – the surrounding hospitals, shrines and schools were wiped out. More than 39,000 people died instantly – fewer than in Hiroshima because the hillsides contained the bomb's shockwave within the narrow valley. Of the 12,000 to 14,000 Japanese Christians living in Urakami, 8500 were killed – including some 50 Catholics waiting in line to confess to Fathers Nishida and Tamaya at the cathedral. Only the brick facade remained of the nearby French convent 'Les Soeurs de l'Enfant Jésus', in the Josei Girls' High School; the bodies of a few Japanese nuns were flung into the grounds and expired in the flames (foreign nuns were interned in Kobe at the time). Rebuilt soon after the war, then abandoned as if cursed, Josei is known locally as 'the school that disappeared in the atomic bombing'.

Within a kilometre's radius of the bomb, 'no living creatures were seen inside or outside buildings'. The blast wave raced through the sewage ducts of the Shiroyama National Elementary School, a three-storey reinforced concrete building, and blew up the concrete cesspool. In the playground, trees half a metre in diameter were ripped up and tossed away, and the pavement ground to powder. Of Shiroyama's 1500 schoolchildren, 1400 ceased to exist, as well as 105 labourers at the Mitsubishi factory located in the school's grounds. In a shelter beneath the playground a lone mother cradled her child, crying, 'Don't die! Oh please don't die!' Both mercifully passed away, according to a witness.

The wooden Prefectural Keiho Junior High School was 'obliterated, original form could not be traced', noted a Nagasaki report. Ten teachers and all 187 children – that is, those who had not been evacuated – were killed instantly. Among the 'Measures Taken Immediately After the Disaster', a surviving teacher rescued the Imperial Portrait and some of the wounded.

The Mitsubishi Steel and Arms Factory at Ohashi was obliterated; but its primary military function lay underground, in the torpedo works that made the torpedoes released at Pearl Harbor. This was relatively undamaged. There were no other significant military works in the area.

The city centre, 3 to 5 kilometres south, experienced comparatively little damage; across the bay the great Mitsubishi Shipyard – the designated 'aiming point' – was not severely damaged; 195 POWs (152 Dutch, 24 Australian and 19 British) were enslaved there. By night they were confined to the Fukuoka POW camps in Saiwaimachi, 1.5 kilometres from the blast, and were marched to the shipyards each morning. Eight POWs died in the atomic blast; seven were Dutch, who were burnt or crushed to death, including the camp leader, whose head was found wedged between two beams under a pile of rubble, recalls Jurgen Onchen, a captured Dutch soldier. The bomb also killed British airman Corporal Ron Shaw, 25, captured after his plane was shot down over Java. The concrete walls of the factory or their cells protected many POWs. In total, over the duration of the war, 113 POWs died in the camp – mostly of illness – of whom 97 were Dutch, 11 Australian and five British. Tasmanian Gunner Ted Howard clearly remembered the bomb: 'A tremendous blue flash' lit up the prison building, then silence, before the walls started collapsing. He threw himself on the floor; 'then everything began to burn and we got out'. The survivors wandered free, 'in tremendous heat through a city that was burned and blasted flat'. For days after the blast, many helped rescue their fellow prisoners, wounded and buried under the rubble. At the time they ignored rumours of radiation poisoning.

The Buddhist Ideguchi family had moved to the Christian quarter because their city home had been demolished to make way for a firebreak. Their allocated residence stood on the gentle slope of the Urakami Valley, in the shade of four large pine trees. Except for a newly built bricked-off bathroom and the sturdy 'Western-style' spare room on the north side, the house was made entirely of wood and paper in the Japanese tradition.

Just before 10am that day nine-year-old Teruo Ideguchi and his two older brothers, Toshi and Masao, returned from a mission to catch white-eyes. By putting a female white-eye in a cage at the end of a branch smeared with mulched leaves, they were able to lure the male bird down to a waiting net. This morning they caught only cicadas: the cicadas would save their little sisters' lives.

At 11am Teruo, his brothers and a friend, Shizuo Iwanaga, were lazing around in the lounge room reading magazines. Teruo lay on the floor; his brothers sat facing north; Shizuo sat opposite them, looking south. Teruo's sisters, Nobuko, six, and Fusako, three, were playing in the garden with the captured cicadas. Nobuko had tied a string around one and flew it into her younger sister's hair, where it got caught. The girls could not unravel the insect and Fusako started to cry. They ran inside to find their mother, who was putting clothes away in the spare room.

Teruo heard a plane descending. There was no air-raid warning, and nobody cared: 'We'd heard planes all week.' He looked out the window. For a split second one of the pine trees appeared perfectly silhouetted against a bright yellow flash. The flash turned pink as the blast wave smashed the house apart and threw him across the room. Teruo landed near the base of the brick wall of the new bathroom, lacerated, under rubble. The wall sheltered him from the house's collapsing roof. 'I felt completely stupefied,' Teruo said. He curiously remembers the sight of 'a *tatami* mat stuck to the ceiling, and a Japanese sword (*katana*) ... stabbed through it'.

His little sisters reached the spare room just before the blast wave struck; its solid walls saved the girls and their mother, who lay lightly injured under fallen tiles. The blast flattened the rest of the house, and every house in the area. Teruo's brothers survived: one lay under a cupboard in the kitchen, with glass shards through his arm; the other was blown outside. Their friend Shizuo lay dead.

The family's maid, aged 20, had just got home with a bag of rations when *Bockscar* began its descent. Persuaded the enemy would not attack Urakami, she walked down the hall to the front door to watch. In a moment of life-saving serendipity she stood in the shelter of the door frame when the bomb detonated: had she walked any faster, she would have died of burns; any slower, the roof would have crushed her.

She dragged herself out of the rubble, ran to the north side of the house and started tearing through tiles and debris to free the family, who were crying out. The mother, her three sons, two daughters and their heroic maid fled north to a train station. Later that day they squeezed into a train carriage, bound for a medical facility in Isahaya, 25 kilometres north. 'Some would be alive next to me one minute, and dead the next,' Teruo recalled in an interview with me.

All Teruo's school friends were killed: of 1581 students at the local Yamazato school, 1300 died; as well as 28 of 32 teachers – most of them in their homes nearby. Like Hiroshima's Honkawa school, Yamazato stood unobstructed in the path of the flash and blast waves. Nineteen-year-old Sadako Moriyama was wandering around the playground at the time, looking for her little brothers, who were chasing dragonflies. At the sound of the plane, she herded them towards the school shelter. They reached the entrance when the shock threw them against the far wall. She blacked out, then regained consciousness to find her burnt brothers weeping at her feet. Somehow the light got in and she saw at the shelter's entrance 'two hideous monsters ... making croaking noises and trying to enter'. They were schoolchildren. Outside four little burnt creatures were sitting, still alive, in a sandbox. She screamed and dashed back in the shelter as the cry '*Mizu! Mizu!*' pursued her from the playground. She passed out.

·

Fifteen-year-old Tsuruji Matsuzoe, who hoped to become a junior high-school teacher after the war, found himself assigned to work in the underground torpedo factory with hundreds of other mobilised children. He finished his night shift at 7am and returned to the factory dormitory, a three-storey concrete building a short walk away. He ate a rice ball and fell asleep on a futon mat. He had eaten well the night before: his Mitsubishi employers had served up white rice, eggs and vegetables, a treat to commemorate the bombing of Pearl Harbor.

A great jolt and gale of intense heat awoke him. 'It felt like someone had poured a bucket of hot embers on me; I thought someone was playing a joke. I was trying to flick the fire off. When I came to I realised that the entire building was collapsing and covered in dust.'

He ran into the hallway: the building seemed on a slant. Hundreds of students were rushing about, trying to escape. Fearing another air raid, he jumped from his bedroom window 3 metres to the ground. He wore only a light *yukata* over his underwear. He ran through the paddies and into the hills to a bomb shelter dug into the road embankment.

He sat in a hot shelter – a mere hole in the hillside – with strange students and farm workers. His arms, chest and right leg were burnt. His right arm started to blister. Unable to bear the pain, he left the shelter to seek medical help. He returned to the dormitory, now ablaze. Dozens of children were trapped inside, crying out. Tsuruji knew several. He identified one voice as that of a friend, Toyosaki, who screamed repeatedly. Teachers and students tried to break in: 'We're going to save you!' they cried. But the sliding paper screens (*shoji*) caught fire, and the timbers were too hot, too heavy. The flames soon consumed the wooden interior and the dormitory collapsed, burying the sound of voices.

A small boy standing next to Tsuruji wept inconsolably throughout this scene. Nobody had noticed him. The boy had failed his school entrance exams and got a job cleaning classrooms and making tea. The students tended to shun him or laugh at him. Here he stood clutching his stomach from which a huge shard of glass protruded, trying to make himself heard: 'Sir, what should I do?' the little boy implored a teacher. 'My stomach is coming out.' The teacher told the boy to lie down and hold his stomach in until rescue teams arrived. An older student tried to reassure him as his life slipped away.

Tsuruji left. His *yukata*, black with dried blood, stuck to his body like a shroud. 'I was getting close to dying of blood loss. I felt faint. I had no energy and felt sick and nauseous. I had no shoes.'

He decided to return to the torpedo factory along a main road that wound up the valley towards Michinoo Station to the north, from which he heard that trains were taking people out of the city. Up this road streamed thousands of people, 'some injured, some naked, some leaning onto each other, holding each others' shoulders and helping them along. Some sat down to rest and then could not stand up again,' he recalled. They advanced on faltering steps, their hands hanging out before them, with the same unholy pathos as the Hiroshimans.

It was mid-afternoon. From the hills he looked down the valley, beyond the approaching crowds, to the underbelly of a dense, reddish black cloud that lingered over Nagasaki like a malign visitation: 'I remember wondering why it was so black, like a black curtain over everything.'

On the way fellow students from the dormitory joined him. As at Hiroshima, social codes and small modesties lingered incongruously among the people. Older students, in line with the authority that entitled them to beat younger boys, ordered Tsuruji to give his *yukata* to their history teacher, a man called Buhei Hishitani. This teacher had a badly burnt head and blue face, Tsuruji noticed, and needed a bandage: 'If I did that, I would have been left in my underwear … I was scared of what would happen to me if I disobeyed, so I just acted as if I had not heard and kept walking. I remember feeling very angry, which was unusual for me. Then Mr Hishitani said "No, that is not necessary." After that the older students didn't say anything.'

The torpedo factory was empty. There were no medicines or doctors. The students dabbed cooking oil – used to lubricate the machines – on their burns and departed. Mr Hishitani and Tsuruji lay down to rest. 'At the time, I didn't realise my teacher was also there.' Tsuruji awoke some time later and resumed his odyssey. He returned to the road leading to Michinoo Station. On the way he joined a line of people queuing to drink from a well. Soon more arrived, many more, pushing in, until hundreds shoved him aside, crying, '*Mizu*.' 'I waited for so long but never reached the front … So I continued to the station.' He later heard that rapid rehydration killed the badly burnt.

Just after 6pm he boarded a train packed with sick and wounded children. They were told to get off at Nagayo and go to the local national elementary school for help. On arrival, Tsuruji collapsed. Local volunteers laid him on the floor of the school gymnasium. Here a medical team, comprising one doctor and three nurses – sent from Ureshino Naval Hospital, 50 or so kilometres east of Nagasaki – performed triage day and night on thousands of arrivals.

'The doctor and nurses were crouching and looking down,' he remembered. 'The nurse was holding a candle and I could see her eerie shadow cast on the roof of the building. I remember that image very well …' He fixed his eyes on the nurse's shadow and grew delirious and started muttering. 'Without meaning to I was raising my voice and acting strangely. That is when I was given the injection … probably something like a tranquilliser … I have no more memories of that night.'

The boy awoke with 'something on my face', irritating, painful. A straw mat called a *mushiro*, used to cover the dead, had been laid over him. 'I raised a finger in the air … someone came past and said, "Hey, there's something moving … Maybe there's someone alive in there?"'

Tsuruji had been cast aside as a corpse, sleeping amid piles of the dead. Volunteers lifted him out and he opened his eyes: all over the floor were *mushiro* mats. Years later he sympathised with the doctors' error: 'Treating so many badly injured people in candlelight ... such mistakes can happen.' The bodies were taken away by horse and cart and cremated.

He spent the next few days at the school clinic. On 11 August his father, alerted by a friend from a neighbouring village, arrived. 'My father was not a very talkative person, but he came up to me and said something simple like, "How are you feeling?" I told him he should have come earlier. Later on I apologised for being so harsh.' They took a train to Kawatana, the family's village, where Tsuruji was admitted to the nearby Shibukawa Hospital. His burns healed and he returned to school in December.

•

A few medical teams survived the blast and acted immediately to help others: Tatsuichiro Akizuki, the only doctor to walk away from the wreckage of Urakami No. 1 Hospital, stayed in the stricken area to treat 70 in-patients and more than 300 seriously wounded who later arrived. He toured the bomb shelters, 'fighting hopelessness to administer treatment to the fatally wounded'. Similarly, Dr Raisuke Shirabe of Nagasaki Medical College and his staff dragged themselves without rest around the smoking rubble, tending to hundreds of wounded; Dr Shirabe continued to treat people for weeks afterwards, despite the loss of his two sons.

Dr Takashi Nagai, the dean of the radiology department of Nagasaki Medical College, set a similar example – despite suffering from leukaemia. That morning, just before the plane approached, he walked to his office at the college, a ferro-concrete building 800 metres from Urakami Cathedral. He began preparing a lecture when an 'invisible fist' smashed through the windows. He lay under rubble with a deep cut to his right temple and mumbled for God's forgiveness: the zealous Catholic convert had intended to confess three sins that afternoon.

Next door, in the X-ray room, Nurse Hashimoto clung to a bookshelf bolted to the floor. The room seemed to sway and bend; the primitive X-ray machines broke up and scattered. Through the window, beyond the smoke and wreckage, the formerly green Mount Inasa glowed red, like a large ember rising out of the bay. She ran down the hall, protected by the heavy walls of the X-ray department, and found five other survivors; they dragged out Dr Nagai and bandaged his head. 'Help the patients!' he cried. With wet hand-towels over their faces the nurses plunged into the smoke-filled wards and led or stretchered the patients out. Dr Nagai hurried downstairs to the underground emergency theatre, now flooded and useless. His illness and blood loss – his bandages now resembled a red turban – overwhelmed him. His colleague Dr Raisuke Shirabe used a novel technique to staunch the blood flow: he pressed a tampon into the wound and sutured the skin over it.

Thousands of near dead staggered up the Urakami Valley towards the hospital. The wounded carried the fatally wounded; children dragged their dying parents; parents clutched their children's bodies. They bumped and shuffled up the hillsides, glancing back at the fires that drew closer, and one by one they collapsed from dehydration or exhaustion.

At the sight Dr Nagai and his team 'started to lose our nerve', he later said. The hospital and medical college were virtually destroyed; the ruins of five auditoriums later revealed the charred remains of hundreds of students sitting at their desks. A heavy roof had fallen on Dr Nagai's First Year; all except for one died like butterflies pinned to a specimen board. Nothing useful remained in the college, no medicines or equipment. Sensing his colleagues' despair, Dr Nagai issued an order: 'Quick, find a *Hi no Maru* [a Japanese flag].' This struck his friend Dr Ogura as absurd, but Dr Nagai insisted: when a flag could not be found, he grabbed a white sheet and smudged a red circle in the centre with his blood-soaked bandages. Dr Ogura tied the flag to a bamboo pole and drove it into a clearing on the hill above the hospital – a rallying point beyond reach of the fires. 'It was so simple an act and yet the psychological effect was profound,' he remembered. Dr Nagai then directed the construction of a field clinic on a stone embankment near the clearing.

Fires, not a firestorm, consumed Urakami; droplets of black rain extinguished the flames. Dr Nagai scoured the lines of refugees coming up the hill for any sign of his wife. When she did not appear his strength broke: 'She's dead, she's dead!' he cried, and resisted a terrible urge to rush headlong into the smoke-filled valley to find her. His legs gave way and he fainted. He awoke on a stretcher in a field, looking up at a thinly curved moon. Men were building shelters for the injured as nurses boiled vegetables in air-raid helmets over open fires. Somewhere people sang, 'Umi Yukaba' (If I Go Away to the Sea), a popular patriotic song. He got up and took the hands of his medical staff; they sat in silence, looking out over the remains of Nagasaki.

●

Truman heard of the success of the Nagasaki mission through cables from Farrell and Groves. His reaction was sombre; gone were the exuberant speeches, fellow back-slapping, mutual congratulations. His mood was sober, defiant – and vengeful. The Japanese people had brought this on themselves, his radio address made clear that day. 'We have used it against those who attacked us without warning at Pearl Harbor ... against those who have abandoned all pretense of obeying international laws of warfare.' Truman drew no distinction between civilian and soldier; mother and murderer; child and monster. In a total racial war, there were no distinctions; the victor wrote the laws of war. The 'Japanese people' had inflicted these atrocities on America, Truman said in tones that suggested he protested too much. 'They' had broken the laws of war; 'they' must pay for the crimes of their masters. He reflected the feelings of mainstream America, for whom 'the Japs' were collectively guilty.

Condemnation of the double bombing was tentative at first. The muted disapproval of the American churches sounded rather like a chastisement of man's folly and an appeal to the better angels of their parishioners than a judgment on the war or nation. Mankind, not America, had erred, consoled their church leaders; mankind had collectively brought this abomination on the world, and the Japanese were complicit in their coming doom.

A few clergymen were specific in their condemnation. The day after Nagasaki, Bishop Garfield Bromley Oxnan, President of the Federal Council of Churches of Christ in America, and John Foster Dulles, President of the Council's Commission on a Just and Durable Peace, warned Americans in the starkest terms to suspend atomic war or risk Armageddon. 'One choice open to us is immediately to wreak upon our enemy mass destruction such as men have never before imagined. That will inevitably obliterate men and women, young and aged, innocent and guilty alike, because they are part of a nation which has attacked us and whose conduct has stirred our deep wrath. If we, a professedly Christian nation, feel morally free to use atomic energy in that way men elsewhere will accept that verdict. Atomic weapons will be looked upon as a normal part of the arsenal of war and the stage will be set for the sudden and final destruction of mankind.'

If atomic power were to be 'a powerful and forceful influence towards the maintenance of world peace', as Truman had said, the time to prove it was now, Bishop Oxnan pleaded. Oxnan and Dulles urged 'a temporary suspension ... of our program of air attack on the Japanese homeland to give the Japanese people an adequate opportunity to react to the new situation'. That 'will require of us great self-restraint. However, our supremacy is now so overwhelming that such restraint would be taken everywhere as evidence not of weakness but of moral and physical greatness ...'

The Reverend Dr Bernard Iddings Bell of Providence, Rhode Island, a prominent cleric, delivered a more blistering critique. His sermon to a noonday service at Trinity Church, Wall Street and Broadway, Lower Manhattan, laid the wrath of God at the window to the American soul. Victory gained by nuclear weapons would be 'at the price of worldwide moral revulsion against us', he declaimed. 'The Orient has long perceived that Anglo-Saxon diplomacy is based not on Christian principles but on canny Imperialistic expediency. Now it has been shown that our methods of war are ... cold-bloodedly barbarous beyond previous experience or possibility.' America annihilated 100,000 persons, most of them civilians, at Hiroshima, he said; and then, in spite of 'universal horror', repeated the performance at Nagasaki.

A minority shared this sentiment, which must be set against the views of the overwhelming majority of Americans at the time, who supported the use of the weapon for a variety of reasons; chiefly the hope that it would end the war, bring the boys home and avenge Pearl Harbor. Prominent among them was the Georgian Senator Richard B. Russell, one of the most influential and well regarded in Washington, whose sentiments reflected a fair swathe of American opinion. He sent this telegram to the President the day after Hiroshima:

Permit me to respectfully suggest that we cease our efforts to cajole Japan into surrendering in accordance with the Potsdam Declaration. Let us carry the war to them until they beg us to accept the unconditional surrender ... If we do not have available a sufficient number of atomic bombs with which to finish the job immediately let us carry on with TNT and firebombs until we can produce them. I also hope that you will issue orders forbidding the officers in command of our air forces from warning Jap cities that they will be attacked ... Our people have not forgotten that the Japanese struck us the first blow in this war without the slightest warning. They believe that we should continue to strike the

Japanese until they are brought groveling to their knees. We should cease our appeals to Japan to sue for peace. The next plea for peace should come from an utterly destroyed Tokyo. Welcome back home. With assurances of esteem.

Richard B Russell US Senator.

'Dear Dick,' the President replied, on 9 August, the day of Nagasaki,

... I know that Japan is a terribly cruel and uncivilized nation in warfare but I can't bring myself to believe that, because they are beasts, we should ourselves act in the same manner.

For myself I certainly regret the necessity of wiping out whole populations because of the 'pigheadedness' of the leaders of a nation and, for your information, I am not going to do it unless it is absolutely necessary. It is my opinion that after the Russians enter into the war the Japanese will very shortly fold up.

My objective is to save as many American lives as possible but I also have a humane feeling for the women and children of Japan.

Sincerely yours,

Harry S Truman.

Truman's replies were carefully calibrated to offset the zealous counsel of church and state. Merciful in answer to the aggression of senators, he threatened a continuation of the atomic blitzkrieg in response to the softness of the clergy: while no one was more troubled than he over the use of atomic weapons, he replied to the Federal Council of Churches, Japan's unwarranted attack on Pearl Harbor and the murder of prisoners greatly disturbed him: 'The only language they seem to understand,' he wrote, 'is the one we have been using to bombard them. When you have to deal with a beast you have to treat him as a beast. It is more regrettable but nevertheless true.'

CHAPTER 20

SURRENDER

Now that the Soviet Union has entered the war to continue
[fighting] would only result in further useless damage and
eventually endanger the very foundation of the empire's existence.

Emperor Hirohito's second edict of surrender to the armed forces, 17 August 1945

AN HOUR BEFORE THE SECOND atomic bomb fell on Nagasaki the Big
Six met in the shelter beneath the Imperial Palace. A tedious debate
about how to surrender in light of the Russian invasion proceeded in the
hot little room; the leaders sank deep in their chairs and the usual
hopeless divisions emerged. 'We can't get anywhere by keeping silent
forever,' noted the unusually outspoken Navy Minister Yonai. The 'peace'
and 'war' factions were split equally over whether: (1) to surrender in line
with the terms of Potsdam on condition that the Emperor be preserved;
or (2) to surrender with four conditions attached: that the Imperial
House remain intact; that Japanese forces be allowed voluntarily to
withdraw; that alleged war criminals be tried by the Japanese government;
and that Japan's mainland territory remain free of foreign occupation. In
short, fantasy vied with delusion for a claim on their minds.

Moderates Suzuki, Togo and Yonai supported the first path;
hardliners Anami, Umezu and Toyoda the second. The latter controlled
the armed forces, whose officer class continued ferociously to resist
any talk of surrender. Nothing of great moment had occurred in

Hiroshima to persuade them of the futility of further defiance; the militarists scorned the weapon as a cowardly attack on defenceless civilians. Towards the end of the interminable discussion – now into its third hour – a messenger arrived with news of the destruction of Nagasaki – by another 'special bomb'. The Big Six paused, registered the news, and resumed their earlier conversation. The messenger, bowing apologetically, was sent on his way. 'No record … treated the effect [of the Nagasaki bomb] seriously,' noted the official history of the Imperial General Headquarters.

In an effort to break the impasse, Cabinet Secretary Sakomizu proposed a full cabinet conference later that day. It began at 2.30pm. For hours, the 16 members (including the Big Six) examined the situation – chiefly the Russian threat – from every perspective, hammered out their arguments, and honed their ancient references and sophistries – as Nagasaki burned. After seven hours the impasse remained and Suzuki declared an intermission. They broke for dinner.

During this epic, the Emperor's inner circle of courtiers – led by Prince Konoe, whose mission to Russia never eventuated – were hatching a secret peace plan of their own to end the war. It hung on a conversation in the Imperial library that afternoon between Marquis Kido and Hirohito. There is no record of the discussion, but according to Japanese sources Kido attempted to persuade Hirohito that His Majesty's survival depended on acceptance of the Potsdam terms, with one condition attached: that America accept the preservation of the Imperial order within the national laws of Japan. The Emperor agreed, and the Imperial nod moved Japan a step closer to the hitherto unthinkable: Hirohito's open intervention to stop the war.

A little later that day Suzuki and Togo, with Secretary Sakomizu taking notes, met in the Prime Minister's office: 'What should we do?' Sakomizu asked.

'How about this?' Suzuki replied (by then, he had heard of Kido's talk with Hirohito). 'Go to the Emperor, report the conferences in detail, and get the Emperor's own decision.'

One did not simply 'go to the Emperor': two nuclear strikes and the Russian invasion were not permitted to disturb the laborious etiquette of being granted an audience with the Imperial presence. The minimum requirement – expected of the Prime Minister – was the dispatch of an official statement outlining the purpose of the meeting.

To describe the progress as glacial is to understate the empire's adherence to form, however ponderous or unhurried: at 11.50pm that night, the Emperor, the Big Six and Baron Kiichiro Hiranuma, an extreme nationalist and President of the Privy Council, met in the Imperial shelter. Each wore a formal morning suit or military uniform, carefully pressed; they carried white handkerchiefs, and sweltered in the badly ventilated shelter just 5.4 by 9 metres. Cabinet Secretary Sakomizu read the Potsdam Declaration; the reading was 'very hard', he later wrote, 'because the contents were not cheerful things to read [to] the Emperor'.

One by one the Big Six gave their opinions, starting with Foreign Minister Togo. Again, they were drearily divided, but Togo's tilt at realism struck a chord: Japan should impose a single condition, not four, he insisted. The withdrawal of troops could be dealt with later, he argued; the future of war criminals did not 'justify the continuation of the war'. Only the fate of the Imperial Household was 'non-negotiable', as 'the basis for the future development of our nation'. Japan should insist only on the preservation of the dynasty.

'I totally agree,' said Navy Minister Yonai.

'I totally disagree,' barked War Minister Anami, leaping to his feet. Japan must fight on, he argued. Japan would 'lose its life as a moral nation' if it accepted '… the annihilation of the Manchurian state'. The four conditions must be met, Anami warned. The War Minister's absolute control of the army fortified his desertion of reality and nobody dared challenge him. He concluded with a death sentence: 'We should live up to our cause even if our hundred million people have to die … I am sure we are well prepared for a decisive battle on our mainland even against the United States.'

'I absolutely agree,' chimed in the equally belligerent Chief of Army General Staff Umezu. 'Although the Soviet entry into the war is disadvantageous … we are still not in a situation where we should be forced to agree to an unconditional surrender.' He insisted on the four conditions 'at the minimum'.

The Soviet Union, the loss of Manchuria, the collapse of the Kwantung Army: these were the threats and disasters that governed debate; these were the forces on which Japanese destiny hinged – in the minds of her leaders. The destruction of Hiroshima and Nagasaki was scarcely mentioned. The wretchedness of the Japanese people impinged little on the samurai elite, spellbound by the whisper of their ancestral exhortation to die with honour: 'The sudden death of ten key men would have meant more than the instant annihilation of ten thousand subjects,' noted the historian Butow. 'Hiroshima and Nagasaki were in another world.'

A long interrogative refrain by the President of the Privy Council revealed the low priority the meeting attached to the atomic bombs. Near the end of a great list of questions about the Soviet invasion and the state of Japan's food supply, between his concerns about air raids in general and the paralysis of public transportation in particular, Baron Hiranuma asked: 'And are you confident in our defense against atomic bombs?'

Poker-faced Umezu, a stranger to understatement, replied, in all sincerity: 'Though we haven't made sufficient progress so far in dealing with air raids, we should expect better results soon since we have revised our tactics. But there is no reason to surrender to our enemies as a result of air raids.'

Hiranuma concluded his cross-examination with a ticking off on a matter of state: contrary to what Togo had earlier suggested, His Majesty was not a constitutional monarch with a 'legal position'. The Foreign Minister had gravely misinterpreted the status of the Imperial

House; the Emperor, in fact, was a living god whose theocratic powers were unrestrained by any law; the Japanese were spiritually bound to preserve this understanding of the Emperor and the *Kokutai* 'even if the whole nation must die in the war'.

Hirohito sat silent throughout. A little after 2am, Prime Minister Suzuki rose, bowed to His Highness and made a statement that changed the course of Japanese history: 'The situation is urgent ... I am therefore proposing to ask the Emperor his own wish [*goseidan* – sacred judgment]. His wish should settle the issue, and the government should follow it.'

Under Japanese custom, the Emperor did not decide anything 'by himself'. He was expected to follow the government's advice rather than suffer the indignity of speaking his mind. Only once, in 1936, had Hirohito been asked to intervene in the affairs of state. Now the Voice of the Sacred Crane was prevailed upon to speak again; what he said would end or prolong the war. The peace faction, however, had laid the groundwork and knew the Emperor's mind. Hirohito leaned forward and said: 'I have the same opinion as the Foreign Minister.' That is, Japan should surrender 'unconditionally' – with the single proviso that the Imperial House be allowed to persist.

'I have been told that we have confidence in our victory but the reality doesn't match our projections,' the Emperor continued. 'For example, the War Minister told me that the defense positions along the coast of Kujukuri Hama would be ready by mid-August but it is not yet ready. Also I have heard that we have no more weapons left for a new division. In this situation, there is no prospect of victory over the American and British forces ... It is very unbearable for me to take away arms from my loyal military men ... But the time has come to bear the unbearable, in order to save the people from disaster ...'

A white-gloved hand wiped away His Majesty's tears. His ministers dutifully followed his lead, and burst into tears. Handkerchiefs appeared. 'We have heard your august Thought,' said Suzuki, through the sobbing. They bowed deeply as Hirohito departed. Suzuki moved that His Majesty's 'personal desire' be adopted as 'the decision of this conference'. The war faction was effectively silenced.

Hirohito had deigned merely to express his feelings, not to instruct his subjects. Nor had the Emperor explicitly mentioned the atomic bombs or their victims. The preservation of the Imperial line occupied his mind – and that issue permeated the debate. And yet, his *goseidan* – sacred judgment – had broken the factional division and set an extraordinary precedent. Suzuki concluded the meeting at 3am, 10 August, Tokyo time, and the secretaries drew up the surrender offer.

At 7am Domei News dispatched Tokyo's formal surrender to Washington via the Swiss Chargé d'Affaires in Berne – by Morse instead of shortwave radio to escape military censors (and the eyes of the Imperial Army, who would refuse to accept it). The Japanese government, the statement said, having failed to achieve a peaceful resolution through the offices of the Soviet Union, was 'ready to accept the terms' of the Potsdam Declaration, on the understanding that it would not 'comprise any demand which prejudices the prerogatives of His Majesty as a Sovereign Ruler'.

•

American radio picked up the message at 7.30am on 10 August – a day, incidentally, when Admiral Halsey's carrier-borne planes subjected Japan to 'the most nerve-wracking demonstration of the whole war' – the sustained obliteration of most of the remaining war factories on the mainland.

The Japanese insistence on a single condition perplexed Truman's cabinet, committed as they were to the mantra of unconditional surrender. The President canvassed his colleagues' views at a meeting that morning. Should they accept the condition? Yes, said Leahy: the Emperor's future was a minor matter compared with delaying victory. Yes, said Stimson, who more persuasively argued that America needed Hirohito to pacify the scattered Imperial Army and avoid 'a score of bloody Iwo Jimas and Okinawas all over China and the New Netherlands [sic]'. Later Stimson gave another, more pressing reason to accept: 'To get the [Japanese] homeland into our hands before the Russians could put in any substantial claim to occupy and rule it.'

No, said Byrnes. He rejected his colleagues' consensus; he saw no reason *openly* to accept the Japanese demand, for which a furious American public would 'crucify' the President. Why, Byrnes, asked, should we offer the Japanese easier terms now the Allies possessed bigger sticks: that is, the atomic bomb and the Soviet army? Yet the Secretary understood the Emperor's value at the peace: the Imperial House may be allowed to exist, he reasoned, but it should *be seen to exist* at America's pleasure, *not* at Japan's insistence.

'Ate lunch at my desk,' Truman jotted down later, mightily pleased with Byrnes' contribution. 'They wanted to make a condition precedent to the surrender ... They wanted to keep the Emperor. We told 'em we'd tell 'em how to keep him, but we'd make the terms.' Here was the first clear admission of a presidential compromise: Washington would tolerate Hirohito's survival as a post-war figurehead in order to tame the Japanese forces. The political arguments that had demanded his head as a war criminal were gossamer on the wind.

The diplomatic challenge was how to frame the concession without seeming weak; in short, how to impose a 'conditional unconditional surrender'? The wily Byrnes had the answer. Not for nothing had Stalin called Byrnes 'the most honest horse thief he had ever met'. Byrnes drafted a compromise that read as an ultimatum. In fact, the 'Byrnes Note', a single sheet of paper, was a little masterpiece of amenable diktat: it gave while appearing to take; it demanded an end to the Japanese military regime while promising the people self-government; it stripped Hirohito of his powers as warlord while re-crowning him 'peacemaker' – all in the service of America.

'From the moment of the surrender,' the Byrnes Note stated, 'the authority of the Emperor shall be subject to the Supreme Commander of the Allied Powers ...' Hirohito 'shall issue his commands to all the

Japanese military, navy and air authorities and to all the forces under their control wherever located to cease active operations and to surrender their arms … The ultimate form of government of Japan shall … be established by the freely expressed will of the Japanese people.'

In offering part of what the Japanese wanted, Byrnes' supple diplomacy clarified, for the first time, Hirohito's post-war role. He framed the concession as a stern demand lest the press and the American people interpreted the Note as a compromise, which is precisely what it was. As he read the draft to cabinet, Byrnes laid special emphasis on 'an American' (MacArthur being the likely choice) as 'top dog commander'. There would be no misunderstandings; Stalin would *not* have a slice of the cake – a point with which Truman fiercely concurred: 'We would go ahead without [the Russians].' It was in America's interests, Truman asserted, that 'the Russians not push too far into Manchuria'. American policy had travelled far in a month: from a strident call for Soviet help in the Pacific at the outset at Potsdam … to this rejection of the Red Army's presence in Manchuria. That partly explained the easier terms Washington was prepared to offer the Japanese: to snare their surrender before the Russians got any deeper into Asia. It was a delicate political dance, with Byrnes in the role of chief choreographer.

The Byrnes Note had an unintended consequence: it dragged the Soviet Union's true agenda into the cold light of day. That afternoon in Moscow, the arrival of the draft Note interrupted a meeting of US Ambassador Harriman, Soviet Foreign Minister Molotov and the British Ambassador Sir Archibald Kerr. Moscow was as yet unaware of the contents; and the revelation that America intended exclusively to command a defeated Japan alarmed Molotov – Stalin's mouthpiece – who intepreted it as a direct threat to Russia's grand design in the East, which was to secure a Soviet seat at the Pacific table; a claim on the territory of the defeated nation; and a springboard for Bolshevism in Asia. Stalin had in fact anticipated America's acceptance of

Moscow's claim to the northern half of Hokkaido and joint command over Japan. Reflecting his boss's wishes, Molotov now proposed that Russia's Marshal Aleksandr Vasilevsky share the command of Japan with MacArthur (in fact, days later Vasilevsky sought permission from Moscow to sieze Hokkaido before the Japanese had time to surrender to the Americans). An appalled Harriman rejected the idea as 'absolutely inadmissible': it would give veto powers to the Soviet government in the choice of the supreme commander. The US had not carried the burden of the Pacific War for four years to yield to the Russians, who had entered the campaign two days ago: 'It was unthinkable that the Supreme Commander could be other than American.' Molotov relented, but an open 'tug of war' over the Pacific spoils had begun.

That night Stimson expressed his satisfaction with the Byrnes Note – it reflected precisely what he had been saying, but its rich irony was not lost on this thoughtful elder. Had he not consistently argued that America needed the Emperor? Had not his Potsdam draft permitted the 'continuance' of the Imperial line – that is, until 'the President and Byrnes struck it out' – to appease those 'uninformed agitators' against Hirohito in the departments of State (chiefly Assistant Secretary Dean Acheson) and War (Assistant Director of the Office of War Information, Archie MacLeish) who knew no more about Japan than Gilbert & Sullivan's *Mikado*? Byrnes' 'wise and careful' Note stood a better chance of being accepted by the Japanese than a more outspoken one, Stimson mused. But why, he wondered, had it not been proposed – and couched in such language – in the first place, in an attempt to end the war sooner and avoid further slaughter? Just how he answered this question is not recorded.

The Byrnes Note flashed to Tokyo, via Switzerland, and the wait began, with palpable anxiety in Washington: 'We are all on edge waiting for the Japs to surrender,' Truman wrote. 'This has been a hell of a day.' That afternoon he met Archbishop Spellman, to discuss the Pope's condemnation of the bomb. He had listened to the Church's views, and his concerns deepened. At a cabinet meeting on 10 August he remarked (according to the diary of Henry Wallace, then Secretary of Commerce) that the prospect of wiping out another 100,000 people 'was too horrible'; the thought of 'killing all those kids' oppressed him.

That day, to Groves' frustration, Truman ordered an end to the atomic bombing – its resumption would require his express permission. The Manhattan Project was then four days ahead of schedule in delivering the next plutonium weapon – which would be available for use 'on target' at the first suitable weather 'after 17 or 18 August', Groves told General Marshall, the Chief of Staff. Beneath this message, Marshall had handwritten: 'It is not to be released on Japan without express authority from the President.' Groves was not alone in wanting further atomic strikes; he represented a vast new war industry, working tirelessly to supply nuclear weapons as required. On 10 August Brigadier General Farrell recommended that Tokyo be added to the list of approved nuclear targets. Six plutonium bombs were expected to be ready by October, with possibly a seventh in November, making one bomb available 'every 10 days' from September, projected Colonel L.E. Seeman, Groves' liaison officer.

Truman's order did not stop the atomic production line. Indeed, Groves, Marshall and their senior officers were the first military strategists openly to contemplate the use of tactical nuclear weapons in a land invasion, the possibility of which they still entertained. A conversation on 13 August between Colonel Seeman and General John Hull, Assistant Chief of Staff in the War Department's Operations Division, reveals the thinking: 'The problem now is whether or not, assuming the Japanese do not capitulate, [we] continue on dropping them every time one is made,' Hull said, 'or whether to … pour them all on in a reasonably short time. Not all in one day, but over a short period … should we not concentrate on targets [that is, enemy troops] that will be of the greatest assistance to an invasion rather than industry, morale, psychology, etc?'

'Nearer the tactical use rather than other use,' said Seeman.

Hull replied, 'That is what it amounts to. What is your own personal reaction to that?'

'I have studied it a great deal,' said Seeman. 'Our own troops would have to be about six miles [10 kilometres] away. I am not sure that the Air Forces could place it within 500 feet [150 metres] of the point we want. Of course, it is not that "pinpoint".'

The risk of radiation to ground troops should not inhibit an invasion, they reasoned: an invading army could enter the battlefield 'certainly within 48 hours', Seeman insisted, but 'it is not something that you fool around with'. To which Hull suggested, 'Should we not lay off awhile, and then group them one, two, three?' They decided to get General Groves' 'slant on the thing'.

•

Meanwhile, to their horror, the Japanese people heard of the Soviet invasion – at the same time as they were expected to endure further atomic attacks. On 10 August Japanese newspapers reported the Soviet declaration of war in banner headlines. The news laid bare the extent of the people's vulnerability and ignorance. On the same day, the English-language *Nippon Times* ran a leader on Hiroshima: 'How can a human being,' it asked – wilfully blind to Japan's own record of war crimes – 'with any claim to a sense of moral responsibility let loose an instrument of destruction which can at one stroke annihilate an appalling segment of mankind? This is not war, this is not even murder, this is pure nihilism. This is a crime against God and humanity ...'

The next day Tokyo laid out the brutal truth about the nation in a statement of Spartan clarity leavened with defiance: 'We cannot but recognise that we are now beset with the worst possible situation ... the people [must] rise to the occasion and overcome all manner of difficulties in order to protect the polity of the Empire.' The same day the War Ministry issued in Anami's name an explosive exhortation to arms: 'Even though we may have to eat grass, swallow dirt and lie in the fields, we shall fight on to the bitter end, ever firm in our faith that we shall find life in death.' The Japanese should follow the example of Shogun Tokimune, who repulsed the Mongol invasion of 1281, 'and surge forward to destroy the arrogant enemy'.

The militants went further: the Japanese spirit would somehow overcome atomic warfare, no less: on the 12th, Tokyo Radio and the national newspapers issued instructions – 'Defenses Against the New Bomb' – on how to withstand the nuclear threat: civilians were told to strengthen their shelters and 'flee to them at the first sight of a parachute' (a reference to the parachute attached to technical instruments dropped in advance of the weapon). The cities of Kyushu should expect to be atom-bombed 'one after another'; Kyushu island's 10 million spiritual weapons (that is, the people) must stand and fight America's 'beastliness'; anyone who read enemy flyers and faltered at his place of duty 'has fallen for the devilish strategy of the enemy'. Gloves, headgear, trousers and long-sleeved shirts made of 'thick cloth' should be worn at all times; 'stay away from window glass even if the shutters are pulled down'; carry emergency air-defence first-aid kits, with burn ointment.

Radio broadcasts promoted the miraculous resurrection of Hiroshima and Nagasaki, whose people had recovered phoenix-like from the ashes: the citizens of Nagasaki were 'rising again all over the city with resolute determination'. The volunteer corps were working with 'tears in their eyes and determination for revenge'. Miss Shizuko Mori, 21, offered a shining example to all Japan. The Nagasaki telephonist had stayed at her post after the blast and, ignoring the deaths of several members of her family, continued to connect the red lights flashing on her console: 'I shall fight through even though I remain the only one alive,' she was reported as saying. Her fellow workers were inspired as though by a miracle; and 'the constantly blinking lights of the dials are shining brilliantly evermore in tribute to the determination of these operators,' Tokyo Radio announced. In a similar spirit of misinformation, the radio declared that streetcars, railways, sanitation and telephone services in the A-bombed cities would resume shortly (indeed, some trains and streetcars did resume within days). For his part in the atomic war effort, Governor Nagano of Nagasaki commissioned the design of a special 'field cap', rather like a ski-cap, with flaps over the ears and a visor over the eyes to protect civilians 'from the terrific blast and high heat' of future atomic bombs.

The Byrnes Note's unmistakable compromise obtruded on this defiant realm like a strange new language; it had the perverse effect, however, of deepening the factional divide. Two hardliners, Umezu and Toyoda, argued at a meeting on the 12th that acceptance would 'desecrate the Emperor's dignity' and reduce Japan to a 'slave nation'. Hirohito chided them for drawing quick conclusions; as did Navy Minister Yonai. But the hardliners mocked Yonai: 'They say that I am a wimp,' Yonai confided later to a colleague. A thoughtful man who had in fact opposed Japan's alliance with Germany, Yonai displayed a novel concern for the people's welfare and welcomed the bombs and the Soviet invasion, 'as, in a sense, gifts from the gods': they would hasten the end and offer Japan a chance to quit the war due to 'domestic circumstances' – without having to say they were defeated on the battlefield. He advocated surrender 'not because I am afraid … of the atomic bombs or Soviet participation in the war. The most important reason is my concern over the domestic situation.'

The informal meeting dragged to another incoherent stalemate. After a brief absence, Togo returned to find that Suzuki had radically changed his mind and now sided with the war faction. The furious Foreign Minister threatened to appeal directly to the Emperor to overrule Suzuki: 'If you persist in this attitude I may have to report independently to the Throne.' The intervention of Kido brought Suzuki back from the brink.

Tokyo dithered for two days. The leaders vacillated over the meaning of Byrnes' wording, chiefly the intent, if any, of the lower case 'g' in 'government': the Byrnes Note used 'ultimate form of government'; a Japanese translation rendered this, 'ultimatum form of *the Government*'. By this, did Washington mean the Imperial institution or just the administrative organs of state, wondered the Big Six.* And what did the 'the' entail? In a rare moment of rationality, Tokyo decided that if Byrnes' formula did not incorporate the Emperor, the 'freely expressed will of the people' would prevail to reinstate Him.

As this debate ground on, American air raids continued: between 10 August (when Tokyo offered its 'conditional' surrender) and 14 August, 1000 B-29 bombers attacked Japanese cities and military facilities, killing an estimated 15,000 people. Groves submitted target lists for a third atomic bomb and six cities were slated for complete destruction: in order of priority, Sapporo, Hakodate, Oyabu, Yokosuka, Osaka and Nagoya. Pacific commanders, in talks with Parsons and Farrell, 'expressly recommended' that the next bomb be dropped on or near Tokyo to maximise its psychological impact.

•

Meanwhile, the Imperial forces were determined unilaterally to sabotage the 'peace process'. Japan's losses to the American air war – in terms of casualties and property destroyed – caused not a tic of anxiety in the minds of the hardliners: Japan would never surrender. While the Imperial Navy and Air Force were virtually non-existent, the chief authority for the war devolved on the Army, which drew on hidden supplies of food and ammunition – despite the near-complete disruption of road, rail and water transport.

The Imperial Army's General Staff moved first to misinform Japan's diplomatic corps in Europe: on 12 August the US intercepted a message from the Vice Chief of the General Staff to military attachés

---

* They argued, too, over the meaning of 'subject to' American power: did it mean 'controlled by' or 'obedient to'?

in Sweden, Switzerland and Portugal, which pledged Japan's determination to fight 'to the bitter end': 'Russia's entrance into the war' posed a major threat to the nation, the cables said. They made no mention of the atomic bombs. The army officers behind these futile gestures of defiance rejected the government's defeatism, and it fell to Anami to curb the insurrectionists – placing him in an impossible situation. Officially, the War Minister obeyed the Emperor's *goseidan*, the sacred intervention, as 'absolutely irreversible', and pledged to punish on charges of sedition anyone in the army who defied it. On a personal level, as an old soldier, he deeply sympathised with these bellicose young officers – many of whom were his protégés – and their plea to fight on.

To quell rising tension between peace and war factions, and attempt to devise a reply to the Byrnes Note, Prime Minister Suzuki convened another epic meeting of the Supreme War Council on the morning of 13 August. The ministers ruminated for five hours, lapsing into arcane digressions – on the nature of 'harmony'; obscure metaphors – 'we should accept in a spirit of a worm that bends itself'; and deep references to samurai glory – for example the Genji defeat of the Heike samurai in the 12th century and the rout of the Toyotomi by the Tokugawa during the Edo shogunate. An Imperial summons briefly interrupted these high deliberations, during which Hirohito urged his commanders to suspend all military action throughout the negotiations.

The meeting resumed. Reality loitered like an unwelcome ghost, laying a chill hand on the more sentient officials: Togo grasped the point of the Byrnes Note insofar as it preserved a shadow of the Emperor – at the people's pleasure; Yonai agreed: 'To much regret ... there is no option left to us but to accept.' Anami's ears pricked up: accepting the Byrnes Note would destroy the *Kokutai*, he snapped. The weight of his conflicting loyalties – to Emperor and army – now plunged the War Minister into a state of incoherent bluster: 'We are still left with some power to fight! ... We should do what we should do.'

Suzuki moved to break the stalemate. Having recovered from his apostasy and regrouped with the peace faction, the Prime Minister felt disposed to accept the Byrnes Note – which, he conceded, changed 'little of substance concerning the Emperor' and offered a 'dim hope in the dark'. In this spirit he resolved to ask Hirohito, again, for another *goseidan*. Mere mortals, helpless in Japan's hour of crisis, appealed for further divine intervention.

In a last desperate bid to buy time, Anami tried to stall the process: he urged Suzuki to delay the next Imperial conference by two days; he needed time to consult with the armed forces. The Prime Minister refused: 'Now is the time to act … there is no more time to waste,' Suzuki warned. Anami abruptly left the room.

Suzuki's doctor, who happened to be present, asked the Prime Minister why he could not wait a few days.

'I can't do that,' Suzuki said. 'If we miss today, the Soviet Union will take not only Manchuria, Korea, Karafuto, but also Hokkaido. This would destroy the foundation of Japan. We must end the war while we can deal with the United States.'

'You know that Anami will commit suicide?' the doctor replied.

'Yes, I know, and I am sorry.'

That night six officers subjected Anami to a detailed outline of their plan; Anami listened and reserved judgment: if he endorsed the coup they plotted, he defied his Emperor in the act of a traitor; if he tried to stop it, he lost his soldiers' faith and risked assassination.

•

On the morning of 14 August, Hirohito prepared to deliver his second sacred intervention at the hastily convened Imperial conference: at 10am the entire cabinet, as well as generals, admirals and their secretaries, were summoned once more to the Imperial bomb shelter. This time, the outcome had been decided in advance in a private meeting between Suzuki and the Emperor.

And so, once more, Japan's rulers gathered in the Imperial presence. They walked through the Fukiage Imperial Gardens to the door of His Majesty's shelter, descended the deep staircase, filed along the tunnel and assembled in the conference room beneath the palace. Some had had little time to dress and wore borrowed neckties or clothes exchanged with their secretaries. A reverential silence and prolonged bows greeted the Emperor, dressed in his marshal's uniform and snow-white gloves. Suzuki opened the proceedings. After deeply apologising for bothering His Majesty again, he summarised the latest events and explained their disunity. Would His Majesty listen once more and offer another sacred judgment?

One by one Anami, Toyoda and Umezu, grave and tearful, rose to plead the case for further resistance. Yonai openly opposed them. He dared to demand – 'bravely', a witness said – that 'the sword we brandished be laid down'. There was little further discussion; but Anami seethed with anger at Yonai, whom he detested – as events would later show.

At 11am His Majesty spoke: 'I hope all of you will agree with my opinion,' he began. 'It is impossible for us to continue the war anymore,' he said. While understanding that disarmament and occupation were 'truly unbearable to the soldiers', he continued, 'I would like to save my people's lives even at my expense. If we continue the war our homeland will be reduced to ashes. It is really intolerable for me to see my people suffering anymore … We, with the nation firmly united, should set out for a future restoration by tolerating the intolerable and bearing the unbearable.'

He offered to read an Imperial Rescript of surrender on public radio: 'I will stand in front of the microphone at any time. As we have not informed people of anything so far, our sudden decision will be very disturbing to them.' He asked the military chiefs to take control of their ranks and the government to draft the edict: 'The above is my idea,' he stressed. From every corner of the room rose the sound of sobbing at His Majesty's 'holy words'. Overcome by grief, the delegates could barely rise from their chairs at the end of it. 'Through the long tunnel back to the surface, in the car, back to the Prime Minister's residence,' recalled Hiroshi Shimomura, director of the Cabinet Information Bureau, 'we could not suppress our tears every time we remembered the scene.'

On 14 August the Imperial government issued a statement to the 'United States, England [sic], Russia and China': 'The Emperor hereby proclaims the acceptance of the terms of the Potsdam Declaration'; he 'is also prepared to turn over all battle and seige equipment' in possession of the Imperial forces to the Allies, and 'to issue all orders required by the High Command of the Allies that may be necessary to carry out the [Potsdam] provisions'. As Grew, Stimson and other senior officials had warned, the Emperor's authority proved critical to the surrender process; but not even Hirohito's intervention would quell the army's more extreme malcontents.

•

The regime's last days degenerated into a peculiarly Japanese farce, marked by plot and counter-plot, a failed coup, an assassination and a dismal case of *seppuku*. Incensed by rumours of surrender, convinced the government had influenced the Emperor against his will, on 14 August several army officers reprised their plan to seize the Imperial Palace and establish martial law under War Minister Anami's authority. That day Anami faced down a group of angry officers who burst into his office and pledged to fight on: 'Over my dead body,' Anami shot back, slamming his swagger stick on the table.

The next day enraged staff officers led by Anami's brother-in-law, Lieutenant Colonel Masahiko Takeshita, announced plans to occupy the war ministries, radio stations and Imperial palace, and 'protect' the Emperor from his poisonous advisers. Anami listened once more to their arguments, but refused to back them, pledging his obedience to the Emperor's decision. The plotters were momentarily stunned. Coup leader Takeshita saw the futility and abandoned the plan, but two others – firebrand officers Major Hatanaka and Lieutenant Colonel Shiizaki – persevered, despite the fact that their commanders were bound to act 'in accordance with the Emperor's sacred decision'.

That afternoon, Cabinet Secretary Sakomizu busily drafted the Imperial Rescript – that is, a document written not on the initiative of the author, but 'written back' in response to a request by the recipients – which would announce the surrender. Radio technicians prepared to record the Emperor's voice. There were delays. The Privy Council exerted its legal right to approve the rescript, and the Japanese cabinet argued furiously over the wording. Anami insisted on changing the phrase 'the military situation is becoming unfavourable day by day' to 'the military situation does not develop in our favour'; he also hoped to insert a reference to the preservation of the *Kokutai* to reassure his troops. At 7pm, after hours of wrangling, the approved text migrated to the desk of the Imperial calligrapher who, brushes poised, began his slow and ancient art. No exigencies – not the bombs, the Russians, nor the wretchedness of the Japanese people – were calamitous enough to hasten the scribe's perfect brush strokes. The scroll received the official imprimatur that night above the signatures of cabinet members – Anami's a hurried scrawl. At 11pm Tokyo telegraphed Hirohito's acceptance of the Byrnes Note to Bern and Stockholm, thence to the four Allied powers. The Emperor repaired to his office to record the rescript to the people.

•

Outside, the cratered streets were deserted. The ruins of Tokyo silhouetted in the curfewed silence stood as an ashen reproach to these proud deliberants who felt little responsibility for the misery their policies had brought to millions. A lone black car wound through the rubble towards the palace to attend the Imperial recording. The occupant, Hiroshi Shimomura, director of the Cabinet Information Bureau, gazed at the stars that appeared through the clearing mist and consoled himself that they would be there, still, in a thousand years, when Japan's defeat and degradation 'would have been long since forgotten'.

Elsewhere in the city the army officers' coup gathered momentum. Concocting a story that the whole army backed the insurrection, the instigators persuaded the commander of the 2nd Imperial Guards Division to join them. A potentially bigger ally, Lieutenant General Takeshi Mori, commander of the 1st Imperial Guards, refused his support – and paid that night with his life: in a burst of rage, Hatanaka shot Mori and forged an order in his name to the seven Imperial Guard Regiments to occupy the palace. The ensuing revolt lost momentum – although not without the leaders' near-successful attempt to broadcast their message to 'fight on' over national radio – when General Shizuichi Tanaka, commander of the Eastern District Army, heard of the plan and, furious at this gross act of insubordination, moved to crush the rebellion.

Throughout this tumult Anami sat quietly drinking *sake* at his official residence, contemplating death. That he had not betrayed the activities of the peace faction to the army's fanatics lent a little dignity to the War Minister's miserable last days. His brother-in-law, Takeshita, arrived, hoping to change Anami's mind and persuade him to lead the rebellion. Anami assured Takeshita that the Eastern Army would destroy the uprising (a prophetic warning: an officer later burst in with news of its failure). The two men yielded to the inevitable.

'I am going to commit *seppuku*,' Anami said. 'What do you think?'

Takeshita had no intention of stopping him; he was more concerned by the amount of *sake* his brother-in-law had drunk. Would Anami be able to hold the dagger properly? Anami assured him that he had a fifth-degree rank in swordsmanship (*kendo*) – and if anything went wrong, Takeshita would assist. They talked a little longer: Anami passed on messages to his wife and colleagues, and then, when asked about his enemy Yonai, in a burst of rage screamed, 'Kill him! Kill him!'

After 3am, alone in his room, Anami swallowed his last mouthful of rice spirit, folded his uniform and put on a white shirt – a gift from Hirohito. Squat on the *tatami* mat, facing the palace, he drew a ritual dagger from its sheath, thrust the blade into his stomach, sliced to the right and up, and disgorged his intestines. He struggled to sever his carotid artery, missed, and collapsed in a pool of blood. Takeshita found him breathing: 'No need to help me. Leave me alone,' were Anami's last words. Takeshita thrust the dagger into Anami's throat and placed his brother-in-law's uniform and last testament on the corpse. The latter had left a written 'humble apology' to the Emperor 'for my great crime' (his part in the army's defeat) and a *haiku* of leaden inconsequence:

Having received great favors
From His Majesty, the Emperor
I have nothing to say to posterity
In the hour of my death

Thus ended the life of the man who personified the spirit of the Imperial Army; in coming days, some 2000 soldiers and civilians would follow his example and destroy themselves. Most, however, bitterly obeyed the ceasefire.

•

Near noon on 15 August Hirohito addressed his subjects. In the devastated cities and countryside, now swollen with refugees, in broken temples and shrines and school halls, among the ruins and makeshift shelters, the people assembled around radios to hear the recording of the Voice of the Sacred Crane. Hirohito's archaic Japanese obscured his intent. What did it mean – 'Our Empire accepts the provisions of their Joint Declaration?' Few understood, or had read, the Potsdam ultimatum. The people's ears keened to hear the scratchy recording (see full text in Appendix 7).

'Indeed,' Hirohito continued, 'We declared war on America and Britain out of Our sincere desire to ensure Japan's self-preservation and the stabilization of East Asia, it being far from Our thought either to infringe upon the sovereignty of other nations or to embark upon territorial aggrandizement.' But 'despite the best that has been done by everyone ... and the devoted service of our one hundred million people, the war situation has developed not necessarily to Japan's advantage, while the general trends of the world have all turned against her interest.'

Here, then, was the staggering rationale for Japanese aggression. Japan had not, in fact, 'surrendered', according to this broadcast; Hirohito never used the word. He conveyed the impression that the Japanese had suffered the loss of a great ideal, that forces beyond their control had thwarted Tokyo's benign motives ... words intended to calm the army, navy and air force whose malcontents remained wedded to insurrection. Herein lay the genesis of the myth of Japanese 'victimhood'.

There was another reason why Tokyo had 'decided' to end the war, the Emperor said. 'The enemy had begun to employ a new and most cruel bomb, the power of which to do damage is indeed incalculable, taking the toll of many innocent lives.' The Emperor, the cabinet, the Big Six had barely mentioned the atomic bomb during their long discussions; if an external threat hastened their actions, it was the Soviet invasion. Hirohito now cited the loss of Hiroshima and Nagasaki as contributing to Japan's 'decision' to lay down its arms. Implicit here was the face-saving influence of a weapon of spectacular power that lent a psychological crutch to the regime and the armed forces in their hour of acute humiliation.

Without Japan's surrender, the Emperor continued, atomic war endangered the very survival of the Japanese people and would possibly 'lead to the total extinction of human civilization'. Japan's capitulation, he implied, had heroically delivered the world from nuclear annihilation. 'The inference,' noted Robert Butow, 'was that Japan, by her own act, was saving the rest of the world' – a grotesque travesty that debauched the history of the Japanese government's responsibility for the outbreak of war, the slaughter of millions of Asian civilians and the torture, starvation and massacre of tens of thousands of Allied prisoners. And it callously denied the claims of the actual victims of the bomb: the ordinary people of Hiroshima and Nagasaki.

•

Shock, relief and sadness resounded in the bombed cities. In Tokyo the troops defied the order, shouted '*Banzai!*' and prayed at the Yasukuni Shrine. Conservatives refused to accept the truth. When he heard the address, Zenchiki Hiraki fell into deep depression. 'He didn't believe it,' his granddaughter said. Zenchiki became stricter and meaner, and lived out his remaining days under American occupation in deep delusions of a Japanese victory.

Hiroshimans wept with a mixture of relief and despair. 'I felt very relieved,' Iwao Nakanishi remembered. 'We no longer risked being bombed; we no longer risked a Russian invasion. But I couldn't have expressed this out loud. The Emperor was regarded as a god, so it was shocking for many people to hear his voice. It was inconceivable that he would speak with a human voice.'

Dr Hachiya, who had observed the victims of Hiroshima's bombing from his hospital stretcher, heard only the phrase 'bear the unbearable'. He wondered what could be more unbearable than what his city had borne. He and his hospital staff were furious at the decision to surrender and turned their wrath on the armed forces: the soldiers, he noticed, had deserted their posts and the police 'hid behind the hospital every time an air-raid sounded'. They were nothing more than cowards, he thought, who deserved 'to commit *harakiri* [the vulgar form for *seppuku*, or ritual suicide] and die!'.

The next day Genshin Takano, Governor of Hiroshima Prefecture, officially notified the people of Hiroshima of the end of the war. They raised their heads from their hovels and mats to hear an extraordinary admonition: somehow, they were to blame for the catastrophe, and owed the Emperor an apology. 'You must share the blame' for the desecration of the *Kokutai*, Takano declared. Had not His Imperial Majesty 'graciously and warmly favoured us ... with an insight into His mind? ... all Japanese people must share the blame [for the 'national hardship'] and apologise with deep reverence to His Majesty ...'

The Governor urged the citizens of Hiroshima Prefecture to 'maintain your pride in being Japanese' – and suggested a few ways they might achieve this:

(1) Do not disrupt law and order and maintain moral rectitude. Refrain from panicking as our food and financial circumstances are secure.
(2) Perform your regular duties and do not abandon your job. Make every effort to increase the food supply.
(3) Do not listen to rumours and be deceived.
(4) Take care of war victims.

Nagasaki listened with tearful faces and spent bodies, exhausted by want. Like millions of their countrymen, the family of Kiyoko Mori, the teenage girl whose teacher had asked her class to list the things they would like to eat, gathered around the radio. The Emperor's strange voice, barely distinguishable in the static, drew sobs of defiance from those who understood its meaning. 'It was the first time we had ever heard the Emperor,' she said. 'We learned that Japan had surrendered and we couldn't believe it. All the adults were crying and saying that it was a lie. But there weren't any air raids after that, so we started to believe it was true.'

Two days later Hirohito issued another rescript – explicitly to the soldiers, sailors and airmen of the Imperial forces. This time he urged them to lay down their weapons and surrender – and gave a single reason: 'Now that the Soviet Union has entered the war,' the Emperor said, 'to continue under the present conditions at home and abroad would only result in further useless damage and eventually endanger the very foundation of the empire's existence.' In the eyes of the Imperial forces, then, the decisive factor in the surrender was the Soviet invasion. The Emperor made no mention of the atomic bomb in his second rescript to the Imperial forces; in their minds they had surrendered to a worthy foe.

•

At 7pm, 14 August 1945, Washington time, Truman announced Japan's 'unconditional surrender' to the White House press room. Hirohito had offered 'full acceptance of the Potsdam Declaration ... In the reply there is no qualification.' The celebrations were long and deep, on this first morning of peace in America in four years: the Allies were victorious; the sacrifices of the dead and wounded not in vain. The fact that humanity had bombed, shot, gassed, starved, tortured or otherwise terminated the lives of more than 50 million people; that a clash of global ideologies loomed over the triumphant power blocs of Western democracy and Soviet communism – understandably these terrible consequences were not invited to the victors' celebrations. In Washington and London grateful minds dwelt on the returning servicemen, families, wives, children ... and God. Truman dedicated Sunday 19 August as a Day of Prayer: 'After the two days' celebration I think we will need the prayer,' he told the buoyant media.

America had not defeated Germany's and Japan's 'grandiose schemes' to enslave the world through strength of spirit, industry and arms alone: God had been America's spiritual guide and comrade in arms, Truman declared; 'God ... has brought us to this glorious day of triumph.' The Day of Prayer – 'thanksgiving for victory' and 'intercession to the Most High' – proceeded in the East Room of the White House in the presence of Washington's highest office holders, community leaders and foreign diplomats.

They sang...

O beautiful for spacious skies,
  For amber waves of grain,
For purple mountain majesties
  Above the fruited plain!
America, America,
  God shed His grace on thee,
And crown thy good with brotherhood
  From sea to shining sea!

... and thanked God for the countless acts of service performed selflessly by men, women and children; for husbands restored to wives, sons and daughters to parents; and 'for the unfaltering witness of Thy Church in many places in every land'. They prayed and wept for the fallen.

·

On the morning of 2 September, General Douglas MacArthur received the Japanese surrender in Tokyo Bay aboard the battleship *Missouri*. The admirals, generals and officials converged on the mother ship at anchor amid 260 vessels representing America, Britain, China, Australia and other participants in the victory. After some last-minute wrangling over who should sign the hated surrender document – the Big Six and cabinet had all resigned – Hirohito authorised his new Foreign Minister, the one-legged Mamoru Shigemitsu, First Class of the Imperial Order of the Rising Sun, and General Yoshijiro Umezu, First Class of the Imperial Order of the Rising Sun and Second Class of the Imperial Military Order of the Golden Kite. Dressed in formal wear and top hats, the Japanese party, led by the hobbling Shigemitsu, advanced across the deck and found their places; MacArthur, Nimitz and Halsey emerged from a hatch.

MacArthur's few words befitted the nobler aspirations of the moment: 'Nor is it for us here to meet, representing as we do the majority of the people of the earth, in a spirit of distrust, malice or hatred. But rather it is for us, both victors and vanquished, to rise to that higher dignity which alone befits the sacred purposes we are about to serve, committing all our people unreservedly to faithful compliance.'

The Japanese and the Allied representatives signed. Their signatures proclaimed the unconditional surrender of all Japanese forces 'wherever situated'; and the immediate liberation of Allied prisoners of war and civilian internees. The Emperor, the government of Japan, 'and their successors' – that is, the Imperial dynasty duly recognised – would carry out the provisions of the Potsdam Declaration 'in good faith'.

In an emphatic demonstration of who now controlled Japan, 400 B-29 Superfortresses and 1500 carrier fighters thundered across the sky. The Japanese people had kept their Emperor and lost an empire. 'We shall not forget Pearl Harbor,' Truman told the American people that day, and 'the Japanese militarists will not forget the USS *Missouri*.' God had been America's witness, he reminded them; the Almighty had 'seen us overcome the forces of tyranny that sought to destroy His civilisation'.

CHAPTER 21

RECKONING

Groves: Radio Tokyo described Hiroshima as a city of death ...
'peopled by [a] ghost parade, the living doomed to die of
radioactive burns'.
Rea: Let me interrupt you here a minute. I would say this. I think
it's good propaganda. The thing is these people got good and burned
– good thermal burns.

General Groves in discussion with Lieutenant Colonel Charles Rea, a Manhattan Project medical
officer, about Japanese reports of radiation poisoning

THE BOMBED CITIES WERE LIKE upended graveyards, 'with not a tomb standing'. Bodies were cremated on the spot and, if possible, the bones, or a bone, turned over to relatives, friends or, in their absence, the city hall. The urgency and scale of the task soon made that impossible. The 'mortuary services' – police and civilian defence teams, themselves usually sick or wounded – spent weeks loading human remains onto wheelbarrows and burning them in mass pyres in local schools and field clinics. Boats trawled the Ota River tributaries and Nagasaki Bay for corpses, hooking them out, towing them ashore. 'We were burning from morning to night,' said a volunteer. Hundreds at a time were incinerated in the great furnaces near Nigitsu Shrine, northeast of Hiroshima Castle. Dr Hachiya wondered whether Pompeii in its last days had resembled the city.

'We didn't see blue sky in Nagasaki for long after,' said one worker. 'The school must be full of the spirits of the many people we burned there.' By 21 August, the Hiroshima cremation teams had disposed of 17,865 corpses out of a total 32,959 eventually burned. A tribute to their thoroughness was the control of infectious diseases: there were only 75 cases of typhoid.

Estimates of total casualties on the days of the explosions vary. The following are the generally accepted figures, although they fail to reflect the ongoing casualties, which had more than doubled in Hiroshima and Nagasaki by the end of 1945 largely due to acute radiation sickness. Indeed, by the end of 1945, 25,000 had died as a result of radiation poisoning or diseases associated with it, out of the 160,000 killed or wounded at Hiroshima. Of the total killed in Hiroshima on 6 August, 20,000 were Korean labourers enslaved by the Japanese after Tokyo annexed the peninsula 40 years earlier, as well as hundreds of Chinese forced labourers. Deaths attributable to radiation exposure continue to this day.

These are the best estimates of casualties on the day of the bomb:

	HIROSHIMA	NAGASAKI
Population	320,000	260,000
Dead	78,000	35,000
Wounded	37,000	30,000
Total	115,000	65,000

The ashes of the cities yielded sad mementoes: *bento* lunchboxes, a charred tricycle, children's school uniforms. Keiko Nagai found her dead sister's aluminium lunchbox melted, with the child's chopsticks still attached to the lid – one item among hundreds of schoolchildren's possessions that had been moved to a nearby temple: 'My mother later prised the bento open and we saw her lunch still there.'* And there

---

* The family later donated the lunchbox to the Hiroshima Peace Museum, where it has become one of the best known exhibits.

were the atomic novelties that would later draw the macabre curiosity of millions of museum-goers: clocks and fob watches stopped at 8.15; the shadow of a human body on the steps of a bank.

•

This reckoning of loss and damage draws on several reports: a British scientific mission sent to Japan in September 1945; Hiroshima's and Nagasaki's own prefectural reports; and the biggest study of the Allied air war on Germany and Japan, the presidentially approved United States Strategic Bombing Survey, set up in 1944 on Roosevelt's directive. In August 1945 Truman instructed the Survey to examine the impact of strategic bombing on Japan. Under the guidance of a committee of respected government officials, including Paul Nitze, Kenneth Galbraith and Franklin D'Olier, a thousand researchers fanned out over the ruined land, interviewing hundreds of survivors, inspecting every aspect of the Japanese war effort, from air-raid shelters and food supplies to sewerage systems and mortuary services. The USSBS's findings on Japanese morale, medical services, defensive manoeuvres, and so on – as well as special reports on Hiroshima and Nagasaki – were not published until June 1946, but the researchers witnessed the atomic bombs' aftermath. While some of the USSBS's conclusions remain a source of keen controversy (see Chapter 23), the researchers drew a detailed and largely accurate portrait of the Japanese experience of the US strategic air war.

'An unbroken expanse of flimsy wooden houses' stood here, observed the British scientific mission to Hiroshima in September 1945. The uranium bomb had destroyed 55,000 buildings out of about 90,000 in greater Hiroshima, the study found. Virtually all were schools, offices, hospitals and homes in the centre of town. Few military structures were destroyed: the castle, which housed a communications centre, the Asahi Munitions Company, the headquarters of the 2nd General Army, a barracks and drill grounds, which contained at the time about 10,000 reservists and supply troops (not the 30,000 to 40,000 as commonly supposed), according to the British mission, of whom nearly 4000 were killed immediately and probably double that by the end of 1945. The bomb left undamaged Hiroshima's vital port and military embarkation point at Ujina on the Ota Delta. The military and industrial plants on the city's periphery, which accounted for 74 per cent of its industrial capacity, were undamaged; and 94 per cent of the workers were unhurt – confounding the original intent to kill 'urban workers'. The factories would have resumed normal production within 30 days of the blast, had the war continued, noted the USSBS. In fact, trains resumed running through the rubble of Hiroshima two days after the blast, on 8 August.

The schools 'completely burned' or 'totally destroyed' in Hiroshima included 21 middle (secondary) schools and 18 grammar schools. Schools partially destroyed included eight junior high schools and six grammar schools.

Of the 12,000 schoolchildren aged between 12 and 17 who worked in the city as mobilised labour – and who had attended these schools – 8500 were killed instantly or within weeks; most of the surviving child labourers were severely wounded and irradiated.

Four hospitals – the Prefectural Hospital, Communications Bureau Hospital, Railroad Hospital, Hijiyama Rest Room and Shima Hospital – were completely wiped out, killing all patients and medical staff. The Tada, Red Cross and Army hospitals were mostly destroyed, along with many private clinics and most of their occupants. The Red Cross Hospital retained some reinforced concrete walls, but the blast wave gutted the interior; 90 per cent of the staff and patients died instantly. The ruins served as a field clinic and a symbol of hope to thousands of survivors.

Most of the city's doctors and nurses were killed or wounded and the makeshift field clinics swiftly overwhelmed:

HIROSHIMA	TOTAL	KILLED IN ATOMIC BOMB
Doctors	298	270
Dentists	152	132
Herb doctors	140	112
Nurses	1780	1654

Of the buildings destroyed, the most prominent were the Imperial Headquarters building, Hiroshima Castle, the Gokoku Shrine, the Hiroshima Gas Company and the Fukuya Department Store (reduced to a gloomy cave filled with the sick and wounded). Among public administration buildings 'completely burned' were the Hiroshima Prefectural Office, City Hall, Maritime Transport Bureau, District Courthouse, police stations and fire stations (east and west) including their rescue operations, communications bureau and post office.

Companies 'completely burned' included the Chugoku Newspaper Co., Hiroshima Broadcasting Station, local branches of Domei News Agency, Bank of Japan, Japan Electric Co., Sumitomo Bank, Geibi Bank and the People's Savings Bank.

Of 9600 subscribers to the Hiroshima telephone system, 8600 or 90 per cent of the lines were scorched, and most of the telephonists electrocuted, burned or crushed. In addition 12 theatres and playhouses and all the city's geisha houses and brothels were wiped out. Most of the prostitutes were reportedly killed.

•

The plutonium bomb dropped on Nagasaki missed its designated target, the Mitsubishi Shipyards, and did little serious damage to the shipyards, city centre and underground torpedo factory. The dockyards would have returned to normal production within three to four months had the war continued. The weapon destroyed Mitsubishi's arms and steelworks, located near the detonation point in Urakami.

Non-military – 'unprofitable' – areas were the worst affected. The bomb completely burned or wrecked the city's hospitals, most schools, and virtually the entire Christian community, according to the first Nagasaki Prefecture's damage report, whose findings the USSBS confirmed.

Among the 18 schools and universities totally destroyed were Nagasaki Medical College, Nagasaki Medical School, Nagasaki School of Pharmacy, Prefectural Keiho Junior High School, Shiroyama National School, Yamazato National School and the Prefectural School for the Deaf. Of the thousands of children, teachers, students, doctors and nurses killed instantly were 2375 secondary school pupils, and several thousand junior pupils.

The bomb obliterated Nagasaki's main hospitals, which were concentrated in Urakami. The Nagasaki University Hospital – 730 metres from ground zero – contained more than 75 per cent of the city's hospital beds. All disappeared in the blast; none of the patients survived. Blast and fire completely flattened the city's Tuberculosis Sanitorium, with the death of everyone inside. In an instant, Nagasaki's medical system ceased to exist. The almost complete destruction of medical facilities and supplies in the two cities added to the total death toll: thousands of the severely wounded died because treatment arrived too late.

In the days after, many dead and injured students of the Nagasaki University and Medical College were returned to their rural communities. In small boats they came across Nagasaki Bay, 20 per vessel, to the villages by the sea. The locals stood on the beach in shock as the vessels unloaded their grim cargo. 'They were dropped off on the shore opposite our house,' recalled Kikuyo Nakamura. 'When we went over to see what was happening, we were asked to help.' The students were 'mangled beyond recognition', lying on the beaches in rows, crying for water. 'We were told not to give them water because if they drank, they would die.' Unable to stand the sound, Kikuyo squeezed water into their mouths out of the towels that hung around their necks. It was a brief reprieve; most were dead by nightfall.

The Buddhist, Shinto and Christan faiths witnessed their 'shrines and churches completely burnt'. The most prominent were: Urakami Cathedral, Nakamichi Church, Gokoku Shrine, Fuchi Shrine, Yamanoo Shrine, Kokuho Fukusai Temple and Honren Temple. The bomb decapitated the statue of Christ hanging over the doors of Urakami Cathedral, and scattered the stone remains of saints among the rubble. Inside, a crowd of Christians at prayer or confession ceased to exist.

The Urakami Orphanage, Urakami Prison, Medical Association Clinic and city crematorium were among the civic buildings obliterated, along with their human occupants. Of 81 prisoners killed, 33 were Chinese and 16 were Korean, jailed for 'theft' and 'espionage'. In addition, 80 per cent of Nagasaki's rice stocks were wiped out.

•

The explosion released several kinds of energy at different speeds, in the trite summary of the USSBS. 'Light, heat, radiation and pressure. The complete bands of radiation, from X- and gamma-rays, to ultraviolet and light rays, to the radiant heat of infra-red rays, travelled with the speed of light.' Gamma rays exposed X-ray film stored in the concrete basement of Hiroshima's Red Cross Hospital, the surveyors found.

'The light and radiant heat accompanying the flash travelled in a straight line and any opaque object, even a single leaf of a vine, shielded objects lying behind it. The duration of the flash was only a fraction of a second but it was sufficiently intense to cause third degree burns to exposed human skin up to a mile ...

'Black or other dark-coloured surfaces of combustible material absorbed the heat and immediately charred or burst into flames ... The heavy black clay tiles which are an almost universal feature of the roofs of Japanese houses bubbled at distances of up to a mile ... The shock waves ... moved out more slowly, that is at about the speed of sound ...'

These shock waves flattened dwellings of all types: brick buildings were levelled at 2225 metres in Hiroshima and 2590 metres in Nagasaki, and traditional Japanese wooden homes were utterly destroyed: 65,000 of Hiroshima's 90,000 buildings were 'rendered unuseable' and the rest partially damaged. Glass windows were blown out at a distance of up to 8 kilometres. But 'nothing was vaporised', the report noted optimistically; and vegetation grew back almost immediately after the explosion.

•

A tremendous spirit of co-operation gradually mobilised Japanese civilians to aid the stricken cities. Many demonstrated to a remarkable degree the quality of self-denial in the interests of common welfare; though no doubt social pressure and the military were on hand to compel the reluctant. Hiroshima's experience of the aftermath, related here, mirrored to a large degree that of Nagasaki.

For days the people received no adequate medical treatment and were forced to rely on emergency relief teams from neighbouring cities, themselves the frequent targets of ongoing conventional bombardment. On 7 August, 33 voluntary first-aid teams arrived in Hiroshima from surrounding communities. They set up simple clinics in the ruins of hospitals, schools and temples; lacking medicines, dressings and clothes, and with little food, they were swiftly overwhelmed. Faced with so many burnt patients, and others who exhibited symptoms of a strange, new illness, 'it was eventually impossible to administer even minimum medical treatment'.

The city's surviving doctors performed miracles in the ruins. At the Hiroshima Communications Hospital, Dr Hachiya felt well enough to go back to work – in filthy, flyblown conditions, prey to disease and exhaustion and surrounded by the moans of the sick and injured. Parents 'crazy with grief' wandered the grounds in search of their children. One mother 'insane with anxiety' circled the hospital shouting her child's name. Dr Hachiya felt an 'animal loneliness' through those long nights of unimaginable torment: 'I became part of the darkness of the night. There were no radios, no electric lights, not even a candle. The only light that came to me was reflected in the flickering shadows made by the burning city. The only sounds were the groans and sobs of the patients. Now and then a patient in delirium would call for his mother, or the voice of one in pain would breathe out the word *eraiyo* – "the pain is intolerable; I cannot endure it!"'

Dr Hachiya had no supplies, not even clothes for the patients: 'I felt ashamed to be as well dressed as I was when I witnessed the misery of the pitiful people around me,' he wrote. 'Here was … a horribly burned young man lying completely naked on a pallet. There was a dying young mother, with breasts exposed, whose baby lay asleep in the crook of her arm with one of her nipples held loosely in its mouth, and a beautiful young girl, burned everywhere except the face, who lay in a puddle of blood and pus.' Others wore rags fashioned out of curtains, sheets or tablecloths scavenged from the ruins.

The Red Cross Hospital performed similar feats under the sole unhurt survivor, Dr Sasaki; only six of the hospital's 30 doctors and 10 of its 200 nurses were fit to work. On the night of the bombing 10,000 victims lay in the hospital grounds. Dr Sasaki and his staff moved among them with bandages and bottles of Mercurochrome, stopping now and then to stitch up the worst lacerations. There was nobody to remove the dead. After 19 hours of this, Dr Sasaki and his exhausted staff lay down to sleep outside, soon to be awoken by 'a complaining circle' around them, crying, 'Doctors! Help us! How can you sleep?'

The wounded experienced agonies of torment: 'One after another the male and female junior high school students were placed nude on top of desks, held down by their parents on either side, while the doctors used scalpels to scrape away the pus. The students, unable to bear the pain, cried and screamed, begging to be killed. There was no way to stop my flood of tears …'

In coming days, food and supplies started flowing in. Police and volunteer groups in surrounding towns organised the dispatch to Hiroshima of rice, salt, matches, candles, clothes, sandals and toilet paper. The army opened stores hitherto denied to the people. In the six days between 12 and 18 August, the city received 130,000 rice meals; eight barrels of pickled plums; 3000 *monpe* (women's uniforms); 249,600 cans of food; 10,000 pairs of straw sandals; and 21,800 towels. People queued in open clearings, amid the wreckage: 'Over 650,000 meals were served during the 10-day period following 9 August 1945. About 98 percent of these were during the first 5 days …' Local bureaucrats were swift to exploit the opportunity. Officials insisted on bribes for the release of donations. Some charged a 10 per cent cut of the value of goods taken to hospitals in official cars – 'Sometimes I even have to give [the officials] extra alcohol, gauze and cotton goods,' said one doctor.

On 14 August, a meagre insurance and postal service resumed and the surviving banks of the local 12 made cash payments to account holders, many of whom had no proof they were customers. One bank transaction occurred on 7 August, 33 on the 8th, 460 on the 15th and 900 transactions on the 16th, across makeshift counters.

Looting and rioting were non-existent in the immediate aftermath due to a deep-rooted fear of the military police and the wartime attitude of self-restraint. Only two arrests for theft were made in Hiroshima between 6 and 21 August; each of the perpetrators received 10 years' hard labour. Looting rose steadily in the last week of August, and later vandals stole supplies designated for hospitals. Within weeks, great numbers of orphans roamed the city in gangs, pilfering, begging and preying on the weak. Japanese self-policing – which was already breaking down – did not delay the draconian re-imposition of law and order. The authorities hastily rebuilt Hiroshima's prison and the 400 or so surviving prisoners – 42 had been killed and 152 severely injured – were back behind bars by 21 August.

The press, too, was soon back on its feet: of the 280 employees of the city's only newspaper, published by the Chugoku News Company,

100 were killed by the bomb and 80 were left injured or missing. Despite this, by 21 August the remaining 100 staff were back at work in the burnt-out shell of the building trying to put out a newspaper. In the interim the city received 210,000 copies of the *Asahi Shimbun* donated from neighbouring cities.

Within days, survivors were erecting small temporary dwellings in the ash and rubble of their previous homes, or living in air-raid shelters. Shops made of lean-tos distributed potatoes and rice. Women cooked the first pans of *okonomiyaki* – (*okonomi*, 'what you like'; *yaki*, 'grilled or fried', usually pancakes and eggs; today the dish is a favourite in Hiroshima) – over open fires. Clerks cleared the charred matter from their offices. Later that year the greatly diminished ranks of schoolchildren helped to clear their smashed classrooms and playgrounds, and resumed lessons.

'It was a tragic sight to see all these 15- and 16-year-old girls who didn't have any hair,' recalled Kiyoko Mori, on returning to school. 'They were wearing black scarves to cover their heads … it was such a shocking sight.' Her class was a fraction of its normal size, but 'we cleaned the whole campus. We cleared all of the broken glass from the window panes, and wiped everything down and classes resumed.' Orphans brought mementoes of their dead families to class: one little girl carried some bones in a tin can, which she always placed on her desk. When her calligraphy teacher asked her to remove it, she began to cry: 'They're the bones of my father and mother.'

•

Weeks after the bomb the apparently healthy and unhurt started dying. They suddenly fell ill with unheard-of symptoms and passed away in the tens of thousands. For some, death was mercifully swift; in others, it would take weeks or months of incremental sickness.

Emiko Okada, who survived the illness, suffered bleeding gums, fatigue, hair loss and other symptoms. 'There were no medicines so my grandmother gave us drinks made from boiled weeds and herbs, to try to get rid of the poisons even though we didn't know what they were.'

One by one they succumbed to terrible nausea, diarrhoea and fever. Dr Hachiya was the first local doctor to study the phenomenon. He noticed the start of strange symptoms in the days after the bomb: some patients produced 'forty to fifty' bloody stools a night, he wrote in his diary. Initially he diagnosed the illness as bacillary dysentery, an understandable error; the doctors did not yet know the bomb had been atomic and none understood the symptoms of radiation sickness.

Death was a daily routine. The causes, however, intrigued Dr Hachiya. Not a single patient exhibited symptoms 'typical of anything we knew'. He began to wonder at the nature of the explosion, the *pikadon*, as the people began calling it: *pika*, bright flash of light; *don*, boom. Perhaps a sudden change in atmospheric pressure caused these curious reactions, he thought.

On 12 August, Dr Hachiya heard the term 'atom bomb' used for the first time, by a naval captain who was passing through town; he also heard a rumour that victims had a very low white blood cell count. He could not prove this because the hospitals' microscopes were smashed. Stories spread that Hiroshima would be uninhabitable for 75 years – a notion swiftly confounded by the sudden appearance of weeds on the twisted train tracks and flowers in the ashen parks. Yet the horrific new illness made many wonder whether the bomb had forever poisoned human life in the Ota Valley.

New symptoms appeared in the patients, many of whom had suffered no injuries or burns, were not directly exposed to the bomb, or who had entered the city the day after the explosion: ulceration of the mouth and throat, balding, bleeding from every orifice, gangrenous tonsillitis, stomatitis and purpura – subcutaneous haemorrhaging manifested by petechiae (pinpoint blood spots) – most common in those closest to the explosion. Petechiae usually heralded death, and patients grew terrified of the spots' appearance. 'We were suffering "spot phobia",' said Dr Hachiya, who scoured every inch of his own body and tugged at his hair at night for signs of onset of the illness. When epilation and petechiae appeared within hours of each other, the victim rarely lived. Some experienced one or the other symptom; some enjoyed a full recovery, with hair regrowth and white blood cell counts rising; others seemed to rally then fall fatally ill. A week after the bomb, the death rate declined – as the burn and blast victims expired; then, towards the end of August, it started rising again. 'So many patients died without our understanding the cause of death that we were all in despair,' Dr Hachiya wrote on 19 August. Japanese doctors were mystified.

The disease manifested itself in another shocking way: the soaring rates of miscarriage in women who had been pregnant during the blast: 'All pregnant women who survived within 3000 feet [900 metres] of the bomb have had miscarriages,' the British study observed. It did not matter whether they were two or nine months' pregnant. Pregnant women between 900 metres and 2000 metres away suffered miscarriages or gave birth to premature infants, who swiftly died. In the ranges of 2000 to 3000 metres only a third of pregnant women gave birth to apparently normal children. Two months after the bomb the incidence of miscarriages, abortions and premature births was 27 per cent compared to the usual 6 per cent.

On 20 August, Dr Hachiya received the microscope he had ordered from a Tokyo hospital and immediately examined the blood of six patients: all had white blood cell counts of about 3000, half the normal count of 6000 to 8000. One had just 200 and died soon after the test. Those closest to the hypocentre had counts of about 1000. Clearly a kind of toxic substance had reduced the white corpuscles, doctors concluded.

An autopsy on a deceased woman helped to solve the mystery: the woman's internal organs were covered in petechiae, and internal bleeding failed to coagulate. Her white corpuscles and her blood platelets were catastrophically low. Dr Hachiya's report on 'radiation sickness' found no relation between the severity of burns and the fall in white blood cells: radiation, not thermal burns, had caused the strange illness.

His diagnosis of Miss Emiko Nishii was typical:

Nishii, Emiko, female, aged sixteen: First seen on 28 August 1945 with complaints of general malaise, petechiae, and inability to sleep. At time of bombing patient was on the second floor of the Central Telephone Bureau, a concrete building five hundred metres from the hypocentre. Immediate onset of dizziness and general weakness; vomited repeatedly. For next three days had nausea and malaise. Gradually recovered appetite but did not recover completely. Patient returned to light work ... Severe epilation [had] appeared on 23 August 1945 and from then on malaise gradually increased ...

Examination: Moderate stature. Nutrition poor ... Numerous petechiae over chest and extremities. Agonized appearance on face. Inner surfaces on eyelids suggest severe anaemia ... Breath sounds over chest weak with dull percussive note over both lung fields ... Pulse weak and rapid with rate of 130/minute, respirations 36, body temperature 104 degrees [Fahrenheit - 40 degrees Celsius] ... Died 29 August 1945, complaining of severe shortness of breath.

Closer diagnosis revealed cytosis – immaturity or abnormal growth in red blood cells – and abnormalities in the internal organs. Thousands of patients died from severe internal bleeding caused by the absence of platelets. In short, gamma radiation had poisoned their entire blood system.

•

Japanese reports of the new medical phenomena piqued American and British interest. Local doctors told the Japanese media of the 'uncanny delayed effects' in those exposed to the bomb and in those who went into the cities days after the explosions. Two weeks later the numbers of dead and wounded were rapidly rising, reported Tokyo Radio. On 22 August the total number killed or wounded in both cities had more than doubled since the first day, to 160,000 in Hiroshima and 60,000 in Nagasaki, where the narrow valley limited the spread of radiation. On 30 August the London *Times* reported the case of a 29-year-old woman who died 19 days after receiving a small bruise at Hiroshima. The post-mortem showed her white corpuscle count at one-tenth of normal, and 'striking changes' to her 'blood-making organs', the liver, spleen, kidneys, lymph glands and marrow. All showed signs of acute exposure to radiation. Thousands were suffering similar symptoms, even though they were a long way from the hypocentre or had entered the city immediately after the explosion.

Japanese doctors sought to broadcast the evidence of 'radiation sickness'. They received a sympathetic ear from doctors in America and Britain, who confirmed that the symptoms described the reaction of the human body to severe exposure to radiation.

Deeply concerned by the emerging medical evidence, American government and military officials sought to counter the Japanese reports, damning them as propaganda. The authority of Dr Robert Oppenheimer, no less, who had earlier rejected the risk of residual radiation, was re-invoked: There was 'every reason to believe', he stated on 8 August, 'based on all of our experimental work and study, and on the results of the test in New Mexico' that 'no appreciable radioactivity [existed] on the ground at Hiroshima and what little there was decayed very rapidly'.

Four days later the Manhattan Project issued a memo to the press, 'Toxic Effects of the Atomic Bomb', which announced, 'No lingering toxic effects are expected in the area over which the Bomb has been used', because the atomic bombs were detonated at such a height 'as to disseminate the radioactive products as a cloud'. This finding echoed the conclusions of the first technical history of the bomb, written by Princeton's Dr Henry Smyth, and released to the American public on

12 August 1945: 'On account of the height of the explosion, practically all the radioactive products are carried upward in the ascending column of hot air and dispersed harmlessly over a wide area.' The 'cloud dispersal theory', however, failed to account for the tendency of clouds to deposit rain; a highly radioactive cloud was apt to produce slimy water droplets filled with radioactive fallout, ash, dust etc – dubbed 'black rain'. 'Strong radioactivities' were found in the sand deposits in northwestern Hiroshima, where most of the black rain fell, according to the USSBS researchers. Nor did the radioactive cloud simply drift away, as it had over the New Mexican desert; Hiroshima's bowl-shaped valley contained the mushroom formation. In Nagasaki, similarly, the hills above the city served as a natural buffer.

A disturbing aspect of Washington's 'toxic' memo was that it contained precise medical guidance on radiation sickness, which the Americans refused to share with Japanese physicians trying to treat thousands of people suffering from it. Gamma rays, the memo noted, were 'very penetrating' and 'of great concern biologically', as they easily entered the most vital parts of the body. It accurately described the symptoms of radiation sickness down to the last details that were then mystifying Dr Hachiya: a sharp reduction in white blood cells resulting in leucopenia and lymphocytopenia, and severe haemorrhaging, with death expected in days or 'postponed for various periods up to 60 days'. Non-lethal doses would induce sterility (both permanent and temporary), hair loss (permanent and temporary) and chronic skin conditions.

The Japanese press reported those very symptoms in rising numbers of people – and American scientists verified the reports, to Washington's alarm. The American media shed its complaisant line and took a close interest. Anxious to establish the truth, Groves phoned Lieutenant Colonel Charles Rea, a medical officer at Oak Ridge Hospital, Tennessee. Groves and Rea were aghast at the possibility that 'Jap propaganda' would elicit American sympathy for the bomb casualties and martyr the nation that brought war to the Pacific.

Groves: 'The death toll at Hiroshima and at Nagasaki ... is still rising,' the [Radio Tokyo] broadcast said. Radio Tokyo described Hiroshima as a city of death ... 'it is peopled by [a] ghost parade, the living doomed to die of radioactive burns'.

Rea: Let me interrupt you here a minute. I would say this. I think it's good propaganda. The thing is these people got good and burned – good thermal burns.

Groves: That's the feeling I have. Let me go on ... 'So painful are these injuries that sufferers plead: "Please kill me",' the broadcast said. 'No one can ever completely recover.'

Rea: This has been in our paper too, last night.

Groves: Then it goes on: 'Radioactivity ... is taking a toll of mounting deaths and causing reconstruction workers in Hiroshima to suffer various sicknesses and ill-health.'

Rea: I would say this ... they just got a good thermal burn, that's what it is. A lot of these people ... don't notice it much. You may get burned and you may have a little redness, but in a couple of days you may have a big blister and a sloughing of the skin ...

Groves: That is brought out a little later on. Now it says here: 'A special news correspondent of the Japs said that three days after the bomb fell there were 30,000 dead and two weeks later the death toll had mounted to 60,000 and is continuing to rise.' One thing is they are finding the bodies.

Rea: They are getting the delayed action of the burn ...

Groves: Now then ... this is the thing I wanted to ask you about particularly – [reads Japanese news report] – 'An examination of the soldiers working on reconstruction projects one week after the bombing showed that their white corpuscles had diminished by half [with] a severe deficiency of red corpuscles.'

Rea: I read that too – I think there's something hokum about that.

Groves: Would they both go down?

Rea: They may, yes – they may, but that's awfully quick, pretty terrifically quick. Of course, it depends – but I wonder if you aren't getting a good dose of propaganda.

Groves: Of course we are getting a good dose of propaganda, due to the idiotic performance of the [American] scientists [who had substantiated the Japanese reports] ...

Rea: I think you had better get the anti-propagandists out.

Groves: We can't you see, because the whole damage has been done by our own people ... The reason I am calling you is because ... I might be asked at any time and I would like to be able to answer.

Rea: ... I would say this right off the bat – anybody with burns, the red count goes down after a while, and the white count may go down too, just from an ordinary burn. I can't get too excited about that.

Groves: We are not bothered a bit, excepting for – what they are trying to do is create sympathy ...

Rea: Let me look it up and I'll give you some straight dope on it.

Groves: This is the kind of thing that hurts us – 'The Japanese ... probably were the victims of a phenomenon that is well known in the great radiation laboratories of America.' That, of course, is what does us the damage ...

Groves' apparent ignorance of the dangers of radiation rang false given his leadership of the Manhattan Project, where he insisted on the highest safety precautions, and his concern about the toxic effects of radiation in salmon; and there was the very public example of the Radium Girls (see Chapter 5). Determined to satisfy the American media, Groves in late August sent a scientific team to Japan headed by Brigadier General Farrell, to investigate the local claims and make absolutely certain of 'no possible ill effects to American troops from radioactive materials'. Farrell's mission was also to protect the occupying forces then entering the cities from any residual radiation, 'although we have no reason to believe that such effects actually exist.'

Meanwhile, Washington pursued another means of playing down the US media's attention to radiation sickness in Japan. The President's press secretary invited friendly reporters to the Trinity site to test residual levels of radiation there. 'This might be a good thing to do,' Charles Ross advised the War Department on 27 August, 'in view of continuing propaganda from Japan that radio activity [sic] in the areas of atomic bomb explosions continues for an indefinite period.' The press tour of Alamogordo proceeded in early September, six weeks after the test; William Laurence, the Manhattan Project's official propagandist, nominally of the *New York Times*, headed the list of friendly reporters.

During this time, one foreign journalist's persistence undermined the US investigation, and put a spur in the side of the world's press. On 5 September, three days after the surrender ceremony, the London *Daily Express* published a front-page article that appeared to confirm the worst fears of gamma radiation and deeply embarrassed the US government. It was the work of an intrepid Australian correspondent, Wilfred Burchett, a young man not yet sullied by his love affair with communism. On 2 September, while most correspondents were attending the ceremony on the *Missouri* in Tokyo Bay, Burchett slipped through MacArthur's media net and travelled 640 kilometres by train to Hiroshima, equipped with seven army rations, an umbrella and a typewriter. The first Western reporter to enter the city found 'the most terrible and frightening desolation', which made a 'blitzed Pacific island seem like an Eden'. After a tense encounter with the Japanese police at their makeshift HQ in the ruins of the Fukuoka Department Store – where he ran the risk of being shot – Burchett received permission from the senior officer, a member of the 'Thought Prosecutors', to tour the wreckage and hospitals. 'Show him what his people have done,' the officer, mistaking Burchett for an American, ordered his staff. That day the Australian pounded out his impressions on his old typewriter sitting on a 'chunk of rubble that had escaped pulverisation'.

'I write this as a warning to the world,' he began portentously – under the misleading headline 'The Atomic Plague' – and took the reader down the long lines of patients who had survived the bomb uninjured and yet who now lay dying from the effects of 'a mysterious illness', which Japanese doctors mistook for malnourishment: 'They gave their patients Vitamin A injections,' Burchett wrote. 'The results were horrible. The flesh started rotting away from the hole caused by the injection of the needle. And in every case the victim died ... They are dying at the rate of 100 a day.' The death toll numbered 53,000, with 30,000 missing – 'certainly dead'. His report was broadly accurate, but failed to identify whether their poisoning resulted from residual radiation in the days after the bomb or whether they were exposed to it at the time of the blast.

Back in Tokyo Burchett attended a press conference where furious US army officers set out to discredit his story (which had been syndicated throughout the world). A 'scientist in brigadier-general's uniform' told him that the bombs were detonated at such a height as to avoid 'residual radiation'. Dirty after his long journey, Burchett rose to his feet and asked if the officer had been to Hiroshima. 'He had not.' Burchett described what he had seen: many cases of people uninjured by the bomb, or who had entered the city immediately after, who were later struck down by radiation sickness. The officer patiently explained that they were victims of blast and burn, 'normal after any big explosion'. What of the fish, turning belly up as they entered a poisoned stretch of the river, weeks later, Burchett asked. 'I'm afraid you've fallen victim to Japanese propaganda,' said the officer. The press conference ended. It was the high point of Burchett's career, when he enjoyed a reputation for credibility. The experience, however, seeded in the young journalist an abiding hatred of America, and he later ruined his reputation by serving as the paid 'consultant' – and mouthpiece – for several odious communist regimes.

Days later, Brigadier General Farrell's mission arrived in Hiroshima accompanied by Laurence, who had dutifully reported the absence, six weeks after the test, of radiation at Trinity, 'giving the lie' to Tokyo's claims of residual radiation in Hiroshima. 'The Japanese claim that people died from radiation,' he quoted Groves as saying. 'If this is true the number was very small,' he wrote on 9 September. From Hiroshima, under the headline, 'No Radioactivity in Hiroshima Ruin', Laurence reported that Farrell's team had found 'no evidence of continuing radioactivity in the blasted area on Sept. 9 … and that there was no danger to be encountered by living in the area …'. That was true of 9 September; but Laurence ignored the effects of radiation during and in the immediate aftermath of the blast. His article was a piece of shameless propaganda that soiled the pages of the *New York Times*. Worst of all, this paid fabricator misreported the full conclusions of the Farrell mission, which essentially confirmed the Japanese claims: 'Summaries of Japanese reports previously sent are essentially correct, as to clinical effects from single gamma radiation dose,' Farrell's team concluded.*

•

Partly in response to Burchett and the prying eyes of the Western media, MacArthur reneged on his promise (of 10 September) – to allow an 'absolute minimum restriction on the freedom of the press' – and imposed on 18 September a censorship regime every bit as rigorous as totalitarian Japan's. The new press code banned anything that 'might directly or by inference disturb public tranquility' or 'convey false or destructive criticism of the Allied Powers'. Hiroshima and Nagasaki were shut off; reports of the bomb disappeared from

---

* Farrell's findings were inconvenient in that they confirmed Japanese reports. On 4 October, Dr Karl Compton made another attempt to quash the radiation reports. He went to Hiroshima and a week later wrote to Truman that American scientists had found no dangerous radioactive 'burning' as an after-effect of the bomb; and not a single 'authenticated' case of 'any person having been damaged by going into the area after the explosion'. In fact, numerous cases of radiation poisoning – including many people who had entered the cities soon after the blasts to rescue survivors – were only then beginning to appear. The very month Dr Compton gave Hiroshima and Nagasaki a clean bill of health, thousands of people started dying from bomb-related illnesses, chiefly leukaemia. They have continued dying for decades.

the local and international press; photographers' films of the cities were confiscated. The newspaper *Asahi Shimbun* was suspended – for branding the bomb 'a war crime worse than an attack on a hospital ship or the use of poison gas' (the *Nippon Times* had earlier called the attack on Hiroshima 'an act of premeditated wholesale murder'). The Domei News Agency was restricted to local news, and Japanese cartoons, novels and poems about the bomb were driven underground. American correspondents were thoroughly tamed, or spiked: MacArthur killed all 25,000 words on Nagasaki written by the Chicago *Tribune*'s George Weller, the first foreign journalist to enter the city. Facts on the atomic casualties died with the lives of victims.

•

The aftermath of the first use of nuclear weapons in war drew hundreds of foreign scientists to the stricken cities, anxious to study the human exhibits of widespread radiation disease. MacArthur's censorship regime permitted their work but not its publication, or its dissemination to Japanese doctors. Farrell's brief mission (referred to above) was the first to arrive; three more American teams, representing the Manhattan Engineer District under Stafford Warren, the US Army Forces, Pacific, under Ashley Oughterson, and the US Navy Bureau of Medicine and Surgery under Shields Warren streamed into the cities in coming weeks. One team of 21 American doctors and four Australian medical corps officers investigated Nagasaki between 20 September and 6 October, and Hiroshima in early October. They examined 900 patients: 432 from Hiroshima and 468 from Nagasaki – visiting the dilapidated Red Cross and Teishin hospitals in Hiroshima; and the Omura, Shinkozen and Isahaya hospitals in Nagasaki. Their findings, and those of other scientific teams, were amalgamated into the Joint Commission for the Investigation of the Atomic Bomb in Japan, which included input from the prominent physicists Dr Robert Serber and Dr Hans Bethe, the Manhattan Project, the USSBS and the British Mission to Japan.

The feature which strikes the reader about the summary of their first report is the curious tone, almost boastful, with which they described the physical damage in what was supposedly a medical assessment: the reinforced concrete smoke stacks designed to resist earthquakes that 'overturned' up to 4000 feet (1200 metres) from 'X' (the hypocentre); the flash charring of wooden telegraph poles up to 13,000 feet (4000 metres) from X; the complete destruction of church buildings with 18-inch (45-centimetre) brick walls 3500 feet (1000 metres) from X. The weapons twisted, ripped, bent, wrecked and melted. The tone seems to be one of clinical satisfaction, if not actual pride, in the bomb's potency. The authors appear to bask in the reflected glory of the discovery of nuclear energy, as though it were an extrapolation of their scientific power rather than the release of phenomena at the heart of nature. 'As intended,' they wrote, 'the bomb was exploded at an almost ideal location over Nagasaki to do the maximum damage to industry ...' (On the contrary, the bomb missed its designated target and landed in the heart of a religious, medical and educational community, which happened to accommodate an underground torpedo factory.) There is not a frisson of concern or regret here at the human cost; not a scintilla of doubt in the doctors' minds that the cities' annihilation was anything other than a necessary military operation. It reads as one imagines aliens might record the effects of their cosmic lasers on vaporised earthlings. One does not expect compassion in what was ostensibly a medical report; yet the authors devote only a few pages of their summary to the human effects of the bomb, written in the desultory tone of an afterthought:

'It seems highly probable that the greatest total number of deaths were those occurring immediately after the bombings.' The experts explained this breathtaking statement of the obvious thus: the people nearest the blast were probably 'killed, as it were, several times over, by each casualty-producing agent separately' (as the British Mission had independently so aptly stated). The 'proper order of importance' of the 'casualty-producing agents' were 'burns, mechanical injury, blast'. That is to say, people were killed threefold. Curiously the authors refrained from adding radiation sickness as the fourth horseman; a likely reason is that their superiors were then arguing that people were no longer dying of radiation poisoning.

Under the subtitle 'Radiation', they described hair loss – epilation – as one of their 'most spectacular findings': 'In many instances the resemblance to a monk's tonsure was striking. In extreme cases the hair was totally lost ...' (regrowth of hair, they discovered, began in some cases as early as 50 days after the blast). Colonel Stafford Warren's party echoed these findings: 'Some were taken ill by the atomic bomb and became bald,' he observed. 'But their hair will come back by and by.'

•

The American investigators were not permitted to share their findings with Japanese physicians then struggling to treat the victims. The local doctors had little if any knowledge of the new illness at a time when American scientific teams were gathering a deep, empirical understanding of radiation sickness. Nor were Japanese doctors allowed to keep their own notes on the illness. The American forces confiscated – 'stole', in the view of the Kyoto Physicians' Association – any local research on radiation sickness:

'Research on the Japanese side hardly progressed at all,' concluded the Association in late 1945. 'This was because their ability to publish their findings was stolen along with their materials.' The US, it claimed, 'did not want to let the world know the reality of the atomic bomb's effect'.

This ban on sharing clinical knowledge severely hampered Japanese medical efforts to save lives. 'To forbid publication of medical matters is unforgivable from a humanitarian standpoint,' noted Professor

Masao Tsuzuki, chair of the medical subcommittee of Japanese doctors in Hiroshima. His views reflected the feelings of many Japanese physicians, furious at their inability to treat the sick and wounded. Nor were the American occupying forces disposed to supply medical equipment. Throughout late 1945 and early 1946, Japanese doctors applied for penicillin and other medical supplies; very little arrived. Observation and testing of the A-bomb victims, not their treatment, were the priorities of the American scientific teams.

Japanese scientists turned to nature, and the dead, as the only legal focus for their experiments. Their clinical fascination tried to measure, for example, the neutron penetration rate through human tissue by grinding the bones of skeletons found at the bomb's hypocentre – as the detonation point came to be known – and testing neutron presence in the skeletal dust. More reassuring were their agricultural and botanical findings: 60 days after the bombings, grasses were sprouting under the detonation points, insects were plentiful and Chinese cabbages, sweet potatoes, radishes and other crops displayed no unusual growth patterns. Flowers were pushing up through the ash: morning glories, day lilies, panic grass and feverfew were visible within weeks. Weeds, too, seemed normal. Buckwheat and 'Welsh onion' soon returned to Urakami. The atomic bombs would prove no hindrance to future farming, Kyoto University's agronomists concluded; in fact, residual radioactivity in the soil appeared to have a stimulating effect on plant growth.

Ants and bugs resumed their natural life cycles, which were only briefly disrupted, found a Kyushu Imperial University study. 'Injuries to them were negligible … even in the vicinity of the central explosion zone.' Soil radiation had had no effect on black carpenter ants, black mountain ants, reticulated ants and others; they seemed to carry on regardless. So too did an array of water insects that returned to frolic in the tanks and boundary ditches and the ponds of Asano Park.

CHAPTER 22

HIBAKUSHA

Maybe I didn't look like a proper woman but my feelings inside have not changed. If I don't like a man, I don't want him. Even if I have bad scars ... I don't have very high ideals but I have some ideals, you see? Inside I have the same hopes, same dreams!

A 19-year-old *hibakusha*, or 'bomb-affected person'

SEIKO IKEDA'S FATHER DID NOT recognise his daughter when he found her on the floor of the village hospital. He came with a neighbour and a stretcher. The little clinic was dark and crowded. 'Seiko, your father is here for you!' he announced in the twilit gloom. He knelt over the children's figures. 'Daddy!' Seiko cried out. He turned towards a vaguely human form strewn on the floor, the hair matted and singed, the face, swollen beyond recognition. The impression was of a little red scarecrow. At first, her father refused to believe this was the daughter who had set off on her first day as a labourer in Hiroshima.

He took the bundle home. For days she suffered a 40-degree Celsius temperature, nausea and diarrhoea; her face and body swelled; her lacerations got infected. 'I was just crying and shouting all day, "It hurts, it hurts."' She received no medical treatment; local doctors set her aside with the hopeless, and reserved their time for the hopeful. 'They even started to prepare my funeral service,' she recalled in an interview with me. Her parents fought to keep their daughter alive, scrounging for nutritious food and bartering for medicines with products from her father's shop.

Her friend Chie returned to the village a few days later. Chie had volunteered to stay in Hiroshima to help the injured and sick. She had no burns or external injuries; her pretty face was unblemished. The village rejoiced; Chie's family wept with joy.

The Ikeda family did what they could to ease their daughter's pain: her father applied a white powder to her face, then a gauze, which had to be changed twice a day to release the pus. Flies laid eggs in her wounds; her sister removed the maggots with tweezers: 'My father would be so upset by the maggots on my face,' Seiko recalled.

Then, within a few weeks, Seiko's fever eased and she sat up in bed. At about the same time, her girlfriend Chie, whose unmarked features had caused such delight, lost her hair. Then Chie's stomach and chest swelled up and purple spots appeared on her body. She bled from her nose, ears, mouth and vagina. The girl begged the doctor to heal her: 'Please help me! I don't want to die yet … I didn't do anything bad, why do I have to die? Help me!' She died, to her family's shock, yet she bore no physical injuries. 'She spent days in Hiroshima helping others,' Seiko said. Chie, who had stayed in the city immediately after the bomb, died of residual radiation poisoning, the existence of which the American authorities continued to deny.

Within three months, Seiko was able to move. The family hid the mirrors. But if she could not see her face, she could feel the changes to her skin – the clawed and matted scar tissue called 'keloids': 'I searched all over for a mirror and found a small one hidden away in the back of a drawer. I was so shocked when I saw myself. My face was bright red, with keloids and skin gathered. The keloids built up and it looked like a liver. My chin was stuck to my neck. My bottom lip was open and the water would come off it when I tried to drink, since I couldn't hold it in. I was so afraid and scared when I saw my face. I lost all my courage.'

She ventured outside to play with other children. They ran away, shouting, 'Red Demon! Red Demon!' Seiko ran home crying. 'My mother would hold me every day, saying: "Who made my daughter's face like this!"' In time her psychological anguish would overtake her physical suffering. She was 13 years old.

Eight months after the bomb Seiko returned to school. An empty space surrounded her on the packed train: 'When my face came near someone else they would shrink away from me. They didn't want my face touching them. I couldn't bear other people watching me like this.' At school, the children fled at the sight of her, pointing and yelling 'monster' and 'demon'. She lost hope: 'I became such a bad girl. I would yell at my parents and say that they should not have taken care of me, they should have left me to die. I envied Chie for dying; I wanted to die too.'

One day, after running away from school, she stole in the back door and overheard her father talking in a low voice to the neighbour. Seiko, he said, 'had a wound on her heart, and she was a bad girl, but since she was so strong she would once again be a good girl'.

'When I heard my father saying this about me,' she recalled, 'it was such a shock. I thought that no one loved me; but my father believed in me despite my behaviour. I was so moved, I started crying while hiding and listening. From then I resolved to do my best to live. Then I started going to school.'

Gloves had protected her hands from the flash. They are beautiful hands, and she displayed them shyly to me, as though offering the only valued part of her body to a careless world. Her face is now slightly scarred, and warm and engaging; modern surgery has reconstructed what had once seemed beyond repair: 'Beauty and looks were important for girls, and my mother brought me up thinking that. Though I wanted to be beautiful I could not [be].' Her mother purchased creams and make-up, and kept applying it, 'saying that I would be beautiful, always encouraging me'.

Her family did not believe the girl would marry. Yet she did, in 1950, at the age of 18, to a cousin, a childhood friend, who had lost one brother in the atomic bomb; another languished in a Soviet prison camp. Her husband cared little for her deformities: 'We just wanted to live together and support each other,' she said. 'So he asked me to marry him.'

Early attempts at plastic surgery failed. But in coming years, Japanese plastic surgeons would learn to apply American techniques and work wonders. Seiko underwent surgery 15 times, and her face gradually improved. 'My doctor told me I was becoming beautiful, but that was not the reason I went through the surgery; I wanted my original face back.' Thanks to the operations and her doctor's determination, Seiko recovered her self-confidence. She learned dress-making and later opened a dress-making school. It prospered. She learned to dance – beautifully. She and her husband soon produced a healthy daughter.

•

There was never any pretence that the foreign medical teams entering Hiroshima and Nagasaki were there to ease the people's suffering. Navy Secretary James Forrestal outlined their experimental role with crystalline clarity in a note to Truman on 18 November 1945. The study of the effect of radiation 'on personnel' – that is, Japanese civilians – he wrote, had started as soon as possible after Japan's capitulation, under the auspices of the army and navy and the Manhattan Project: 'Preliminary surveys involve about 14,000 Japanese who were exposed to the radiation of atomic fission. It is considered that the group and others yet to be identified offer a unique opportunity for the study of the medical and biological effects of radiation which is of utmost importance to the United States.' The scientists' express instructions were not to treat the people; rather, to experiment on them. 'The Japanese had sole responsibility for treating bomb victims,' noted the USSBS, 'though the American forces did provide some medical supplies …' Late in 1945, Japanese doctors in Hiroshima and Nagasaki continued to work in ruins, had no plasma or blood, and only a negligible quantity of vital drugs.

In 1946 it was evident to the Joint Commission for the Investigation of the Atomic Bomb (the peak body representing the US Armed Forces and the Manhattan Project teams) that long-range investigations of the survivors were required. On 25 November 1947 America's highest scientific body, the National Research Council of the National Academy of Sciences, received a presidential directive 'to undertake a long-range continuing study of the biological and medical effects of the atomic bomb on man'. The directive shifted the experiment from military to civilian scientists because the danger of radiation went 'beyond the scope' of the armed forces to 'humanity in general' – not only in war but 'in peaceful industry and agriculture'. The directive did not mention treatment: prolonging life, easing pain, were neither the intentions nor the by-products of the job.

Whether the patients – or more accurately the exhibits – lived or died was immaterial to the foreign doctors' charter. *How* the victims lived or died; *whether* their conditions improved or deteriorated; *whether* they suffered from cancer at some distant date or reproduced it in their children; such were the questions of cold scientific inquiry. In short, irradiated Japanese civilians were to serve as American laboratory rats. Herein lay a benefit – future rationalists would argue – of dropping the bomb on a city: to harvest scientific data about gamma radiation. The doctors were not expected to show a duty of care – though in practice, incidental to the experiment, many did. They were human, after all. The degree of care depended on the attitude and resources of the particular physician but it was not part of his job description. In this sense, the presidential directive to America's peak medical research body prescribed a flagrant violation of the Hippocratic Oath: 'I will provide regimens for the good of my patients and … never do harm to anyone.'

The research proceeded under the auspices of a new body set up under MacArthur's jurisdiction named the Atomic Bomb Casualty Commission (ABCC), which drew on, and vastly expanded, the work of the pioneering medical teams that had first entered the cities. The initial aim of the ABCC's Hiroshima laboratory, which moved from the Red Cross Hospital to a purpose-built facility on Hijiyama hill – was to 're-examine' Japanese records, autopsy records and patient medical histories (the patients had no right to privacy); to 'collect' and examine survey cases; and to 'obtain' photographs of victims, according to the ABCC's charter.

They were to prod, probe and test the irradiated human relics of the first atomic bomb. 'Does radiation produce long range effects in human beings? Finding the answer to this question is the purpose of the ... ABCC,' wrote Lieutenant Colonel Carl F. Tessmer, the first director of the commission. 'Treatment of patients is not undertaken by the ABCC because such matters properly are in the hands of Japanese physicians in Kure, Hiroshima and Nagasaki.' Such matters may have been in their hands had Japanese doctors access to American medical research on radiation sickness. In any case, many Japanese doctors did not help themselves, or their patients: wary of foreign rivals, some refused American help even when it was offered and were strongly opposed to Americans treating Japanese people. And some had no qualms about studying – when they might also be treating – their countrymen and accepted jobs at the ABCC.

•

About 100,000 Japanese adults and 80,000 children who had been exposed to the atomic bombs participated in these wider experiments in some form. They were not coerced; the ABCC offered incentives such as nutritious rations, fresh water and candy. Many 'patients' were furious when they later discovered that they had been used as part of a vast experiment, and claimed they were misled. The arrival of American doctors, who showed them kindness and understanding, fostered the hope of treatment for their and their children's ongoing sickness. They were now callously disabused. Stories of the humiliation of Japanese patients and the insensitivity of foreign scientists proliferated – for example, of naked women being examined by crowds of male physicians; of the ABCC jeep arriving at family funerals 'to ask if they could dissect the body … it would be good for the society as a whole'.

Some foreign teams gave the impression of being quack apothecaries with a 'morbid lust for corpses', harvesting Japanese cemeteries for victims of the bomb. There is some official evidence for this: Dr Stanley Finch, Chief of Medicine at the ABCC in 1960–62, wrote of the 'base population' for a pathology study in Hiroshima–Nagasaki consisting of a 'sub-set of persons' who were candidates for post-mortem studies. In other words, the ABCC chose some patients precisely *because* they had no chance of living, the better to study the effect of radiation on the dying. To their families' horror, many 'patients' did not come out of the lab on Hijiyama hill alive. 'Autopsy rates as high as 45 per cent in the early 1960s,' Finch added, 'have provided information of great value'; his clinical contentment suggested to journalist Wilfred Burchett a vision of the 'professional body snatchers', who valued the dead and dying over the living.

That grim conclusion must be set against the fact that the ABCC's scientists did draw valuable medical findings from their dazed and defeated sample. What galled the local population was the lack of any concurrent duty of care for the people under the microscope. 'Why won't the American doctors help us?' was a constant local refrain. Their point was reasonable: the Americans were far better resourced and experienced to treat radiation sickness than local doctors, on whose flimsy and ill-equipped clinics fell that terrible responsibility.

The ABCC's earliest experiments were devoted to the genetic effects, if any, of radiation on the development of children – including foetuses. Under the guidance of Dr James Neel, the study examined more than 71,000 pregnancies in the two cities between 1948 and 1953. Neel used rice ration registration forms to identify subjects: pregnant women were allowed an extra rice ration from the fifth month of pregnancy. The study found no relationship between the parents' exposure to radiation and subsequent rates of genetic mutation and stillbirths in children conceived after the bomb. This confounded expectations. Indeed, one Hiroshima politician, convinced of future hereditary deformities, proposed that bomb survivors submit to a voluntary program of sterilisation: 'I had my wife sterilised because I don't want abnormal children,' he said. 'We should set them [the A-bomb survivors] aside and not mix them with the rest of the population.'*

Foetuses exposed to the bomb were, however, severely affected, especially those irradiated prior to the 25th week of gestation. Many of those who survived to birth were born with smaller head sizes (microcephaly), severe mental retardation, stunted development and anaemia. The worst affected infants were placed in psychiatric institutions. Their afflictions were not genetic malformations, as was commonly supposed – rather, the baleful influence of radiation on a developing foetus. Another dramatic early finding was the sharply increased rate of leukaemia in A-bomb survivors. After a two to three year period of latency, the number of cases peaked in 1950–52.

•

They were called the *hibakusha* – literally, 'bomb-affected people' – a neutral term that pointedly did not connote 'survivor' or 'victim'. For years they existed in a nether world, the flotsam of official indifference and the jetsam of American experimentation. To Japanese society, they were untouchable, the people you did not employ or let your son or daughter marry. Many were refused compensation, jobs, love, family – shunned to the extremities of a community unable to bear the hideous after-effects of total war; their scars were painful reminders of the disgrace Japan had brought upon herself. A red-hot iron pressed against the bare skin would have had the same penetrating effect as

---

* The question of hereditary genetic mutation remains the focus of experimental study in Japan and America.

the flashburn, searing deep into the flesh, according to Dr Tomin Harada. The resulting wounds took months to heal, leaving the victim's face contoured in thick keloids – derived from the Greek word for crab claws – which had the segregating power of leprosy. The afflicted were refused entry to public baths in case they contaminated the water, and compelled to work in nocturnal jobs out of private shame and public revulsion. The keloid-scarred women who staffed one nightly pinball parlour 'dreaded the daylight … because they had such hideous burn scars as a result of the *pikadon* [explosion]'. These girls were obliged to hang their clothes and wash their plates in separate areas to prevent the 'contamination' of healthy employees.

For a brief period the *hibakusha* seemed to occupy the place vacated by the *burakumin*, the untouchables of Japanese society, whose Hiroshiman ghettos were destroyed in the bomb. The comparison is inapt, as the *burakumin*'s strong identity, dating back to their ancestors, the *eta*, readily reformed in fresh ghettos, confounding the hopes that the bomb had blown away entrenched discrimination. The *burakumin* were segregated by class and occupation. The *hibakusha*, however, shared no attributes of class, religion or culture – only common exposure to the bomb. Gamma rays did not discriminate.

The A-bomb survivors responded to society's repugnance with deep anxiety and shame – as though it were somehow *their* fault that they wore the mark of a defeated nation. The most miserably scarred became the elephant men and women of Japanese society. Playground cruelty knew no restraint in the presence of such disfigurement; teenage *hibakusha* were taunted to the edge of suicide. They hid themselves away, stayed indoors, and shielded their faces in masks.

The *hibakusha*'s awareness of the improbability of love, marriage, even friendship – that ordinary jobs were unobtainable, that Japanese society shunned them – were preludes to an ocean of loneliness. Many younger victims, denied the normal hopes of adulthood, experienced a common death wish. One young man, aged 26, his face covered in keloids, tried to end his life several times after his marriage proposals were rebuffed. Thirty per cent of *hibakusha* have experienced suicidal feelings since the war, with the figure rising to 70 per cent among those with the worst physical deformities, according to a statistical study by Tadashi Ishida.

'Nobody's going to marry those Nagasaki girls,' said one woman from a village in Nagasaki Prefecture. 'Even after they reach marrying age, nobody's going to marry them. Ever since the Bomb fell, everybody's calling them "the never-stop people". And the thing that never stops is their bleeding. Those people are outcasts – damned Untouchables. Nobody's going to marry one of them ever again.' Gossip condemned houses, suburbs and whole villages of survivors as untouchable. One rural community near Nagasaki feared it would become 'a village of bleeders'. Kawauchi village in the Asa district became known as 'Widows' Village' – *Goke-mura* – when all 75 wives living there became instant widows after the bomb killed their husbands, then labouring in Hiroshima.

Their fear of cancer – or confirmation of it – drove many to suicide: one girl happened to see 'myeloid leukaemia' on her medical chart and promptly hanged herself. Whenever he heard such stories the great Japanese writer Kenzaburo Oe, himself exposed to the bomb, felt relieved that Japan 'is not a Christian country. I feel an almost complete relief that a dogmatic Christian sense of guilt did not prevent the girl from taking her own life. None of us survivors can morally blame her. We have only the freedom to remember the existence of "people who do not kill themselves in spite of their misery".' In time, many *hibakusha*'s resistance to illness faded, and in subsequent decades tens of thousands succumbed to radiation-related cancers, usually leukaemia.

Kikuyo Nakamura had her uterus and ovaries removed at the age of 25 years. She lost her hair as a result of anti-cancer treatment and wears a wig. She experienced little discrimination because 'almost everybody else in Nagasaki of my age has been exposed to radiation'. Her baby son, Hiroshi, later developed leukaemia, for which she blames her exposure to radiation, despite the fact that the link between *hibakusha* and second-generation medical conditions is not proven.

'But when I asked my doctor why my child developed leukaemia, he told me that it was because he had been fed my breast milk.'

Hiroshi grew up and married but did not tell his wife that his mother was a *hibakusha*. When his wife found out she screamed at her mother-in-law, 'The doctor told me that you gave my husband this disease!' The two women could not live together: 'Every time she looked at me, she felt angry,' Kikuyo recalls. Hiroshi's wife soon moved out and divorced him; he died soon after.

Stronger souls resisted the condemnation of the post-war society: 'Maybe I didn't look like a proper woman,' said one spirited 19-year-old girl, her face rent with scars, in the 1950s, 'but my feelings inside have not changed.' Social attitudes that expected her to feel grateful for a man's attention enraged her. 'If I don't like a man, I don't want him. Even if I have bad scars ... I don't have very high ideals but I have some ideals, you see? Inside I have the same hopes, same dreams!'

•

Not all experienced misery. Thousands recovered and lived relatively happy lives, found jobs, married and had healthy children despite rumours of possible deformities in the second generation. Their experiences vary widely, of course; but their personal stories convey the range of consequences of the bomb better than statistics or medical analyses. Their determination to live as comfortably and happily as possible confounds the agenda of those who, usually foreigners, seek to impose an unwelcome martyrdom on the A-bombed cities. Here are a few examples.

Of 165 Japanese people who experienced both atomic bombs, Tsutomu Yamaguchi is the only officially recognised survivor 'twice over'. That extraordinary coincidence for these people did not, of course, necessarily make their ordeal any more trying than that of a person acutely exposed to a single bomb. During the war Yamaguchi lived with his young family in Nagasaki, where he worked as an engineer with Mitsubishi shipyards. On 6 August he visited Hiroshima on a business trip. Three kilometres from the blast, he sustained facial burns and spent the night in an air-raid shelter before returning home two days later, where he experienced the second nuclear attack. He, his wife and baby son survived without injury.

After the surrender Yamaguchi worked as a translator for the US forces and then became a teacher. He broke his silence about his past when his son, six months old at the time of the Nagasaki bombing, died of cancer, aged 59. His loss turned Yamaguchi into a vocal supporter of nuclear disarmament (and a key participant in the documentary *Niju Hibaku* – Twice Bombed, Twice Survived).

His mother's will probably saved her son Iwao Nakanishi. In late August 1945 the 15-year-old's seemingly healthy body broke down with the usual bomb-related symptoms. Lacking medicine, she sold her kimonos, *obi* (sash) and Japanese trinkets to the occupying troops in exchange for nutritious food – butter and canned meat, condensed milk and chocolate. At a time when women were too scared or shy to approach foreign soldiers, many of them Australians and New Zealanders, this drew nasty gossip.

Iwao lived. So would his younger sister for a while. Unhurt in the blast, she grew into a beautiful young woman. In 1951, aged 18, she came second in the Miss Hiroshima beauty contest. Her exposure to residual radiation is believed to have caused the cancer that killed her several years later.

Iwao joined a Japanese company and married, aged 27. His wife's family were 'very worried' and asked for his medical reports. A year later his wife gave birth to a healthy son. For the rest of his life Iwao experienced intermittent illnesses linked to bomb exposure; he survived prostate cancer. Today he works as a volunteer guide at the Peace Museum in Hiroshima. He studied the history of the war and, ashamed of Japanese war crimes, visited the Nanking Massacre

Memorial Museum recently on a 'cultural exchange': it was 'the best I can do as an act of atonement'.

Tsuruji Matsuzoe, the boy in the torpedo tunnel who had been consigned to the piles of dead in a Nagayo school, graduated from his teachers' training college in Nagasaki in 1949. His health returned, though his hands remain badly scarred, and three operations failed to heal his crippled right hand. He suffers from a recurring pulmonary disease, which has required years of treatment. He briefly worked as a national elementary school teacher, and later became a newspaper journalist. 'I did not feel particularly discriminated against. However, when I married, I did not mention it. My wife did not mind at all, but her family were not happy when they found out.'

On 11 August, in the ashes of their home, Dr Takashi Nagai discovered the bones of his wife's body beside her rosary beads; he buried her in a full Catholic ceremony. Over the next few years he resumed teaching at the university, and writing his book *The Bells of Nagasaki*. Completed on the first anniversary of the bomb, it was initially suppressed but eventually became a bestseller. Seriously ill with leukaemia, Dr Nagai spent the last few years of his life confined to bed in a small hut built near the site of Urakami Cathedral, where he received visits from Emperor Hirohito and an envoy of the Pope. He died in 1951 and 20,000 people attended the funeral of the man who became known as 'the Saint of Urakami'.

Tomiko Matsumoto's uncle retrieved her from a makeshift clinic in the school assembly hall in Fuchucho, a village in the Aki district, near Hiroshima. The 13-year-old's facial burns shocked her family. Without any medicines or painkillers, her grandmother used herbs and grated cucumber juice on the wounds.

One day Tomiko's grandmother produced a black box containing the bones of her mother and three-year-old brother; nothing was found of her second-youngest brother who died instantly while playing outside. To this day, he remains one of thousands of people whose bodies are unaccounted for. Tomiko's father, notwithstanding his injuries and exposure to black rain while searching for her, survived.

Father and daughter recovered and returned to Hiroshima, he limping, drawing her along in a two-wheeled cart. They built a small, makeshift shelter in the ashes of their former home. There they lived, on pumpkin roots, rice balls and grass. 'People said that nothing would grow in Hiroshima for 75 years,' she later said, 'but grass started to appear among the burnt-out ruins. We would pick the grass and eat it.'

Tomiko returned to school in 1946. A disused military barracks at the foot of Hijiyama served as a temporary classroom for the 30 surviving children in her form. She wore a hat over her bald head, and long sleeves to hide her keloids, 'even in summer'.

'We all had keloid scars. There was one girl whose fingers had been stuck together with burns and she couldn't separate them.' Soon many of her school friends died of leukaemia and other diseases. 'I hoped to be able to return to some semblance of a normal life after the war ended, but I was terribly wrong.' One night in May 1948 her father, almost completely bedridden with radiation sickness, killed himself. 'Every day, he would say, "I want to die, I want to die",' she recalled.

She lived off the kindness of strangers and her own wits. After class, she would go around the city collecting scrap metal, which she sold for rice or vegetables. Her relatives sent money for school fees, and she graduated in 1950. She began to look for work in Hiroshima – but 'nobody would give us jobs'. Tomiko found work in 1952, at a sweet shop, where she remained for six months. She survived, and lives today in Hiroshima, a robust 78-year-old, and an outspoken advocate for peace.

Taeko Nakamae's father, a soldier, came looking for her on a bicycle; student relief workers had seen her name on a list of survivors at Kanawa Island near Uji. He had already organised the funeral for her younger sister, Emiko, who died of burns at Koi Primary School. The beautiful young teacher who saved Taeko also passed away, on 30 August: 'When I heard that,' Taeko said, 'I wished that I had died in her place. She not only saved me but she had gone all over Hiroshima saving her students.'

Her family kept her face bandaged and the mirrors hidden. 'Even when I cried and asked them what my injuries were, they continued to hide it from me.' In October, feeling better, she found the strength to remove her bandages and look at herself in a mirror: 'When I saw my face, I lamented my teacher for helping a student with such a terrible ugly face – as a 15-year-old girl. I felt that as a woman my face was the most important thing …'

She underwent three plastic surgery operations. She later fell in love but left the relationship to avoid causing the man or his family trouble. Though not openly discriminated against, she remembers lots of rumours about *hibakusha* having deformed children. 'I resolved to be a working woman instead' – and she got a job as a clerk at the national railway company. In 1963, aged 33, she married. To please her husband, she resolved to undergo a fourth operation on her face, but 'he told me that he married me – even though I had scars – just the way I was, and so I shouldn't go through any more pain'.

•

In the years after the A-bomb, survivors hoped to receive government support. Their cities were battlefields; they were legitimate casualties of war. Yet they were denied any medical recognition or compensation for more than a decade.

In October 1945, the Japanese domestic law that guaranteed compensation for war victims ceased to exist. Neither *hibakusha* nor victims of conventional bombing received a cent's worth of medical care. 'There were no appeals at the United Nations, nothing was done by the government for 11 years,' said Nori Tohei, co-chairperson of *Nihon Hidankyo* (the Japanese Confederation of A- and H-bomb Sufferers Organisation).

For nearly a decade, the world would hear nothing of the human repercussions of the atomic bombings. The occupying forces not only ignored the A-bomb survivors' medical complaints; they refused to recognise their existence. American censors forbade media references to the atomic bomb or its effects. The Japanese media, under American control, readily complied. The *hibakusha* were a low priority in a nation that shunned such dreadful reminders of a disastrous war. Not until 1952, when the occupation ended, were Japanese reporters able to write about the atomic bomb. With the media's tiresome love of sensation, they cast a lurid eye over these wretched young women whom they dubbed 'Keloid Girls' or 'A-bomb Maidens', cruelly reinforcing the girls' marital status. While reports of 'atomic freaks' boosted newspaper sales, the government ignored the issue and refused to accept the *hibakusha*'s medical complaints. Indeed, the A-bomb survivors became a national irritant: why should they receive special treatment, complained the victims of conventional firebombing. Nobody accepted the peculiar, ongoing horror of radiation exposure as a special case, a long-term medical issue. Indeed, it took another atomic bomb to provoke any interest.

On 1 March 1954, America detonated a hydrogen bomb, the world's first thermonuclear weapon, on Bikini Atoll, in the Marshall Islands. The explosion, the first in a series of thermonuclear tests, yielded energy equivalent to 15 megatonnes of TNT – or about 600 to 800 Hiroshima bombs. Extensive radioactivity saturated the atoll and neighbouring islands. Of the 290 people unintentionally exposed to radiation, 239 were inhabitants of three nearby atolls (of whom 46 died between 1954 and 1966); 28 were American observers on Rongerik Island; and 23 were crewmen of a Japanese fishing boat, the *Daigo Fukuryu Maru* (*Lucky Dragon No. 5*), one of whom died. Among the non-human casualties, millions of irradiated fish were rendered inedible.

Most upsettingly for the *hibakusha*, the American director of the Atomic Bomb Casualty Commission offered immediate medical treatment to the crew of the *Lucky Dragon*. So either the commission was lying when it claimed it had no authority to treat A-bomb casualties; or it meant to use the fishermen for similar experiments under the pretence of treating them. The case provoked a public outcry in Japan – dozens of Japanese articles appeared under headlines such as, 'We Won't be Treated as Guinea Pigs' – and stirred the ire of the *hibakusha*, whose claims had gone unheard for almost a decade. The disease that dared not speak its name had killed a Japanese fisherman; now it would bear responsibility for the chronic illnesses and deaths of hundreds of thousands.

'This was when the broad citizens' movement in Japan against nuclear weapons started, and the *hibakusha* gained attention,' said Nori Tohei. 'Not until 1954,' noted a committee of Japanese scientists, 'did the Japanese government adopt any official policies to help the A-bomb victims. The immediate cause was the groundswell of public concern ... provoked by the damages to crewmen of the Japanese fishing vessel.'

The explosive publicity drew foreign sympathy and hopes of recovery: US doctors had developed new plastic surgical techniques which promised to restore some of a patient's original likeness. In 1955 the American philanthropist Norman Cousins organised, with the assistance of a Japanese cleric, Reverend Kiyoshi Tanimoto, a visit to the United States of some 25 Hiroshima women with severe A-bomb scarring. The 'Hiroshima Maidens' arrived to a lavish New York reception and the welcoming smile of the star turn, a local debutante called Candis, aged 18. Candis's beauty and long white ball gown set in brutal relief the disfigured faces and drab clothes of the vanquished Japanese. Their smiling American hosts arranged an excruciatingly implausible A-bomb 'reunification' on the popular daytime television show *This Is Your Life* (dedicated to the life of Reverend Tanimoto) of the Maidens and Captain Robert Lewis, assistant pilot of the *Enola Gay*, whose plane had delivered their misery. Two Maidens appeared on the program behind screens: 'To avoid causing them any embarrassment,' the presenter explained, 'we'll not show you their faces.' If the producers felt any qualms over this exhibition, they did not let it intrude on the show's constructive intent: to raise $60,000 in donations to finance the Maidens' tour and plastic surgery.

Surgeons at New York's Mount Sinai Hospital performed 127 operations on the women. The methods – including 'Z-plasty', 'defatting', 'split skin graft', 'scar bridle' – were not completely successful. The women received tattooed eyebrows and grafts of skin taken from their thighs, arms and stomachs. Doctors removed the worst of their keloids. Simply being able to blink, or open and close their mouths, transformed their lives. One girl's lidless left eye, open and weeping for 10 years, received a new eyelid; another woman, named only Hiroko T., underwent 12 operations, including a 'tubed pedicle' – a grafting technique that involved sewing down a long flap of skin – after which she found she could eat through her mouth for the first time since the bomb. Asked what she would like to eat she indicated 'a hot dog'. On their return to Japan, the Maidens' medical improvements were not instantly perceptible to a nation hungry for miracles; but the trip performed a useful public relations role.

•

The publicity surrounding the *Lucky Dragon* and the Maidens' tour compelled the Japanese government to act. In 1957 a new Medical Law granted a 'health passbook' to people who could prove they had been exposed to atomic radiation. Initially, it entitled them to twice-yearly free medical checks and a percentage of burial costs. For the first time the Japanese government recognised the *hibakusha* as medical casualties of war – that is, wounded by a weapon of war.

Having a passbook did not mean the government offered to pay for their medical treatment or provide other form of compensation. In 1963, the Tokyo District Court rejected – on the grounds that an individual could not act alone – the case of an atomic bomb victim who tried to sue for compensation. The decision marched in lockstep with the government's unwillingness to accept the scale of radiation sickness, partly to avoid embarrassing American interests in Japan. The *hibakusha* took their cue that only mass action would work and organised as a collective force. Over the years, after countless legal cases, their medical complaints gradually won recognition.

In time the Japanese government placed them in categories. There were those directly exposed to the bomb; those exposed to residual radiation (that is, who entered the city within two weeks after the bomb fell); those exposed while treating victims; and those not yet born who received radiation poisoning *in utero*. Sub-categories further delineated the claimants – for example, according to their proximity to the hypocentre; people within a 1-kilometre radius were the most seriously affected, of course. But that did not automatically qualify them as *hibakusha* – sometimes local councils in the bombed cities defined them differently. A sympathetic mayor might help citizens less eligible than those living nearer the hypocentre.

Being *hibakusha*, carrying a passbook, still did not generate government-paid medical assistance. The new Atomic Bomb Survivors Relief Law recognised as bomb-related a very narrow range of cancers: just 2000 people, or about 0.6 per cent, of the 300,000 A-bomb survivors, received government-assisted treatment between 1945 and 2000. The rest got nothing because their peculiar ailments were not 'officially approved' illnesses.

In time, mounting evidence of a far wider range of medical problems prompted the *hibakusha* to launch a series of class actions. In 2003, the first cases came before the Japanese courts; in the next six years *Nihon Hidankyo*, the peak body representing radiation victims, won 19 cases representing 306 claimants, 60 of whom died during the proceedings. These high-profile victories forced the government to add liver, thyroid and other conditions to the official list of bomb-related diseases, and to accept that the *hibakusha*'s medical ailments were linked to their exposure, directly or indirectly, to the atomic bombs.

In March 2009, nearly 70 years after Groves, Oppenheimer, Farrell, the *New York Times* and teams of US experts dismissed the risk of widespread radiation sickness, Japanese authorities had designated precisely 235,569 Japanese people as atomic bomb sufferers, and granted them a health passbook. Their average age was 75.9 years. That year the passbook entitled the holder to, among other benefits, free medical checks twice a year (including cancer checks) and a state subsidy of 90 per cent of their medical fees. Most passbook holders have had cancer; most will die of it.

On 6 August 2009 the *hibakusha* won another victory: the Japanese government unconditionally surrendered to the atomic bomb victims. After losing 19 straight cases over the right to certification of people seeking recognition as sufferers of bomb-related illnesses, Tokyo granted unconditional medical relief to every case that succeeded at the first hearing. A compensation fund is being planned. The age and vulnerability of the plaintiffs, and the government's and claimants' deep reluctance to relive the horror of the atomic bombs through the Japanese courts, partly explained the legal capitulation. The then Prime Minister, Taro Aso, signed the agreement after attending the commemorative ceremony marking the 64th anniversary of the Hiroshima bombing, thus ending the 306 plaintiffs' six-year-long legal battle. 'Considering that the plaintiffs are aging,' Aso said, 'and they have fought this legal battle so long, we have decided to introduce the new policies to bring relief to them swiftly.' Tokyo recently extended the medical compensation to foreign nationals – Koreans and prisoners of war – who were exposed to the atomic bombs: survivors are urged to contact the Japanese government through their embassies.

On hearing of the breakthrough, Haruhide Tamamoto, a 79-year-old plaintiff, told the *Japan Times*: 'A certificate … means the government admits that it started a war and caused this atrocity. Being dead without receiving one is an absolute tragedy.' Another plaintiff, Kamiko Oe, 80, said: 'I once held grudges against the government, but my hard feelings went away today.' The government added an apology, which may be read as a symbolic act of restitution after 65 years of neglect. 'Lawsuits have been drawn out, A-bomb survivors have aged, and their illnesses have worsened,' said then Chief Cabinet Secretary Takeo Kawamura. 'By extending its thoughts to A-bomb survivors' sufferings, which cannot be described in words ... the government apologizes.'

•

Some of the worst cases triumphed over the most appalling circumstances, to lead happy and fulfilling lives. They tend to share a striking absence of self-pity. As Anne Chisholm has recorded in her book *Faces of Hiroshima*, Hiroko T., whose deformities were so acute she wore a face mask for years, grew into a lively, quick-witted woman utterly free of morbid self-consciousness or self-pity; she later married an ex-marine who read about her in the papers and courted her for 10 years before she finally said 'yes' (at first she angrily interpreted his affection for pity). Hiroko has had 29 facial operations – between finding work as a shop-owner, a dressmaker and a fashion saleswoman. Had she never despaired, Chisholm asked. 'So many of the others had thought about killing themselves,' Hiroko replied. 'I, never! If I think back on my life, I think I was really a lucky girl.'

'My best subject was mathematics,' said Miyoko Matsubara, a pupil at Hiroshima Girls' Commercial School in Danbara. 'I wanted to become a bank employee.' Miyoko came from a poor family and hoped to find financial security. Her school was progressive: it believed girls should have a trade and get a job. On 6 August she was among 250 mobilised children working on a demolition site at Tsurumicho. Her flashburns were so severe no one would employ her after she graduated. She quickly abandoned her dream of working in a bank and getting married. She spared Japan the trouble of setting eyes on her: she shut herself away as a live-in carer at an orphanage for the blind. For eight years, from morning to night, a group of sightless children were among the few creatures on earth who valued Miyoko's existence.

A friend persuaded her to join the Nagarekawa Methodist Church, where Reverend Tanimoto's mission worked to help the worst affected. Her new faith, and just as likely, the influence of this enlightened pastor, 'put my heart at ease'. He recommended her for a job at the Peace Museum before he passed away. For years she underwent plastic surgery, and recent advances have partly remoulded her face. Her personal warmth shines through the residual scarring: 'My younger brother and my niece and several other relatives are all bank employees now,' she said with a sigh, when I spoke to her in 2009.

•

Yet some cases were so severe, they at first appeared beyond the reach of medical science or even humanitarian care. A close friend of Miyoko Matsubara has endured 66 surgical operations in an effort to rebuild her broken body. So badly burned and smashed up, these people live solely, it seems, because they can: a stoic rebuttal to those who, 65 years ago, set them aside on triage fields to die.

In 2009 I visited a nursing home in the suburbs of Hiroshima built exclusively for *hibakusha*: the Kurakake Nozomi-en (Nursing Home for A-bomb Survivors) is devoted to treating the full range of physical and psychological problems associated with exposure to the atomic bomb. The president, Dr Nanao Kamada, showed me over the facility. 'In a general nursing home they cannot mention the atomic bomb, but here,' he said, 'they can speak freely about their psychological problems.' The patients were having lunch as we entered. The upward gaze of the ward seemed surprised by the sight of a Western visitor – 'Why is he here, to study us?' their eyes seemed to say. Some were psychologically damaged, mute, expressionless, with no outward physical signs of bomb exposure, only a dark and abiding memory; others were severely deformed, their bodies twisted, dessicated and tiny, their faces scarred and wrenched in extreme directions. One or two waved from their wheelchairs, smiling. The effort lent a strange sense of hope – that nobody here takes for granted the use of their hands or the movement of their lips. A source of happiness here is being able to smile.

·

About a month after the bombings, some 16,000 child evacuees waited in the temples and shrines around Hiroshima for their parents to collect them. About 5000 of these children were not yet aware they were orphans. In time a succession of strange uncles and aunts and cousins would arrive; or their badly scarred mother or father. Some children, not recognising their parents, would run away in terror.

One unclaimed third-grade girl recounted her experience: 'The friends I had been living with were gradually taken back to Hiroshima – today one, the next day two – by their fathers and brothers. It was saddest when the time came for my best friend to go. Just before she left, when I should have been there to share her joy, I hid instead in the shadow of the old temple and wept ... no-one came for me ...'

On 10 August Shoso Kawamoto, the Hiroshiman schoolboy evacuated with his classmates to Kamisugi village in the countryside, sat similarly unclaimed in the temple near Hiroshima. He thought his entire family were dead and 'cried tears of joy and couldn't speak, I was so happy', when his elder sister Tokie, 15, arrived to collect him. The day after the explosion, Tokie, apparently unhurt, had returned to the site of their home where she found the burnt remains of their mother, younger brother and sister locked in an embrace. She could not find their father and second sister.

Shoso and Tokie took a train to the outskirts of Hiroshima to resume the search. Not a building interrupted their view from West

Hiroshima Station to the mountains. After fruitless hours spent wandering the ruined city, they collected a little ash near their home and buried it in Numatacho, a suburb where their relatives lived.

Their whole family, they presumed – with the exception of their eldest brother then in Manchuria – were dead. Shoso and Tokie went to their relatives' house in Numatacho. Their uncles and aunts wanted to adopt them separately but Tokie refused. 'My sister,' said Shoso, 'was adamant that we wouldn't be separated as we were the only survivors of our family.' The two children returned to Hiroshima and lived together in a corner of the partly destroyed train station. The strong-willed Tokie got an administrative job in the railway company, which let them wash in the station bathroom. They scavenged for food – potatoes and rice balls – with the help of friends. 'Looking back I realise what a wonderful older sister she was,' Shoso said. Hundreds of these atomic street kids would die of cold and starvation in the coming winter; the survivors, like Tokie and Shoso, got jobs; or formed gangs that roamed the cities, thieving and begging, and growing up to become *Yakuza*-style gangsters.

Within a year Tokie succumbed to radiation sickness and Shoso's uncle placed him in the care of a nearby village head, Rikiso Kawanaka, who gave the boy a job in the family's soy sauce store. 'Mr Kawanaka told me that if I worked hard for 10 years he would build me a house,' Shoso said. 'So I worked very hard for the next 10 years and he built me a house.'

Shoso fell in love with a local girl, whose father refused to allow the marriage: 'You were in Hiroshima,' he said. 'You must have been exposed to radiation. You probably won't live long either, so I can't possibly give my daughter to you.'

Shoso walked away in shock. His house, his job, meant nothing to the girl's father. 'Even after working so hard for all that time, I couldn't marry the woman I loved, so I quit my job.' He returned to Hiroshima and joined a gang – many of them A-bomb orphans, like

Shoso – who took care of him. 'My experience had taught me that hard work doesn't pay off. So I just hung out with gangsters and lived a low life.'

In his 30s he abandoned the life of a gangster, moved to Okayama and started a food production business. In 1995, a fellow A-bomb orphan invited him to the 50th anniversary of the atomic attack. Shoso, then almost 60, decided to sell up and return to Hiroshima. Today he derives great happiness guiding school groups around the Peace Museum.

'Did you subsequently marry?' I asked him.

'No, I didn't. I didn't want to experience that pain again.'

'You never saw her again?'

'Never.'

'Have you ever been sick yourself in any way? Do you suffer any radiation effects?'

'No, thankfully I haven't experienced any radiation effects at this stage.'

•

Mitsue Fujii's eyes blaze with anger at the memory of the short life of her mother, Shizue. After the family returned empty-handed to Hiraki village, Zenchiki, Mitsue's grandfather, literally worked Shizue to death. In addition to her domestic and farm chores, she was responsible for nursing his two surviving daughters who had returned from Hiroshima severely injured; every day she washed them and placed cucumber slices on their burns.

Meanwhile, anxious to find out whether her husband – Zenchiki's son – had survived the war, Shizue plied returning soldiers for news. They tried to sound optimistic – 'He'll come back one day' and 'I think he's alive' – and averted their eyes from the woman's imploring gaze. One day, a year after the surrender, she received a standard government notice informing her that her husband was dead. There was no ceremony, funeral or compensation. The news mortified the old man, and left Shizue helpless: 'From that point on, my mother had no future,' her daughter Mitsue said.

Two years later, aged 35, Shizue died of illnesses linked to her exposure to residual radiation. Her three children survive her. 'I remember feeling pure hatred towards my grandfather,' said Mitsue, now 70. 'My grandfather only cared about his own children, and he considered his daughter-in-law disposable.'

When their grandfather died shortly after, the three children stayed on the farm – in happiness. To avoid being sent to orphanages, Mitsue's elder brother, Hisao, then 16, insisted on raising her, then 10, and their little sister. 'To me, my older brother is my father,' Mitsue says. The three children helped each other: 'We never fought, we just worked hard.' They gathered wood and vegetables, cooked, worked in the fields. Their neighbours were very kind, and offered food and a regular bath. Mitsue, eager for an education, sometimes lied to her brother and slipped away to school in Hiroshima.

At 19, Hisao married and his wife came to live with them: 'She was like a mother to me,' Mitsue recalled. 'They were so kind. I was so happy. I wanted to live with them forever.' But she had to find work. In her teens, Mitsue got a job as a trainee hairdresser and lived in the salon. A highly intelligent child, she studied in her spare time, to the anger of her boss who ordered her to stick to hairdressing. For nine years she cut hair and suffered recurring illnesses linked to her exposure to residual radiation. She was constantly being told that '*hibakusha* are weak', '*hibakusha* cannot have children'.

Doctors measured her white blood cell count at half that of a normal person. Her illness ruined her hopes of marriage. So she saved money to open a salon. Within a few years, however, she met her future husband – a man who, like Mitsue, had lost his mother to the bomb and his father in the war.

The couple started a family with deep anxiety: 'When I was pregnant, I was in constant fear that my child would be born without arms or legs, or have some other deformity,' she said. Mitsue 'aches with sorrow' over the memory of her mother but she has found happiness and is now a cheerful grandmother, with five healthy grandchildren. Fifteen years ago, aged 55, she resumed her studies. Today, she is a volunteer worker in the Hiroshima Peace Museum.

CHAPTER 23

WHY

*... this deliberate, premeditated destruction was
our least abhorrent choice.*

Henry Stimson, Secretary of War in the Truman administration, defending the decision to use the
bomb (*Harper's Magazine*, February 1947)

IN THE IMMEDIATE AFTERMATH OF the bombing, American consciences
were settled: the weapon had avenged Pearl Harbor and Japanese
atrocities, avoided a land invasion, saved hundreds of thousands of
American lives and ended the war – so believed an emerging consensus.
The targets were 'military', Washington repeatedly assured the public.
The media caressed the bomb as the saviour of mankind – only 1.7 per
cent of 595 newspaper editorials in 1945 opposed the use of the atomic
bomb.

The press and public mutually reinforced their satisfaction at a job
well done. Asked whether they approved or disapproved of the atomic
strikes, 85 per cent of Americans said in a Gallup Poll published on
26 August 1945 they approved. The responses of men and women,
young and old, middle- and working-class, fetched the same result.
Curiously 50 per cent of Americans said, in the same poll, that they
were against the use of poison gas – even if gassing the Japanese would
have reduced American casualties (40 per cent of men and a larger
percentage of women supported the use of gas). The reasons were

possibly connected with the ghastly memory of mustard gas used in World War I and the emerging horror of the Nazi death camps and gas chambers. Atomic bombs were seen as spectacular new weapons that somehow inflicted a cleaner, quicker death. That perception gradually cooled as the public learned the truth about the destruction of civilian life, and the facts of radiation poisoning; two years after the war, the number of respondents who approved of the bomb had halved, according to a similar poll.

Letters to the editor of August 1945 in America and Britain conveyed the full range of feelings, from ardent approval of the weapon to moral outrage at the wanton destruction of civilian life. The London *Times* registered the angst of clergymen, politicians and artists. There were soul-searchers – 'shall [we] not lose our souls in the process of using these new bombs?'; those disgusted – 'a few months ago, we were expressing horror at the inhumanity of the Germans' use of indiscriminate weapons ... Must we not therefore now apply this criticism to ourselves?'; pulpit pounders – the bomb, claimed Sir William Beveridge, had 'obliterated' any distinction between combatants and civilians as targets for attack and exacted too great a price for peace; and George Bernard Shaw, who wrote, 'We may practise our magic without knowing how to stop it, thus fulfilling the prophecy of Prospero.' Prince Vladimir Obolensky, a Russian aristocrat then living in London, challenged the emerging consensus that the bomb 'brought the Japanese war to a magic end', as he wrote in *The Times* on 14 August 1945, 'The belief that it has saved millions of the allies' lives is a misconception ... In reality, Japan has been brought down by the interruption of her sea communications by Anglo-American air and sea power, and a danger of a Soviet thrust across Manchuria, cutting the Japanese armies in Asia from home.'

The bomb provoked extreme reactions in American church leaders: provincial firebrands extolled atomic power as a heavenly thunderbolt with which the Almighty had endowed His American disciples to smite the wicked – echoing Truman's description of the bomb as the 'most powerful weapon in the arsenal of righteousness'. Many religious leaders, however, quietly registered their Christian disapproval of the mass killing of noncombatants. This intensified as the truth about the effects of the bombs emerged. The Federal Council of Churches was among the most vociferous, branding the atomic bombing of Japan 'morally indefensible'; in so doing, America had 'sinned grievously against the law of God and the Japanese people'. The sermon jarred in a country where most people despised the Japanese and believed in God.

Several church leaders condemned the bombings as war crimes. On 29 August an influential magazine, the *Christian Century*, published an article headlined, 'America's Atomic Atrocity', which strummed the nerve of moral outrage and provoked a stream of letters that shared the author's revulsion at America's impetuous adoption of 'this incredibly inhuman instrument': the atomic bomb had placed the United States in 'an indefensible moral position', the magazine concluded, and landed the blame squarely on the shoulders of the US government and, by extension, the American people: it was a collective crime against humanity: *we* dropped the bomb on 'two helpless cities'; *we* destroyed more lives than the US lost in the entire war; *we* crippled America's reputation for justice and humanity. The *Christian Century's* philippic ended with an appeal to Nagasaki's devastated Catholic community for understanding and forgiveness, and a rallying cry to action: 'The churches of America must dissociate themselves and their faith from this inhuman and reckless act of the American government ... They can give voice to the shame the American people feel concerning the barbaric methods used in their name.'

And despite Truman's efforts to appease the Holy See, the Vatican's disgust presaged a long accretion of Catholic condemnation. Father John Siemes, a priest who experienced Hiroshima, alerted the Vatican to the horror on the ground. 'The crux of the matter,' Siemes concluded, 'is whether total war in its present form is justifiable, even when it serves a just purpose. Does it not have a material and spiritual evil ... which far exceed[s] whatever good might result?' The Vatican, by its actions, believed it did. Leaders of the Anglican Church shared that sentiment. The Dean of St Albans, England, rather churlishly refused a service to commemorate Victory in the Pacific because 'victory was clinched by the atomization of a quarter of a million Japanese'.

•

As the facts of the destruction filtered back to Los Alamos in August and September, the earlier exuberance of the Manhattan Project's scientists and engineers turned introspective and, by stages, morose. Some found themselves reflecting guiltily on what they had done. The nuclear reckoning preoccupied the experts in ways they had not foreseen: the 'questionable morality' of dropping the bomb without warning 'profoundly disturbed' many, and their moral qualms deepened after Nagasaki, observed Edward Teller: 'After the war's end,' he wrote, 'scientists who wanted no more of weapons work began fleeing to the sanctuary of university laboratories and classrooms.'

In Oppenheimer we encounter a man who seemed to reflect the median temperament, rather like a psychic bellwether who captures the emotional impulses of those around him. On 16 October, his last day on the 'Hill', Los Alamos held a farewell ceremony in Oppenheimer's honour. Upon accepting his Certificate of Appreciation, Oppenheimer addressed the mesa's entire workforce (each of whom later received a sterling silver pin stamped with a large 'A' and a small 'BOMB', in recognition of his or her service):

'It is our hope,' Oppenheimer said, 'that in years to come we may look at this scroll, and all that it signifies, with pride. Today that pride must be tempered with a profound concern. If atomic bombs are to be added as new weapons to the arsenals of a warring world, or to the arsenals of nations preparing for war, then the time will come when mankind will curse the names of Los Alamos and of Hiroshima. The peoples of this world must unite or they will perish. This war, that has ravaged so much of the earth, has written these words. The atomic bomb has spelled them out for all men to understand. Other men have spoken them, in other times, of other wars, of other weapons. They have not prevailed. There are some, misled by a false sense of human history, who hold that they will not prevail today. It is not for us to believe that. By our works we are committed, committed to a united world, before this common peril, in law, and in humanity.'

On 25 October 1945 Truman received Oppenheimer in the Oval Office; the physicist had requested the meeting in an effort to persuade the President to support international controls on nuclear weapons. Truman disarmed Oppenheimer by asking when the latter thought the Russians would develop a nuclear weapon; Oppenheimer replied that he did not know, to which Truman interjected: 'Never!' Sensing a lack of urgency in the US leader, and perhaps a little overwhelmed by their first meeting, Oppenheimer confided, 'Mr President, I feel I have blood on my hands.' The remark infuriated Truman who bluntly replied (as he later told David Lilienthal, chairman of the Atomic Energy Commission), that 'the blood is on my hands, let me worry about that', smoothly ejected the physicist, and instructed Dean Acheson never to bring 'that son of a bitch in this office ever again'.

Oppenheimer meant that he wore the blood of future casualties of nuclear war; not the blood of the Japanese. The 'cry baby scientist', as Truman later dismissed him (itself a rather infantile remark) cared not for the Japanese – who were 'poor little people', the collateral damage of a war they had brought on themselves; he wept for the Western victims of a future nuclear Armageddon. Oppenheimer's remark alluded to his responsibility for the deaths of millions of individuals in some distant apocalypse, which would be traced to Little Boy and Fat Man. Hiroshima and Nagasaki served as terrible, if necessary, examples of what the bomb *might* do; he did not think of them as avoidable tragedies in their own right. He quickly cast forward – as though he dared not look back – to a world where, he dreamed, global controls on nuclear weapons would entrench a lasting peace. In later years he alluded to a collective sense of regret for the general horror of war. He spoke of the 'numbing and indifference' World War II had imbued in mankind; he warned that 'we have made a very grave mistake' in contemplating the massive use of the weapon; and that 'in some sort of crude sense ... the physicists have known sin'. He did not define the nature of his 'sin'. Oppenheimer's mind was impervious to the probes of an ordinary conscience. He felt a terrible responsibility for *what might happen*; not for what he had helped to do.

His great speech of 2 November 1945 to the Association of Los Alamos Scientists (ALAS) – the spirit of whose acronym he did not share – was notable for what it did not say. The man who bore most responsibility for developing the weapon did not name Hiroshima or Nagasaki; they were already part of a fading past. He made one oblique reference, however, that suggested niggling regret. 'There was a period immediately after the first use of the bomb,' he told the 500 members of ALAS, 'when it seemed most natural that a clear statement of policy, and the initial steps of implementing it, should have been made; and it would be wrong for me not to admit that something may have been lost, and that there may be tragedy in that loss.'

He called for noble goals – the shared exchange of atomic knowledge, the creation of a world fraternity of nuclear scientists, and the abolition of nuclear weapons; he spoke of the 'deep moral dependence' of mankind during the 'peril and the hope' of the nuclear era; he cited Lincoln's Gettysburg Address. He charmed and pullulated and a kind of statesman was born. These were, however, the impossible ideals of a very clever problem-solver, not a moral visionary. Grand gestures could not change the fact that Oppenheimer had personally recommended a nuclear attack on a city of civilians without warning. This happened; the rest was wistful dreams. He later insisted – perhaps it was unbearable to think otherwise – that the atomic bombs 'cruelly, yet decisively ended the Second World War'.

Scandal and tragedy punctuated the rest of Oppenheimer's life. The recipient of several job offers, from Harvard, Princeton and

Columbia, Oppenheimer chose initially to return to Berkeley. In the event, however, he would serve in Washington as an adviser and contributor to the Acheson-Lilienthal Report on international atomic controls. His acme as chairman of the powerful General Advisory Committee to the US Atomic Energy Commission came tumbling down with the suspension of his security clearance in 1954: the inquiry dredged up his pre-war associations with communists, notably his late lover Jean Tatlock and his friend Haakon Chevalier, who claimed the scientist had been a member of a communist cell. The witch hunt, of course, was a 'travesty of justice', and had more to do with Oppenheimer's refusal to support the hydrogen bomb project than any serious truck with Reds. That Groves should be among those who refused to clear him would surely have embittered a lesser man; but Oppenheimer rode the inquiry with dignity and circumspection. He retained his job as director of the Institute of Advanced Study, was later rehabilitated and retired with a certain honour. He died on 18 February 1967.*

If Oppenheimer lacked the moral courage directly to recognise his role in the destruction of Hiroshima and Nagasaki, the facts of the atomic bombs jolted other scientists, as if from a fitful sleep, to a keener awareness of a terrible new reality. Many lapsed into torments of self-accusation and spent much of their lives expiating guilt. Physicist Mark Oliphant, for example – who had played a critical role in persuading the US to build the bomb – 'could hardly believe the early reports of the incineration of Hiroshima ... for he had not really come to grips with the possibility that a civilised and reputedly Christian nation was capable of such a deed'. His denial was hardly credible: what on earth did he suppose he had been working on, if not a bomb that would, if the chance arose, be used? Unlike Oppenheimer, Oliphant ran the gauntlet of guilt and later damned himself: 'During the war I worked ... on nuclear weapons so I, too, am a war criminal.'

---

* Of 4,756,705 American citizens screened for loyalty to the state between 1947 and 1952 just 560 were dismissed or refused employment as 'disloyal', and fewer than 20 charged with actual sedition or treason.

Szilard and the Franck Committee imposed, to a lesser degree, a similar self-indictment. Their consciences were less harmful, as they had persisted with their protest weeks before the nuclear destruction of the two Japanese cities. Many of these dissident scientists, however, had been all too eager to drop the bomb on Germany – casting a shadow over the consistency of their moral position: were they opposed to the nuclear destruction of all innocent civilians, or just non-German ones? Were their principles absolute or relative? Many were Jews who had lost family members in the Nazi round-ups, and their personal loss clearly influenced their support for the destruction of Germany. At the time, however – 1941–44 – they were unaware precisely of what had become of their families and friends. The Soviets liberated the first Nazi death camp in July 1944. The fact is, many émigré scientists turned decidedly cool on the bomb when Japan loomed in the sights of the Target Committee; Szilard in particular had adopted 'diametrically opposite positions' in relation to Germany and Japan.

James Conant, the chemist who drove the S-1 program, manifested a third expression of the scientists' moral dilemma: pride and an utter lack of remorse were the hallmarks of Conant's response to the atomic bombs. Conant disdained those scientists who 'paraded their sense of guilt' about the bomb. The moral conflict over Hiroshima 'hardly existed in my mind', said the man who had led America's mustard gas research program during World War I and played a vital role in the development of proto-napalm dropped on Japanese cities.

Conant understood the gravity of a nuclear arms race; his answer was to ratchet up America's nuclear arsenal to force concessions from the Kremlin. Within weeks of Hiroshima he advised the War Department to prepare for nuclear war. Fear of an atomic conflagration should not render the American public insensible or hysterical, he counselled – for that may lead them to reject the new weapon. Fear must be managed, distilled and drip-fed to the people, rather like a doctor treating a man with diabetes. 'The physician, therefore,' Conant wrote, 'had to frighten the patient sufficiently in order to make him obey the dietary rules; but if he frightened him too much, despondency might set in – hysteria if you will – and the patient might overindulge in a mood of despair, with probably fatal consequences.'

Conant's arms race had limits: he drew the line at the creation of the first hydrogen bomb, which he opposed in his capacity as a member of the General Advisory Committee on thermonuclear power. He saw in its hideous potential the capacity 'to destroy far more than military objectives might ever justify' – surely a self-deceiving view after the events of 6 and 9 August. He reconciled his opposition to the superbomb on the grounds that nations at peace had no moral case for building such a weapon. 'Let us freely admit,' he said in 1943, with reference to general advances in the technology of weapons of mass destruction, 'that the battlefield is no place to question the doctrine that the end justifies the means' – that is, in war anything goes, mustard gas, napalm etc. 'But let us insist, and insist with all our power, that this same doctrine must be repudiated in times of peace.' In old age, fretful over his role in the bomb, Conant conceded that it had been a 'mistake' to destroy Nagasaki.

In Edward Teller we encounter a fourth response: the utter rejection of any controls on nuclear arms. Teller was the apotheosis of the 'warrior-scientist', a man who gave his working life to the hydrogen bomb, and who saw in the megatonne dawn over Eniwetok the harbinger of the American century. Teller, on whom Stanley Kubrick partly modelled the character of Dr Strangelove, argued that America must equal or exceed the Soviet balance of terror – to assure the continuation of world peace. Negotiation, compromise, the preservation of the species, the ideal of a shared humanity: none had any traction on his argument that only by matching the Soviet arsenal could peace be assured. The planet was doomed, Teller believed, unless America subscribed to the logic of mutually assured destruction (MAD), in which bigger and more devastating bombs were the sole currency. Peace would only prevail in a world in which each side possessed the power to annihilate the other. In this, he was prophetic. The grim truth is that posterity has thus far judged him, and the exponents of MAD, partly correct, insofar as mankind has avoided a nuclear war through the assurance of mutual annihilation; that does not mean, of course, that it will not happen, and the dire uncertainty and immense expenditure of maintaining the balance of mutually assured death has turned the minds of enlightened leaders to the policy of nuclear disarmament.

•

And the politicians? After the war, the politicians stuck to their guns. Truman, Stimson and Byrnes argued that the bomb alone had ended World War II and saved hundreds of thousands, if not a million, American lives. Stimson travelled a difficult road to this position. The War Secretary had repeatedly bemoaned his countrymen's indifference to the firebombing of Europe and Japan and the 'appalling lack of conscience and compassion' the war had brought. His colleagues on the Interim Committee on nuclear energy, formed in June 1945, were at least consistent: having approved the firebombing of noncombatants, how then could they oppose the nuclear-bombing of noncombatants? The atomic bombs were a continuation of the existing strategy of civilian extermination to break the people's 'morale'.

If words alone be his judge, Stimson was consistent: he had objected to targeting civilians – in private talks with Truman – and maintained on 16 May 1945 that the 'same rules of sparing the civilian population should be applied as far as possible to the use of any new [atomic] weapons'. His actions betrayed this apparent conviction. He raised no objection when the Interim Committee proposed the nuclear destruction of an urban area. In so doing, he was reduced to accepting the grotesque casuistry that 'workers' homes' represented a military target. His fixation with the temples and shrines of Kyoto – as if they were mankind's last link with a dying civilisation – brought to a point all the frustrations of a man unable to reconcile conscience with action. Tens of thousands of civilians would die regardless of whether the bomb fell on shrines in Kyoto or workers' homes in Hiroshima.

Stimson erased this truth from his mind, whose elasticity would stretch to one last act in the story of the bomb – as the draught horse for its defence. This eminent statesman, of unimpeachable public example, who had done more than any high official to question (if not oppose) the bomb, would now serve as the official mouthpiece for the arguments deployed *in favour* of using it. Stimson's role was to 'silence the chatterers' – the scientists, journalists and clerics whose shrill denunciations of the bombing of Hiroshima and Nagasaki had put the White House in the witness box. Their case gathered momentum in 1946: *The Bulletin of the Atomic Scientists*, for example, published on 1 May the Franck Committee's report opposing the bomb, which the influential radio broadcaster Raymond Swing gave national airplay. Einstein, in a front-page article in the *New York Times* (19 August), 'deplored' the use of the weapon. John Hersey's article 'Hiroshima' – which occupied an entire issue of the *New Yorker* (on 31 August) – showed Americans what it meant to experience a nuclear attack through the lives of six survivors, and had a huge impact on perceptions of the nuclear weapons (notwithstanding the reporter Mary McCarthy's demolition of Hersey's article as a 'human interest story' that treated the bomb as an earthquake or other natural disaster and failed to consider why it was used, who was responsible, and whether it had been necessary).

Prominent official voices joined the backlash. In 1945 Truman extended the United States Strategic Bombing Survey (USSBS) to the Pacific, in order to record the effectiveness of the air war over Japan. Its findings, which appeared in July 1946, diametrically opposed Truman's case for the bombs. The USSBS argued that the weapons were unnecessary, and that Japan had been effectively defeated long before their use. That much the military commanders already knew. But the study went further, speculating that Japan would have surrendered 'certainly prior to 31 December 1945, and in all probability prior to 1 November 1945 ... even if the atomic bombs had not been dropped, even if Russia had not entered the war, and even if no invasion had been planned or contemplated'. This seems unlikely – Tokyo refused to yield and the Russian invasion, as we have seen, played a decisive role in Japan's surrender – and the USSBS's conclusions have been heavily criticised. At the time, however, the report certainly added to the growing unease over nuclear weapons.

Washington insiders, chiefly Conant, were concerned at the cumulative effect of these voices and proposed a reply. The result was a long article, in Stimson's name, sourced to a memorandum from his assistant, Harvey Bundy, and written largely by Bundy's precociously clever son, McGeorge. Groves, Conant and several senior officials edited the draft. The article first appeared in the February 1947 issue of *Harper's Magazine*, reappeared in major newspapers and magazines, and was aired on mainstream radio. It purported to be a straight statement of the facts, and quickly gained legitimacy as the official case for the weapon. The *Harper's* article (and a parallel piece in the *Atlantic Monthly* by Karl Compton) reinforced in the American mind the tendentious idea that the atomic bomb saved hundreds of thousands (perhaps several millions, Compton claimed) of American lives by preventing an invasion of Japan. The article's central plank was that America had had no choice. There was no other way to force the Japanese to surrender than to drop atomic weapons on them. By this argument, the atomic bombings were not only a patriotic duty but also a moral expedient:

In the light of the alternatives which, upon a fair estimate, were open to us I believe that no man, in our position and subject to our responsibilities, holding in his hands a weapon of such possibilities for accomplishing this purpose and saving those lives, could have failed to use it and afterwards looked his countrymen in the face.

The decision to use the atomic bomb brought death to over a hundred thousand Japanese. No explanation can change that fact and I do not wish to gloss over it. But this deliberate, premeditated destruction was our least abhorrent choice. The destruction of Hiroshima and Nagasaki put an end to the Japanese war. It stopped the fire raids, and the strangling blockade; it ended the ghastly specter of the clash of great land armies.

Editors and the public warmly approved: here, they felt, was an honest justification for this horrific weapon; the A-bomb did good, in the end. The *Harper's* article put the American mind at ease, slipped into national folklore, and the Stimsonian spell appeared to tranquillise the nation's critical faculties on the subject. Only the *Washington Post* made a serious attempt at a critique. It trenchantly argued that, contrary to Stimson's claim, clear evidence was available of Japan's terminal weakness before the bombs; and that his 'apologia' would 'not altogether remove the feeling that the use of the bomb put upon us the mark of Cain'.

The *Harper's* article was profoundly flawed. Stimson had not intended to deceive the American public, but the omissions and selective use of facts deployed in his name had that effect. The essay made no mention of the long debate over the role of the Emperor and Japan's last (and only persuasive) offer to surrender on condition that the Emperor be preserved (a condition Washington, in the end, accepted). Nor did it mention the opposition of senior officials to bombing a city without warning – a target that only the most wilfully self-deceived could construe as 'military'; or the Soviet Union's role in the timing of the bomb; or the USSBS's (contested) claim that a defeated Japan would have surrendered without the bomb or an American invasion. Most erroneously it argued that a land invasion of Japan and the atomic weapons were mutually exclusive – a case of 'either-or'. This flawed nexus ignored the fact that Truman and senior military advisers had all but abandoned the land invasion by early July 1945, irrespective of whether Trinity bathed Alamogordo in neutrons.

Basic errors of fact compounded these sins of omission. The article was plain wrong, for example, to claim that the 'direct military use' of the bomb had destroyed 'active working parts of the Japanese war effort'. Nobody on the Target Committee pretended that 'working men's homes' were military targets whose destruction would seriously hamper the enemy's fighting ability. In any case, more than 90 per cent of Hiroshima's war-related factories were on the city's periphery. On Conant's nod, the committee had clearly recommended that the bomb be dropped on the heart of a city – that is, on noncombatants. The priority was not the destruction of 'workers' homes' (though their presence served a useful public relations role); it was to shock Japan into submission by annihilating a city.

As to Stimson's claim that America used the bomb reluctantly – 'our least abhorrent choice' – suggesting that Washington and the Pentagon had wrestled painfully with alternatives, the facts demonstrate precisely the opposite. Everyone involved expected, indeed hoped, to use the bomb as soon as possible, and gave no serious consideration to any other course of action. The Target and Interim Committees swiftly dispensed with alternatives – for example, a warning, a demonstration, or attacking a genuine military target. Indeed, Byrnes rejected these over lunch in the Pentagon – arguing that a warning imperilled the lives of Allied POWs whom the Japanese would move to the target area (the US Air Force had shown no such restraint in the conventional air war, which daily endangered POWs). As well as this, he argued, a demonstration might be a dud (unlikely, given Trinity's success, and the fact that Manhattan scientists saw no need even to test the gun-type uranium bomb used on Hiroshima); that they had only two bombs (untrue – at least three were prepared for August, and several in line for September through to November); and that there were no military targets big enough to contain the bomb. In fact, Truk Naval Base was considered and rejected; no other military target was seriously examined; only Kokura, a city containing a large arsenal, came close to that description, and the attempt to bomb it was abandoned due to the weather.

The nuclear attacks were an active choice, a desirable outcome, not a regrettable or painful last resort, as Stimson insisted. The administration never seriously considered any alternative; its members focused on how, not whether, to use atomic weapons. Every high office-holder believed the bomb would be dropped if Trinity proved successful. 'I never had any doubt it should be used,' Truman said on many occasions. 'The decision,' wrote Churchill later, 'whether or not to use the atomic bomb to compel the surrender of Japan was never an issue.' Groves dismissed Truman's role as inconsequential. 'Truman's decision,' the general wrote, 'was one of non-interference – basically a decision not to upset the existing plans.'

In this frame, a complete Japanese surrender at an awkward time – that is, after Trinity's success and before the bombs arrived on Tinian – would have frustrated any hope of using the weapons. This is not to impute sinister motives to any man, whose heart and mind we may never truly know; simply to assert that Washington and the Pentagon were absolutely determined to use the two atomic bombs. 'American leaders did not cast policy in order to avoid using the atomic weapons,' in the historian Barton Bernstein's view. The phrase 'our least abhorrent choice' grossly misrepresents a gung-ho, indeed diabolically zealous, enterprise.

Stimson's least persuasive claims were that the atomic bombs ended the war and prevented up to a million American casualties. While the bombs obviously contributed to Japan's general sense of defeat, not a shred of evidence supports the contention that the Japanese leadership surrendered *in direct response to* the atomic bombs. On the contrary, Tokyo's hardline militarists shrugged as the two irradiated cities were added to the tally of 66 already destroyed, and overrode the protests of the moderates. They barely acknowledged the news of Nagasaki's destruction. Nor would a nuclear-battered Japan consider modifying its terms of 'conditional surrender': the leaders clung stubbornly to that central condition – the retention of the Emperor – to the bitter end. In fact, state propaganda immediately after Hiroshima and Nagasaki girded the nation for a continuing war – *against a nuclear-armed America*.

A regime that cared so little for its people except insofar as they served as cannon fodder in a last miserable act of national *seppuku*; a nation so fearful of the Soviet Union that it sent message after message imploring their intervention in the dying months of war; a people so steadfast in their refusal to yield that they actually prepared to defend their cities against further atomic bombs – this was not a country easily shocked into submission by the sight of a mushroom cloud in the sky (and it is worth remembering that, the day after, Tokyo had no film or photographs of the bomb; only US pamphlets and military reports claiming it had been used).

A greater threat than nuclear weapons – in Tokyo's eyes – drove Japan finally to accept the surrender: the regime's suffocating fear of Russia. The Soviet invasion on 8 August crushed the Kwantung Army's frontline units within days, and sent a crippling loss of confidence across Tokyo. The Japanese warlords despaired. Their erstwhile 'neutral' partner had turned into their worst nightmare. The invasion invoked the spectre of a communist Japan, no less.

Russia matched iron with iron, battalion with battalion. This was a war that Tokyo's samurai leaders understood, a clash they respected – in stark contrast to America's incendiary and atomic raids, which they saw as cowardly attacks on defenceless civilians.

Americans are not alone in seeing the past through their national prism; their pervasive power, however, enables them to project a decidedly American impression of what happened – or should have happened – onto the rest of the world. Photos of the mushroom cloud over Hiroshima impressed most Americans, who could readily imagine the pulverisation of St Louis or Dallas or Chicago. How on earth would the little yellow people endure such a fate? Other realities, however, prevailed in Tokyo and Moscow – a ground war of immense military dimensions and far-reaching political implications, which had a more exacting influence on Tokyo's decision to surrender than the death of two more cities.

•

What of Truman? The President knew the script well. On 6 August he told Eben Ayers, his impressionable White House press official, that military advisers had warned a million men might be needed to invade Japan, with casualties of 25 per cent. Truman estimated that if Hiroshima's population were 60,000 then 'it was far better to kill 60,000 Japanese than to have 250,000 Americans killed'. He added: 'I therefore ordered the dropping of the bomb on Hiroshima and Nagasaki.'

In fact, Truman had approved a decision that had been made for him in the extraordinary confluence of American industrial wealth, European scientific brilliance, Japanese war crimes, Russian imperialism, the incendiary campaigns over Germany and Japan, and the acute geo-political pressures of the last year of the war, which met in the minds of Churchill and Roosevelt at Hyde Park. Only a character of unearthly will, vast authority and transcendent moral vision could have resisted the fatal momentum of the atomic project. However great a president, Truman was not that character.

The President never lost any sleep over 'his' decision, he often claimed; he would have done it again, he said on several occasions. He told a Mrs Klein, years after the war, that the bomb saved a quarter of a million American and Japanese boys. As a result, 'I never worried about the dropping of the bomb. It was just a means to end the war.' He told a 1958 CBS *See It Now* TV report that he had 'no qualms' about ordering the use of the atomic weapon on Japan – prompting a letter of protest from Hiroshima City Council, in reply to which Truman attempted to justify the use of the bomb as revenge for Pearl Harbor.

The council responded: 'Do you consider it a humane act to try to justify the outrageous murder of two hundred thousand civilians of Hiroshima, men and women, young and old, as a countermeasure for the surprise attack [on Pearl Harbor]?' Nagasaki Municipal Assembly weighed in: 'We deeply regret the war crimes committed by our nation during the last World war ... Nevertheless, we cannot remain silent in the face of your persistent attempt to justify the atomic raids.'

Again on 5 August 1963, in a letter to Irv Kupcinet of the *Chicago-Sun Times*, Truman claimed that the bomb saved 125,000 American and 125,000 Japanese 'youngsters', and avenged Pearl Harbor. Oddly, lower down in the same letter, he doubled the number of lives he reckoned he saved: 'I knew what I was doing when I stopped the war that would have killed a half a million youngsters on both sides if those bombs had not been dropped. I have no regrets and, under the same circumstances, I would do it again.' He protested too much, it seems. He later told Paul Tibbets not to lose any sleep over the mission. 'It was my decision. You had no choice.' Selflessly determined to shoulder the burden alone, Truman moved to silence the worm of conscience in others.

The simple fact is, Truman never presented the bomb as an alternative to invasion until after the war. He had always resisted the invasion of Japan regardless of whether the bomb worked. The prospect of several hundred Okinawas on the shores of Kyushu horrified him. He expected the naval blockade, the air war and – at least until mid-July – the Russians would together finish the job. Marshall, Stimson, Leahy, Eisenhower and Halsey all came to believe this, to a greater or lesser extent.

The President was too smart a politician – with a genuine desire to protect American lives – to risk political suicide through the loss of so many young men against a regime that everyone in power in Washington knew was, for all practical purposes, defeated. In this context, the bomb was not a substitute for an invasion for the simple reason that Truman had no intention of approving one. He could not say this after the war, because that would have emasculated his claim that the bomb saved up to 'a million' lives.

•

The magic number is hard to source. At various times, from 1947 on, Truman, Stimson and Byrnes argued that the bomb saved a quarter of a million, a million and millions of lives. How they arrived at these spectacular figures is a question of deep and continuing controversy. On one occasion, Byrnes contended that the weapon rescued 'some millions [of Japanese people] who would have perished [under] incendiary bombs [used] against a people whose air force had been destroyed'. By ending the war, Byrnes implied, the bomb also ended the firebombing missions that were steadily exterminating the Japanese people. Only a peculiarly gymnastic political mind could have managed this somersault of present expediency over past reality: the bomb was not built to deliver the Japanese people from the wrath of Curtis LeMay.

Years after the war the President sourced his casualty estimates to General Marshall who, Truman said, had privately warned him that invading Japan would cost 'as much as a million' dead and wounded 'on the American side alone' (despite the fact that Marshall had predicted 31,000 casualties at the meeting on 18 June 1945). Truman appears to have drawn on several other sources – not least former President Herbert Hoover's warning in a memo on 15 May 1945 that an invasion would cost between '500,000 and 1,000,000' American *fatalities* – a death rate of such astonishing scale as to discredit this analysis, regardless of Hoover's access to 'intelligence briefings'. US military experts – notably General George Lincoln – derided Hoover's calculation as deserving 'little consideration'; Hoover's upper limit in fact implied *total casualties* (that is, including wounded and missing) of between three and five million. That the comparatively weak and dispirited Japanese might kill or wound virtually every American soldier several times over was ludicrous: the US commanded the air over Japan and ringed the islands in steel; their armies were well trained and equipped and the enemy in an abject state.

Nonetheless, the American press and public fixed on the magic number – 1,000,000 American dead – as a monument to the justification of the atomic destruction of Hiroshima and Nagasaki. It is worth restating that Truman and the Joint Chiefs approved the *planning* of an invasion of Kyushu – *not* its execution – at the 18 June meeting on the modest assumption of 31,000 battle casualties (dead, wounded and missing). These numbers, no less dreadful to the family of one of those killed, hardly made invasion desirable or advisable, of course – especially if a credible alternative (blockade, bombardment, Russian intervention) would force the surrender. The point is that Americans – from ordinary citizens to serious thinkers – continue to justify the bomb with a blizzard of unreliable casualty estimates for an invasion Truman had not approved and was desperate to avoid. America's collective conscience – unable, perhaps, to bear any other interpretation – seems to have fixed on this version of the past.

There were other ways of saving American lives – if we accept this as Washington's most urgent imperative – each admittedly fraught with risks, but surely worth consideration. These were by clarifying the terms of 'unconditional surrender' and the status of Hirohito, whose continuation in a symbolic role every sentient Washington official knew would be vital in managing post-war Japan; by offering to test the bomb before the eyes of the United Nations – with Japanese observers in attendance – as the Franck Committee proposed, which, although fraught with practical problems, had the advantage of demonstrating America's civilised restraint; or seriously encouraging Russia to enter the war and retaining Moscow's signature on the

Potsdam Declaration (a course which, after Trinity, was the least appealing option for Truman and Byrnes). Truman chose none of these alternatives, or variations of them. With Byrnes at his ear, he rejected the counsel of Grew and Stimson, who recommended the gift of the Emperor as the quickest way to end the war. He did not actively solicit Russian help. If he went to Potsdam, as he said, to get the Russians into the Pacific, he left Potsdam determined to keep them out. 'Truman,' observed historian Barton Bernstein, 'did nothing substantive at Potsdam to encourage Soviet intervention and much to delay or prevent it.' Everything changed with Trinity, of course; the first nuclear-armed President believed he had no further need of his erstwhile ally.

These actions may not have resulted in Japan's surrender, of course. The point is, if ending the war and saving lives were Truman's chief aim, why did he not at the very least explore these alternatives? The historian Gar Alperovitz answered that 'saving lives was not the very highest priority' – ascribing an unmerited callousness to the American leadership. More accurately, the United States sought to end the war *on its own terms*, and in accordance with the wishes of the people, who would never forgive Pearl Harbor. The bomb, Truman believed, handed him that opportunity. He felt that it removed the need to engage the Russians or appease the Japanese. America would win the war on her terms – without Russian help – and *proudly*.

After the war Truman tried to amend the record to project this noble sentiment. It is a fact, for example, that the Japanese were given no warning of the first bomb. In 1959 Truman emphatically denied this to an astonished Richard Hewlett, author of the official history of the US Atomic Energy Commission. 'Certainly,' Truman said, 'the Potsdam Declaration did not contain such a warning but the Japanese had been warned through secret diplomatic channels by way of both Switzerland and Sweden ... this warning told the Japanese that they would be attacked by a new and terrible weapon unless they would surrender.' Asked whether he had copies of the cables, Truman replied that they could be found in the files of the State Department or CIA. No record of any such warning has been found.

A less clear-cut issue divided Truman and Byrnes after the war – and has vexed historians ever since: whether the bomb was used as an emphatic expression of American power to deter Russian aggression in Europe and Asia. Absolutely not, Truman told Richard Hewlett; the President had always claimed that his primary motive in Potsdam was to get the Russians into the war. Byrnes, on the other hand, believed the bombs had a definite role in 'managing' the Soviet Union, and told US News in the 1960s that they were dropped to end the war 'before Russia got in'. In a later interview with Fred Freed of NBC, Byrnes undermined Truman's public position: 'Neither the President nor I were anxious to have [the Russians] enter the war after we had learned of the successful test.'

No documentary evidence exists to support the blackest reading of Byrnes' role – that he engineered Japan's refusal to surrender until the bombs were ready. It is intriguing to speculate, nonetheless, whether the Japanese would have surrendered had they received the Byrnes Note *before* the atomic bombs fell. Remember that Tokyo's 'peace faction' had repeatedly requested the Emperor's continuation as the sole precondition for surrender. After the bombs, they doggedly stuck to this condition; Hiroshima and Nagasaki made little dent in their resolve on this issue. They surrendered only *after* the Russians invaded and *after* the Byrnes Note effectively met Tokyo's condition. In this light, it seems likely that Japan, in receipt of the Byrnes Note, and in the grip of the Russian onslaught, would have surrendered without the use of atomic bombs.*

---

* Byrnes' true role in the delivery of the bomb will never be known – thanks, in part, to his obsession with secrecy which extended well after the war. Nine years later, in June 1954, the State Department's Historical Division wrote to Byrnes in pursuit of crucial information about his role at the Potsdam Conference; the Secretary's colleagues had insisted that he 'kept some kind of personal record of the meeting'. Alas, Byrnes replied that he had kept no personal record at Potsdam, which is odd given his copious note-taking at Yalta (although his assistant, Walter Brown, kept a diary). Undaunted, the State Department a month later sent Byrnes a long list of questions about Potsdam: would he divulge the discussions of 'atomic matters' with Stalin and the British? Did he have a copy of Grew's draft declaration calling for Japan's surrender (as State could not properly identify its copy of the draft)? He could offer neither.

In a similar spirit of hypothetical inquiry, what might have happened had America not used the atomic bombs? One probable scenario is this: within weeks Russia would have crushed Manchuria, and Japan – crippled by the naval blockade, subjected to constant conventional aerial attack, and fearful of the communist advance – would have surrendered to the more acceptable enemy, America. No US invasion would have been necessary. The total casualties, Allied and Japanese, of this scenario are unknowable.

•

The chiefs of the armed forces took a different line on the bomb. The bomb dismayed many of America's most senior soldiers. There was something deeply abhorrent to the traditional commander's mind about the air war on Japanese civilians. Deliberate, indiscriminate slaughter of noncombatants – unless a woman or child with a bamboo spear may be classed as a combatant – did not figure in their conception of warfare. The new world order reflected this view: of what use were the Hague and Geneva Conventions if not to protect the innocent, preserve the lives of prisoners and limit unnecessary cruelty? In defying them, Japan and Germany had waged a bestial campaign against the very essence of what distinguishes the species as human. Surely America, the self-appointed standard-bearer of God-fearing morality, would not follow them?

The use of nuclear weapons was unconscionable and/or militarily unnecessary – so argued Generals Eisenhower and MacArthur, and Admirals Halsey, King and Leahy. They each condemned and opposed, with varying degrees of emphasis, the nuclear attacks – on strategic and/or moral grounds. But they only did so publicly *after* the bombs fell, which diminished their arguments – a case of wanting to look virtuous after the event. Pride also played a part in their objections to the bomb: their own forces – sea, conventional air , and land troops – would have defeated Japan without atomic weapons, they contended, to a greater or lesser degree. They believed the war would have ended anyway, by blockade, bombardment and/or Russian help (only MacArthur, in the last weeks, held a candle for the invasion).*

Leahy was the most emphatic opponent of the bomb: for him, dropping it on Japan was an act of inhumanity unworthy of a Christian country. 'It is my opinion,' he wrote in his memoirs, 'that the use of this barbarous weapon at Hiroshima and Nagasaki was of no material assistance in our war against Japan. The Japanese were already defeated and ready to surrender [on their terms, Leahy omits to say] … in being the first to use it we had adopted an ethical standard common to the barbarians of the Dark Ages … wars cannot be won by destroying women and children.' Halsey, the feared Commander of the 3rd Fleet, publicly declared on 9 September that the 'first atomic bomb was an unnecessary experiment … It was a mistake ever to drop it.' He dismissed the weapon as a 'toy' the scientists 'wanted to try out'. MacArthur, perhaps the last of the gentlemen soldiers, was implacably opposed; his aide, Brigadier General Bonner Fellers, described America's strategic air offensive – in its incendiary and atomic forms – as 'one of the most ruthless and barbaric killings of noncombatants in all history'. Generals Arnold and LeMay were arrogantly dismissive of the new weapon, which merely continued, they believed, what they had started. For them, incendiaries and conventional explosives would have eventually forced Japan to surrender, a campaign LeMay proudly portrayed in terms of the body-count scorecard that he would later apply in Vietnam. 'We scorched and boiled and baked to death more people

---

* The historian Robert Maddox's spirited attempt to show that these commanders did not really mean what they said is unconvincing. If their words have, at times, been taken out of context or misapplied, they certainly did not believe the bomb, by itself, ended the war or obviated an invasion. However, it appears true, as he states, that none took his argument to Truman before the bombs were dropped. Nobody, it seems, had the courage of their convictions openly to oppose the use of the weapons until after the war; Maddox is also right to attack the least credible of the 'revisionist' views that Truman and Byrnes deliberately prolonged the war in order to use the atomic bombs.

in Tokyo on that night of 9–10 March than went up in vapor at Hiroshima and Nagasaki.'*

Eisenhower was adamantly opposed, after the event. He famously stated in his post-war memoirs that 'in [July] 1945' – he was then Supreme Allied Commander in Europe – '

> Secretary of War Stimson, visiting my headquarters in Germany, informed me that our government was preparing to drop an atomic bomb on Japan ... During his recitation of the relevant facts, I had been conscious of a feeling of depression and so I voiced to him my grave misgivings, first on the basis of my belief that Japan was already defeated and that dropping the bomb was completely unnecessary, and secondly because I thought that our country should avoid shocking world opinion by the use of a weapon whose employment was, I thought, no longer mandatory as a measure to save American lives. It was my belief that Japan was, at that very moment, seeking some way to surrender with a minimum loss of 'face'.

In a *Newsweek* interview in 1963, Eisenhower again alluded to what he told Stimson, that 'the Japanese were ready to surrender and it wasn't necessary to hit them with that awful thing'.

Japan's smarter military commanders conceded this. They fought on in the hope of extracting conditions at the surrender. Japan's most

---

* Curtis LeMay displayed a capacity for self-deception that consumed his post-war years: 'We were going after military targets,' LeMay later claimed to universal disbelief. 'No point in slaughtering civilians for the mere sake of slaughter.' A drill press discovered in the roasted wreckage of 'tiny houses' informed LeMay that the 'entire population ... worked to make those airplanes or munitions of war ... men, women and children. We knew we were going to kill a lot of women and kids when we burned that town. Had to be done.' If the urban areas played 'at least some role in Japanese war production', according to Gordon Daniels, a scholar of strategic bombing, no clear demarcation line divided Japanese industrial and residential areas. True enough as a statement of the obvious: Japanese cottage industries nestled among the populace. Every urban area contributed to the war effort in that Japanese civilians helped to feed, equip and care for their servicemen – as did civilians in all countries locked in mortal combat. But LeMay and Daniels (and their apologists) entirely miss the point: a lathe in a Japanese backyard was a US public relations convenience after the event; it was never the target. The ordinary people ('civilian morale'), the workers' homes, were always the targets – of both incendiary and nuclear bombs – as the Target Committee had made abundantly clear.

famous soldier, General Tomoyuki Yamashita, the 'Tiger of Malaya' – a thickset, bullet-headed giant of a man – summed up this attitude at the highest level during an interview in his prison cell in Manila in March 1946: 'Our cause was lost even before you had recourse to atomic bombs and long-range bombers. We were fatally handicapped by lack of … material resources.'

•

The American press helped to cement the myth that Japan surrendered *in direct response to* the nuclear attacks. The media reached this consensus *after* the war. In the months before the surrender reporters sang a very different tune: then, newspapers and radio consistently portrayed Japan as starving, vanquished, reduced to sending out desperate peace feelers, and so on. The contrast between the two Japans is startling, and suggests an indoctrinated or confused press (as historians Uday Mohan and Sanho Tree show in their essay 'The Construction of Conventional Wisdom'). Which was the true Japan – the pathetic, hungry, defeated adversary the press portrayed during the last months of the war, or the resilient, still threatening nation whose surrender would require a land invasion or nuclear holocaust (or both)? Neither portrayal was entirely accurate, but the former closer to the truth. That's not to say the US media all bought the government's post-war line. The *Washington Post* was a constant thorn in Truman's side. And intriguingly, the most strident critique of the newly minted 'orthodox' view – that the atomic bomb ended the war and saved hundreds of thousands of lives (itself a revision of the facts) – came from the right-wing press, in magazines such as *The Freeman*, *National Review* and *Human Events*. Truman's refusal to modify the terms of 'unconditional surrender', observed Forrest Davis in *The Freeman*, came 'little short of being a high crime and one that may return unmercifully to plague us'. These days, few agree with *The Freeman*: from US tabloids to Pulitzer-prizewinning historians, the consensus is that the atomic bombs alone won the war and were our least abhorrent choice.

The stifling debate between those who claim the bombs were dropped as a warning to Russia, the first blow of the Cold War ('revisionist'), or used to avoid a land invasion and save a million US lives, as Truman argued after the war ('orthodox'), little helps our understanding of the complex reality of what actually happened in 1945. In its extreme form each side presents a simplistic, partisan reading of the past, which superimposes a story that best reflects the beliefs of the authors but little serves the pursuit of truth. Indeed, the very terms 'revisionism' and 'orthodoxy' seem interchangeable, depending on the context, timing and characters involved. Truman, for example, expressed what has become the 'orthodox' view of the bomb after the war; his feelings were more complex before the armistice – in flux, amenable and, under Byrnes' influence, shaded with 'revisionist' tendencies. For his part, Byrnes sounded robustly 'revisionist' in his view of the bomb's role in managing or subduing the Soviets. And Stimson's 'orthodoxy' was not fully formed until he put his name to an article in 1947. Minds, moods, change – for all kinds of reasons. To ascribe a single, static motive to the behaviour of an individual or government little helps us to understand the complex interplay of relationships, thoughts and feelings that drive human history.

•

The most that may be said in defence of the bombing of Hiroshima – in strictly military and political terms – is that it bounced the Russians into the war a week or so earlier than Moscow planned. Whether that justified the destruction of a city is a different question. There appears to be no justification – in military or political terms – for Nagasaki. The two atomic strikes did, however, furnish Tokyo's leaders with a face-saving expedient, to surrender to the more acceptable enemy. Far better to capitulate to a nuclear-armed, democratic America than to a vengeful, communist Russia.

Perversely the bombs offered the Japanese leadership an unintended propaganda tool: it let them present the surrender as the act of a martyred nation, forced to yield, as Hirohito said, to a 'new and most cruel bomb … taking the toll of many innocent lives'. This characterisation cruelly overshadowed the genuine claims of the victims of Japanese war crimes, and was utterly false. Tokyo did not surrender to protect the Japanese people from the weapon; the leadership had shown not the slightest duty of care towards these 'innocent lives'. For the Japanese leaders, the bombs served another purpose, unwelcome in Washington: Tokyo was able to surrender without conceding defeat on the battlefield, where it mattered most to the samurai mind. In this sense, the Imperial forces were able to capitulate with military honour intact. Little Boy and Fat Man saved the faces of a people for whom 'saving face' meant more than saving their lives.

In 1945 America occupied and started to rebuild the smoking disgrace of Japan – and turned its shoulder to the nuclear winter. The splitting of the atom, to paraphrase Einstein, changed everything except 'the way men think'. Truman later expressed a similar spirit: 'The human animal and his emotions change not much from age to age. He must change now or he faces absolute and complete destruction and maybe the insect age or an atmosphere-less planet will succeed him.' It reflected his new understanding of the bomb, not as a weapon of war but as a clear and present danger to humankind.

•

This book is less interested in finding villains than understanding the bloody acts to which German and Japanese aggression had impelled the Allies to resort. Those acts were unconscionable, and unworthy of the civilised world. But even if we accept Truman's later defence of his decision – and no doubt he sincerely wished to end the war as soon as possible and save lives – the question is whether good intentions alone justify the flouting of war conventions and the massacre of ordinary people.

The bomb gave America's tooth-for-a-tooth sensibility the power to scatter a billion molars. They used it, without warning, in an attempt to extract 'unconditional surrender' from a defeated foe, 'manage' Russian aggression in Europe and Asia, and avenge Pearl Harbor, as Truman and Byrnes said. The bomb achieved none of those goals (unless two destroyed cities is accepted as proportionate punishment for Pearl Harbor): Tokyo surrendered with its sole condition intact, and Russia continued to stamp and snort and foment communist revolution around the world – and would soon rush to join the nuclear arms race.

Taken together, or alone, the reasons offered in defence of the bomb do not justify the massacre of innocent civilians. We debase ourselves, and the history of civilisation, if we accept that Japanese atrocities warranted an American atrocity in reply.

DEAD HEAT

... dehumanise a belt across the Korean Peninsula by surface radiological contamination ... broadcast the fact to the enemy ... that entrance into the belt would mean certain death or slow deformity to all foot soldiers ... and, further, that the belt would be regularly recontaminated until such time as a satisfactory solution to the whole Korean problem shall have been reached ...

Al Gore, Tennessee senator, father of Al Gore jnr, Vice President 1993–2001

It is generally recognised throughout the world that the United States has no aggressive designs on any other nation and is, therefore, the safest possessor of the [nuclear] secret.

US Joint Chiefs of Staff

WITHIN A MONTH OF JAPAN'S surrender the Pentagon had written the death sentence of America's next enemy: the destruction of 66 Soviet cities of 'strategic importance' would require an 'optimum' 204 atomic bombs, advised a US Army Air Force study in September 1945. Of unspecified kilotonnage, the new bombs would far exceed the power of the one that destroyed Hiroshima (equal to about 16 kilotonnes of TNT), they noted, and obliterate most of Russia's population and industry – chiefly its capacity to refine oil and produce aircraft and tanks.

The Army Air Forces thus commended to Washington the first official calculation of the atomic bombs needed 'to insure our national security'. 'It is obvious,' Major General Lauris Norstad, Deputy Chief of Air Staff at Army Air Forces HQ, told Groves on 15 September 1945, that America and Russia 'will be the outstanding military powers' for the next 10 years. The destruction of Russia's capability to wage war must serve as the 'basis upon which to predicate the US atomic bomb requirements'.

It may not be necessary to remove 66 Soviet population centres, conceded the Army Air Forces Study; wiping out 15 'first priority' cities may have the same impact, it calculated. 'The primary objective for the application of the atomic bomb is ... the simultaneous destruction of these 15 first priority targets,' Norstad advised Groves. The prime targets and the estimated number of atomic bombs required to destroy each of them were: Moscow – six, Leningrad – six, Tashkent – six, Novosibirsk – six, Nizhni Tagil – five, Sverdlovsk – five, Kazan – five, Gorki – four, Chelyabinsk – three, Tbilisi – three, Kuibyshev – three, Magnitogorsk – three, Stalinsk – three, Saratov – two, Molotov – two, Baku – two, Omsk – two, Grozny – one (see full list, Appendix 9).

In this scenario, a massed nuclear air raid would kill, wound or displace a population of 10,151,000, destroy the cities' industrial and military facilities, and devastate about 600 square kilometres of urban area. The question that troubled Groves was this: were so many atomic bombs required? Recognising that the weapon could not be regarded as 'just another bomb' – 'these bombs are very expensive, cannot be produced in mass, require special storage conditions' – Norstad replied that three 'well-placed' nuclear strikes on a single target 'would throw a modern city of any size into chaos and definitely incapacitate it for an appreciable period of time'. To inflict this on 15 Soviet cities, the general recommended the production of '39 bombs as a minimum'.

'It is not essential,' Groves concurred, 'to get total destruction of a city in order to destroy its effectiveness. Hiroshima no longer exists as a city even though the area of total destruction is considerably less than total.' Fewer bombs were therefore needed, he advised – and approved the bomb production plan.

In Moscow meanwhile, Soviet military planners dealt in theory, not practice. Russian scientists were scrambling to meet Stalin's highest military priority: the construction of a Soviet nuclear weapon. Only then would the Russians join the new arms race and place New York and Washington at the top of their list of priority enemy targets.

•

Relief at peace in Europe soon yielded to the long, painful process of mopping up and the fraught realisation of what had been done. The continent smouldered and starved and struggled to assimilate a dark new reality: tens of millions dead and displaced; the revelation of the Nazi death camps; and the shadow of the Soviet Union over Western Europe. In Asia the Japanese, their cities destroyed and the *Kokutai* humiliated, sought to emulate the strange new system – democracy – espoused by the leader of the occupying forces, General MacArthur. In China, Mao Tse-tung's tribes of peasant guerrillas were on the brink of victory in a civil war with Chiang Kai-shek's Nationalists. In Moscow, the Russians looked hungrily out on a planet that, after their successful land grab at Yalta and Potsdam, seemed theirs for the taking. The only mote in the eye of the Kremlin's vision was the prospect of a nuclear-armed America.

The lines of this binary world cracked, heaved and slowly cooled. Well before Churchill descried an 'iron curtain' descending on Eastern Europe, the President, with James Byrnes at his ear, contemplated a planet split along political and economic lines. In such a world the bomb would serve, they hoped, as a lever to prise concessions from their expansionist former ally.

Nobody declared the Cold War; the confrontation did not 'begin' at a fixed point. Distrust festered. Cumulative acts of deceit drained the political will required to arrest the downward trajectory of Soviet–American relations. A series of hard-edged incidents – Soviet perfidy after Yalta, Russia's sequestration of Eastern Europe, and mutual distrust at Potsdam – prodded the belligerents into their corners and cumulatively led to the Soviet–American stand-off at the very birth of the nuclear age. In this atmosphere of spiralling mutual ill-will, both sides rushed to build and were prepared to use nuclear arsenals in anticipation of a future atomic confrontation. Fear of the weapon fixated Moscow; the possession of it empowered America. The former hungered to acquire nuclear power; the latter hastened to develop, hide and protect it.

•

The vigilant, if powerless, Stimson had warned of the breakdown in US–Soviet relations and the risk of a nuclear arms race. He now grimly watched these events unfold. And while the gentleman within him abstained from *Schadenfreude*, the knight errant continued to tilt at hopes of a diplomatic breakthrough. Nonetheless, his warnings to the White House precisely anticipated the state of the planet 10 years hence. In a long letter to the President on 11 September 1945, Stimson warned that any attempt to freeze Russia out of the atomic secret and use nuclear power as a 'direct lever' – a diplomatic weapon – against Moscow would provoke extreme hostility. Unless the Soviets were invited to join the Anglo-American nuclear partnership, Stimson wrote, they would resort to a 'feverish' attempt to build a bomb 'in what will in effect be a secret armament race of a rather desperate character'. The Russians, he added, may already have commenced. In fact, they were well advanced, thanks largely to the classified information gleaned from Los Alamos by Fuchs and other lesser spies.

The horror of nuclear warfare imposed trust on an otherwise distrustful relationship, reasoned Stimson with the wistful idealism of old age. The world, he believed, had no choice: 'The chief lesson I have learned in a long life,' he wrote to Truman, 'is that the only way you can make a man trustworthy is to trust him; and the surest way to make him untrustworthy is to distrust him and show your distrust. If the atomic bomb were merely another though more devastating military weapon to be assimilated into our pattern of international relations ... we could then follow the old custom of secrecy and nationalistic military superiority relying on international caution to prescribe the future use of the weapon as we did with gas. But I think the bomb instead constitutes merely a first step in a new control by man over the forces of nature too revolutionary and dangerous to fit into the old concepts ... Since the crux of the problem is Russia, any contemplated action leading to the control of this weapon should be primarily directed to Russia.'

Stimson proposed that Washington approach Moscow with a bipartisan plan for control of the bomb. The two powers would jointly limit the production of nuclear weapons and guide atomic energy to humanitarian and peaceful uses. Perhaps America might even 'impound what bombs we now have', on the condition that the Russians and British agreed – and no government used the weapon without the approval of all three. Above all, the pact should be American-inspired and delivered, with British backing.

Stimson's proposal struck a decent note for posterity when, in the twilight of his career, it was safe to do so. He had nothing to lose; his better angel had simply written a time capsule. But a year of Soviet perfidy had weaned Truman's mind off any illusion that Moscow could be trusted. The barriers were being thrown up even as Stimson dreamed of breaking them down. In early September the White House asked editors and broadcasters to self-censor in the interests of national security: to withhold all information or editorial comment on the scientific processes, formulas and mechanics employed in the operation of the atomic bomb; and the location, procurement and consumption of uranium stocks.

Later that month, the President canvassed his senior cabinet members on Stimson's ideas: should America share the science of the atomic bomb with Russia? Most agreed with the War Secretary. Extraordinary as it may seem, in 1945–46 the hope of avoiding a nuclear arms race eclipsed the fear of the Soviet Union. Somehow the two powers had to get along, should be compelled to co-operate; the stakes of failing to do so were too high. Vannevar Bush, director of the Office of Scientific Research and Development, crisply defined the issues in a memo to the President: 'Down one path lies a secret arms race on atomic energy; down the other international collaboration and possibly ultimate control. Both paths are thorny but we live in a new world and have to choose.' A secret race to build atomic bombs would lead to 'a very unhappy world'. Rogue regimes would develop nuclear weapons 'underground'. Bush urged Truman to act swiftly to heal the divisions at Potsdam: 'We lost something when we could not make the move [to share the secret with Russia] until the bomb actually exploded. Now no unnecessary time should be lost.'

Truman's friend Dean Acheson, Assistant Secretary of State, broadly agreed. Acheson urged the abandonment of 'the policy of secrecy', which was 'both futile and dangerous'. In its place America should establish 'conditions' that would 'govern' the exchange of atomic secrets, and work with Russia and other nations to build a system of international control of nuclear weapons that would 'prevent a race toward mutual destruction'.

Most nuclear scientists agreed. Einstein, Fermi, Oppenheimer, Szilard and Bethe were vocal supporters of shared scientific exchange. The scientists, however, grossly underestimated the complexity of managing a politically divided world on the threshold of a nuclear arms race. Einstein's reply to Truman naively presumed that the bomb might end wars of 'nationalism' and clearly misjudged the enduring power of the idea of the sovereign state. 'The most important thing we intellectuals can do,' Einstein wrote, 'is to emphasize over and over again the establishment of a solidly built world-government and the abolition of war preparations (including all kind of military secrecy) by the single states.'

Byrnes and the military establishment were having none of this. The Joint Chiefs of Staff were adamant that America's nuclear secrets be cloaked in the strictest secrecy for several reasons, some of them persuasive: sharing the bomb would accelerate, not curb, an arms race at a time of international political division; American cities were believed to be particularly vulnerable to atomic attack; international controls on atomic power were not in place; and the world would interpret any sharing of atomic secrets as weakness. 'It is generally recognised throughout the world,' the Joint Chiefs less convincingly concluded, 'that the United States has no aggressive designs on any other nation and is, therefore, the safest possessor of the secret.'

Self-interest played an obvious part: the leaders of the US military-industrial complex saw the Soviet lag as a golden opportunity for unilateral US nuclear development, and pointedly *not* the moment to bring Russia into the atomic tent. Transparency was not only naive and irresponsible, they argued; it would deny American arms makers the chance to exploit their head start.

Truman went through the arguments and made up his mind. He sided with Byrnes and the military establishment. Moscow could not be trusted to abide by Stimsonian notions of mutual co-operation. America would keep its atomic secrets for another reason high in Byrnes' mind: the presence of a US nuclear threat would place Washington in a commanding position in any future negotiations with the Russians.

•

Byrnes soon had a chance to put this theory to the test, during the meeting of the Council of Foreign Ministers in London in September 1945, convened to negotiate the details of Potsdam. This first act of nuclear diplomacy got off to a disarming start when Molotov joked about whether Byrnes had 'an atomic bomb in his side pocket'. Byrnes replied, 'You don't know southerners. We carry our artillery in our pocket. If you don't cut out all this stalling [in the talks] ... I am going to pull an atomic bomb out of my hip pocket and let you have it.' As the nervous laughter subsided, Molotov simply ignored Byrnes' implied threat, and carried on with Moscow's territorial demands in Europe and Asia. Then at a cocktail party that evening the Russian let slip, 'You know we have the atomic bomb.' A colleague gruffly interrupted – the commissar had overstepped the mark – and escorted Molotov from the room.

Byrnes dismissed the comment as light provocation. Privately, however, he was deeply concerned. News of this exchange went straight to Groves via the American embassy in London. To the Secretary of State's chagrin, the tacit threat of the bomb had demonstrated not the slightest leverage over Soviet action. Manifestly, it would not make Russia 'more manageable' in Europe, as Byrnes had hoped. It had the opposite effect; the Soviet Union turned its back on America's atomic revolver and sauntered off into the sunset. The Russians, Byrnes quipped a few weeks later, were 'stubborn, obstinate and they don't scare'.

Ironically, it was Byrnes, and not his Russian counterpart, who weakened in coming talks. During his mission to Moscow that December, the Secretary adopted – without Truman's approval – a compromising line with Stalin. This act of insubordination marked the beginning of his fall from grace; Truman felt that Byrnes was deciding foreign policy without informing the White House.

'I had been left in the dark about the Moscow conference,' Truman told Byrnes in a letter that year. He urged Byrnes in subsequent correspondence to take a much harder line with the Soviet Union, chiefly over Moscow's burgeoning troop presence in Iran. 'Unless Russia is faced with an iron fist and strong language another war is in the making,' Truman wrote. 'Only one language do they understand – "how many divisions do you have?" I do not think we should play compromise any longer ... I am tired of babying the Soviets.' Byrnes reverted to a tougher approach, but Truman had lost confidence in his Secretary and Byrnes felt compelled to resign, with bitterness, in 1947.

•

In early October 1945 the President announced publicly what had been agreed in private, that America would not share its nuclear technology with the world: if the other nations were to 'catch up ... they will have to do it on their own hook, just as we did', he told reporters from the porch of Linda Lodge, Reelfoot Lake, Tennessee, where he was on holiday. His decision heralded the new American policy of 'monopoly and exclusion' in relation to nuclear weapons, and answered the hopes of the defence industry, which had most to gain from US investment in a nuclear arsenal.

The new Atomic Energy Commission, which enveloped the Manhattan Engineer District, was charged with responsibility for developing America's nuclear industry. Thus began a series of atomic weapons tests in Nevada and the Pacific launched in tandem with a new strategic outlook. In 1946–47 the concept of the pre-emptive strike – the 'anticipatory counterattack' – insinuated itself into America's highest military and policy-making realm.* Commanders preached to an audience of nuclear converts, confident that America would be the world's only nuclear power for many years.

Their voices were uncompromising, their plans gargantuan. Groves and LeMay – who had been earlier dismissive of the weapon in the context of the Pacific War – were in the vanguard of this call to nuclear arms. LeMay demanded a fundamental rethink of US military policy that would pre-empt the future threat of Soviet missiles. The next war, he warned, would be atomic and the danger, once Russia had the bomb, ever-present. 'When these bombs fall on our industrial

---

* The Pincher Plan, for example, envisaged a nuclear attack on Russia sometime between the summers of 1946–47, with massive civilian casualties.

heart – that rich and highly developed area that lies between Boston and Baltimore and Chicago … the war may already be over.'

In reply, US air power must be stronger than the world's strongest bully, LeMay argued. America, he declared, would serve as the world's policeman. Against atomic weapons 'American air power must protect not only our own nation but every other nation. It can do this only by becoming so strong that no other nation dare attack us … When it is that, the American Army Air Force will be the custodian of the peace of the world.' LeMay's words had already acquired lethal substance. By November 1946, the new B-36 heavy bombers in production at Fort Worth, Texas, were capable of delivering atomic bombs to any city on earth. 'Inhabited Regions of World Now in Range of B-36's Atomic Bombs,' the press release announced. The B-36s were 'designed for a normal range of 10,000 miles' carrying '10,000 pounds of bombs' without the need to refuel.

In 1947 the concept of the pre-emptive nuclear strike received the imprimatur of the Joint Chiefs of Staff 'Evaluation Board', a presidium of four military heads, three top industrialists and a scientist. They sought to overturn the US tradition of 'never striking until we are struck'. 'An enemy armed with atomic weapons,' the board stated in its preliminary report on the Bikini Atoll tests, 'might do irreparable damage before we could strike back.' The Chiefs therefore 'directed' the President, as Commander in Chief, *'to launch an attack against any nation when in his opinion and in that of his cabinet, that nation is preparing an atomic attack on the United States'*. The pre-emptive nuclear strike was the centrepiece of the Joint Chiefs' final report, 'The Evaluation of the Atomic Bomb as a Military Weapon', released on 30 June 1947, which enshrined the concept of 'offensive defence' in American military planning. A few clauses capture the thinking:

- 'Offensive measures will be the only effective measures of defense.'
- The President should be empowered 'to order atomic bomb

retaliation when such retaliation is necessary to prevent or
frustrate an atomic weapon attack upon us'.

- An adequate program of defence must possess enough atomic
  bombs to 'deter a potential enemy from attack' or, 'if he plans an
  attack, overwhelm him and destroy his will and ability to make
  war before he can inflict significant damage on us'. In such a war,
  'the element of surprise ... will be the only assurance of success';
  the lack of it 'may be catastrophic'.
- 'The effects of radiation upon living organisms,' the Joint Chiefs
  added, 'were untreatable, nor is there any reason to hope that
  prophylaxis [immunisation] may be found.' If this seemed to
  undermine the central idea that it was possible to 'win' a global
  nuclear war, the Joint Chiefs swiftly disabused their readers: on the
  contrary, the bomb's very toxicity was 'of decisive importance in
  war'; its radioactive power, a desirable attribute of offensive action.

The military chiefs, with the support of their industrial and scientific
partners, concluded that America must produce atomic weapons and
fissionable materials 'in such quantities and at such a rate of production
as will give to it the ability to overwhelm swiftly any potential enemy'.
The nuclear arsenal would be a psychological weapon as much as a
physical threat: unused, its terrifying *potential* would conquer an
enemy's will to resist. And therein lay a further reason to stockpile it.
Nobody who read this document mistook 'potential enemy' for anyone
other than the Soviet Union; nor did anyone doubt that Moscow, had
it then possessed nuclear weapons, would have pursued a mirror image
of the policy.

In 1948, US defence chiefs, assuring themselves that Russia did not
have the bomb, planned future military operations around the concept
of the pre-emptive strike. One scenario, 'Plan Trojan', for example,
earmarked 30 Soviet cities for complete incineration. A more extensive
map of devastation superseded it when LeMay, head of Strategic Air
Command, incorporated nuclear weapons into his arsenal of conventional

and incendiary bombs. The man Washington credited with killing, wounding or dehousing some two million Japanese civilians in the firestorms of 1945 regarded nuclear terror as an extension of his incendiary campaign. LeMay's War Plan 1-49, drawn up in March 1949, envisaged dropping the entire US nuclear stockpile – then some 133 bombs – over Russia in a pre-emptive act of annihilation. The plan aimed to knock the Soviet Union out of any future conflict before Moscow acquired the bomb and envisaged some seven million casualties, including three million dead. On this occasion, LeMay's ideas were not adopted – shelved, however, rather than rejected.

•

The High Noon of the Cold War arrived sooner than Truman or Byrnes dared anticipate. Washington had worked on the assumption that the Russians were years, if not decades, away from producing a bomb. General Groves himself had reckoned it would take the Soviets at least 10 years. In 1949, news from Kazakhstan corrected Groves, astonished Washington and pressed the Pentagon to accelerate America's nuclear program. On 29 August the Soviet Union detonated an atomic device, according to leaked accounts of the explosion, which obliged Truman to speak. On 23 September he informed the American people that they were 'entitled' to know that 'within recent weeks an atomic explosion occurred in the USSR'.

The news astonished – 'flattened', according to one report – politicians, scientists and members of the public. That night Moscow Radio ominously reported that America, on the eve of the Anglo–American–Canadian Atomic Conference, continued to assert a 'monopoly' over the production of atomic bombs. The report, in the context, was interpreted as a taunt. Only the stock market rejoiced: the Russian explosion reversed a recent, sharp decline in shares on the assumption that a new arms race would cut unemployment and prevent another recession.

Panic-stricken statements in the US Senate presaged all-out nuclear conflict. America and Russia would now enter a 'mad armament race' that would lead to 'inevitable war', declared Senator John Sparkman, the Alabaman Democrat, who appealed for an international system of controls through the United Nations, coupled with inspections run by the Security Council. Many speakers echoed his plea.

The military affected an easy calm of business as usual: the Pentagon was silent; the Joint Chiefs of Staff did not meet. Some overplayed the *sang-froid*: after he heard the news, General Omar Bradley, Chairman of the Joint Chiefs, took time off for a round of golf. Defence heads did make one insistent comment: that the Russian explosion had not relied on 'any information stolen or attained from the United States' – drawing attention to the possibility that it had.

The news gravely upset the expectations of Groves, who had assured Washington in December 1945 and many times since – most recently just weeks before the Soviet detonation – that it would take the Russians 10 to 20 years to build an atomic bomb. The event diminished the general, who tried to present its premature arrival as nothing out of the ordinary. The US had been preparing for the eventuality for some time, he said. He would not lose any sleep over it. In any case the public need not fear, he said. 'We are organized in a quiet, orderly way.' He cited the new medical schools at the Walter Reed Hospital, which had been set up to 'indoctrinate' the medical profession in handling an atomic attack, expected as early as 1950. In Groves' mind the Russian bomb simply hastened his call for a nuclear-armed America so powerful that its supremacy would deter a would-be aggressor.

The newly formed Central Intelligence Agency lent impetus to the rush to re-arm. Asked to assess the Russian nuclear threat, the CIA drew the grimmest conclusions. It sheeted home the start of the arms race to autumn 1945 – in that strange limbo between Potsdam and the attacks on Hiroshima and Nagasaki, when, according to the CIA, the Soviet Union decided on an all-out effort to produce atomic bombs, 'in the greatest number possible'. Since that time, Moscow had acquired the technical knowledge and capacity to employ 'Nagasaki-type' fissile weapons – plutonium bombs – and decided in 1949 to develop thermonuclear devices, the CIA concluded.

.

A kind of nuclear fever gripped the American imagination in the early 1950s. The wonder, horror and glamour of the nuclear age awed and strangely thrilled the public. The advent of nuclear power led millions to dream of a world of limitless energy as cheap as 'water, sunlight and air', according to a Columbia University dean; of the age of Prospero, when humanity would finally conquer its most terrible foe – the weather – and make deserts blossom, the arctic habitable and the climate bend to human needs. The flipside of fear was public infatuation with the atomic age, a kind of patriotic fetish for all things nuclear. This acquired a totemic, curiously sexual dimension. Nuclear paraphernalia mirrored the national preoccupation with the bomb: mushroom clouds appeared on posters, car stickers, T-shirts; fashion magazines depicted 'anatomic woman' – a model sunbathing by a swimming pool in the caress of a radioactive cloud. Such were the symptoms of a society obsessed by intimations of its own annihilation. The government encouraged the public to prepare for nuclear Armageddon, which official propaganda sold with a neighbourly touch: schoolchildren were trained to 'duck and cover' beneath their desks to escape the Soviet holocaust; civil defence brochures explained how to evacuate the cities, feed millions of refugees, bury the dead (Britain, too, produced in 1950 'Atomic Warfare – Manual of Basic Training' in civil defence). Russian bombs were no match for America's robust community spirit, these policies defiantly announced.

While American neighbourhoods rallied to meet the threat, the Pentagon was busy refining its plans for all-out atomic war. In late 1949, Truman took a close interest in the development of the 'superbomb', the hydrogen weapon many times more powerful than

Little Boy or Fat Man, already crude antiques by comparison. A war involving several thermonuclear bombs deeply concerned the President: would it not release vast amounts of radioactive agents into the atmosphere? Could humanity survive such an event?

Sumner Pike, acting chairman of the Atomic Energy Commission (AEC), put the President's mind at rest: 'Dear Mr President,' Pike wrote in a letter of 7 December 1949. 'We have now had several independent calculations made and have concluded that this danger is not serious. Very conservative assumptions indicate that about 500 [hydrogen] bombs could be exploded before the danger point would be reached. More reasonable assumptions lead to a figure of 50,000 bombs.'

While not entirely reassured, Truman decided to proceed with the hydrogen bomb: on 31 January 1950 he directed the AEC, in co-operation with the Defense Department, to go ahead with plans to determine the technical feasibility of a thermonuclear weapon.

The superbomb drew the unwelcome scrutiny of the media, which were less amenable to the self-imposed restraints of wartime censorship. The Alsop brothers – Joseph and Stewart – of the *Washington Post* regularly provoked anxiety in the White House over alleged breaches of security. Their report 'Pandora's Box 1' of 2 January 1950 dared to suggest that faceless officials, unaccountable to the taxpayer, were deciding the fate of the earth: 'Thus dustily and obscurely,' they wrote, 'the issues of life and death are settled nowadays – dingy committee rooms are the scenes of the debate; harassed officials are the disputants; all the proceedings are highly classified, yet the whole future hangs, perhaps, upon the outcome. It will no doubt cause irritation, it may probably provoke denials, to bring the present debate out of its native darkness. Yet this must be done, since deeper issues are involved, which have been far too long concealed from the country.'

•

The first chance to test the Cold Warriors' willingness to use nuclear weapons came in 1950–51, with the outbreak of the Korean War. Commanders, politicians and prelates urged Truman to A-bomb the Chinese and North Koreans. Many senators and congressmen backed a nuclear strike. The horror of the Korean 'meat grinder' moved Democrat Al Gore of Tennessee, father of the future Vice President and environmental campaigner, to propose a 'cataclysmic' atomic intervention. Having served on the Appropriations Committee of the Atomic Energy Commission, Gore claimed to understand the consequences of his decision: even so, he wrote, the 'tragic situation' in Korea 'demands some dramatic and climactic use of some of these immense weapons'. Any attack on American-occupied Japan would justify the use of the atomic bomb, he added.

He recommended that Truman

... dehumanise a belt across the Korean Peninsula by surface radiological contamination ... broadcast the fact to the enemy, with ample and particular notice that entrance into the belt would mean certain death or slow deformity to all foot soldiers; that all vehicles, weapons, food, apparel entering the belt would become poisoned with radioactivity, and, further, that the belt would be regularly recontaminated until such time as a satisfactory solution to the whole Korean problem shall have been reached. This ... would be, I believe, morally justifiable under the circumstances.

Church leaders and their parishioners were among the loudest exponents of a nuclear solution to the war. Some had much bigger game in mind: 'Your Excellency,' the Reverend Kenneth Eyler of the Wesleyan Methodist Church of East Michigan addressed the President, 'As a minister of the Gospel and a Bible-believing Christian ... there is much that has been bothering me lately. This war in Korea. Why is it we fuss around at the fringe instead of getting at the heart of the matter? ... You know as well as I do where this whole matter lies. That is in MOSCOW. I would rather see Moscow destroyed than our boys die in Korea at the hands of the Chinese Red ... You can use the Atom bomb. Don't pay attention to these liberal and modernist preachers ...'

Albert Sheldon, associate director of the God's Way Foundation (whose letterhead listed 'God' as its director), told the President: 'If and when you decide it is necessary to use atomic bombs, we hope and pray that they will be dropped first of all on Moscow ... Please, *please*, do not allow any atomic bombs to be dropped on Chinese territory ... until we have done our best to knock Russia cold.'

A great many ordinary people similarly urged the President to use the bomb: pawn shop owner W.D. Westbrook, who used to buy mules from Truman's father, pleaded, 'for all our sakes', to drop the bomb: 'Your one bomb stopped the Japs. Several will end it all, all over the world. The world is waiting.'

And there was General MacArthur, the commander on the ground, who demanded up to 50 nuclear weapons be dropped on Manchuria to thwart the massed Chinese attacks on his positions; at the time MacArthur enjoyed huge public support.

At the other extreme were many ordinary Americans who failed to see the hand of God in the creation of a radioactive wasteland on the Korean Peninsula. Clara Bergamini lost her son on Guam:

> I and my whole family are appalled at this talk in government circles about whether or not we should use the atom bomb on our foes in Korea. We should not have used the hideous weapon in Japan, and believe me, the crime we perpetrated on helpless non-combatants there will inevitably have to be paid for by us.
>
> My husband and I with two children were interned for these years by the Japanese, in the Philippines, and we lost everything we owned, including our eldest son, who was killed with the marine assault on Guam. We would rather go through all that again, and lose our only surviving son, than turn loose this atomic horror on anyone.

Restraint proved the better part of valour. Truman refused to approve the use of nuclear weapons on Korea, defying MacArthur's plea for deliverance. (Later, Henry Kissinger, Secretary of State in the Nixon administration, rammed home the message, according to historian Richard Rhodes: he prevented the South Koreans from building a bomb by threatening to withdraw US forces completely from the Peninsula.) The President's gravest fear was that a nuclear strike would ignite a global nuclear confrontation between the two power blocs.

•

His restraint notwithstanding, the Korean War supercharged America's domestic atomic program. In 1952 Congress prevailed upon the President to build a thermonuclear arsenal; the influential Joint Committee on Atomic Energy recommended the mass production of hydrogen bombs. Senator Brien McMahon made the case 'in the defense of peace' in a letter to Truman of 30 May 1952: it seemed likely, he wrote, that several kinds of H-bombs were deliverable by 1954, at a fraction of the cost of the $2 billion spent on the Manhattan Project; some of the new weapons 'very possibly' had the explosive power of 'tens of millions of tons of TNT'. The sheer size of the superbomb limited its use to strategic targets: whole cities, ports, and huge industries. McMahon urged Truman, however, to consider whether 'tactical H-bombs' might be developed, for use against ground troops; and if so, whether hydrogen bombs may become 'our primary nuclear weapon':

'So long as the arms race continues, the ineluctable logic of our position leaves us without choice except to acquire the greatest possible firepower in the shortest possible time … overwhelming American superiority in H-bombs may well be the decisive means of keeping open the future for peace …'

Truman promised, in his reply, to refer McMahon's views to the Special Committee of the National Security Council where they 'will be given the careful study which they deserve'.

The detonation of the world's first hydrogen bombs at the Pacific atoll of Eniwetok marked the beginning of an arms race with the Soviet Union in a new family of nuclear weapons. The era Churchill called the 'peace of mutual terror' had begun. Never in history, the British leader declared, had nations fewer inducements to start a war. Commentators invoked a flurry of metaphors to describe the global deadlock: like two men holding a cocked pistol, for example, at each other's heads with the triggers connected by an invisible wire, so that when one fired they both died. Oppenheimer suggested two scorpions in a bottle, locked in a fight to the death.

By the mid-1950s the Cold War had exceeded the projections of mutually assured destruction; America and the Soviet Union were now building nuclear arsenals powerful enough to annihilate the other several times over. In 1955, the US Strategic Air Command calmly contemplated destroying 75 per cent of the population of 118 Russian cities, in a war in which total Soviet fatalities were expected to exceed 60 million. 'The final impression,' remarked a shocked witness at one briefing, 'was that virtually all of Russia would be nothing but a smoking, radiating ruin at the end of two hours.' These were no longer the scenarios of military planners; they were seen as likelihoods.

Such projections were beyond the comprehension of most people; few realise how close America and the Communist Bloc have come to all-out nuclear war. The French pleaded for nuclear weapons to save their garrison at Dien Bien Phu in 1954, which Washington refused. Kennedy's and Khrushchev's restraint avoided nuclear combat during the 1962 Cuban Missile Crisis, the closest the world has been to a global atomic confrontation. Later, in Vietnam, America chose to fight a 'limited' – that is, non-nuclear war – and lose, rather than risk an atomic clash with North Vietnam's communist sponsors, China and the Soviet Union. The Soviet Union similarly resisted the temptation to use atomic weapons in Afghanistan, and lost that war too.

In such a world, Washington and Moscow viewed any move to dismantle their nuclear arsenals as suicidal – a point lost on campaigners for nuclear disarmament, who failed to appreciate the impossibility of unilateral action. The bombs were here to stay, for as long as Russia posed a threat to America's survival, and America to Russia's. The weapons could not be uninvented, shoved back inside Pandora's Box.

Overawed by these events, shocked by the scale of the arms race, many people sought refuge in protest – or dark satire. Kurt Vonnegut's 1963 novel *Cat's Cradle*, for example, sought absolution from the nuclear nightmare in the blackest sci-fi comedy. And *The Report from Iron Mountain* (1967) purported to be an official report on the doings of a government-appointed Special Study Group that met in various secret locations, for example Iron Mountain, to discuss the prospect of peace on earth. According to *The Report*, a bestselling spoof by 'John Doe' (Leonard Lewin) – a member of the panel – the Iron Mountain Special Study Group concluded that, even if lasting peace could be achieved, 'it would almost certainly not be in the best interests of society to achieve it.' A state of chronic war – the 'war system' – should be maintained because it bequeathed all the economic benefits, political stability and progress mankind could desire. To author Leonard Lewin's astonishment, officials in Washington and the Pentagon took his report seriously, bearing as it did the hallmark of the official mindset and written in familiar jargon. Only 'normal people' – ordinary Americans – got the joke: 'Those who thought it might be a hoax were people without any background in the subject,' remarked Lewin.

•

The nuclear arms race was probably beyond satire. The United States and the Soviet Union were deadly serious, and nobody doubts they would have used nuclear weapons at the height of the Cold War had their survival been seriously threatened. By the 1980s they were capable of destroying the world many times over. The 'republic of insects and grass' of Jonathan Schell's portentous 1982 vision, *The Fate of the Earth*, seems a rather benign place (for cockroaches, at least) beside the smoking cinder America and Russia would have left the planet had they engaged in nuclear war. The superpowers held their fire, thanks largely to the post-war restraint of Truman, Eisenhower, Kennedy, Khrushchev and Gorbachev – until the war of economic stealth waged by Ronald Reagan and Margaret Thatcher worked its lethal ministry and defeated Soviet communism in the late 1980s. By then, the superpowers had run a dead heat – a stalemate of mutually assured annihilation culminating in President Reagan's 'Star Wars' program that conceived of a vast missile shield across the United States. And by then, the Soviet Union was effectively bankrupt.

The real weapon in the Cold War was economic. The nuclear program had cost America US$5.5 trillion and was, in the long run, unaffordable. The arms race boiled down to the question: which political system could afford to maintain such destructive power? (Even today, America's nuclear arsenal costs US$50 billion a year to maintain.) The answer lay in the rubble of the Berlin Wall: the collapse of the Soviet Empire in 1989 led both sides eventually to agree, after a series of marathon negotiations, to dismantle part of their nuclear arsenals (in 2009 the US possessed about 10,000 nuclear warheads, and Russia 12,500. Adding the 900 or so of China, Britain, France, India, Pakistan, Israel and North Korea, the world remains quite capable of blanketing the atmosphere in nuclear waste).

The Soviet Union and America did not reduce their arsenals as a humanitarian gesture: Moscow simply ran out of money. In this light, the oft-cited 'good consequences' of Hiroshima and Nagasaki, as a warning to the world, seem elusive. Today, every tin-pot regime wants the bomb. Even if we accept the dubious notion that the bombing of Hiroshima and Nagasaki has acted as a deterrent – and pulled the world back from the brink of nuclear war in Korea, Cuba and Vietnam – Washington obviously did not drop the weapon for that reason. This is a dangerous example of reading history backwards: consoling in hindsight, it utterly misrepresents the motives and thinking of the time.

This is not the counsel of despair, however: a glimmer of hope exists in the steady gains achieved by the proponents of nuclear non-proliferation, whose efforts in the past decade have bequeathed a safer world, with fewer nuclear states than once feared: in 1960 President Kennedy said he expected to see 50 nuclear powers emerge within four years; today there are fewer than 10. Some naively believe that in a distant, nuclear-free world any nation that dares attempt to build nuclear weapons will be charged with a crime against humanity. That is unlikely and unwise: it is inconceivable that America, Russia or the emerging Chinese superpower would destroy their nuclear deterrents in obedience to international law. For who would risk living in a world at the mercy of nuclear pirates?

•

In contemplating a way forward, wiser heads return to the past, and try honestly to confront what did and did not happen in Hiroshima and Nagasaki nearly 70 years ago. Let us dispense with the easy myths: the bomb did not 'shock Japan into submission'; the bomb did not save a million American servicemen; the bomb did not, of itself, end the war. Little Boy and Fat Man were two, obviously big, components of the immense American and Soviet war machinery ranged against Japan. More critical than the bombs in forcing Tokyo to surrender were the US naval blockade, in its sustained damage to the Japanese economy, the Russian invasion of Japanese-occupied Manchuria and – the clincher for Tokyo – the continuation of the Imperial dynasty, granted by Washington on 11 August 1945. These events would have occurred regardless of whether the bombs were dropped; together they persuaded Tokyo to surrender, chiefly by empowering the moderates over the militarists and assuring Hirohito a role in the post-war world. The truth is, the bombs were militarily unnecessary, as the plain facts show and the leading American commanders confirmed after the war: Eisenhower, MacArthur,

Halsey, Leahy and LeMay all, to varying degrees (and, in some cases, with a fair dollop of personal ego thrown in), dismissed the weapon as having little impact on the end of the war. At least two of these men condemned the bombs as inhumane, utterly barbaric.

And yet, from this great distance, academics, politicians and millions of people presume to know better than the commanders on the ground. To this day they labour under the delusion – however comforting – that the destruction of two cities composed mostly of civilians was critical in ending the war. They rarely ask why LeMay spent the best part of six months massacring innocent people, instead of targeting the vestiges of Tokyo's war machine; i.e. kamikaze airfields, coal ferries, underground factories (a job Halsey's aircraft resumed with lethal effect in July 1945). The atomic attacks were the culmination of this process of deliberate civilian annihilation. Millions of Christian Americans routinely defend this act as somehow OK in the 'context' of total war. This is the special pleading of armchair generals and the ignorant, not of the men who were there.

But the chief prop in the dismal defence of the bomb is the utter fallacy that it 'saved a million American lives' (implying 3–4 million *casualties*, or the death or wounding of *every American serviceman* in a putative land invasion, which would not have gone ahead in any case) – as well as thousands of prisoners of war. Paul Fussell offers the most eloquent expression of this sentiment in his essay, 'Thank God for the Atom Bomb'. He fails to see that the soldiers would have thanked God for the Russian invasion, the Byrnes Note and the US naval blockade, had they known that this combination actually ended the war.

At a time of war, people will applaud any story their government feeds them. Americans continue to swear blind that the bombs alone ended the war; that they were America's 'least abhorrent' choice. These are plainly false propositions, salves to uneasy consciences over what was actually done on 6 and 9 August 1945 when, under a summer sky, without warning, hundreds of thousands of civilian men, women and children felt the sun fall on their heads.

APPENDIX 1

TIMELINE – HIROSHIMA NAGASAKI

1920s – Ernest Rutherford's experiments at Cambridge discover the basic composition of the atom: a nucleus of positively charged protons surrounded by orbiting negatively charged electrons.

February 1932 – James Chadwick proves the existence of the neutron, a neutral particle that binds with the proton in the nucleus.

12 September 1933 – On a London street Leo Szilard conceives of a chain reaction of neutrons colliding with atomic nuclei, dislodging more neutrons and so on, releasing huge amounts of energy. He speculates about the possibility of harnessing the energy for weapons.

21 December 1938 – Otto Hahn submits a paper to *Naturwissenschaften* showing the first clear evidence of nuclear fission – the 'splitting' of the nucleus of the atom.

26 January 1939 – Niels Bohr announces the discovery of fission at a physics conference at George Washington University in Washington, DC.

29 January 1939 – Robert Oppenheimer conceives of the construction of an atomic bomb, based on the energy released through fission of the uranium atom.

31 August 1939 – Bohr co-publishes with John Wheeler a theoretical analysis of uranium fission, showing that U235 is more fissile than U238.

1 September 1939 – Germany invades Poland, initiating the outbreak of World War II.

11 October 1939 – Alexander Sachs delivers a letter signed by Albert Einstein (and drafted by Szilard) to President Franklin D. Roosevelt, warning of the risk that Germany may be developing nuclear weapons and urging America to take action. Roosevelt appoints an 'Advisory Committee on Uranium'.

10 May 1940 – Germany attacks Holland, Belgium and France.

1 July 1940 – The United States' newly created National Defense Research Council (NDRC), headed by Vannevar Bush, takes over responsibility for uranium research.

26 February 1941 – Scientists Glenn Seaborg and Arthur Wahl demonstrate the presence of the highly radioactive element 94, which they later name plutonium.

March 1941 – The Maud Committee, drawing on the work of Otto Frisch and Rudolf Peierls, issues a report that explains fast fission in bomb design and the radiation risks. The committee sends it to the US, where Lyman Briggs, chairman of the Uranium Committee, locks it in his safe.

28 March 1941 – Joseph Kennedy, Glenn Seaborg and Emilio Segrè demonstrate slow fission in a plutonium sample, showing its potential in bomb material.

May 1941 – Dr Ernest Lawrence of the Radiation Laboratory, University of California, reports that plutonium would be highly fissionable. Tokutaro Hagiwara at the University of Kyoto announces the possibility of an atomic fusion explosion, the first such mention of an 'implosion' weapon.

18 May 1941 – Emilio Segrè and Glenn Seaborg show that plutonium is a better prospect than uranium for use in a nuclear weapon.

15 July 1941 – The contents of the Maud Report on the practical development of an atomic bomb reaches Vannevar Bush, following pressure from Mark Oliphant and Ernest Lawrence. Bush delays action.

3 September 1941 – Prime Minister Winston Churchill and British Chiefs of Staff approve the development of a nuclear weapon in a program to be codenamed 'Tube Alloys'.

7 December 1941 – The Japanese attack Pearl Harbor.

8 December 1941 – The US declares war on Japan.

11 December 1941 – The US joins the war against Germany and Italy.

18 December 1941 – The first meeting on America's S-1 (atomic) project is held – charged with research and development of fission weapons.

April 1942 – Enrico Fermi's team completes the building of a subcritical experimental pile in the Stagg Field squash courts at Chicago University, in order to test the world's first slow-fission nuclear reaction.

August 1942 – The Manhattan Engineer District (Manhattan Project) created to co-ordinate all work on S-1.

17 September 1942 – (Then) Colonel Leslie Groves, a military engineer, appointed to command the Manhattan Project. Shortly promoted to Brigadier General, Groves is given top security clearance and emergency procurement powers.

18 September 1942 – On behalf of the Manhattan Project, Groves purchases 1250 tons of high quality Belgian Congo uranium ore then stored on Staten Island.

15 October 1942 – Groves appoints Oppenheimer head of Project Y, the new bomb laboratory to be based at Los Alamos.

December 1942 – Roosevelt approves a further $400 million in Manhattan Project funding (almost five times the previous estimate).

2 December 1942 – Fermi's group achieves the world's first 'slow' chain reaction at Chicago University.

18 February 1943 – Construction begins at Oak Ridge on the vast electromagnetic U235 separation plant.

31 May 1943 – Construction begins on the gaseous diffusion uranium enrichment plant at Oak Ridge.

June 1943 – Navy Captain William 'Deak' Parsons begins gun-assembly research at Los Alamos, as Ordnance Division leader.

July–November 1943 – Massive expansion of Oak Ridge, Hanford and Los Alamos, with many new arrivals, including Johann (John) von Neumann, a pioneer of the implosion bomb-detonation method; and Norman Ramsey, who will select and modify aircraft for delivering the atomic bombs. The physicists Niels Bohr, Otto Frisch, Rudolf Peierls, James Chadwick, William Penney and German émigré Klaus Fuchs, the Soviet spy, join the Manhattan Project.

20 July 1944 – Los Alamos channels further resources into the plutonium detonation system, based on implosion.

August 1944 – The US Air Force begins modifying 17 B-29s for combat delivery of nuclear weapons.

20 January 1945 – Curtis LeMay takes charge of the XXI Bomber Command in the Marianas, with a fleet of 345 B-29 aircraft.

February 1945 – Fleet Admiral Chester Nimitz, Commander in Chief, Pacific Ocean Areas, is informed of the atomic bomb project. Tinian Island is selected as the base of operations for the atomic attack.

13 February 1945 – Dresden burned in a massive Allied air raid, killing or wounding more than 100,000.

9–10 March 1945 – LeMay's aircraft firebomb Tokyo, killing at least 100,000 people and seriously wounding 41,000.

11–18 March 1945 – US aircraft firebomb Nagoya, Osaka and Kobe, burning 40 square kilometres of residential area and killing 50,000 people.

12 April 1945 – President Roosevelt dies of a brain hemorrhage.

13 April 1945 – War Secretary Henry Stimson informs President Harry Truman of the existence of the atomic bomb project.

25 April 1945 – Stimson and Groves give Truman first in-depth report on the Manhattan Project.

27 April 1945 – The first meeting of the Target Committee, established to choose targets for the atomic bomb, shortlist Hiroshima, Niigata, Kokura and Nagasaki, being the only big cities not yet burned or destroyed.

8 May 1945 – V-E Day: the unconditional surrender of Germany.

25 May 1945 – US defence chiefs start planning Operation Olympic, the invasion of Kyushu, Japan, scheduled for 1 November.

28 May 1945 – The Target Committee meets again listing as atomic targets Kyoto, Hiroshima and Niigata. Byrnes tells an astonished Szilard that the atomic bomb will make Russia 'more manageable' in Europe.

30 May 1945 – War Secretary Henry Stimson orders Kyoto, the ancient capital of Japan, to be removed from the atomic target list.

1 June 1945 – The Interim Committee, set up to oversee post-war controls on nuclear weapons, recommends that the atomic bomb be dropped as soon as possible, without warning, on the urban heart of a large Japanese city.

Early June 1945 – The Scientific Panel of the Interim Committee (Robert Oppenheimer, Enrico Fermi, Ernest Lawrence and Arthur Compton) meet at Los Alamos to prepare a report on the possible non-military demonstration of the bomb. Within a few days they reject the idea, and recommend direct military use of the bomb, without warning, on a Japanese urban area.

10 June 1945 – The 509th Composite Group charged with responsibility for dropping the atomic bombs begins arriving on Tinian Island.

June 1945 – LeMay estimates that his XXI Bomber Command will completely destroy Japan's 60 most important cities by 1 October 1945.

11 July 1945 – Japanese Foreign Minister Shigenori Togo cables Ambassador Naotake Sato in Moscow asking him to explore the possibility of the Soviet Union acting as an intermediary in surrender negotiations.

15 July 1945 – Truman, Secretary of State James Byrnes and their staff arrive in Potsdam for the Three Power summit with Britain and the Soviet Union.

16 July 1945 – Trinity, the world first's atomic explosion, occurs at 5:29:45am at Alamogordo, New Mexico.

23 July 1945 – The latest target list – Hiroshima, Kokura, and Niigata, in order of priority – is sent to Stimson in Potsdam; Nagasaki is later added. The bomb-making schedule estimates that a second Fat Man plutonium bomb will be ready by 24 August, with three more available in September, and seven or more atomic bombs available per month thereafter.

24 July 1945 – Truman tells Stalin that America possesses a 'new weapon of unusual destructive force' (he does not divulge that it is 'atomic'). Stalin is already aware of the Manhattan Project, through Fuchs and other spies. Groves drafts the final directive authorising the use of atomic bombs as soon as available and weather permitting. The target list is Hiroshima, Kokura, Niigata and Nagasaki.

26 July 1945 – Truman issues the Potsdam Declaration, requiring the unconditional surrender of the Japanese armed forces; the USS *Indianapolis* delivers the Little Boy uranium bomb parts to Tinian.

28 July 1945 – The Japanese government responds to the Potsdam ultimatum with *mokusatsu* – 'killing it with silence'.

4 August 1945 – Tibbets informs the 509th Composite Group that they are to drop immensely powerful bombs on Japan; he does not reveal the nature of the weapon.

5 August 1945 – Tibbets christens his aircraft *Enola Gay* after his mother; Little Boy is loaded on the plane.

6 August 1945 – At 8:16:02am Hiroshima time Little Boy explodes at an altitude of 1850 feet (560 metres), 550 feet (160 metres) from the target, the Aioi Bridge, over the centre of the city, instantly killing more than 70,000 people.

8 August 1945 – Russian Foreign Minister Molotov announces that the Soviet Union will be at war with Japan the next day. At midnight, 1.5 million Red Army soldiers invade Japanese-occupied Manchuria.

9 August 1945 – At 3.47am, Bockscar departs Tinian, bound for Kokura Arsenal. After three failed runs on Kokura, the aircraft turns towards Nagasaki, the only secondary target in range. At 11:02 (Nagasaki time) Fat Man explodes at 1950 feet (580 metres) near the perimeter of the city; the yield is 19 to 23 kilotonnes, instantly killing at least 30,000 people.

9–11 August 1945 – Japan's leaders refuse to surrender, and continue to disagree over the surrender terms they are willing to accept. Their primary concern is the Soviet invasion of Japanese-occupied territory. State propaganda urges the Japanese public to prepare for atomic warfare.

11 August 1945 – Truman and Byrnes send Japan an amended ultimatum, the 'Byrnes Note', which acknowledges the Emperor's existence in Japanese society, but insists that America will circumvent and control his powers.

12–15 August 1945 – Tokyo fiddles while the nation burns; Japan's 'Big Six' rulers attend a series of meetings with the Emperor that culminate in the decision to surrender – notwithstanding an 11th hour coup attempt by army officers.

15 August 1945 – Hirohito accepts the terms of the Potsdam Declaration – as conditioned by the Byrnes Note. His 'surrender' speech does not use the word 'surrender'. On the contrary, he tells the Japanese people merely that the war has not necessarily turned to Japan's advantage.

17 August 1945 – Hirohito delivers second 'surrender' speech, to the armed forces, and urges them to lay down their weapons; he blames the Russian invasion and makes no mention of the atomic bomb.

2 September 1945 – General Douglas MacArthur and other Allied commanders receive Japan's conditional surrender aboard the battleship USS *Missouri*. The Emperor remains in a figurehead role.

3 October 1945 – President Truman asks Congress for atomic energy control legislation; a special Senate Committee on Atomic Energy is created on 29 October.

November 1945 – Truman and the Prime Ministers of the United Kingdom and Canada announce agreement on principles of international control of atomic energy.

1946–1989 – The Soviet Union and United States together build nuclear arsenals containing thousands of nuclear weapons, capable of destroying the world several times over.

May 2009 – President Barack Obama announces his desire for a nuclear-free world, the fourth president publicly to do so.

APPENDIX 2

HYDE PARK AGREEMENT
ATOMIC ENERGY

10 Downing Street, Whitehall – Tube Alloys

Aid Memoires [sic] of Conversation between the President and the Prime
Minister at Hyde Park September 18, 1944.

1. The suggestion that the world should be informed regarding tube alloys with a
view to an international agreement regarding its control and use is not accepted.
The matter should continue to be regarded as of the utmost secrecy, but when
a bomb is finally available it might perhaps, after mature consideration, be used
against the Japanese, who should be warned that this bombardment will be
repeated until they surrender.

2. Full collaboration between the United States and the British Government
in development of tube alloys for military and commercial purposes should
continue after the defeat of Japan unless and until termination by joint
agreement.

3. Inquiry should be made regarding the activities of Professor Bohr and steps
taken to insure that he is responsible for no leakage of information, particularly
to the Russians.

F.D.R.
W.S.C.

NOTE: A copy of this aid memoirs [sic] was left with President Roosevelt;
another copy was given to Admiral Leahy to hand to Lord Cherwell.

APPENDIX 3

POTSDAM PROCLAMATION, 26 JULY 1945

(1)   We – the President of the United States, the President of the National Government of the Republic of China, and the Prime Minister of Great Britain, representing the hundreds of millions of our countrymen, have conferred and agree that Japan shall be given an opportunity to end this war.

(2)   The prodigious land, sea and air forces of the United States, the British Empire and of China, many times reinforced by their armies and air fleets from the west, are poised to strike the final blows upon Japan. This military power is sustained and inspired by the determination of all the Allied Nations to prosecute the war against Japan until she ceases to resist.

(3)   The result of the futile and senseless German resistance to the might of the aroused free peoples of the world stands forth in awful clarity as an example to the people of Japan. The might that now converges on Japan is immeasurably greater than that which, when applied to the resisting Nazis, necessarily laid waste to the lands, the industry and the method of life of the whole German people. The full application of our military power, backed by our resolve, will mean the inevitable and complete destruction of the Japanese armed forces and just as inevitably the utter devastation of the Japanese homeland.

(4)   The time has come for Japan to decide whether she will continue to be controlled by those self-willed militaristic advisers whose unintelligent calculations have brought the Empire of Japan to the threshold of annihilation, or whether she will follow the path of reason.

(5)   Following are our terms. We will not deviate from them. There are no alternatives. We shall brook no delay.

(6)   There must be eliminated for all time the authority and influence of those who have deceived and misled the people of Japan into embarking on world conquest, for we insist that a new order of peace, security and justice will be impossible until irresponsible militarism is driven from the world.

(7)   Until such a new order is established and until there is convincing proof that Japan's war-making power is destroyed, points in Japanese

territory to be designated by the Allies shall be occupied to secure the achievement of the basic objectives we are here setting forth.

(8)   The terms of the Cairo Declaration shall be carried out and Japanese sovereignty shall be limited to the islands of Honshu, Hokkaido, Kyushu, Shikoku and such minor islands as we determine.

(9)   The Japanese military forces, after being completely disarmed, shall be permitted to return to their homes with the opportunity to lead peaceful and productive lives.

(10)   We do not intend that the Japanese shall be enslaved as a race or destroyed as a nation, but stern justice shall be meted out to all war criminals, including those who have visited cruelties upon our prisoners. The Japanese Government shall remove all obstacles to the revival and strengthening of democratic tendencies among the Japanese people. Freedom of speech, of religion, and of thought, as well as respect for the fundamental human rights shall be established.

(11)   Japan shall be permitted to maintain such industries as will sustain her economy and permit the exaction of just reparations in kind, but not those which would enable her to re-arm for war. To this end, access to, as distinguished from control of, raw materials shall be permitted. Eventual Japanese participation in world trade relations shall be permitted.

(12)   The occupying forces of the Allies shall be withdrawn from Japan as soon as these objectives have been accomplished and there has been established in accordance with the freely expressed will of the Japanese people a peacefully inclined and responsible government.

(13)   We call upon the government of Japan to proclaim now the unconditional surrender of all Japanese armed forces, and to provide proper and adequate assurances of their good faith in such action. The alternative for Japan is prompt and utter destruction.

APPENDIX 4

July 17, 1945

A PETITION TO THE PRESIDENT OF THE UNITED STATES

Discoveries of which the people of the United States are not aware may affect the welfare of this nation in the near future. The liberation of atomic power which has been achieved places atomic bombs in the hands of the Army. It places in your hands, as commander in chief, the fateful decision whether or not to sanction the use of such bombs in the present phase of the war against Japan.

We, the undersigned scientists, have been working in the field of atomic power. Until recently we have had to fear that the United States might be attacked by atomic bombs during this war and that her only defense might lie in a counterattack by the same means. Today, with the defeat of Germany, this danger is averted and we feel impelled to say what follows:

The war has to be brought speedily to a successful conclusion and attacks by atomic bombs may very well be an effective method of warfare. We feel, however, that such attacks on Japan could not be justified, at least not unless the terms which will be imposed after the war on Japan were made public in detail and Japan were given an opportunity to surrender.

If such public announcement gave assurance to the Japanese that they could look forward to a life devoted to peaceful pursuits in their homeland and if Japan still refused to surrender our nation might then, in certain circumstances, find itself forced to resort to the use of atomic bombs. Such a step, however, ought not to be made at any time without seriously considering the moral responsibilities which are involved.

The development of atomic power will provide the nations with new means of destruction. The atomic bombs at our disposal represent only the first step in this direction, and there is almost no limit to the destructive power which will become available in the course of their future development. Thus a nation which sets the precedent of using these newly liberated forces of nature for purposes of destruction may have to bear the responsibility of opening the door to an era of devastation on an unimaginable scale.

If after this war a situation is allowed to develop in the world which permits rival powers to be in uncontrolled possession of these new means of destruction, the cities of the United States as well as the cities of other nations will be in continuous danger of sudden annihilation. All the resources of the United States, moral and material, may have to be mobilized to prevent the advent of such a world situation. Its prevention is at present the solemn responsibility of the United States—singled out by virtue of her lead in the field of atomic power.

The added material strength which this lead gives to the United States brings with it the obligation of restraint and if we were to violate this obligation our moral position would be weakened in the eyes of the world and in our own eyes. It would then be more difficult for us to live up to our responsibility of bringing the unloosened forces of destruction under control.

In view of the foregoing, we, the undersigned, respectfully petition: first, that you exercise your power as commander in chief, to rule that the United States shall not resort to the use of atomic bombs in this war unless the terms which will be imposed upon Japan have been made public in detail and Japan knowing these terms has refused to surrender; second, that in such an event the question whether or not to use atomic bombs be decided by you in the light of the considerations presented in this petition as well as all the other moral responsibilities which are involved.

Leo Szilard plus 69 signatories

APPENDIX 5

FINAL DIRECTIVE AUTHORISING USE OF
ATOMIC WEAPONS AGAINST JAPAN

War Department
Office of the Chief of Staff
Washington 25, D.C.
25 July 1945

TO: General Carl Spaatz
        Commanding General
        United States Army Strategic Air Forces

1. The 509 Composite Group, 20th Air Force will deliver its first special
   bomb as soon as weather will permit visual bombing after about 3 August
   1945 on one of the targets: Hiroshima, Kokura, Niigata and Nagasaki. To
   carry military and civilian scientific personnel from the War department
   to observe and record the effects of the explosion of the bomb, additional
   aircraft will accompany the airplane carrying the bomb. The observing planes
   will stay several miles distant from the point of impact of the bomb.
2. Additional bombs will be delivered on the above targets as soon as made
   ready by the project staff. Further instructions will be issued concerning
   targets other than those listed above.
3. Dissemination of any and all information concerning the use of the weapon
   against Japan is reserved to the Secretary of War and the President of the
   United States. No communiques on the subject or releases of information
   will be issued by Commanders in the field without specific prior authority.
   Any news stories will be sent to the War Department for special clearance.
4. The foregoing directive is issued to you by direction and with the approval
   of the Secretary of War and of the Chief of Staff, USA. It is desired that you
   personally deliver one copy of this directive to General MacArthur and one
   copy to Admiral Nimitz for their information.

THOS. T. HANDY
General, G.S.C.
Acting Chief of Staff

APPENDIX 6

PRESS RELEASE BY THE WHITE HOUSE, 6 AUGUST 1945.

The White House
Washington, D.C.

IMMEDIATE RELEASE

STATEMENT BY THE PRESIDENT OF THE UNITED STATES

Sixteen hours ago an American airplane dropped one bomb on Hiroshima and destroyed its usefulness to the enemy. That bomb had more power than 20,000 tons of T.N.T. It had more than two thousand times the blast power of the British "Grand Slam" which is the largest bomb ever yet used in the history of warfare.

The Japanese began the war from the air at Pearl Harbor. They have been repaid many fold. And the end is not yet. With this bomb we have now added a new and revolutionary increase in destruction to supplement the growing power of our armed forces. In their present form these bombs are now in production and even more powerful forms are in development.

It is an atomic bomb. It is a harnessing of the basic power of the universe. The force from which the sun draws its power has been loosed against those who brought war to the Far East.

Before 1939, it was the accepted belief of scientists that it was theoretically possible to release atomic energy. But no one knew any practical method of doing it. By 1942, however, we knew that the Germans were working feverishly to find a way to add atomic energy to the other engines of war with which they hoped to enslave the world. But they failed. We may be grateful to Providence that the Germans got the V-1's and V-2's [sic] late and in limited quantities and even more grateful that they did not get the atomic bomb at all.

The battle of the laboratories held fateful risks for us as well as the battles of the air, land and sea, and we have now won the battle of the laboratories as we have won the other battles.

Beginning in 1940, before Pearl Harbor, scientific knowledge useful in war was pooled between the United States and Great Britain, and many priceless helps to our victories have come from that arrangement. Under that general policy the research on the atomic bomb was begun. With American and British scientists working together we entered the race of discovery against the Germans.

The United States had available the large number of scientists of distinction in the many needed areas of knowledge. It had the tremendous industrial and financial resources necessary for the project and they could be devoted to it without undue impairment of other vital war work. In the United States the laboratory work and the production plants, on which a substantial start had already been made, would be out of reach of enemy bombing, while at that time Britain was exposed to constant air attack and was still threatened with the possibility of invasion. For these reasons Prime Minister Churchill and President Roosevelt agreed that it was wise to carry on the project here. We now have two great plants and many lesser works devoted to the production of atomic power. Employment during peak construction numbered 125,000 and over 65,000 individuals are even now engaged in operating the plants. Many have worked there for two and a half years. Few know what they have been producing. They see great quantities of material going in and they see nothing coming out of these plants, for the physical size of the explosive charge is exceedingly small. We have spent two billion dollars on the greatest scientific gamble in history – and won.

But the greatest marvel is not the size of the enterprise, its secrecy, nor its cost, but the achievement of scientific brains in putting together infinitely complex pieces of knowledge held by many men in different fields of science into a workable plan. And hardly less marvellous has been the capacity of industry to design, and of labor to operate, the machines and methods to do things never done before so that the brain child of many minds came forth in physical shape and performed as it was supposed to do. Both science and industry worked under the direction of the United States Army, which achieved a unique success in managing so diverse a problem in the advancement of knowledge in an amazingly short time. It is doubtful if such another combination could be got together in the world. What has been done is the greatest achievement of organized science in history. It was done under high pressure and without failure.

We are now prepared to obliterate more rapidly and completely every productive enterprise the Japanese have above ground in any city. We shall destroy their docks, their factories, and their communications. Let there be no mistake; we shall completely destroy Japan's power to make war.

It was to spare the Japanese people from utter destruction that the ultimatum of July 26 was issued at Potsdam. Their leaders promptly rejected that ultimatum. If they do not now accept our terms they may expect a rain of ruin from the air, the like of which has never been seen on this earth. Behind this air attack will follow sea and land forces in such numbers and power as they have not yet seen and with the fighting skill of which they are already well aware.

The Secretary of War, who has kept in personal touch with all phases of the project, will immediately make public a statement giving further details.

His statement will give facts concerning the sites at Oak Ridge near Knoxville, Tennessee, and at Richland near Pasco, Washington, and an installation near Santa Fe, New Mexico. Although the workers at the sites have been making materials to be used in producing the greatest destructive force in history they have not themselves been in danger beyond that of many other occupations, for the utmost care has been taken of their safety.

The fact that we can release atomic energy ushers in a new era in man's understanding of nature's forces. Atomic energy may in the future supplement the power that now comes from coal, oil, and falling water, but at present it cannot be produced on a basis to compete with them commercially. Before that comes there must be a long period of intensive research.

It has never been the habit of the scientists of this country or the policy of this Government to withhold from the world scientific knowledge. Normally, therefore, everything about the work with atomic energy would be made public.

But under present circumstances it is not intended to divulge the technical processes of production or all the military applications, pending further examination of possible methods of protecting us and the rest of the world from the danger of sudden destruction.

I shall recommend that the Congress of the United States consider promptly the establishment of an appropriate commission to control the production and use of atomic power with the United States. I shall give further consideration and make further recommendations to the Congress as to how atomic power can become a powerful and forceful influence towards the maintenance of world peace.

APPENDIX 7

SURRENDER SPEECHES OF EMPEROR HIROHITO

Rescript to Japanese people, 15 August 1945

To our good and loyal subjects:

After pondering deeply the general trends of the world and the actual conditions obtaining in our empire today, we have decided to effect a settlement of the present situation by resorting to an extraordinary measure.

We have ordered our government to communicate to the governments of the United States, Great Britain, China and the Soviet Union that our empire accepts the provisions of their joint declaration.

To strive for the common prosperity and happiness of all nations as well as the security and well-being of our subjects is the solemn obligation which has been handed down by our Imperial ancestors and which we lay close to the heart.

Indeed, we declared war on America and Britain out of our sincere desire to insure Japan's self-preservation and the stabilisation of East Asia, it being far from our thought either to infringe upon the sovereignty of other nations or to embark upon territorial aggrandisement.

But now the war has lasted for nearly four years. Despite the best that has been done by everyone– the gallant fighting of our military and naval forces, the diligence and assiduity of out servants of the State and the devoted service of our 100,000,000 people – the war situation has developed not necessarily to Japan's advantage, while the general trends of the world have all turned against her interest.

Moreover, the enemy has begun to employ a new and most cruel bomb, the power of which to do damage is, indeed, incalculable, taking the toll of many innocent lives. Should we continue to fight, it would not only result in an ultimate collapse and obliteration of the Japanese nation, but also it would lead to the total extinction of human civilization.

Such being the case, how are we to save the millions of our subjects, nor to atone ourselves before the hallowed spirits of our imperial ancestors? This is the reason why we have ordered the acceptance of the provisions of the joint declaration of the powers.

We cannot but express the deepest sense of regret to our allied nations of East Asia, who have consistently co-operated with the Empire toward the emancipation of East Asia.

The thought of those officers and men as well as others who have fallen in the fields of battle, those who died at their posts of duty, or those who met death [otherwise] and all their bereaved families, pains our heart night and day.

The welfare of the wounded and the war sufferers and of those who lost their homes and livelihood is the object of our profound solicitude. The hardships and sufferings to which our nation is to be subjected hereafter will be certainly great.

We are keenly aware of the inmost feelings of all of you, our subjects. However, it is according to the dictates of time and fate that we have resolved to pave the way for a grand peace for all the generations to come by enduring the [unavoidable] and suffering what is unsufferable. Having been able to save and maintain the structure of the Imperial State, we are always with you, our good and loyal subjects, relying upon your sincerity and integrity.

Beware most strictly of any outbursts of emotion that may engender needless complications, of any fraternal contention and strife that may create confusion, lead you astray and cause you to lose the confidence of the world.

Let the entire nation continue as one family from generation to generation, ever firm in its faith of the imperishableness of its divine land, and mindful of its heavy burden of responsibilities, and the long road before it. Unite your total strength to be devoted to the construction for the future. Cultivate the ways of rectitude, nobility of spirit, and work with resolution so that you may enhance the innate glory of the Imperial State and keep pace with the progress of the world.

Rescript to Japanese armed forces, 17 August 1945

To the officers and men of the Imperial forces:

Three years and eight months have elapsed since we declared war on the United States and Britain. During this time our beloved men of the army and navy, sacrificing their lives, have fought valiantly on disease-stricken and barren lands and on tempestuous waters in the blazing sun, and of this we are deeply grateful.

Now that the Soviet Union has entered the war against us, to continue the war under the present internal and external conditions would be only to increase needlessly the ravages of war finally to the point of endangering the very foundation of the Empire's existence.

With that in mind and although the fighting spirit of the Imperial Army and Navy is as high as ever, with a view to maintaining and protecting our noble national policy we are about to make peace with the United States, Britain, the Soviet Union and Chungking.

To a large number of loyal and brave officers and men of the Imperial forces who have died in battle and from sicknesses goes our deepest grief. At the same time we believe the loyalty and achievements of you officers and men of the Imperial forces will for all time be the quintessence of our nation.

We trust that you officers and men of the Imperial forces will comply with our intention and will maintain a solid unity and strict discipline in your movements and that you will bear the hardest of all difficulties, bear the unbearable and leave an everlasting foundation of the nation.

APPENDIX 8

CASUALTY TABLES

Total Number of Casualties due to the Atomic Bomb, Hiroshima, as at 10 August 1946

Distance from Hypocentre (km)	Killed	Severely injured	Slightly injured	Missing	Not injured	Total
Under 0.5	19,329	478	338	593	924	21,662
0.5–1.0	42,271	3,046	1,919	1,366	4,434	53,036
1.0–1.5	37,689	7,732	9,522	1,188	9,140	65,271
1.5 – 2.0	13,422	7,627	11,516	227	11,698	44,490
2.0–2.5	4,513	7,830	14,149	98	26,096	52,686
2.5–3.0	1,139	2,923	6,795	32	19,907	30,796
3.0–3.5	117	474	1,934	2	10,250	12,777
3.5–4.0	100	295	1,768	3	13,513	15,679
4.0–4.5	8	64	373		4,260	4,705
4.5–5.0	31	36	156	1	6,593	6,817
Over 5.0	42	19	136	167	11,798	12,162
TOTAL	118,661	30,524	48,606	3,677	118,613	320,081

Casualties among Pupils and Students in Hiroshima

Name of School	Number of cases	Distance from Hypocentre (km)	Exposed Condition	Survived	Killed	Mortality Rate (%)
Second Hiroshima Prefectural High School (first year)	308	0.6	Outdoors	0	308	100.0
Hiroshima Municipal Girls' High School (first year)	277	0.6	Outdoors and in shade	0	277	100.0
Hiroshima Municipal Girls' High School (second year)	264	0.6	Outdoors and in shade	0	264	100.0
Hiroshima Municipal Shipbuilding School (first year)	200	0.8	Outdoors	0	200	100.0
First Hiroshima Prefectural Girls' High School (fourth year)	40	0.8	Indoors	0	40	100.0
First Hiroshima Prefectural High School (third year)	40	0.9	Outdoors	0	40	100.0
First Hiroshima Prefectural Girls' High School (first year)	250*	0.9	Outdoors	0	250*	100.0
First Hiroshima Prefectural High School (first year)	150*	1.2	Outdoors	0	150*	100.0
First Hiroshima Prefectural High School (first year)	150*	1.2	Indoors	17	133*	88.6
Second Hiroshima Prefectural Girls' High School (second year)	37	1.3	Outdoors	1	36	97.2
Hiroshima Prefectural Commercial High School (second year)	44	1.3	Outdoors	0	44	100.0
Third Primary School (boys)	72	1.3	Outdoors	1	71	98.6
Third Primary School (girls)	139	1.3	Outdoors and in shade	68	71	51.0
First Hiroshima Prefectural High School (third year)	75	1.6	Outdoors	74	1	1.3
Second Hiroshima Prefectural Girls' High School (first year)	71	2.1	Outdoors	70	1	1.4
Second Hiroshima Prefectural Girls' High School (second year)	40*	2.1	In shade	40*	0	0
Hiroshima Prefectural Commercial High School (first year)	250*	2.1	Outdoors and in shade	Over 233	Below 17	6.8%§
Hiroshima Prefectural Commercial High School (second year)	200*	2.1	Outdoors and in shade	Over 185	Below 15	7.5§
Second Hiroshima Prefectural High School (second year)	200*	2.3	Outdoors	200*	0	0
Hiroshima Municipal Shipbuilding School (first year)	140	3.5	Indoors	139	1	0.7

*Approximate

§ Including those dead at home and other places

Comparison of adjusted atomic bomb survivor mortality rates by type of malignant neoplasm with all-Japan figure (per 100,000 of total 1951 population)

Disease Type / Population	Digestive tract organs and peritoneum		Respiratory tract organs		Breasts	Uterus and other female organs	Male organs	Urinary organs		Lymphatic tissue and hematopoietic tissue	
	Male	Female	Male	Female	Female	Female	Male	Male	Female	Male	Female
All Japan (1951)	68.2	43.9	4.2	2.0	3.3	19.9	0.7	1.1	0.7	3.1	2.0
Atomic bomb survivors (1951–55)	83.1	37.6	7.8	3.6	3.4	19.8	1.2	1.4	0.7	9.7	8.7

APPENDIX 9

Pentagon's Estimated Bomb Requirements for Destruction of Russian Strategic Areas, September 1945

City	Area of City in sq. miles	No. of Bombs
Moscow	110.0	6
Leningrad	40.4	6
Tashkent	28.9	6
Baku	7.0	2
Novosibirsk	22.0	6
Gorki	13.5	4
Sverdlovsk	20.2	5
Chelyabinsk	11.5	3
Tbilisi	12.7	3
Omsk	6.6	2
Kuibyshev	12.6	3
Kiev	64.4	6
Lvov	20.0	5
Kazan	20.0	5
Alma Ata	13.1	4
Kharkov	30.1	6
Riga	40.0	6
Saratov	8.8	2
Koenigsberg	37.8	6
Odessa	28.7	6
Rostov-on-Don	14.4	4
Dnepropetrovsk	9.2	3
Stalino	7.1	2
Yaroslavl	14.0	4
Ivanovo	16.2	4
Archangel	11.0	3
Khabarovsk	10.0	3
Tula	8.1	2
Molotov	5.7	2
Astrakhan	4.8	1

City	Area of City in sq. miles	No. of Bombs
Magnitogorsk	10.0	3
Vladivostok	10.0	3
Stalingrad	20.3	5
Ufa	10.8	3
Irkutsk	11.5	3
Vilna	20.0	5
Voronezh	17.0	5
Izhevsk	7.5	2
Chkalov	10.2	3
Grozny	1.3	1
Stalinsk	10.8	3
Nizhni Tagil	17.3	5
Penza	5.8	2
Minsk	4.2	1
Kirov	5.3	2
Tallinn	16.0	4
Kemerovo	5.0	2
Ulan Ude	22.3	6
Komsomolsk	5.0	2
Murmansk	4.0	1
Belostok	6.0	2
Vitebsk	3.9	1
Zlatoust	5.6	2
Makhach Kala	1.8	1
Syzran	5.4	2
Chimkent	13.4	4
Batum	3.9	1
Kovrov	1.8	1
Orsk	4.8	2
Kamensk	4.0	1
Brest Litovsk	4.5	1
Gurev	4.0	1
Sterlitamak	3.1	1
Ishimbaevo	4.0	1
Neftedag	4.0	1
Ukhta	4.0	1
TOTAL – 66 CITIES	901.3	204

APPENDIX 10

DRAFT OF THE POTSDAM ULTIMATUM TO JAPAN

Like the final version, it threatened 'prompt and utter destruction' unless Japan surrendered. In this draft, James Byrnes, US Secretary of State, has struck off the names of Russia and China, who were allied to America at the time. The removal of the Soviet Union's name ensured that Japan continued to think of Moscow as neutral, and reduced the chances of Tokyo surrendering just weeks before the bomb was dropped on Hiroshima.

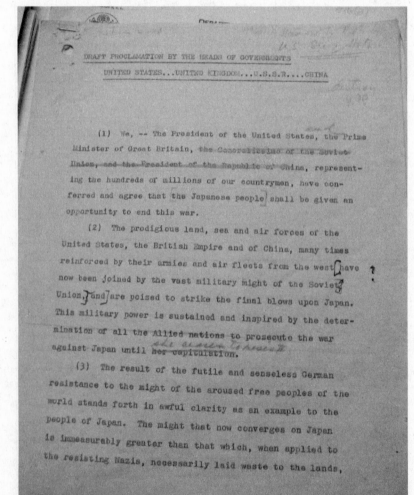

ENDNOTES

CHAPTER 1 WINTER 1945

Stalin expression: Byrnes, J., *Speaking Frankly*, p. 44.

Distrust between Anglo-America and the Soviet Union: McCullough, D., *Truman*, Touchstone, p. 371.

Development of extraordinary new weapon: Correspondence ('Top Secret') of the Manhattan Engineer District, 1942–46, NARA RG77, microfilm publication 1109, roll 3: entry 1 subfile, 16a – summary of facts relating to breach of Quebec Agreement.

Bomb use and control: Atomic Energy, 1944–65, Byrnes Papers.

Churchill softening: See Sherwin, M., *A World Destroyed*, appendices.

Stimson outlook: quoted in Malloy, S. L., *Atomic Tragedy*, pp. 4, 6.

Sharing the secret of the atomic bomb with Russia: Stimson Papers, Diary, 3 December 1944.

Churchill's shock: quoted in Sherwin, M., *A World Destroyed*, p. 110.

Purpose to destroy German Nazism: Byrnes, James F., Yalta Conference, February 1945, Brown Papers, box 13, folder 13.

Unconditional surrender of Japan: Toland, J., *The Rising Sun*, p. 438.

Germans and Japanese must abandon militarism: Butow, R., *Japan's Decision to Surrender*, p. 137.

Churchill's doubts about policy's extension to Japan: quoted in Thorne, C., *Allies of a Kind*, p. 372.

Relaxation of surrender terms: quoted in Hasegawa, T., *Racing the Enemy*, . 36.

Russian entry in the Pacific War: Orwell, G., *Diaries*, (ed. Davison, P.).

London and Washington desired Russian entry: Hasegawa, T., *Racing the Enemy*, p. 33.

Stalin's careful line: Toland, J., *The Rising Sun*, p. 66.

Moscow's successive breaches: Hasegawa, T., *Racing the Enemy*, p. 19.

Stalin's price for Pacific entry: Thorne, C., *Allies of a Kind*, p. 526.

Stalin's timely denunciation: Butow, R., *Japan's Decision to Surrender*, pp. 41–42.

Top secret 'Protocol' signed: Sherwin, M., *A World Destroyed*, pp. 85–89.

Byrnes unaware of deal: Byrnes, J., *Speaking Frankly*, p. 43.

Their secrecy: Leahy, W., *I Was There*, p. 318.

Stalin and the 'Polish Door': Byrnes Papers, box 67, folder 12, transcript draft A, 9III.

Churchill and the Poles: Leahy, W., *I Was There*, p. 309.

Soviet breach of Yalta and Churchill–Roosevelt exchange: Correspondence, Roosevelt, Stalin, Churchill, Truman, Yalta to Potsdam, 1944–45, Byrnes Papers, box 7, folder 5; Byrnes Papers, box 67, folder 12, transcript draft A, 9III; see also Brown Papers.

The Japanese will not crack: Joseph Grew, lecture, quoted in Rhodes, R., *The Making of the Atomic Bomb*, p. 520.

Japanese atrocities: *Japan Times*, quoted in Toland, J., *The Rising Sun*, p. 301.

Rabaul, 1942: See Ham, P., *Kokoda*.

American prisoners: Frank, R., *Downfall*, pp. 160–163.

Japan's biological warfare unit: Tanaka, Y., *Hidden Horrors*, pp. 139–144.

Japanese cholera strategy: Tanaka, Y., *Hidden Horrors*, p. 144.

Allies' view of Japanese war crimes: Tanaka, Y., *Hidden Horrors*, p. 132; see also Ham, P., *Kokoda*.

Epidemic of the Japanese: See Dower, J., *Japan in War and Peace – Selected Essays*.

Media portrayal of racial war: Bird, K. & Lifschultz, L. (eds), *Hiroshima's Shadow*, p. 27.

Japanese prisoners: normal human beings: quoted in Dower, J., *Japan in War and Peace*, p. 258.

Western governments and Japanese hatred: Bird, K. & Lifschultz, L. (eds), *Hiroshima's Shadow*, p. 27.

Japanese Americans interned: Dower, J., *Japan in War and Peace*, p. 258.

General contempt for the race: Thorne, C., *Allies of a Kind*, pp. 167–168, 722.

Australia's formal policy of hatred: Ham, P., *Kokoda*, p. 296.

Japan's prison camps in previous wars: See Tanaka, Y., *Hidden Horrors*.

Japanese 'spirit' and conquest: Togo, S., *The Cause of Japan*, p. 8.

Russia wounded: Hasegawa, T., *Racing the Enemy*, p. 10.

Emperor's divinity: *The Essence of the Kokutai*, quoted in Alperovitz, G., *The Decision to Use the Atomic Bomb*, p. 35.

The roots of this belief: Herschel Webb, in Mitchell, R., *Censorship in Imperial Japan*, p. 107.

Hirohito: power of the sun: Hasegawa, T., *Racing the Enemy*, p. 4.

Hirohito and Japan's military expansion: For a persuasive case against Hirohito, see Bix, H.P., *Hirohito and the Making of Imperial Japan*.

Hirohito saying nothing: Shillony, B., *Politics and Culture in Wartime Japan*, p. 40.

Aggressive interpretations of Hirohito's words: Butow, R., *Japan's Decision to Surrender*, p. 33.

Japanese hatred of the West: quoted in Shillony, B., *Politics and Culture in Wartime Japan*, p. 146.

Japan Steel's letters of encouragement: Poster: *Seisanryoku Zokyo Soshinguntaikai* [Marching Parade for Boosting Productivity] (1933–34).

Japanese response to propaganda: Shillony, B., *Politics and Culture in Wartime Japan*, pp. 145, 147.

Monitored 'dangerous thinkers': Jansen, M., *The Making of Modern Japan*, p. 505.

Reporter dispatched to China: Cook, H.T. & Cook, T.F., *Japan at War*, p. 222.

Few challenged censorship laws: Shillony, B., *Politics and Culture in Wartime Japan*, pp. 12–13.

Violaters arrested: Dower, J., *Japan in War and Peace*, p. 111.

Responses to the war effort: Shillony, B., *Politics and Culture in Wartime Japan*, pp. 71–72, 100, 118.

Heard only good news: Fumie Katayama interview, Hiroshima, 2008; Kiyomi Igura, interview, 2008

Kiyosawa, K., *A Diary of Darkness*, entries 15 December 1943, 6 January 1944, 8 March 1944, 20 April 1944; pp. 154, 176, 249.

Tojo's resignation statement: Shillony, B., *Politics and Culture in Wartime Japan*, p. 67.

Japan in misery: Kiyosawa, K., *A Diary of Darkness*, p. 228, p. 304.

Japanese situation and open surrender: Butow, R., *Japan's Decision to Surrender*, pp. 23–25.

Togo caution: Togo, S., *The Cause of Japan*, pp. 21–21.

Twelve years later: Butow, R., *Japan's Decision to Surrender*, pp. 38–39.

Konoe's verdict: quoted in Hasegawa, T., *Racing the Enemy*, p. 37.

CHAPTER 2 TWO CITIES

Hiraki family: Mitsue Fujii (nee Hiraki), interview, Hiroshima, August 2009.

Great famine of 1934: Ackerman, E., *Japan's Natural Resources*, p. 3.

Pre-war Showa income: Hane, M., *Peasants, Rebels and Outcasts*, p. 31.

Utter destitution: Nagatsuki, T., *The Soil*, p. 47.

Cannibalism and infanticide: Hane, M., *Peasants, Rebels and Outcasts*, pp. 3–4.

A little hope in wartime: Hearn, L., *Lafcadio Hearn's Japan*, p. 68.

Diet enacted the National Mobilisation Law: Shillony, B., *Politics and Culture in Wartime Japan*, p. 2.

Three badges of pride: Shinzo Kiuchi and Motoki Sonoike, 'The Outline History of Hiroshima City', NARA RG 77, Reports Pertaining to the Effects of the Atomic Bomb.

Water metropolis: Ogura, T., *Letters from the End of the World*, p. 37.

Nakajima entertainment district: Hearn, L., *Peasants, Rebels and Outcasts*, p. 72.

Rigid daily routine: Cook, H.T. & Cook, T.F., *Japan at War*, chapter 8.

The fall of average daily ration: Coox, A., *Japan: The Final Agony*, p. 51.

Ministry of Health and Welfare's nutritional standard: Ienaga, S., *The Pacific War*, p. 193.

Severe conditions in Hiroshima: Coox, A., *Japan: The Final Agony*, pp. 51–52; Ienage, S., *The Pacific War*, p. 193; Kiyosawa, K., *A Diary of Darkness*, p. 25.

'Lifestyle reform' policy: The Force that Supports the Home Front (Japanese wartime brochure).

Coupons for *monpe*: Cook, H.T. & Cook, T.F., *Japan at War*, chapter 7.

Hunger conquered female vanity: The Force that Supports the Home Front (Japanese wartime brochure).

Samurai tradition: Jansen, M., *The Making of Modern Japan*, p. 39.

Samurai thrived on war: Jansen, M., *The Making of Modern Japan*, p. 108.

Samurai's end: Turnbull, S.R., *The Samurai*, p. 287.

Exotic Nagasaki: Jansen, M., *The Making of Modern Japan*, pp. 64–65.

The destination of traders and buccaneers: Nagasaki Testimonial Society, 'A Journey to Nagasaki' (brochure), p. 6.

Minminzemi cicada: quoted in Hearn, L., *Lafcadio Hearn's Japan*, p. 79.

Dutch traders: Jansen, M., *The Making of Modern Japan*, pp. 84–85.

Fascination and suspicion of the West: Shinji, T., *Listening to the Wishes of the Dead*, p. 24.

Expelling white foreigners: quoted in Jansen, M., *The Making of Modern Japan*, pp. 266–267.

Challenge, emulate and oppose the West: Macpherson, W.J., *The Economic Development of Japan, c. 1868–1941*, p. 31.

Christianity in Nagasaki: Jansen, M., *The Making of Modern Japan*, pp. 28, 67.

Church property confiscated: http://www.cbcj.catholic.jp/eng/ehistory/table02.htm

Japanese Christians persecuted: Jansen, M., *The Making of Modern Japan*, pp. 67, 77.

Troops storm the Christian ghetto: Nagasaki Testimonial Society, '*A Journey to Nagasaki* (brochure)', p. 5.

Ban on Christianity withdrawn: http://www.cbcj.catholic.jp/eng/ehistory/table02.htm

Hostility to Christianity renewed: Kazuhiro Hamaguchi, interview, Nagasaki, 2009.

Yukio Mishima: Mishima, Y., *Yukio Mishima on Hagakure*, p. 27.

Marriage of Catholicism and Shinto: http://www.cbcj.catholic.jp/eng/ehistory/table02.htm

Japanese priests and the war: http://www.cbcj.catholic.jp/eng/ehistory/table02.htm

Demand for prostitutes' services: Kiyosawa K., *A Diary of Darkness*, p. 202.

Takashi Nagai and his ideals: Shinji, T., *Listening to the Wishes of the Dead*, pp. 24–26.

Nagai, the Moriyamas and Christianity: Glynn, P., *A Song for Nagasaki*, pp. 39, 56.

Nagai baptised: Shinji, T., *Listening to the Wishes of the Dead*, p. 26.

Faith and medical science: Glynn, P., *A Song for Nagasaki*, p. 83.

Value of rice and wheat: Ackerman, E., *Japan's Natural Resources*, p. 85.

Grass, roots and berries: Kazuhiro Hamaguchi, interview, Nagasaki, 2009.

'Child soldier' killed off: Tsuruji Matsuzoe, interview, Nagasaki, 2009.

Mass casualties of air raids: Yamazaki, J. & Fleming, L., *Children of the Atomic Bomb*, p. 1.

Hand of God in Nagasaki's preservation: Teruo Ideguchi, interview, Nagasaki, 2009.

Mitsubishi shipbuilding plant attacked: Shillony, B., *Politics and Culture in Wartime Japan*, p. 139.

Ships constructed in secret: Nagasaki Testimonial Society, 'A Journey to Nagasaki' (brochure), p. 7.

CHAPTER 3 FEUERSTURM

Mitchell and strategic air power: LeMay, C., *Superfortress*, p. 14.

First civilian victims of 'strategic bombing': Tuchman, B., *The Guns of August*, p. 9.

German Zeppelins and Gotha heavy bombers: Pape, R., *Bombing to Win*, p. 59.

Giulio Douhet's: http://en.wikipedia.org/wiki/Giulio_Douhet

Air strategy against civilians: Douhet, G., *The Command of the Air*, pp. 21, 52, 54, 142, 150, 153, 159.

Effectiveness of precision bombing: British Cabinet Papers, PREM 3 – Papers concerning Defence & Operational Subjects, 1940–45.

British government's new policy and RAF's new aims: Pape, R., *Bombing to Win*, p. 261.

Cherwell's obsession with destroying enemy homes: Pape, R., *Bombing to Win*, p. 261. See also Dehousing Memorandum, http://en.wikipedia.org/wiki/Dehousing#Contents_of_the_dehousing_paper

Francis Vivian Drake and area bombing: de Seversky, A., *Victory Through Air Power*, p. 171.

Harris and air raid experiment: Harris, A., *Bomber Offensive*, p. 119; 'Bomber Harris was unfairly blamed for terror raids', *The Independent*, 26 October 1997, p. 9.

Operation Gomorrah: quoted in Rhodes, R., *The Making of the Atomic Bomb*, p. 471.

Feuersturm: quoted in Harris, A., *Bomber Offensive*, p. 174.

'Terror-bombing' and Goebbels: Hessel, P. (2005) *The Mystery of Frankenberg's Canadian Airman*, Toronto: James Lorimer & Co., p. 107, and Kochavi, A. J., (2005) *Confronting Captivity: Britain and the United States and their POWs in Nazi Germany*, Chapel Hill: University of North Carolina Press, p. 172.

Basil Liddell Hart: quoted in Sayle, M., 'Did the Bomb End the War?' in Bird, K. & Lifschultz, L. (eds), *Hiroshima's Shadow*, p. 25.

Inferno in Germany: Harris, A., *Bomber Offensive*, pp. 176, 261.

Military targets and main casualties: de Seversky, A., *Victory Through Air Power*, p. 171.

A raging furnace: quoted in Irving, D., *The Destruction of Dresden*, p. 146.

Kurt Vonnegut, a prisoner of war: quoted in Selden, 'The Logic of Mass Destruction' in Bird, K. & Lifschultz, L. (eds), *Hiroshima's Shadow*, p. 54.

Churchill and Bomber Command: Webster, C. & Frankland, N., *The Strategic Air Offensive Against Germany 1939–1945*, Vol. 3, pp. 112, 117.

US Air Force: Pape, R., *Bombing to Win*, p. 8.

General Spaatz's restraint: Rhodes, R. *The Making of the Atomic Bomb*, p. 310.

US Air Force options: LeMay, C., *Superfortress*, p. 121.

Experimental raid: Roberts, A. & Guelff, R., (eds.), *Documents on the Laws of War*, pp. 122, 126.

Low-level incendiary raids: LeMay, C., *Superfortress*, pp. 48, 122.

Superfortresses and the firestorm: Final Report Covering Air-Raid Protection and Allied Subjects in Japan, No. 11, February 1947, NARA RG243, box 96, p. 231–232.

Mouth of hell, Tokyo ablaze: LeMay, C., *Superfortress*, p. 123.

Smell of human flesh: quoted in Hastings, M., *Nemesis*, p. 320.

Spot fires ignited: Frank, R., *Downfall*, p. 9.

Destroyed property: http://www.cbcj.catholic.jp/eng/ehistory/table02.htm

Air-inflicted body damage: Frank, R., *Downfall*, pp. 18, 234. Final Report Covering Air-Raid Protection and Allied Subjects in Japan, No. 11, February 1947, NARA RG243, box 96.

Greater casualties: Selden, M., 'The Logic of Mass Destruction', in Bird K. & Lifschultz, L. (eds), *Hiroshima's Shadow*, pp. 56–57.

General Arnold praised LeMay: quoted in Frank, R., *Downfall*, p. 67.

With every weapon in LeMay's arsenal: quoted in Hastings, M., *Nemesis*, p. 320; Pape, R., *Bombing to Win*, p. 92.

LeMay's XXI Bomber Command: Final Report Covering Air-Raid Protection and Allied Subjects in Japan, No. 11, February 1947, NARA RG243, box 96.

US pilots dropped pamphlets: Frank, R., *Downfall*, p. 77.

Japanese people ordered not to read: Hastings, M., *Nemesis*, p. 57.

Bureaucratic indifference: Magic Diplomatic Summary, Nos 6, 115.

Euphoric media hailed the incendiary campaign: Boyer, P., *By the Bomb's Early Light*, p. 213.

Some US officials demurred: Malloy, S.L., *Atomic Tragedy*, p. 119.

Barbarity of air campaign: See Hastings, M., *Nemesis*.

US Air Force abandoned pretence: Frank, R., *Downfall*, p. 189.

Terror-bombing could not defeat: de Seversky, A., *Victory Through Air Power*, p. 189.

Civilian areas were 'unprofitable' targets: Blackett, P.M.S., *Studies of War*, p. 7;' see also Blackett, P.M.S., *Fear, War and the Bomb*, pp. 20–24, 194–197 for statistics on bombing in Eurpope.

Memorandum to US Secretary of War: 11 June 1945, Preliminary Review of Effectiveness of the Combined Bomber Offensive in the European Theater of Operations, President's Secretary's File, Truman Library, subject file 1940–53: Cabinet File, box 136.

Destruction of factories, communication posts and transport lines: Harris, A., *Bomber Offensive*, p. 208.

Harris applauded precision bombing of vital facories: Harris, A., *The Organization of Mass Bombing Attacks*, p. 217.

Aircrews had skills and technology: Lectures and Articles on Air Power, in the Strategic Bomber Offensive Against Germany.

Japan's preserved war infrastructure: Summary Report (Pacific War), Washington, DC, 1 July 1946, NARA RG243, box 93; see also Batchelder, R., 'Changing Ethics in the Crucible of War' in Baker, P. (ed.), *The Atomic Bomb*, p. 135.

Insurrection unthinkable: Pape, R., *Bombing to Win*, pp. 25, 331.

De Seversky on air war objective: De Seversky, A., *Victory Through Air Power*, pp. 158–167.

Harris belatedly acknowledged truth: Harris, A., *Bomber Offensive*, p. 78.

Japan refuses to yield: Summary Report (Pacific War), Washington, DC, 1 July 1946, NARA RG243, box 93.

Mystery why Hiroshima and Nagasaki spared: Miyoko Watanabe, interview, Hiroshima, 2009.

LeMay's Tinian Island cable: Correspondence ('Top Secret') of the Manhattan Engineer District, 1942–46, Events Preceding and Following the Dropping of the First Atomic Bombs at Hiroshima and Nagasaki, NARA RG77, entry 1, roll 1, file 5, subfile 5c – Preparation and Movement of Personnel and Equipment to Tinian.

CHAPTER 4 PRESIDENT

Roosevelt ill: Associated Press, 13 April 1945.

Harry Truman sworn in: *Washington Post*, 13 April 1945.

Truman inherited great burden: McCullough, D., *Truman*, Touchstone, p. 354.

Truman sees as a nobody: *New Republic*, April 1945.

Truman out of his depth: McCullough, D., *Truman*, Touchstone, p. 355.

Truman dismissed as a failure: *Time*, quoted in McCullough, D., *Truman*, Touchstone, p. 320.

Roosevelt excluded Vice President: Hasegawa, T., *Racing the Enemy*, pp. 49–50.

Truman's cabinet meeting: *Washington Post*, 13 April 1945.

Stimson reveals secret: Documents Pertaining to the Atomic Bomb, the Atomic Bomb Collection, Truman Library, box 2; see also Miscamble, W., *From Roosevelt to Truman*, p. 91.

Truman's letter to Schwellenbach: quoted in McCullough, D., *Truman*, Touchstone, pp. 289–290.

Stimson blocked investigations: Documents Pertaining to the Atomic Bomb, the Atomic Bomb Collection, Truman Library, box 2; see also McCullough, D., *Truman*, Touchstone, p. 291.

Stimson's patience exhausted: Stimson Papers, Diary, 13 March 1944.

Roosevelt's casket: Associated Press, 14 April 1945.

Roosevelt's post-war world: Miscamble, W., *From Roosevelt to Truman*, p. 85.

Truman's task awed him: quoted in Miscamble, W., *From Roosevelt to Truman*, p. 87.

How could he succeed a man worshipped?: Truman, H., *Memoirs of Harry S. Truman*, pp. 16–17; also quoted in Miscamble, W., *From Roosevelt to Truman*, p. 87.

Day after his swearing in: Hasegawa, T., *Racing the Enemy*, p. 50.

President met challenge head on: Miller, M., *Plain Speaking*, p. 19.

Truman's first Congressional address: McCullough, D., *Truman*, Touchstone, p. 354.

America will continue to fight: General: Memorandum to Historical Files to September–October, 1946, Historical File, 1924–53, Truman Library, box 188.

Stimson's memo: General Documents, Atomic Bomb Collection, Truman Library, Henry Stimson to Harry S. Truman, 24 April 1945.

Control of the weapon: Documents Pertaining to the Atomic Bomb, the Atomic Bomb Collection, Truman Library, box 2.

Details of the operation: quoted in Hasegawa, T., *Racing the Enemy*, p. 67.

Truman's fact-finding mission: Stimson Papers, Diary, 25 April 1945.

Truman's easy style: Ayers Papers, Diary 1941–53, Truman Library, box 19.

Media fascination with Truman: General: Memorandum to Historical Files to September–October, 1946, Historical File, 1924–53, Truman Library, box 188.

Truman read Roosevelt's foreign policy initiatives: Miscamble, W., *From Roosevelt to Truman*, p. 100.

Stalin's letter to Roosevelt: Correspondence, Roosevelt, Stalin, Churchill, Truman, Yalta to Potsdam, 1944–45, Byrnes Papers, box 7, folder 5.

Secret intelligence report: quoted in McCullough, D., *Truman*, Touchstone, p. 372.

Tougher line with Moscow: quoted in Miscamble, W., *From Roosevelt to Truman*, p. 97.

Truman received Molotov: Molotov Conferences – 1945, President's Secretary's File, Historical File, Truman Library. Accounts of the Molotov meeting vary, but this exchange is a generally accepted version.

'Carry out your agreements': Truman, H., *Memoirs of Harry S. Truman*, p. 82.

President wrote of meeting: Molotov Conferences – 1945, President's Secretary's File, Historical File, Truman Library. According to another account, Truman dismissed Molotov more politely, but the effect was the same; see McCullough, D., *Truman*, Touchstone, p. 376.

Truman's tough line: Harriman, A. & Abel, E., *Special Envoy to Churchill and Stalin 1941–1946*, pp. 453–54.

Leahy's assertion: Leahy, W., *I Was There*, p. 353.

Reissue ultimatum to Japan: Germany – Surrender, 8 May 1945, President's Secretary's File, Historical File, Truman Library, box 195.

Terms of surrender changed: Dower, J., *Japan in War and Peace*, p. 342.

Caution against the destruction of Hirohito: Alperovitz, G., *The Decision to Use the Atomic Bomb*, pp. 38, 41.

Who *is* Harry Truman?: See McCullough, D., *Truman*, Touchstone.

Humble farmer from Missouri: Ayers Papers, Diary 1941–53, Truman Library, box 19,16 May 1945.

Common everyday man: quoted in McCullough, D., *Truman*, Touchstone, p. 360.

Truman grew up in harsh times: Miller, M., *Plain Speaking*, p. 360.

Truman overcame deep prejudices: Truman, H.S., *Dear Bess*, p. xi.

Value of people for what they did and said: Miller, M., *Plain Speaking*, p. 128.

Civil liberties policy: quoted in McCullough, D., *Truman*, Touchstone, p. 247.

Truman given almost nothing: quoted in McCullough, D., *Truman*, Touchstone, p. 325.

Ascribed early success to luck: County Judge File: 3 December 1930 to 7 March 1951, President's Secretary's File, Longhand Notes File, 1930–55, Truman Library, box 281.

Nothing of Uriah Heep: Miller, M., *Plain Speaking*, p. 19.

Truman read history and echoed wisdom of historians: Miller, M., *Plain Speaking*, pp. 26-27, 69, 214.

Truman's personal philosophy: Alperovitz, G., *The Decision to Use the Atomic Bomb*, p. 504.

Soviet–American relations: Stalin to Truman, 24 April FRUS 1945, V, pp. 263–264; Miscamble, W., *From Roosevelt to Truman*, p. 125.

The atomic question: Stimson Papers, Diary, 14–15 May 1945.

Meeting with Stalin should be delayed: Joseph Davies Papers.

Harry Hopkins' message: County Judge File: 3 December 1930 to 7 March 1951, President's Secretary's File, Longhand Notes File, 1930–55, Truman Library, box 281.

Hopkins and Stalin: President's Secretary's File, Truman Library, historical file, box 195.

Soviets in negotiable frame of mind: Correspondence, Roosevelt, Stalin, Churchill, Truman, Yalta to Potsdam, 1944–45, Byrnes Papers, box 7, folder 5; see also Sherwin, M., *A World Destroyed*, p. 155.

Truman refused Churchill's private meeting: County Judge File: 3 December 1930 to 7 March 1951, President's Secretary's File, Longhand Notes File, 1930–55, Truman Library, box 281.

Truman urged patience: Agreed Declaration by US, United Kingdom and Canada, Advisory Committee to Atomic Energy, President's Secretary's File, Truman Library, subject file 1940–53: Cabinet File, box 174.

Prospect of the atomic bomb: Miscamble, W., *From Roosevelt to Truman*, p. 127.

CHAPTER 5 ATOM

Truman expected not to meddle: Truman, Truman, H., *Memoirs of Harry S. Truman*, p. 296.

Atoms were not unchanging: Oppenheimer, J.R., *The Constitution of Matter*, pp. 2–3.

Joseph Thomson demonstrates atom: Clark, R.W., *The Greatest Power on Earth*, p. 10.

Radioactive atoms: Bizony, P., *Atom*, p. 15.

Atom particles: Bizony, P., *Atom*, p. xv.

Fundamental composition of atom: Rhodes, R., *The Making of the Atomic Bomb*, p. 49.

Alpha particles: Clark, R.W., *The Greatest Power on Earth*, p. 35.

Ridicule of atomic transmutation: quoted in Clark, R.W., *The Greatest Power on Earth*, p. 11.

World vanish in smoke: Clark, R.W., *The Greatest Power on Earth*, p. 12.

Military application: quoted in Rhodes, R., *The Making of the Atomic Bomb*, p. 44.

Danish physicist Niels Bohr: Clark, R.W., *The Greatest Power on Earth*, p. 12.

Bohr's biographer's first thought: Pais, A., *Niels Bohr's Times, in Physics, Philosophy and Polity*, p. 5.

Deep paradox: Pais, A., *Niels Bohr's Times, in Physics, Philosophy and Polity*, pp. 137–138.

Electron's own orbit: Bizony, P., *Atom*, p. 39.

Light quanta-electrons behaviour: Pais, A., *Niels Bohr's Times, in Physics, Philosophy and Polity*, p. 231.

Definition of scientist: Rhodes, R., *The Making of the Atomic Bomb*, p. 113.

Physicist discuss nature of matter: See Oppenheimer, J.R., *The Constitution of Matter*.

Precise position of particle: Bizony, P., *Atom*, pp. 60–61.

Nature seemed a random event: Clark, R.W., *The Greatest Power on Earth*, p. 27.

Symptoms of compulsive gambler: Bizony, P., *Atom*, p. 61.

Einstein attacked quantum theory: Rhodes, R., *The Making of the Atomic Bomb*, p. 133.

Rutherford's findings: *The Times*, 12 September 1933, 'The British association – breaking down the atom'.

Szilard irritated by Rutherford's dismissal: *Bulletin of the Atomic Scientists*, December 1992, p. 17.

Szilard and *The World Set Free*: See Jewish Virtual Library, and Lanouette, W. & Szilard, B., *Genius in the Shadows*.

Novel anticipated: quoted in Rhodes, R., *The Making of the Atomic Bomb*, p. 24; see also Wells, H.G., *The World Set Free*.

Szilard's thought: quoted in Rhodes, R., *The Making of the Atomic Bomb*, p. 28.

The light turned green: Lanouette, W. & Szilard, B., *Genius in the Shadows*, pp. 133–135.

Disintegrating nuclei: Oppenheimer, J. R., *Letters and Recollections*, p. 159.

Lawrence's atom smasher: quoted in Clark, R.W., *The Birth of the Bomb*, p. 7.

A distant dream: quoted in Rhodes, R., *The Making of the Atomic Bomb*, p. 227.

Szilard visits Einstein: Lanouette, W. & Szilard, B., *Genius in the Shadows*, p. 199.

Germans were further advanced: Clark, R.W., *The Birth of the Bomb*, p. 7.

Rutherford's alchemic predictions: *Naturwissenschaften*, 6 January 1939.

Fission process: *Nature*, 11 February 1939.

Spectacular little explosion: quoted in Clark, R.W., *The Birth of the Bomb*, pp. 14–15, 46.

Physicist urged Washington to act: Williams, R.C. & Cantelon, P.L. (eds), *The American Atom*, pp. 12–14.

CHAPTER 6 THE MANHATTAN PROJECT

Practical means of making an atomic bomb: Margaret Gowing, British official historian, quoted in Bundy, M., *Danger and Survival*, p. 24.

Uranium isotope and radiation: Frisch–Peierls Memorandum, quoted in 'On the Construction of a Super–Bomb' in Williams, R.C. & Cantelon, P.L. (eds), *The American Atom*, pp. 14, 15.

Extraction of fissile uranium: Maud Report, quoted in 'On the Construction of a Super–Bomb', in Williams, R.C. & Cantelon, P.L. (eds), *The American Atom*, p. 22.

Nuclear blast force: Maud Report, quoted in Hershberg, J., *James B. Conant*, p. 149.

Cheaper than conventional explosives: Williams, R.C. & Cantelon, P.L. (eds), *The American Atom*, p. 22.

Roosevelt approval development of bomb: Correspondence ('Top Secret') of the Manhattan Engineer District, 1942–46, NARA RG77, microfilm publication 1109, roll 3: entry 1, file 25D – 'Top Secret' Corespondence, subseries 1.

Briggs and Maud Report: Kelly, C. (ed.), *The Manhattan Project*, p. 60; see also Cockburn, S. & Ellyard, D., *Oliphant*.

The fate of Maud: Kelly, C. (ed.), *The Manhattan Project*, p. 62.

Oliphant's persuasiveness: Rhodes, R., *The Making of the Atomic Bomb*, p. 372.

Development of S–1: Correspondence ('Top Secret') of the Manhattan Engineer District, 1942–46, NARA RG77, microfilm publication 1109, roll 3: entry 1, file 25D — 'Top Secret' Corespondence, subseries 1.

Compton defined science: Groueff, S., *Manhattan Project*, pp. 33–34.

Groves, L., *Now It Can Be Told*, p. 4; see also Groueff, S., *Manhattan Project*, p. 12.

Groves' self worth: quoted in Kelly, C. (ed.), The Manhattan Project, p. 119.

Portrait of Groves: Lawren, W., *The General and the Bomb*, pp. 43, 46, 62; see also Norris, R.S., *Racing for the Bomb*.

A young captain: See Norris, R. S., *Racing for the Bomb*; also Kelly, *The Manhattan Project*, p. 118.

Cost contract: Correspondence ('Top Secret') of the Manhattan Engineer District, 1942–46, NARA RG77, entry 1, roll 1, file 2: Production, Operations, Raw Materials and Construction.

Groves' security clearance: Correspondence ('Top Secret') of the Manhattan Engineer District, 1942–46, NARA RG77, entry 1, roll 1, file 5, subfile 5b – Directives, Memorandums etc. to and from the Chief of Staff, Secretary of War.

King wrote to Nimitz: C., Kelly, *The Manhattan Project*, pp. 322–323.

Russia, a future enemy: quoted in Sherwin, M., *A World Destroyed*, p. 62.

Byrnes' review: Correspondence ('Top Secret') of the Manhattan Engineer District, 1942–46, NARA RG77, microfilm publication 1109, roll 3: entry 1, file 20 – Miscellaneous.

General Somervell joked: Groves, L., *Now It Can Be Told*, pp. 70, 102.

Groves and 'Oppie': Groves, L., *Now It Can Be Told*, pp. 70, 102, 63.

Oppenheimer, a clever young man: Oppenheimer, J. R., *Letters and Recollections*, pp. 28–29.

Young Oppenheimer and Proust: quoted in Bird, K. & Sherwin, M., *American Prometheus*, p. 51.

Oppenheimer's intellectual talents and depression: Oppenheimer, J. R., *Letters and Recollections*, pp. 41, 70, 74, 94, 165.

Bohr, a security risk: Kelly, C. (ed.), *The Manhattan Project*, p. 102.

Fermi and 'atomic pile' at Stagg Field: Fermi's Own Story, quoted in Kelly, C. (ed.), *The Manhattan Project*, pp. 82, 84; see also Fermi, L., *Atoms in the Family*, pp. 197–198; and Segrè, E., *Enrico Fermi, Physicist*.

Compton call to Conant: The Manhattan Project: An Interactive History. US Department of Energy: http://en.wikipedia.org/wiki/US_Department_of_Energy.

Oppenheimer charmed staff: Conant, J., *109 East Palace*: quoted in Kelly, C. (ed.), *The Manhattan Project*, p. 134.

Groves' and Oppenheimer's ambition: Norris, R.S., *Racing for the Bomb*; quoted in Kelly, C. (ed.), *The Manhattan Project*, p. 139.

The Los Alamos Primer: Rhodes, R., *The Making of the Atomic Bomb*, p. 460.

Bomb's process and risks: Serber, R., *The Los Alamos Primer*, pp. 27, 33–35.

Oak Ridge, secret city: Italy: General to Korea, Japan, President's Secretary's File, Truman Library, subject file 1940–53, Cabinet File: box 158.

Groves' argument: Groueff, S., *Manhattan Project*, p. 66.

Two conditions and construction of Hanford: Internal memo, To All Employees of E. I. DuPont de Nemours & Company, Delaware, 24 August 1945.

Effects of radiation: Groueff, S., *Manhattan Project*, p. 152–153.

Dangers to humans: Groves, L., *Now It Can Be Told*, p. 87.

Electrical conductors: Groves, L., *Now It Can Be Told*, p. 107.

Conant on Oak Ridge: *New York Times*, 29 September 1945, p. 6.

Celebrity scientists use aliases: Fermi, L., *Atoms in the Family*, quoted in Kelly, C. (ed.), *The Manhattan Project*, p. 240.

Codenames and forbidden words: Marshak, R., *Secret City*, quoted in Kelly, C. (ed.), *The Manhattan Project*, p. 167.

Long-term assignments: Lawren, W., *The General and the Bomb*, p. 142.

Screened and cordoned-off employees: Groves, L., *Now It Can Be Told*, p.102.

Dutiful employees: Italy: General to Korea, Japan, President's Secretary's File, Truman Library, subject file 1940–53, Cabinet File: box 158.

Separating uranium isotopes: quoted in Kelly, C. (ed.), *The Manhattan Project*, p. 206.

Unions were banned: Groves, L., *Now It Can Be Told*, p. 100.

Fuchs came highly recommended: http://www.lanl.gov/history/atomicbomb/trinity.shtml, Klaus Fuchs' Questionnaire Form

Fuchs' research: See Holloway, D., *Stalin and the Bomb*; see also Kelly, C. (ed.), *The Manhattan Project*, p. 262.

Fuchs informed Moscow: Teller, E., *The Legacy of Hiroshima*, pp. 27–28; See also Teller, E. & Shoolery, J., *Memoirs*.

Fears of Soviet penetration: Conant, J., *109 East Palace*, p. 160.

Bush and Conant warn Stimson: Memorandum from Vannevar Bush and James B. Conant, Office of Scientific Research and Development, to Secretary of War, 30 September 1944, Background on the Atomic Project, NARA; document 1; Records of the Army Corps of Engineers, Manhattan Engineer District, NARA RG77, Harrison–Bundy Files (H–B Files), folder 69.

CHAPTER 7 SPRING 1945

Playground of the poor: Hearn, L., *Lafcadio Hearn's Japan*, p. 76.

Strange dread: Kohji Hosokawa, interview, Hiroshima, 2009.

Japan's divine destiny: *Keys to the Japanese Heart and Soul*, p. 71.

Treatment of Shizue Hiraki by Zenchiki Hiraki: Mitsue Fujii (Hiraki), interview, Hiroshima, 2009.

'Children's Battalion': Hane, M., *Peasants, Rebels and Outcasts*, p. 81.

Shizue's work: The Force that Supports the Home Front (Japanese wartime brochure).

Orders in Emperor's name: Iwao Nakanishi, interview, Hiroshima, 2009.

Food shortage: The Effects of Strategic Bombing on Japanese Morale, Morale Division, June 1947, NARA RG243; p.17.

Rice harvest: Coox, A., *Japan: The Final Agony*, p. 52.

Food production and consumption: Ackerman, E., *Japan's Natural Resources*, p. 138.

Potato staple and meagre rations: Nagasaki woman, interviewed by Yuta Hiramatsu, 2009

Commodities exhausted: Coox, A., *Japan: The Final Agony*, p. 54.

1945 blockade: Ackerman, E., *Japan's Natural Resources*, p. 218.

Wood replaced metal and plastics: Coox, A., *Japan: The Final Agony*, p. 55.

Coal production: Ackerman, E., *Japan's Natural Resources*, p. 210.

Manpower non-existent: Nagasaki woman, interviewed by Yuta Hiramatsu, 2009.

Scheme for Strengthening Domestic Preparedness: Introduced on 22 September 1944, The Force that Supports the Home Front (Japanese wartime brochure).

Women and labour drive: The Force that Supports the Home Front (Japanese wartime brochure).

Munitions factories: The Japanese Wartime Standard of Living And Utilization of Manpower; Manpower, Food And Civilian Supplies Division, January 1947, NARA RG243, box 106. Compulsory mobilisation of women under *Joshi Kinro Teishin Tai*, p. 75.

National Mobilisation Law's new ruling: Kohji Hosokawa, interview, Hiroshima, 2009.

Schools converted into munitions factories: The Force that Supports the Home Front (Japanese wartime brochure).

Pine oil collected: Coox, A., *Japan: The Final Agony*, p. 54.

Pine oil lauded as saviour: The Force that Supports the Home Front (Japanese wartime brochure).

Yoko Moriwaki's Diary: Hosokawa, H. & Kamei, H., *The Diary of Yoko Moriwaki*.

Shoso's evacuation: Shoso Kawamoto, interview, Hiroshima, 2009.

Nagasaki's rural shelter limited: Field Report Covering Air-Raid Protection and Allied Subjects in Nagasaki, Japan; Civilian Defense Division, March 1947, NARA RG243, box 94.

Teenage labour: Tsuruji Matsuzoe, interview, Nagasaki, 2009.

Kiyoko making torpedoes: Kiyoko Mori, interview, Nagasaki, 2009.

Volunteer Neighbourhood Associations: The Japanese Wartime Standard of Living And Utilization of Manpower; Manpower, Food And Civilian Supplies Division, January 1947, NARA RG243, box 106, Compulsory mobilisation of women under *Joshi Kinro Teishin Tai*, p. 75.

Fire-fighting facilities in pathetic state: Civilian Defense Report No. 1, Hiroshima, Japan Field Report,15 November 1945, NARA RG243, box 95. Yamano Yukio, interview, then acting chief of Hiroshima's West Side Fire Department.

Sand balls: Civilian Defense Report No. 1, Hiroshima, Japan Field Report,15 November 1945, NARA RG243, box 95, pp. 22–23.

Hiroshima and Nagasaki's siren systems: Field Report Covering Air-Raid Protection and Allied Subjects in Nagasaki, Japan; Civilian Defense Division, March 1947, NARA RG243, box 94; as atomic bomb survivors later told the USSB and author.

Nagasaki's shallow trenches: Field Report Covering Air-Raid Protection and Allied Subjects in Nagasaki, Japan; Civilian Defense Division, March 1947, NARA RG243, box 94.

Ineffectual Japanese rescue services: Final Report Covering Air-Raid Protection and Allied Subjects in Japan, No. 11, February 1947, NARA RG243, box 96.

Electronically guided B-29s: Final Report Covering Air-Raid Protection and Allied Subjects in Japan, No. 11, February 1947, NARA RG243, box 96.

Weight of the cities' defence: Final Report Covering Air-Raid Protection and Allied Subjects in Japan, No. 11, February 1947, NARA RG243, box 96.

Second General Army headquarters: Frank, R., *Downfall*, p. 165.

CHAPTER 8 THE TARGET COMMITTEE

Target Committee: The Decision to Drop the Atomic Bomb on Japan, Truman Library; Harry S. Truman Student Research File (b file).

Scientists unaware decision to use bomb predated Germany's surrender: Malloy, S. L., *Atomic Tragedy*, p. 57.

Choice of local targets: http://www.lanl.gov/history/atomicbomb/victory.shtml; see also Groves, quoted in Frank, R., *Downfall*, p. 254.

Debate of the details: Notes on Initial Meeting of Target Committee, 2 May 1945, Top Secret, MED Records, Top Secret Documents, NARA RG77, file no. 5d (copy from microfilm), document 4.

Oppenheimer ran through agenda: Correspondence ('Top Secret') of the Manhattan Engineer District, 1942–46, NARA RG77, entry 1, roll 1, file 5, subfile 5c – Preparation and Movement of Personnel and Equipment to Tinian.

Cities' military attributes: Atomic Bomb – War Department, Memo on Hiroshima as 'Army City', President's Secretary's File, historical file, Truman Library, box 193.

Bomb should be dropped on large urban centre: The Decision to Drop the Atomic Bomb on Japan, Truman Library; Harry S. Truman Student Research File (b file).

Likely effects of radiation: Memorandum from J.R. Oppenheimer to Brigadier General Farrell, 11 May 1945, MED Records, Top Secret Documents, NARA RG77, file no. 5g (copy from microfilm), document 5.

Captain Deak Parsons' reason: quoted in Malloy, S.L., *Atomic Tragedy*, p. 61.

Target must be a city centre: Bundy, M., *Danger and Survival*, p. 67.

Parsons rejected noncombat demonstration: quoted in Malloy, S.L., *Atomic Tragedy*, p. 61.

No doubt where first atomic bomb would fall: Documents Pertaining to the Atomic Bomb, the Atomic Bomb Collection, Truman Library, box 2.

Kyoto's attributes: Correspondence ('Top Secret') of the Manhattan Engineer District, 1942–46, NARA RG77, microfilm publication 1109, roll 3: entry 1, file 25D – 'Top Secret' Correspondence, subseries I.

Stimson order Kyoto's removal from list: Atomic Energy Commission, 'Mr Stimson's "Pet City" – The Sparing of Kyoto', by Otis Cary, Papers of R. Gordon Arneson, Truman Library, box 1.

Stimson adamant with Groves: quoted in Sherwin, M., *A World Destroyed*, p. 230.

Groves' proprietorial control: See Groves, L., *Now It Can Be Told*.

Kyoto preserved from conventional attack: Correspondence ('Top Secret') of the Manhattan Engineer District, 1942–46, NARA RG77, entry 1, roll 1, file 5, subfile 5b – Directives, Memorandums etc. to and from the Chief of Staff, Secretary of War etc.

Interim Committee's members: The Decision to Drop the Atomic Bomb on Japan, Notes of the Interim Committee Meeting, Thursday, 31 May1945, MED Records, NARA RG77, H– files, box 1, folder 100.

Oswald C. Brewster's letter: Documents Pertaining to the Atomic Bomb, the Atomic Bomb Collection, Truman Library, box 2; see also: The Decision to Drop the Atomic Bomb on Japan, Notes of the Interim Committee Meeting, Thursday, 31 May 1945, MED Records, NARA RG77, H–B files, box 1, folder 100.

Stimson's portentous note: Documents Pertaining to the Atomic Bomb, the Atomic Bomb Collection, Truman Library, box 2.

Bomb's potential: Rhodes, R., *The Making of the Atomic Bomb*, p. 643.

Ernest Lawrence's idea killed: Baker, P.R., (ed.), *The Atomic Bomb*, p. 19.

Stimson's military damage objective: Truman, H., *Memoirs of Harry S. Truman*, p. 297.

Kyoto must not be bombed: Documents Pertaining to the Atomic Bomb, the Atomic Bomb Collection, Truman Library, box 2.

Total war had debased everyone: Bernstein, B., *The Atomic Bomb*, p. 146.

Recommended use of gas: Assistant Secretary of War John J. McCloy, Memorandum of Conversation with General Marshall 29 May 1945, Office of the Secretary of War, Formerly Top Secret Correspondence of Secretary of War Stimson ('Safe File'), July 1940–September 1945, NARA RG107, box 12, S-1.

Idea of bomb tormented Stimson: quoted in Malloy, S. L., *Atomic Tragedy*, p. 64.

Most desirable target: The Decision to Drop the Atomic Bomb on Japan, Notes of the Interim Committee Meeting, Thursday, 31 May 1945, MED Records, NARA RG77, H–B files, box 1, folder 100.

Committee agreed on atomic bomb use: The Decision to Drop the Atomic Bomb on Japan, Notes of the Interim Committee Meeting, Thursday, 31 May 1945, MED Records, NARA RG77, H–B files, box 1, folder 100.

Most Japanese cities: quoted in Hasegawa, T., *Racing the Enemy*, pp. 90–91.

Report to soothe dissent: Documents Pertaining to the Atomic Bomb, the Atomic Bomb Collection, Truman Library, box 2.

President Truman's important speech: The Decision to Drop the Atomic Bomb on Japan, Truman Library; Harry S. Truman Student Research File (b file).

Scientist reported to Washington: Documents Pertaining to the Atomic Bomb, the Atomic Bomb Collection, Truman Library, box 2.

'Immediate use' of bomb recommended: The Decision to Drop the Atomic Bomb on Japan, Truman Library; Harry S. Truman Student Research File (b file).

Truman's decision critically influenced by scientists: Truman, H., *Memoirs of Harry S. Truman*, p. 296.

Dissenting scientists and the Franck report: Documents Pertaining to the Atomic Bomb, the Atomic Bomb Collection, Truman Library, box 2.

Gnawing of self-doubt felt: Documents Pertaining to the Atomic Bomb, the Atomic Bomb Collection, Truman Library, box 2.

The weapon would help contain Russia: Lanouette, W., 'Three Attempts to Stop the Bomb', in Bird, K. & Lifschultz, L. (eds), *Hiroshima's Shadow*, p. 104.

CHAPTER 9 JAPAN DEFEATED

Kamikaze airman's mission: Takehiko Ena, personal statement, circa end 1945.

Takehiko Ena's 'last meal': Takehiko Ena, interview, August, 2009.

Farewell poem to the *Kokutai:* Hiroshima City Archives, *Hiroshima Konjaku:* '80 *Hiroshima-shi Seirei Shitei Toshi Kinen* [Hiroshima Today and Long Ago: In Commemoration of Hiroshima Becoming a Government Ordinance Designated City in 1980] pp. 27–28. This is a propaganda poster that exalts the deaths of persons in the military forces (1944). Used in the poster are the words of the farewell poem recited by the kamikaze.

Ena's crew: Summary Report (Pacific War), Washington, DC, 1 July 1946, NARA RG243, box 93.

Engine failure: Ena Takehiko, personal statement, circa end 1945.

The Ryukus and Okinawas: Dept of State Bulletin xii, April 1945, Brown Papers, box 50, folder 2.

Okinawa's relations with mainland Japanese: Dept of State Bulletin xii, April 1945, Brown Papers, box 50, folder 2.

Stench of human detritis: See Hastings, M., *Nemesis*, pp. 414–416.

Battles at Okinawa: See Hastings, M., *Nemesis* and Frank, R., *Downfall* on Okinawa.

Estimates range from 30,000 to 160,000 civilian dead: See Saburo Ienaga, *The Pacific War*, p. 199.

Okinawa people longed to surrender: See Hastings, M., *Nemesis*, pp. 414–416.

Land invasion: quoted in Frank, R., *Downfall*, p. 132; see also County Judge File: 3 December 1930 to 7 March 1951, President's Secretary's File, Longhand Notes File, 1930–55, Truman Library, box 281.

Japan had lost the air war: Potsdam, Germany Trip, *Augusta Press*, July 1945, President's Secretary's File, Truman Library, subject file 1940–53, Cabinet File, box 140.

Japan's aircraft after Okinawa: Francillon, R., *Japanese Aircraft of the Pacific War*, pp. 15, 36, 93.

Japan had lost the sea war: Butow, R., *Japan's Decision to Surrender*, p. 41.

Warships: Summary Report (Pacific War), Washington, DC, 1 July 1946, NARA RG243, box 93.

Japanese ground forces defeated: Magic Diplomatic Summary, 6 June 1945.

Homeland reduced to virtual standstill: Frank, R., *Downfall*, p. 80.

Hiroshima's port ceased operation: Summary Report (Pacific War), Washington, DC, 1 July 1946, NARA RG243, box 93.

Military reinforcements from China cut off: Butow, R., *Japan's Decision to Surrender*, p. 94.

Severance of communication: Magic Diplomatic Summary, 1210, 17 July 1945.

Blockade critical contribution to defeat of Japan: Hastings, M., *Nemesis*, p. 289.

Merchant shipping lost: Frank, R., *Downfall*, pp. 78, 156.

US submarines and Japanese merchant navy: Hastings, M., *Nemesis*, pp. 288, 290, 302. See also Summary Report (Pacific War), Washington, DC, 1 July 1946, NARA RG243, box 93.

Japan's economic defeat: Butow, R., *Japan's Decision to Surrender*, p. 11.

Desperate acts of the kamikaze: Summary Report (Pacific War), Washington, DC, 1 July 1946, NARA RG243, box 93.

D.C. Ramsey on Japanese aircraft: County Judge File: 3 December 1930 to 7 March 1951, President's Secretary's File, Longhand Notes File, 1930–55, Truman Library, box 281.

Allies concentration of troops, ships and planes: 3rd Report to the President, the Senate and House of Representatives, by the Director of War Mobilization and Reconversion, 1 July 1945, Japan the Road to Tokyo and Beyond, President's Secretary's File, Truman Library, box 158.

US bombers: Hastings, M., *Nemesis*, p. 339.

General LeMay's crews and raids: Frank, R., *Downfall*, p. 77.

Firebombing wiped out: Summary Report (Pacific War), Washington, DC, 1 July 1946, NARA RG243, box 93; see also Ackerman, E., *Japan's Natural Resources*.

US Navy multiplied: 3rd Report to the President, the Senate and House of Representatives, by the Director of War Mobilization and Reconversion, 1 July 1945, Japan the Road to Tokyo and Beyond, President's Secretary's File, Truman Library, box 158.

Halsey's precision raids: Frank, R., *Downfall*, p. 157.

Russia's abrogation of the Neutrality Pact: Magic Diplomatic Summary, 25, 28 May 1945.

Old samurais' resources: Frank, R., *Downfall*, p. 184.

Japan possessed aviation fuel for a few thousand planes: The Decision to Drop the Atomic Bomb on Japan, Truman Library; Harry S. Truman Student Research File (b file).

No reinforcements forthcoming: Drea, E.J., *MacArthur's ULTRA*; see Chapter 8.

Tokyo knew nation was defeated: Hayashi, S., *Kogun*, pp. 176–182.

Fight to the last man: 'The Little White House' (pamplet).

Meeting of the Supreme Council for the Direction of War: Butow, R., *Japan's Decision to Surrender*, pp. 80, 96–97.

Suzuki played the militarist's line: Feis, H., *The Atomic Bomb and the End of World War II*, p. 183.

Suzuki's views accepted: Butow, R., *Japan's Decision to Surrender*, p. 102.

Counterplan to end death wish: Butow, R., *Japan's Decision to Surrender*, pp. 47, 49, 84, 114.

Signals intelligence: Van Der Rhoer, E., *Deadly Magic*, pp. 12–13, 185 (Marshall letter).

Tokyo's peace talks interpreted as psychological warfare: Magic Diplomatic Summary, undated, April 1945, and 25 June 1945.

Bagge's discussions with Togo: Butow, R., *Japan's Decision to Surrender*, pp. 55–57.

Magic Diplomatic Summary, 11 June 1945.

Sato struggled to convey dark reality: Magic Diplomatic Summaries: 14, 26, June 1945, 3, 7, 9, 10, 12, 13, 15, 16 July 1945.

Good offices of Soviet Union: quoted in Butow, R., *Japan's Decision to Surrender*, p. 130.

Emperor's intervention astonished American intelligence: See Potsdam Papers, NARA, quoted in Alperovitz, G., *The Decision to Use the Atomic Bomb*, p. 238.

General Weckerling's interpretations: John Weckerling, Deputy Assistant Chief of Staff, G-2, to Deputy Chief of Staff, 'Japanese Peace Offer', 13 July 1945, Ultra NARA RG165, Army Operations OPD Executive file 17, item 13.

CHAPTER 10 UNCONDITIONAL SURRENDER

Moderates' terms for Japanese surrender: See Hasegawa, T., *Racing the Enemy*, pp. 51–53.

Truman pressed to ease terms: Leahy, W., *I Was There*, pp. 384–385.

Poll 1945: Sigal, L., *Fighting to a Finish*, p. 95.

Washington Post: quoted in Bird, K. & Lifschultz, L. (eds), *Hiroshima's Shadow*, p. 13.

A harmless compromise: quoted in Hasegawa, T., *Racing the Enemy*, p. 53.

Military reason for retaining Emperor: Frank, R., *Downfall*, p. 219.

Truman's regard for MacArthur: County Judge File: 3 December 1930 to 7 March 1951, President's Secretary's File, Longhand Notes File, 1930–55, Truman Library, box 281.

Admirals King and Leahy: Skates, J.R., *The Invasion of Japan*, pp. 18, 25.

Nimitz told King: Hastings, M., *Nemesis*, p. 481.

Invasion circumstances: Minutes of Meeting Held at the White House on Monday, 18 June 1945 at 1530, Records of the Joint Chiefs of Staff, Central Decimal Files, 1942–45, NARA RG218, box 198. This source for all minutes quoted in this chapter.

Estimates debate: Truman, H., *Memoirs of Harry S. Truman*.

Kyushu 'different' from Okinawa: quoted in Frank, R., *Downfall*, p. 141.

Leahy challenged the formula: Minutes of Meeting Held at the White House on Monday, 18 June 1945 at 1530, Records of the Joint Chiefs of Staff, Central Decimal Files, 1942–45, NARA RG218, box 198.

Invasion plan fading: Hastings, M., *Nemesis*, p. 498.

Higher casualties: Feis, H., *The Atomic Bomb and the End of World War II*, p. 9.

Invasion plan costly and unnecessary: there is solid consensus for this view, from various sides of the debate; see Hastings, M., *Nemesis*, Hasegawa, T., *Racing the Enemy*, Skates, J.R., *The Invasion of Japan*, p. 256.

McCloy spoke: quoted in Alperovitz, G., *The Decision to Use the Atomic Bomb*, pp. 68, 73.

The weapon nobody dared name: Minutes of Meeting Held at the White House on Monday, 18 June 1945 at 1530, Records of the Joint Chiefs of Staff, Central Decimal Files, 1942–45, NARA RG218, box 198. The bomb is probably referred to as 'certain other matters' in the minutes.

Political solution: McCullough, D., *Truman*, Touchstone, p. 401.

McCloy suggests bomb ultimatum: There are various versions of his actual words at the meeting; quoted in Alperovitz, G., *The Decision to Use the Atomic Bomb*, pp. 73, 503; see also Hasegawa, T., *Racing the Enemy*, p. 105.

Japan's military capability, July 1945: The Decision to Drop the Atomic Bomb on Japan, Truman Library; Harry S. Truman Student Research File (b file).

Qualifications according to US Sixth Army estimates: quoted in Giangreco, D.M., *Hell to Pay*, pp. 205–210; see also G-2 Estimate of Enemy Situation with Respect to Kyushu, US Sixth Army, 1 August 1945.

Soviet Union wild card: The Decision to Drop the Atomic Bomb on Japan, Truman Library; Harry S. Truman Student Research File (b file).

Invasion, a supporting weapon: Skates, J.R., *The Invasion of Japan*, p. 243.

Developments led to setting aside Olympic: See Frank, R., *Downfall*, pp. 211–213.

Stimson's private talk: Stimson Papers, Diary.

Stimson expressed his abhorrence: Stimson Papers, Diary.

Grew advised America should clarify 'unconditional surrender': Documents Pertaining to the Atomic Bomb, the Atomic Bomb Collection, Truman Library, box 2.

Preservation of the throne: quoted in Hasegawa, T., *Racing the Enemy*, p. 98.

Intelligence committee lent weight to deliberations: The Decision to Drop the Atomic Bomb on Japan, Truman Library; Harry S. Truman Student Research File (b file).

Bard's belief: Memorandum from George L. Harrison to Secretary of War, 28 June 1945, Top Secret, enclosing Ralph Bard 'Memorandum on the Use of S-1 Bomb', 27 June 1945, NARA, RG77, document 23.

Stimson's protest to President: Documents Pertaining to the Atomic Bomb, the Atomic Bomb Collection, Truman Library, box 2.

Potsdam Declaration: Truman, H., *Memoirs of Harry S. Truman*, see chapter 26.

Japan to retain Hirohito as powerless head of state: Documents Pertaining to the Atomic Bomb, the Atomic Bomb Collection, Truman Library, box 2; see also Hasegawa, T., *Racing the Enemy*, pp. 117–118.

State department refused to entertain ideas about retaining Emperor: Brown Papers, Diary, Brown Papers, see also Robertson, D., *Sly and Able*, p. 417.

State department news: Secretary's Staff Meetings Minutes, 1944–47, Saturday Morning, 7 July 1945, NARA RG353.

America freed herself: Memorandum from R. Gordon Arneson, Interim Committee Secretary, to Mr Harrison, 25 June 1945, MED Records Nara RG77, see also Documents Pertaining to the Atomic Bomb, the Atomic Bomb Collection, Truman Library, box 2.

Grew and Hirohito's fate: Potsdam: Agenda and Documents, July–August 1945, Byrnes Papers, box 1, folder 10.

Split in Washington: Assistant Secretary of War John J. McCloy to Colonel Stimson, 29 June 1945, Top Secret, Office of the Secretary of War, Formerly Top Secret Correspondence of Secretary of War Stimson ('Safe File'), July 1940–September 1945, RG107, box 8, Japan (after 7 December 1941).

Byrnes sailed east: Japanese Surrender – 14 August 1945, President's Secretary's File, Truman Library, historical file, box 195.

CHAPTER 11 TRINITY

USS *Augusta*: Potsdam, Germany Trip, President's Secretary's File, Truman Library, subject file 1940–53, Cabinet File, box 140.

Conference postponed: Sherwin, M., *A World Destroyed*, pp. 191, 193; Conant, J., *109 East Palace*, p. 231.

Truman insisted on later date: Truman, M.S., *Harry S. Truman*, p. 260.

Stimson shared Byrnes' faith: Sherwin, M., *A World Destroyed*, pp. 193–194; Stimson Papers, Diary.

Conniving Secretary of State: County Judge File: 3 December 1930 to 7 March 1951, President's Secretary's File, Longhand Notes File, 1930–55, Truman Library, box 281.

Truman's priority: *Chicago Daily Tribune*, 9 August 1945.

Russian support would hasten victory: Miscamble, W., *From Roosevelt to Truman*, pp. 188–189.

Possibility of 'constitutional monarchy': See Potsdam Papers in Feis, H., *The Atomic Bomb and the End of World War II*.

Soviet Union as a signatory: Byrnes, J., *All in One Lifetime*, p. 296; quoted in Malloy, S. L., *Atomic Tragedy*, p. 123.

Cordell Hull conversation: Brown Papers, Diary.

Brynes feared Red Army: Byrnes, J., *Speaking Frankly*, p. 208.

Stalin saw accord with China: Feis, H., *The Atomic Bomb and the End of World War II*, p. 72.

Brynes and Truman hope: Byrnes, J., *All in One Lifetime*, p. 298.

President ate lunch with sailors: President's Trip to Berlin Conference, Byrnes Papers, box 16, folder 12.

Red Star: Potsdam, Germany Trip, *Augusta Press*, July 1945, President's Secretary's File, Truman Library, subject file 1940–53, Cabinet File, box 140.

Truman read in the *Augusta Press*: *Augusta Press*, 8–12 July.

'Little White House': The Decision to Drop the Atomic Bomb on Japan, Truman Library, Harry S. Truman Student Research File (b file).

Soviet hosts supplied German furniture: *New York Times*, 14–15 July 1945.

Advised not to discuss confidential matters: The Decision to Drop the Atomic Bomb on Japan, Truman Library, Harry S. Truman Student Research File (b file).

Hidden microphones: Hasegawa, T., *Racing the Enemy*, p. 132.

Hint of crime: 'The Little White House' (pamphlet).

Churchill called on Truman: Truman at Potsdam – His Secret Diary: Notes by Harry S. Truman on the Potsdam Conference, 16 July 1945, Truman Library.

Churchill praised Truman: Churchill, W., *The Second World War*, Vol. 6. *Triumph and Tragedy*, p. 630.

Ruined city: Truman at Potsdam – His Secret Diary: Notes by Harry S. Truman on the Potsdam Conference, 16 July 1945, Truman Library.

German men nowhere to be seen: Brown Papers, Diary.

President sank into his Papers: Hasegawa, T., *Racing the Enemy*, p. 133.

Japan's 'peace proposal': Van Der Rhoer, E., *Deadly Magic*, p. 181.

Words of the Emperor: Alperovitz, G., *The Decision to Use the Atomic Bomb*, p. 233.

'Do not watch for the flash directly': Laurence, W., 'Part 1 – Drama of the Atomic Bomb Found Climax in July 16 Test', *New York Times*, 26 September 1945.

Probable brilliance of the explosion: Correspondence of the Manhattan Engineer District, 1942–46, NARA RG77, entry 1, roll 1, file 4: Trinity Test at Alamogordo 16 July 1945.

Plutonium weapon: http://www.lanl.gov/history/atomicbomb/trinity.shtml; see Rhodes, R., *The Making of the Atomic Bomb*, for full description of the plutonium bomb assembly.

If the reaction failed: Plaque at Trinity site.

Inclement conditions risk: Groves, L., *Now It Can Be Told*, pp. 291–292.

General refused to postpone test: quoted in Alperovitz, G., *The Decision to Use the Atomic Bomb*, p. 148.

Cars and lead-lined trucks: Memo, Colonel Stafford Warren to Groves, 21 July 1945, Correspondence of the Manhattan Engineer District, 1942–46, NARA RG77, entry 1, roll 1, file 4: Trinity Test at Alamogordo 16 July 1945

Mood wavered: Kelly, C. (ed.), *The Manhattan Project*, pp. 294–295.

If the count reached zero: Groves, L., *Now It Can Be Told*, p. 296.

Bets on the force of the blast: Rhodes, R., *The Making of the Atomic Bomb*, pp. 656, 664; Clark, R.W., *The Greatest Power on Earth*, pp. 197–198.

As the deadline approached: Kelly, C. (ed.), *The Manhattan Project*, pp. 294–295.

Countdown continued: Walker, S., *Shockwave*, pp. 61–62.

Billions of neutrons: Rhodes, R., *The Making of the Atomic Bomb*, pp. 670, 673.

Conant witnessed the hills: Conant, J., *109 East Palace*, pp. 231–233.

Nuclear dawn visible in Santa Fe: Sherwin, M., *A World Destroyed*, p. 223.

'What was that?': Kelly, C. (ed.), *The Manhattan Project*, p. 310.

Centre of the fireball: quoted in Rhodes, R., *The Making of the Atomic Bomb*, p. 672.

The purple afterglow and everyone's response: Correspondence of the Manhattan Engineer District, 1942–46, NARA RG77, entry 1, roll 1, file 4: Trinity Test at Alamogordo 16 July 1945, Thought of E.O. Lawrence.

Professor Kistiakowsky and Chadwick observe: Clark, R.W., *The Greatest Power on Earth*, p. 199.

Some felt personally affronted: quoted in Rhodes, R., *The Making of the Atomic Bomb*, p. 672.

Others felt touched by the divine: Laurence, W., 'Part 1 – Drama of the Atomic Bomb Found Climax in July 16 Test', *New York Times*, 26 September 1945.

Conant wept: Conant, J., *109 East Palace*, p. 234.

Bainbridge snorted and Oppenheimer quoted from the Bhagavad Gita: http://www.lanl.gov/history/atomicbomb/trinity.shtml.

Project leader adopted a strut: Walker, S., *Shockwave*, p. 69.

Rabi mused: quoted in Rhodes, R., *The Making of the Atomic Bomb*, p. 672.

Defied God's creation: Leslie R. Groves to Henry Stimson, 18 July 1945, Original Trinity Report, Truman Library.

Blast shifted Fermi's confetti: http://www.lanl.gov/history/atomicbomb/trinity.shtml.

Words were inadequate: quoted in Walker, S., *Shockwave*, p. 67.

Radioactive fallout: Lamont, L., *Day of Trinity*, pp. 235–236.

Generals mercifully glib: Groves, L., *Now It Can Be Told*, p. 298.

Groves' humility: Leslie R. Groves to Henry Stimson, 18 July 1945, Original Trinity Report, Truman Library.

Colonel Stafford Warren: Correspondence of the Manhattan Engineer District, 1942–46, NARA RG77, entry 1, roll 1, file 4: Trinity Test at Alamogordo 16 July 1945, Memo, Colonel Stafford Warren to Groves, 21 July 1945.

Drafts prepared to cover all eventualities: Groves, L., *Now It Can Be Told*, p. 301.

Report on the test: Leslie R. Groves to Henry Stimson, 18 July 1945, Original Trinity Report, Truman Library.

Excited chatter irritated Groves: Groves, L., *Now It Can Be Told*, p. 303.

Franck's committee launch petition: Documents Pertaining to the Atomic Bomb, the Atomic Bomb Collection, Truman Library, box 2.

Oppenheimer banned draft being circulated: Lanouette, W., 'Three Attempts to Stop the Bomb' in Bird, K. & Lifschultz, L. (eds), *Hiroshima's Shadow*, p. 107.

Signatories damaged their case: Documents Pertaining to the Atomic Bomb, the Atomic Bomb Collection, Truman Library, box 2.

Szilard's final version of petition: Lanouette, W., 'Three Attempts to Stop the Bomb', in Bird, K. & Lifschultz, L. (eds), *Hiroshima's Shadow*, pp. 109–110.

Petition never reached Truman: Documents Pertaining to the Atomic Bomb, the Atomic Bomb Collection, Truman Library, box 2, see Lanouette, W., 'Three Attempts to Stop the Bomb', in Bird, K. & Lifschultz, L. (eds), *Hiroshima's Shadow*.

Unwelcome distraction: Alice Kimball Smith, quoted in Lawren, W., *The General and the Bomb*, p. 235.

Szilard evicted: Lanouette, W., 'Three Attempts to Stop the Bomb', in Bird, K. & Lifschultz, L. (eds), *Hiroshima's Shadow*, p. 111.

CHAPTER 12 POSTDAM

News from Alamogordo: Correspondence ('Top Secret') of the Manhattan Engineer District, 1942–46, Events Preceding and Following the Dropping of the First Atomic Bombs at Hiroshima and Nagasaki, NARA RG77, entry 1, roll 1, file 5.

Stimson withdrew to diary: Stimson Papers, Diary.

Harrison's further news and Stimson's reply: Correspondence ('Top Secret') of the Manhattan Engineer District, 1942–46, Events Preceding and Following the Dropping of the First Atomic Bombs at Hiroshima and Nagasaki, NARA RG77, entry 1, roll 1, file 5.

Churchills' delight and conclusions: quoted in Hasegawa, T., *Racing the Enemy*, p. 141; Stimson Papers, Diary.

Fire destruction prophesied: quoted in Alperovitz, G., *The Decision to Use the Atomic Bomb*, p. 250.

President told the Generalissimo: Truman at Potsdam – His Secret Diary: Notes by Harry S. Truman on the Potsdam Conference, 16 July 1945, Truman Library.

President wrote to his wife: Truman, H.S., *Dear Bess*, p. 519.

Flags shared rooftops: Mee, C., *Meeting at Potsdam*, p. 30.

Destiny of European continent: McCullough, D., *Truman*, Touchstone, p. 421.

200 reporters refused entry: *New York Times*, 19 July 1945.

Comforts of home set for Big Three: *Stars and Stripes*, 17 July 1945.

Americans trying to disentangle themselves from Stalin: Brown Papers, Diary.

Diplomatic stick against further Soviet incursions: Correspondence, Roosevelt, Stalin, Churchill, Truman, Yalta to Potsdam, 1944–45, Byrnes Papers, box 7, folder 5; see also Byrnes, J., *All in One Lifetime* and *Speaking Frankly*.

To tell Stalin of the discovery: Mee, C., *Meeting at Potsdam*, p. 87.

No more talk of compromise: McCullough, D., *Truman*, Touchstone, p. 425.

Stalin's responses to Konoe peace mission: Mee, C., *Meeting at Potsdam*, p. 92.

Japan to surrender exclusively to America: Truman Diary, quoted in Frank, p. 242.

Stimson recommends actions to Byrnes: Stimson Papers, Diary.

Stimson's account of atomic test: Stimson Papers, Diary.

Truman reviews military strategy: Truman, H., *Memoirs of Harry S. Truman*, p. 415.

News from Pentagon: Correspondence ('Top Secret') of the Manhattan Engineer District, 1942–46, Events Preceding and Following the Dropping of the First Atomic Bombs at Hiroshima and Nagasaki, NARA RG77, entry 1, roll 1, file 5, subfile 5E – Terminal Cables.

Truman determined at Cecilienhof: Stimson Papers, Diary.

Soviet zone of Poland occupation: quoted in Mee, C., *Meeting at Potsdam*, pp. 126–129.

Truman forced Stalin on defensive: Stimson Papers, Diary.

Stalin's reply: Mee, C., *Meeting at Potsdam*, pp. 130–132.

Truman told Russians where to get off: Truman, H.S., *Dear Bess*, p. 519.

Generalissimo hosts dinner: Truman, H.S., *Letters Home*, pp. 192–193.

Stimson's 'pet city' to return to target list: Correspondence ('Top Secret') of the Manhattan Engineer District, 1942–46, Events Preceding and Following the Dropping of the First Atomic Bombs at Hiroshima and Nagasaki, NARA RG77, entry 1, roll 1, file 5, subfile 5E – Terminal Cables.

Stimson's blunt reply: Stimson Papers, Diary.

Chosen cities and strike conditions: Colonel John Stone to General Arnold, The Decision to Drop the Atomic Bomb on Japan, Truman Library, Harry S. Truman Student Research File (b file).

Russians not needed in Pacific War: Stimson Papers, Diary.

Churchill declared atomic bomb is Second Coming: Feis, H., 'The Secret that Travelled to Potsdam', *Foreign Affairs*, January 1960, Truman Library, box 19, folder 1 – State Dept, USC Material (1945, 1960).

Churchill had worked himself into a great euphoria: quoted in Mee, C., *Meeting at Potsdam*, p. 166.

Russians on warpath again: Stimson Papers, Diary.

Exact dates expected for S-1: Stimson Papers, Diary.

Critical statement cut from Potsdam: quoted in Hasegawa, T., *Racing the Enemy*, p. 146.

Stimson urged Truman to reassure Japanese: Stimson Papers, Diary.

President formally approved use of bomb: McCullough, D., *Truman*, Touchstone, p. 442.

Groves sent cable: Groves, L., *Now It Can Be Told*, p. 309.

Truman rubber-stamped plan: Stimson Papers, Diary.

Truman's 'Potsdam Diary' reflections: Truman Diary, quoted in Frank, R., *Downfall*, p. 245.

Polish delegation present case: Mee, C., *Meeting at Potsdam*, p. 137.

Gist of S-1 revealed to Stalin: Feis, H., 'The Secret that Travelled to Potsdam', *Foreign Affairs*, January 1960, Truman Library, box 19, folder 1 – State Dept, USC Material (1945, 1960).

Truman nonchalantly told Stalin: Brown Papers, Diary.

Stalin glad to hear it: Feis, H., 'The Secret that Travelled to Potsdam', *Foreign Affairs*, January 1960, Truman Library, box 19, folder 1 – State Dept, USC Material (1945, 1960); The Decision to Drop the Atomic Bomb on Japan, Truman Library; Harry S. Truman Student Research File (b file), box 1, Introductory.

Stalin ordered work sped up on Russian bomb: Mee, C., *Meeting at Potsdam*, p. 178; Conant, J., *109 East Palace*, p. 237; Volkogonov, D., *Stalin*.

Political atmosphere degenerated: Byrnes Papers, Forrestal Diaries, box 12, folder 13.

The Declaration's power: Potsdam Proclamation, see Appendix 3.

Stalin comprehensively outmanoeuvred: quoted in Hasegawa, T., *Racing the Enemy*, p. 161.

Byrnes explained situation to Molotov: Foreign Ministers, 27 July, Byrnes Papers, box 16.

Chaplain Northen led prayer: The Decision to Drop the Atomic Bomb on Japan, Truman Library, Harry S. Truman Student Research File (b file).

Truman sidestepped: quoted in Hasegawa, T., *Racing the Enemy*, p. 163.

A travesty of the truth: Report on the Tripartite Conference of Berlin, 2 August 1945, Byrnes Papers, box 16, folder 10.

Potsdam conference agreed: Mee, C., *Meeting at Potsdam*, p. 134.

CHAPTER 13 MOKUSATSU

Council's self-denial: Magic Diplomatic Summary, 17 July 1945.

Horai: quoted in Hearn, L., *Lafcadio Hearn's Japan*, p. 128.

Moscow's dismissive reply: Itoh, T. (ed.), *Sokichi Takagi*, diary entry for 20 July 1945, pp. 916–917.

Sato's impassioned appeal to Tokyo: Magic Diplomatic Summaries, 20, 22 July 1945.

Council decided to fortify Konoe's role: Itoh, T. (ed.), *Sokichi Takagi*, diary entry for 20 July 1945, pp. 916–917.

Togo told his errant diplomat: Magic Diplomatic Summary, 22 July 1945.

Japan's refusal to yield: Naval Historical Center, Operational Archives, James Forrestal Diaries, entry 24 July 1945, 'Japanese Peace Feelers'.

Imperial Mission: Magic Diplomatic Summaries, 23, 25 July 1945.

The Atlantic Charter: Alperovitz, G., *The Decision to Use the Atomic Bomb*, p. 395.

State Department isolated: Minutes 1945, Byrnes Papers, box 18, folder 14.

Tokyo and US media confused: Alperovitz, G., *The Decision to Use the Atomic Bomb*, p. 404.

Japan's position: *New York Times*, 23 July 1945.

Peace mediators and intermediaries: Magic Diplomatic Summaries, 21–27 July 1945.

Grew attempted to dismiss 'peace feelers': quoted in Frank, R., *Downfall*, p. 115.

Offer made front-page headlines: *International Herald Tribune*, 27 July 1945.

Tokyo Radio: *Stars and Stripes*, 27 July 1945.

Leaflets on the streets: Watanabe Miyoko, interview, Hiroshima, 2009.

We knew the words Potsdam Declaration: Watanabe Miyoko, interview, Hiroshima, 2009.

Togo and Suzuki disliked: The Pacific War Research Society, *Japan's Longest Day*, p. 15.

Two points on declaration: Hasegawa, T., *Racing the Enemy*, p. 162.

The hateful document: quoted in Butow, R., *Japan's Decision to Surrender*, p. 141.

Mokusatsu: quoted in The Pacific War Research Society, *Japan's Longest Day*, p. 17.

Togo's dismay: *Yomiuri Hochi*, 28 July 1945.

'The government intends to *mokusatsu*': *Asahi Shimbun*, 28 July 1945; quoted in Hasegawa, T., *Racing the Enemy*, p. 167.

The report quoted sources: quoted in Butow, R., *Japan's Decision to Surrender*, p. 146.

Government exhorted factory supervisors: Ogura, T., *Letters from the End of the World*, p. 41.

Haragei: Sigal, L., *Fighting to the Finish*, p. 47.

Unspoken communication: Sayle, M., 'Did the Bomb End the War'?, quoted in Bird, K. & Lifschultz, L. (eds), *Hiroshima's Shadow*, p. 32.

Suzuki fronted the Japanese media: quoted in Alperovitz, G., *The Decision to Use the Atomic Bomb*, p. 407.

Various translations: quoted in The Pacific War Research Society, *Japan's Longest Day*, p. 17.

Tokyo's statement: *New York Times*, 30 July 1945.

Truman seized on salient point: quoted in Hasegawa, T., *Racing the Enemy*, p. 169.

Sato injected some reality: Magic Diplomatic Summary, 29 July 1945.

Big Six study of Potsdam: Magic Diplomatic Summary, 29 July 1945.

Sato and Kase persuade Tokyo to reason: Magic Diplomatic Summaries, 30 July, 1, 2 August 1945.

Would Sato resume efforts?: Magic Diplomatic Summary, 2 August 1945.

Sato's reply and plea: Magic Diplomatic Summary, 5 August 1945.

CHAPTER 14 SUMMER 1945

Japan lay in ruins: Dower, J., *Japan in War and Peace*, p. 122.

Colossal devastation: Hastings, M., *Nemesis*, p. 342.

Ancient samurai poems: 'Umi Yukaba' [Across the Ocean], by 8th-century poet Yakamochi Otomo, in Dower, J., *Japan in War and Peace*, p. 102.

Warning of their defeat: Matsubara Miyoko, interview, Hiroshima, 2009.

Widow's one-off payment: Hiroshima City Archives, *Hiroshima Konjaku: '80 Hiroshima-shi Seirei Shitei Toshi Kinen* [Hiroshima Today and Long Ago: In Commemoration of Hiroshima Becoming a Government Ordinance Designated City in 1980].

Lenient censorship: Hiromi Hasai, interview, Hiroshima, 2009.

Chiyoka was dong his duty: Mitsue Fujii (Hiraki), interview, Hiroshima, 2009.

Letters from her her husband: Kikuyo Nakamura, interview, Nagasaki, 2009.

Child's awareness: Hosokawa, H. & Kamei, H. (eds), *The Diary of Yoko Moriwaki*.

Midori heard Nagai's news in silence: *Crossroads* (journal).

Fate accepted as God's will: Glynn, P., *A Song for Nagasaki*, p. 91.

Religious consolations: http://www.cbcj.catholic.jp/eng/ehistory/table02.htm

Army officials berated Catholics: Glynn, P., *A Song for Nagasaki*, p. 91.

Medical facilities and aid in dire state: Records of the USSBS, office of the Chairman Pacific Service, Final Report Covering Air-Raid Protection and Allied Subjects in Japan, No. 11, February 1947, Medical, p. 74, Mortuary Services, p. 87, NARA RG243, box 96; Effects of Atomic Bombs on Health and Medical Services in Hiroshima and Nagasaki, No. 12, Medical Division, March 1947, NARA RG243, box 94 esp p. 26; Civilian Defense Report No. 1, Hiroshima, Japan Field Report,15 November 1945 (conducted 10–21 October 1945), NARA RG243, box 95.

Posters exhorted children to worship squads and kamikazes: Hiroshima City Archives, *Hiroshima Konjaku: '80 Hiroshima-shi Seirei Shitei Toshi Kinen* [Hiroshima Today and Long Ago: In Commemoration of Hiroshima Becoming a Government Ordinance Designated City in 1980].

Instil in the young a sense of suicidal revenge: Hiroshima City Archives, *Hiroshima Konjaku: '80 Hiroshima-shi Seirei Shitei Toshi Kinen* [Hiroshima Today and Long Ago: In Commemoration of Hiroshima Becoming a Government Ordinance Designated City in 1980].

Yukio Katayama enlisted: Yukio Katayama, interview by Yuta Hiramatsu.

Mesmerised by kamikaze heroics: Iwao Nakanishi, interview, Hiroshima, 2009.

Expected to give their lives: Hiroshima City Archives, *Hiroshima Konjaku: '80 Hiroshima-shi Seirei Shitei Toshi Kinen* [Hiroshima Today and Long Ago: In Commemoration of Hiroshima Becoming a Government Ordinance Designated City in 1980].

Students marched home singing: Ibuse, M., *Black Rain* (Bester. J., trans.), p. 193.

CHAPTER 15 TINIAN ISLAND

Stimson and Harrison cable communication about weapon's readiness: Correspondence ('Top Secret') of the Manhattan Engineer District, 1942–46, NARA RG77, entry 1, roll 1, file 5 – Events Preceding and Following the Dropping of the First Atomic Bombs at Hiroshima and Nagasaki.

Truman had no hand in the order's creation: Hasegawa, T., *Racing the Enemy*, p. 152.

Truman's later claim: See Truman, H., *Memoirs of Harry S. Truman* (the atomic decision).

Additional bombs and targets: Correspondence ('Top Secret') of the Manhattan Engineer District, 1942–46, NARA RG77, entry 1, roll 1, file 5, subfile 5b – Directives, Memorandums etc. to and from the Chief of Staff, Secretary of War, General Thomas T. Handy to General Carl Spaatz, 26 July 1945; MED records, Top Secret Files No. 5, NARA RG77.

Japanese kept prisoners of war: CPM, MID Confidential Publication, 20 December 1944 – 'Location and known strength of POW camps and civilian assembly centers in Japan and Japanese ...'

American airmen imprisoned in Hiroshima Castle: Correspondence ('Top Secret') of the Manhattan Engineer District, 1942–46, NARA RG77, entry 1, roll 1, file 5, subfile 5b – Directives, Memorandums etc. to and from the Chief of Staff, Secretary of War etc.

Nuclear weapons accompany land invasion: Correspondence ('Top Secret') of the Manhattan Engineer District, 1942–46, NARA RG77, entry 1, roll 1, file 5, subfile 5b – Directives, Memorandums etc. to and from the Chief of Staff, Secretary of War etc.

Destructive power of the bomb: Memorandum from Major General L. R. Groves to Chief of Staff, 30 July 1945.

Manhattan Project devolved on Tinian: Correspondence ('Top Secret') of the Manhattan Engineer District, 1942–46, NARA RG77, entry 1, roll 1, file 5, subfile 5b – Directives, Memorandums, etc. to and from the Chief of Staff, Secretary of War etc.

Farrell supervising 'vial of wrath': Correspondence ('Top Secret') of the Manhattan Engineer District, 1942–46, NARA RG77, entry 1, roll 1, file 5, subfile 5b – Directives, Memorandums etc. to and from the Chief of Staff, Secretary of War etc.

Fast-talking officer: Italy: General to Korea, Japan, President's Secretary's File, Truman Library, subject file 1940–53, Cabinet File: box 158.

509th Composite activated: Correspondence ('Top Secret') of the Manhattan Engineer District, 1942–46, NARA RG77, entry 1, roll 1, file 5 – Events Preceding and Following the Dropping of the First Atomic Bombs at Hiroshima and Nagasaki, subfile 5c – Preparation and Movement of Personnel and Equipment to Tinian

Lieutenant Colonel Paul Tibbets' soapbox: quoted in Walker, S., *Shockwave*, pp. 81–82.

Orange projectiles: Correspondence ('Top Secret') of the Manhattan Engineer District, 1942–46, NARA RG77, entry 1, roll 1, file 5 – Events Preceding and Following the Dropping of the First Atomic Bombs at Hiroshima and Nagasaki, subfile 5c – Preparation and Movement of Personnel and Equipment to Tinian

Airmen had not experienced training like this: Walker, S., *Shockwave*, p. 81.

'Pumpkins' dropped on countryside: Correspondence ('Top Secret') of the Manhattan Engineer District, 1942–46, NARA RG77, entry 1, roll 1, file 5 – Events Preceding and Following the Dropping of the First Atomic Bombs at Hiroshima and Nagasaki, subfile 5c – Preparation and Movement of Personnel and Equipment to Tinian.

Yoko saw B-29 overhead: Hosokawa, H. & Kamei, H., *The Diary of Yoko Moriwaki*.

SeaBees' construction: Correspondence ('Top Secret') of the Manhattan Engineer District, 1942–46, NARA RG77, entry 1, roll 1, file 5 – Events Preceding and Following the Dropping of the First Atomic Bombs at Hiroshima and Nagasaki, subfile 5c – Preparation and Movement of Personnel and Equipment to Tinian.

Hundreds of missions: Walker, S., *Shockwave*, pp. 83.

LeMay's airforce took umbrage at the 509th Composite Group: Correspondence ('Top Secret') of the Manhattan Engineer District, 1942–46, NARA RG77, entry 1, roll 1, file 5 – Events Preceding and Following the Dropping of the First Atomic Bombs at Hiroshima and Nagasaki, subfile 5c – Preparation and Movement of Personnel and Equipment to Tinian.

509th's perks: Walker, S., *Shockwave*, p. 83.

509th's attitude unfortunate: Correspondence ('Top Secret') of the Manhattan Engineer District, 1942–46, NARA RG77, entry 1, roll 1, file 5 – Events Preceding and Following the Dropping of the First Atomic Bombs at Hiroshima and Nagasaki, subfile 5c – Preparation and Movement of Personnel and Equipment to Tinian.

Tibbets' honesty: quoted in Walker, S., *Shockwave*, pp. 90–91.

Tibbets' private air force: quoted in Walker, S., *Shockwave*, pp. 90, 93–94.

Tibbets a capable man: Correspondence ('Top Secret') of the Manhattan Engineer District, 1942–46, NARA RG77, entry 1, roll 1, file 5 – Events Preceding and Following the Dropping of the First Atomic Bombs at Hiroshima and Nagasaki, subfile 5c – Preparation and Movement of Personnel and Equipment to Tinian.

Tibbets addresses crews: quoted in Walker, S., *Shockwave*, p. 165.

Parsons speaks: quoted in Walker, S., *Shockwave*, pp. 168–169.

12-man crew of delivery plane: Bock, F., *Commemorative Booklet for 50th Anniversary Reunion, 509th Composite Group*, August 1995.

Poor weather pushes back departure date: NARA, Documents 50a–c: Weather delays; Document 50b: CG 313th Bomb Wing, Tinian cable APCOM 5130 to War Department, 4 August 1945.

Chaplain blesses mission: quoted in Walker, *Shockwave*, pp. 200–201.

Sweeney attended Mass: Sweeney, C., Antonucci, J. & Antonucci, M., *War's End*, p. 157.

Lewis' pilot log: Lewis, Robert A., Notes Taken During Mission of the *Enola Gay* to Bomb Hiroshima, 6 August 1945, Truman Library, Atomic Bomb Collection, box 1.

About to start the bomb run: Lewis, Robert A., Notes Taken During Mission of the *Enola Gay* to Bomb Hiroshima, 6 August 1945, Atomic Bomb Collection, box 1.

Crew thought they had dropped a dud: Lewis, Robert A., Notes Taken During Mission of the *Enola Gay* to Bomb Hiroshima, 6 August 1945, Truman Library, Atomic Bomb Collection, box 1.

Jeppson's reaction: Lewis, Robert A., Notes Taken During Mission of the *Enola Gay* to Bomb Hiroshima, 6 August 1945, Truman Library, Atomic Bomb Collection, box 1.

Black boiling nest: Kelly, C. (ed.), *The Manhattan Project*, p. 330.

Farrell decoded note: The Decision to Drop the Atomic Bomb on Japan, Truman Library, box 1, Harry S. Truman Student Research File (b file).

Lewis wrote his last line: Lewis, Robert A., Notes Taken During Mission of the *Enola Gay* to Bomb Hiroshima, 6 August 1945, Truman Library, Atomic Bomb Collection, box 1.

Party program: quoted in Walker, S., *Shockwave*, p. 281.

Strike was tremendous: The Decision to Drop the Atomic Bomb on Japan, Truman Library, box 1, Harry S. Truman Student Research File (b file).

Groves thought about casualties: Groves, L., *Now It Can Be Told*, p. 324.

Parsons wrote to his father: Christman, A., *Target Hiroshima*, p. 195.

CHAPTER 16 AUGUSTA

Presidential Party relieved to be away from Berlin: County Judge File: 3 December 1930 to 7 March 1951, President's Secretary's File, Longhand Notes File, 1930–55, Truman Library, box 281, esp 17 June, 4 July 1945 [E813–866], 5 August 1945, four pages meeting with King George VI, 10 August 1945, 13 pages on atomic bomb, surrender, Washington business.

Truman not eager to see Soviet delegation: Brown Papers, Diary, entries July–August 1945, Byrnes, James F., Potsdam: Minutes, July–August 1945, box 10, folder 12.

President dined with King George VI: President's Trip to Berlin Conference, Byrnes Papers, box 16, folder 12.

Leahy expressed doubt about bomb: President's Trip to Berlin Conference, Byrnes Papers, box 16, folder 12; see Alperovitz, G., *The Decision to Use the Atomic Bomb*, and others.

Japanese seeking a peaceful solution: Brown Papers, Byrnes, James F., Potsdam: Minutes, July–August 1945; Brown Papers, Diary, box 10, folder 12.

Truman briefed on Magic intercepts: Outgoing Correspondence to Robert, J. Donovan, Article for the *Washington Post* Regarding Harry S. Truman 1974, Elsey Papers, Truman Library, box 113: chronological file 1952–53. For further evidence of Truman's access to Magic summaries, see White House staffer George Elsey's letter of 25 July 1945 to Commander Tyree, which makes clear the intelligence was shown to the President.

King grabbed Byrnes' arm: Meeting Notes, 3 August 1945, Potsdam: Minutes, July–August 1945, Byrnes Papers, box 18; another version of the King's comment: 'I will send out a ship for it', Brown Papers, Diary.

Tokyo radio report: Potsdam, Germany Trip, President's Secretary's File, Truman Library, subject file 1940–53, Cabinet File, box 140; morning *Augusta Press*, 5 August 1945.

Tokyo defied warnings: Potsdam, Germany Trip, President's Secretary's File, Truman Library, subject file 1940–53, Cabinet File, box 140.

Frank Graham brought a cable to Truman: President's Trip to Berlin Conference, Byrnes Papers, box 16, folder 12; see also Admiral Edwards to Admiral Leahy Re: Dropping of Bomb, 6 August 1945, Atomic Bomb: Hiroshima and Nagasaki (C358), President's Secretary's File, Truman Library, subject file 1940–53, Cabinet File, box 173.

Another cable arrived: Admiral Edwards to Admiral Leahy Re: Dropping of
    Bomb, 6 August 1945, Atomic Bomb: Hiroshima and Nagasaki (C358),
    President's Secretary's File, Truman Library, subject file 1940–53, Cabinet
    File, box 173.
The President's announcement: Ross Papers, Truman Library.
Officers expressed hope: President's Trip to Berlin Conference, Byrnes Papers,
    box 16, folder 12.
Truman greatly moved: Truman, H., *Memoirs of Harry S. Truman*.
Surles warned Ayers: Ayers Papers, Diary 1941–53, entries July–December
    1945, and 1 January 1944–31 December 1946, Truman Library, box 19.
Press statement on the bomb: Correspondence ('Top Secret') of the Manhattan
    Engineer District, 1942–46, Events Preceding and Following the Dropping
    of the First Atomic Bombs at Hiroshima and Nagasaki, NARA RG77, entry
    1, roll 1, file 5, subfile 5b – Directives, Memorandums etc. to and from the
    Chief of Staff, Secretary of War, etc; see also Groves, L., *Now It Can Be Told*,
    pp. 329–330.
Ayers tells the story: Ayers Papers, Diary 1941–53, entries July–December 1945,
    and 1 January 1944–31 December 1946, Truman Library, box 19.
Augusta broadcast of President's statement: Atomic Bomb: Hiroshima and
    Nagasaki, President's Secretary's File, Truman Library, subject file 1940–53,
    Cabinet File, box 173 Atomic Bomb: Hiroshima and Nagasaki one page
    (twice done) of Stimson's message to Truman re bomb, 6 August 1945.
Charlie Ross' contemplative mood: Ross Papers, Truman Library.
Important to have key men in places: President's Trip to Berlin Conference,
    Byrnes Papers, box 16, folder 12; Brown Papers, Diary, box 2, folder 1.
Program of entertainment: President's Trip to Berlin Conference, Byrnes Papers,
    box 16, folder 12.
President awoke to effusive media: Potsdam, Germany Trip, *Augusta Press*, July
    1945, President's Secretary's File, Truman Library, subject file 1940–53,
    Cabinet File, box 140.
Groves launched own press offensive: President's Secretary's File, Truman
    Library; subject file 1940–53, Cabinet File, box 173, Press Releases.
Americans heard of vast secret: President's Secretary's File, Truman Library;
    subject file 1940–53, Cabinet File, box 173, Press Releases.
'New Age Ushered ... Hiroshima is Target': *New York Times*, 7 August 1945.
Laurence's claims: Laurence, W., *Dawn Over Zero*, Knopf, p. 187.
Hanson Baldwin's palliative: *New York Times*, 7 August 1945.
Vatican deplored atomic attack: *Observattore Romano*, quoted in *New York Times*,
    8 August 1945.
Sombre British response: Press Releases, President's Secretary's File, Truman
    Library, subject file 1940–53, Cabinet File, box 173.

Speeches of congratulations: President's Trip to Berlin Conference, Byrnes Papers, box 16, folder 12.

President sought to mollify the Vatican: Ayers Papers, Diary 1941–53, Truman Library, box 19.

Oppenheimer's call from Groves: Groves Personal Papers; see also notes in Lawren, W., *The General and the Bomb*, and Walker, S., *Shockwave*, p. 298.

Scientists assembled to hear the news: Walker, S., *Shockwave*, p. 298; Lawren, W., *The General and the Bomb*, p. 250.

Los Alamos employee wrote to his parents: James Hush's Letter Regarding the Atomic Project, Atomic Bomb Collection, box 1; Los Alamos employee to parents letter, Atomic Bomb Collection, box 1

Security remained paramount: Internal memorandum to Los Alamos personnel first announcing the nature and purpose of the Manhattan Project – 6 August 1945, signed by Commanding Officer – Los Alamos, Col. G. R. Tyler of the Corp of Engineers, NARA, RG77, MED Records.

Factory workers celebrated: Internal memo to MED personnel, NARA, RG77, MED Records.

Scientists' exuberance quickly faded: Oppenheimer, J. R., *Letters and Recollections*, p. 292.

Szilard's petition and scientists' moral opposition: Lanouette, W., 'Three Attempts to Stop the Bomb', in Bird, K. & Lifschultz, L. (eds), *Hiroshima's Shadow*, pp. 559–560.

Dissenting voices irritated White House: County Judge File: 3 December 1930 to 7 March 1951, President's Secretary's File, Longhand Notes File, 1930–55, Truman Library, box 281.

CHAPTER 17 HIROSHIMA, 6 AUGUST 1945

Alert shrugged off: Iwao Nakanishi, interview, Hiroshima, 2009.

Moriwaki's pledge: Hosokawa, H. & Kamei, H., *The Diary of Yoko Moriwaki*.

Classmates assembled and put on their *monpes*: *The Minami Alumni 80th Anniversary Commemorative Magazine*, 1 March 1982.

Ground temperature: Hachiya, M., *Hiroshima Diary*, p. 114.

Buildings destroyed: Hiroshima Peace Memorial Museum manual, *The Outline of Atomic Bomb Damage in Hiroshima*.

Taeko fainted: Taeko Nakamae, interview, Hiroshima, 2009.

Troops were badly scorched: Hachiya, M., *Hiroshima Diary*, p. 15.

In village of Haraki: Mitsue Fujii (Hiraki), interview, Hiroshima, 2009.

Cadets at Etajima Military Academy: Ibuse, M., 'The Crazy Iris', in Oe, K., (ed.), *Fire From the Ashes*, p. 21.

Water became poison: Osada, A. (ed.), *Children of Hiroshima*, p. 229.

Deadly oases: Sugimine, H. & F. (eds), *Doctors' Testimonies of Hiroshima*, p. 84.

Boy ran away in tears: Osada, A. (ed.), *Children of Hiroshima*, p. 176.

Fire reservoirs filled to brim with dead: Hachiya, M., *Hiroshima Diary*, p. 19.

Seiko's first day as war labourer: Seiko Ikeda, interview, Hiroshima, 2009.

Why is it already night?: Hersey, J., *Hiroshima*, p. 27.

Strange, dirty people: Glynn, P., *A Song for Nagasaki*, p. 96.

Voices cried out: Seiko Ikeda, interview, Hiroshima, 2009.

Line of walking and crawling wounded: Ogura, T., (ed.), *Letters From the End of the World*, p. 58.

Truck stopped for Seiko: Seiko Ikeda, interview, Hiroshima, 2009.

Local photographer took photos: Cook, H.T. & Cook, T.F., *Japan at War*, pp. 391–394.

Iwao gave water to those who pleaded: Iwao Nakanishi, interview, Hiroshima, 2009.

Tomiko remembers with terrible clarity: Tomiko Nakamura, interview, Hiroshima, 2009.

She heard voices yelling: Osada, A. (ed.), *Children of Hiroshima*, p. 132.

The rest of Tomiko's friends passed away: Tomiko Nakamura, interview, Hiroshima, 2009.

Blouses and headbands used for wounds: Yoshiko Kajimoto, interview, Hiroshima, 2009.

'Black rain' pelted: Takashi Morita, interview, Peace Boat, January 2009.

Katsuzo tasted the droplets: Katsuzo Oda, 'Human Ashes', in Oe, K. (ed.), *Fire From the Ashes*, p. 75.

Yoshiko bumped into her father: Yoshiko Kajimoto, interview, Hiroshima, 2009.

Teachers and students showed courage: Taeko Nakamae, interview, Hiroshima, 2009.

A teacher's vain attempt: Osada, A. (ed.), *Children of Hiroshima*, p. 58.

Another emerged from a collapsed building: Kenji Kitagawa, interview, Hiroshima, 2009.

Soldiers and military police helpless: USSBS; see Takaki. R., *Hiroshima: Why America Dropped the Atomic Bomb*, p. 46.

Korean prince: Takashi Morita, interview, Peace Boat, January 2009.

Jeep full of American service men: Takashi Morita, interview, Peace Boat, January 2009; see also Sherwin, M., *A World Destroyed*, p. 232.

Thought only of his girlfriend: Takeshi Inokuchi, interview, Hiroshima, 2009.

Yoko pleaded for her mother: Hatsue Ueda, letter to a friend.

At the sight of Yoko's body: Kohji Hosokawa, statement & interview, Hiroshima, 2009.

Witnesses remember walking wounded:, H. & F. (eds), *Doctors' Testimonies of Hiroshima*, p. 93.

Shock turned to stupefaction: Yoshiko Kajimoto, interview, Hiroshima, 2009.

Echoes of deference and duty persisted: Hersey, J., *Hiroshima*, p. 38.

Tamika Hara shuddered: Tamiki Hara, 'Summer Flower', in Oe, K. (ed), *Fire From the Ashes*, p. 45.

Some ran careless of their wounds: Osada, A. (ed.), *Children of Hiroshima*, p. 309.

Ogura heard her shrill cries: Ogura, T., *Letters from the End of the World*, pp. 64–65.

'Jellyfish cloud': Ibuse, M., *Black Rain*, pp. 53, 56.

Red Cross Hospital doctor: Hersey, J., *Hiroshima*, pp. 20–21.

Asano library: Tamiki Hara, 'Summer Flower' in Oe, K. (ed.), *Fire From the Ashes*, p. 51.

Shocked patients: Hiroshima City Archives, *Hiroshima Konjaku: '80 Hiroshima-shi Seirei Shitei Toshi Kinen* [Hiroshima Today and Long Ago: In Commemoration of Hiroshima Becoming a Government Ordinance Designated City in 1980], p. 34.

Feeling of dreadful loneliness: Hachiya, M., *Hiroshima Diary*, pp. 2–3.

Crowds begged for treatment: Hachiya, M., *Hiroshima Diary*, p. 11.

Yoko's brother: Kohji Hosokawa, interview, Hiroshima, 2009.

Thousands wounded bereft of hope: Sugimine, H. & F. (eds), *Doctors' Testimonies of Hiroshima*, pp. 100–101.

Issued identification slips: Tamiki Hara, 'Summer Flower', in Oe, K. (ed.), *Fire From the Ashes*, p. 48.

Children with dead or dying parents: Ogura, T., *Letters from the End of the World*, pp. 71–72.

No beds left: Hiromi Hasai, interview, Hiroshima, 2009.

Scene of wretched humanity: Osada, A. (ed), *Children of Hiroshima*, p. 322.

Piece of charcoal: Sugimine, H. & F. (eds), *Doctors' Testimonies of Hiroshima*, p. 95.

Delirium and shouting: Osada, A. (ed.), *Children of Hiroshima*, pp. 110, 128.

The animals: Jungk, R., *Children of the Ashes*, p. 42.

Ena reached Itsukaichi Station: Takehiko Ena, interview, Hiroshima, 2009.

Survivors told airmen of huge bomb: Tamiki Hara, 'Summer Flower', in Oe, K. (ed.), *Fire From the Ashes*, p. 42.

Relief effort had barely begun: Takehiko Ena, statement, Hiroshima, 2009.

Zenchiki's concern: Mitsue Fujii (Hiraki), interview, Hiroshima, 2009.

Shizue dragged her mortified children: Osada, A. (ed.), *Children of Hiroshima*, p. 94.

Shizue lost her way: Mitsue Fujii (Hiraki), interview, Hiroshima, 2009.

Buddhist incantation to the dead: Tamiki Hara, 'Summer Flower', in Oe, K. (ed.), *Fire From the Ashes*, p. 51.

CHAPTER 18 INVASION

American flyers: Report on Overseas Operations – Atomic Bomb – By General T. Farrell & Flash Burn by Dr R. Serber, Reports Pertaining to the Effects of the Atomic Bomb, Records of the Office of the Commanding General, Manhattan Project, NARA RG77, box 90.

Evacuate your cities: The Decision to Drop the Atomic Bomb on Japan, Truman Library, Harry S. Truman Student Research File (b file).

12 mid-size cities warned: *The Times*, London, 6 August 1945.

B-29s flew incendiary raids: Potsdam, Germany Trip, *Augusta Press*, 7 August 1945, President's Secretary's File, Truman Library, subject file 1940–53, Cabinet File, box 140.

Garbled message: Magic, 10 August 1945.

Government ignorance: Seiji Hasegawa, 'Hakai no Zenya' [The Night Before the Collapse], *Fujin Koron*, August 1947.

Togo and Suzuki study broadcast: *Gaimusho* [Ministry of Foreign Affairs] (ed.), *Shusen Shiroku* [The Historical Records of the End of the War].

Togo sought confirmation: Interview with Sakomizu, *Oni Review*, Italy: General to Korea, Japan, President's Secretary's File, Truman Library, subject file 1940–53, Cabinet File, box 158.

US naval blockade: USSBS, Japan's Struggle to End the War.

An honourable deliverance: Interview with Sakomizu, *Oni Review*, Italy: General to Korea, Japan, President's Secretary's File, Truman Library, subject file 1940–53, Cabinet File, box 158.

Disseminate all known facts: Interview with Sakomizu, *Oni Review*, Italy: General to Korea, Japan, President's Secretary's File, Truman Library, subject file 1940–53, Cabinet File, box 158.

Investigation delay: *Gaimusho* [Ministry of Foreign Affairs] (ed.), *Shusen Shiroku* [The Historical Records of the End of the War].

Anami's one bomb belief: 'Doomsday', *Time* magazine, 7 August 1995.

Anami's defiant nihilism: Interview with Sakomizu, *Oni Review*, Italy: General to Korea, Japan, President's Secretary's File, Truman Library, subject file 1940–53, Cabinet File, box 158.

Samurai honourable course: Yamamoto, T., *Bushido, The Way of the Samurai*, p. 13.

Domei's international statement: *New York Times*, 8 August 1945, p. 7.

Emperor to accept Potsdam: Transcript of Foreign Minister Togo's Testimony, 'Shusen ni saishite' [At the time of the end of the war], September 1945, National Security Archive.

Manchurian border threat: Magic, 3 August 1945.

Japanese underestimated Russia's resolve: Hasegawa, T., *Racing the Enemy*, p. 85.

Boys drafted: Magic, 26 July 1945.

Stalin's paranoia: Hasegawa, T., *Racing the Enemy*, p. 186.

Stalin feared loss of prizes: Magic, circa late July 1945.

Togo, Sato and Soviet intentions: Magic, 8 August 1945.

Takagai and Yonai: Itoh, T. (ed.), *Sokichi Takagi*, diary entry for Wednesday, 8 August 1945, pp. 923–924.

Russia imposed its name on Potsdam: Feis, H., *The Atomic Bomb and the End of World War II*, pp. 126–127.

Japan's darkest fears: Garthoff, R.L., *Soviet Military Policy*, p. 174.

Stalin described onslaught to Harriman: Memorandum of Conversation, 'Far Eastern War and General Situation', 8 August 1945, Harriman Papers, Chron File August 5–9 1945, Library of Congress Manuscript Division, box 181.

Soviet propaganda: Hastings, M., *Nemesis*, p. 531.

Japanese resistance overwhelmed: Glantz, D.M., *August Storm*, p. 1.

Japanese troops neither warned nor equipped: Hastings, M., *Nemesis*, p. 534.

Russians captured: Hayashi, S., *Kogun*, p. 175.

Truman's 'terrible responsibility': Stimson Papers, Diary.

Snap press conference: General: Memorandum to Historical Files to September–October, 1946, Historical File, 1924–53, Truman Library, box 188.

Senator Alexander Wiley: *New York Times*, 9 August 1945.

James Byrnes' statement: Dept of State Bulletin xiii, July–September 1945, Brown Papers, box 50, folder 2.

Truman briefed his press officers: Ayers Papers, Diary 1941–53, Truman Library, box 19

Chinese Nationalists: Garthoff, R.L., *Soviet Military Policy*, pp. 174–175.

Suzuki's reply: Interview with Sakomizu, *Oni Review*, Italy: General to Korea, Japan, President's Secretary's File, Truman Library, subject file 1940–53, Cabinet File, box 158.

Emperor accepts Potsdam terms with condition: Hasegawa, T., *Racing the Enemy*, p. 197.

Soviet declaration, a deeper impression: Sigal, L., *Fighting to a Finish*, p. 226; see also Pape, R., *Bombing to Win*, p. 121.

Secret of atomic power: General: Memorandum to Historical Files to September–October 1946, Historical File, 1924–53, Truman Library, box 188.

CHAPTER 19 NAGASAKI, 9 AUGUST 1945

General Somervell wrote: Correspondence ('Top Secret') of the Manhattan Engineer District, 1942–46, NARA RG77, entry 1, roll 1, file 5 – Events Preceding and Following the Dropping of the First Atomic Bombs at Hiroshima and Nagasaki, file 5, subfile 5b – Directives, Memorandums etc. to and from the Chief of Staff, Secretary of War.

Groves' strategic plan: Groves, L., *Now It Can Be Told*, p. 342.

Perception in minds of Japanese: See Frank, R., *Downfall*.

People of Nagasaki vaguely aware: Toland, J., *The Rising Sun*, p. 799.

In mock *haiku*: Glynn, P., *A Song for Nagasaki*, p. 92.

Flyers before Nagasaki's destruction: The Decision to Drop the Atomic Bomb on Japan, Truman Library, Harry S. Truman Student Research File (b file).

Professor Ryokichi Sagane: Toland, J., *The Rising Sun*, p. 800.

Sweeney gathered his crew: Chinnock, F.W., *Nagasaki*, p. 14.

Sweeney's reactions to mishaps: Sweeney, C., Antonucci, J. & Antonucci, M., *War's End*, pp. 205, 209–210.

Laurence's notes: Laurence, W., 'Atomic Bombing of Nagasaki Told by Flight Member', *New York Times*, 9 September 1945.

Bockscar: Sweeney, C., Antonucci, J. & Antonucci, M., *War's End*, pp. 212–213, p. 218.

Leonard's conscious thought: Cheshire, L., *Where is God In All This?*, p. 65.

An hallucinogenic vision: Laurence, W., 'Atomic Bombing of Nagasaki Told by Flight Member', *New York Times*, 9 September 1945.

Captain Charles Albury: Top Secret, Cable APCOM 5445 from General Farrell to O'Leary [Groves' assistant], 9 August 1945, Attack on Nagasaki, National Security Archive.

Bockscar barely reached Okinawa: Documents Pertaining to the Atomic Bomb, The Atomic Bomb Collection, Truman Library, box 4.

Sirens approached – and questions: Sweeney, C., Antonucci, J. & Antonucci, M., *War's End*, p. 226.

Ashworth's misgivings: Cable CMDW576 to COMGENUSASTAF, for General Farrell to O'Leary [Groves' assistant], 9 August 1945, Attack on Nagasaki, National Security Archive.

Ashworth wondered: Sweeney, C., Antonucci, J. & Antonucci, M., *War's End*, p. 228.

Fat Man landed: Top Secret L COMGENAAF 20 Guam cable AIMCCR 5532 to COMGENUSASTAF Guam, 10 August 1945, NARA, RG77, Tinian Files, April–December 1945, box 20, envelope G Tinian Files, National Security Archive.

Farrell cabled Groves: Top Secret L COMGENAAF 20 Guam cable AIMCCR 5532 to COMGENUSASTAF Guam, 10 August 1945, NARA, RG77, Tinian Files, April–December 1945, box 20, envelope G Tinian Files, National Security Archive.

Fat Man's explosion damage: Glynn, P., *A Song for Nagasaki*, p. 95; Chinnock, F.W., *Nagasaki*, p. 108.

Japanese nuns: Nagasaki Testimonial Society, *A Journey to Nagasaki* (brochure), p. 25.

Lone mother cradled child: Burke-Gaffney, B. (trans.) *et al.*, *The Light of Morning*, p. 26.

Keiho Middle School: Final Report of the Atomic Bomb Investigation Group at Hiroshima and Nagasaki, Records of the Office of the Commanding General, Manhattan Project: Reports Pertaining to the Effects of the Atomic Bomb, NARA RG77, box 90.

Mitsubishi shipyard: Groves, L., *Now It Can Be Told*, p. 343.

Jurgen Onchen recalls: Nagasaki Testimonial Society, *A Journey to Nagasaki* (brochure), p. 14.

POW deaths: POW Research Network Japan.

Tasmanian Gunner Ted Howard: 'POWs Felt Atomic Bomb's Blast', *Sydney Morning Herald*, 15 October 1945, p. 4.

Many helped rescue wounded: See Clarke, H.V., *Last Stop Nagasaki*, for POWs' memories of the bomb, pp. 96–109.

Teruo heard a plane descending: Teruo Ideguchi, interview, *Nagasaki*, August 2009.

Yamazato school: Nagasaki Testimonial Society, *A Journey to Nagasaki* (brochure), p. 22.

'Two hideous monsters': Glynn, P., *A Song for Nagasaki*, p. 96.

Intense heat awoke Tsuruji: Tsuruji Matsuzoe, interview, Nagasaki, August 2009.

Mr Hishitani: Tsuruji Matsuzoe, interview, Nagasaki, August 2009.

Tsuruji admitted to hospital: Tsuruji Matsuzoe, interview, Nagasaki, August 2009.

Tatuichiro Akizuki: Nagasaki Testimonial Society, *A Journey to Nagasaki* (brochure), p. 16.

Dr Shirabe tended wounded: Burke-Gaffney, B. (trans.) *et al.*, *The Light of Morning*, pp. 45–81.

Dr Shirabe's novel technique: Shirabe, R., *A Physician's Diary of the Atomic Bombing and its Aftermath*, p. 7.

Thousands of near dead staggered: Glynn, P., *A Song for Nagasaki*, p. 101.

Hospital and medical school virtually destroyed: Glynn, P., *A Song for Nagasaki*, pp. 99–100, 102.

Dr Nagai's wife: Glynn, P., *A Song for Nagasaki*, p. 102.

'Umi Yukaba': Nagasaki Testimonial Society, *A Journey to Nagasaki* (brochure), p. 11.

Truman's mood: Hasegawa, T., *Racing the Enemy*, pp. 201–202.

Bishop Oxnan urged: Documents Pertaining to the Atomic Bomb, the Atomic Bomb Collection, Truman Library, box 2, 'Oxnan, Dulles ask halt in bomb use', unnamed US daily newspaper, 10 August 1945.

Senator Richard B. Russell's telegram: Richard Russell to Harry S. Truman, 7 August 1945: RE: Japan Should be Beaten to Dust, Forced to Beg, telegram, White House Official File, Truman Papers.

President's reply: Harry S. Truman to Richard Russell, 9 August 1945, White House Official File, Truman Papers.

Unwarranted attack on Pearl Harbor: White House Official File, Truman Papers, box 1527: Atomic Bomb.

CHAPTER 20 SURRENDER

Hopeless divisions: Butow, R., *Japan's Decision to Surrender*, p. 160.

Another 'special bomb': Hasegawa, T., *Racing the Enemy*, p. 204.

Secretary Sakomizu: Interview with Sakomizu, *Oni Review*, Italy: General to Korea, Japan, President's Secretary's File, Truman Library, subject file 1940–53, Cabinet File, box 158.

Opinions on surrender conditions: Hoshina, Z., *Daitoa Senso Hishi*: [Secret History of the Greater East Asia War], excerpts from Section 5, 'The Emperor Made *Goseidan* [Sacred Decision] – the Decision to Terminate the War', pp. 139–149.

Wretchedness impinged little on the samurai elite: Butow, R., *Japan's Decision to Surrender*, p. 166.

Hiranuma asks about defence: Hoshina, Z., *Daitoa Senso Hishi* [Secret History of the Greater East Asia War], excerpts from Section 5, 'The Emperor Made *Goseidan* [Sacred Decision] – the Decision to Terminate the War', pp.139–149.

Suzuki's statement: Interview with Sakomizu, *Oni Review*, Italy: General to Korea, Japan, President's Secretary's File, Truman Library, subject file 1940–53, Cabinet File, box 158.

Emperor speaks: Hoshina, Z., *Daitoa Senso Hishi* [Secret History of the Greater East Asia War], excerpts from Section 5, 'The Emperor Made *goseidan* [Sacred Decision] – the Decision to Terminate the War', pp.139–149.

Another version of Hirohito's speech: 'I think I should tell the reasons … to continue the war means nothing but the destruction of the whole nation … So to stop the war on this occasion is the only way to save the nation from destruction and to restore peace in the world. Looking back at what our military headquarters have done, it is apparent that their performance has fallen far short of the plans expressed. I don't think this discrepancy can be corrected in the future. But when I think about my obedient soldiers abroad and of those who died or were wounded in battle, about those who have lost their property or lives by bombing in the home land; when I think of all those sacrifices I cannot help but feel sad. I have decided that this war should be stopped … in spite of this sentiment and for more important considerations.' Interview with Sakomizu, *Oni Review*, Italy: General to Korea, Japan, President's Secretary's File, Truman Library, subject file 1940–53, Cabinet File, box 158.

Handkerchiefs appeared: Hasegawa, T., *Racing the Enemy*, p. 213.

His Majesty's 'personal desire': Butow, R., *Japan's Decision to Surrender*, p. 176.

Admiral Halsey's carrier-borne planes: Butow, R., *Japan's Decision to Surrender*, p. 181.

Should they accept the condition?: Stimson Papers, Diary.

Byrnes rejects consensus: Brown Papers, Diary.

Truman pleased with Byrnes: County Judge File: 3 December 1930 to 7 March 1951, President's Secretary's File, Longhand Notes File, 1930–55, Truman Library, box 281.

Byrnes Note: Japanese Surrender – 14 August 1945, President's Secretary's File, Truman Library, historical file, box 195.

Stalin wouldn't have a slice of the cake: Wallace Papers.

Alexandr Vasilevsky: Miscamble, W., *From Roosevelt to Truman*, p. 240.

Burden of the Pacific War: 'Japanese Surrender Negotiations', 10 August 1945, Papers of W. Averell Harriman, Chron File, 10–12 August 1945; Memorandum of Conversation, Library of Congress Manuscript Division, box 181.

Stimson's satisfaction with Byrnes Note: Stimson Papers, Diary.

Palpable anxiety in Washington: County Judge File: 3 December 1930 to 7 March 1951, President's Secretary's File, Longhand Notes File, 1930–55, Truman Library, box 281.

Truman's concern deepened: Wallace Papers, Diary.

Truman's express permission: Atomic Energy Commission, Thermonuclear Weapons to R. Gordon Arneson, Tape Recording, Interim Committee On Atomic Energy – Notes of Meetings, Papers of R. Gordon Arneson, Truman Library, box 1.

Recommended Tokyo be added to the list: Correspondence ('Top Secret') of the Manhattan Engineer District, 1942–46, Events Preceding and Following the Dropping of the First Atomic Bombs at Hiroshima and Nagasaki, NARA RG77, entry 1, roll 1, file 5, subfile 5b – Directives, Memorandums etc. to and from the Chief of Staff, Secretary of War.

Plutonium bomb projections: Telephone conversation transcript, General Hull and Colonel Seaman [sic] – 1325 – 13 August 1945, Marshall Papers

Risk of radiation to ground troops: Telephone conversation transcript, General Hull and Colonel Seaman [sic] – 1325 – 13 August 1945, Marshall Papers.

Leader on Hiroshima: quoted in Shillony, B., *Politics and Culture in Wartime Japan*, pp. 107–108.

Tokyo laid out brutal truth: *Nippon Times*, 11 August 1945, quoted in Butow, R., *Japan's Decision to Surrender*, p. 182.

War Ministry's exhortation to arms: *Nippon Times*, 12 August 1945, quoted in Butow, R., *Japan's Decision to Surrender*, p. 183.

Japanese spirit: Various Japanese newspaper reports, August 1945.

Yonai's concern: Itoh, T. (ed.), *Sokichi Takagi*, diary entry for 12 August 1945, pp. 916–917.

Suzuki sided with war faction: quoted in Butow, R., *Japan's Decision to Surrender*, p. 195.

Big Six argued over meaning: Butow, R., *Japan's Decision to Surrender*, p. 198.

American air raids continued: Hasegawa, T., *Racing the Enemy*, p. 234.

Target list for third atomic bomb submitted: Frank, R., *Downfall*, p. 303.

US intercepted message: Magic Diplomatic Summary, 13 August 1945.

Anami to curb insurrectionists: Itoh, T. (ed.), *Sokichi Takagi*, diary entry for 12 August 1945, pp. 916–917

Suzuki moved to break stalemate: *Gaimusho* [Ministry of Foreign Affairs] (ed.), *Shusen Shiroku* [*Historical Record of the End of the War*], The Cabinet Meeting over the Reply to the Four Powers (13 August), Vol. 5, pp. 27–35.

Anami tried to stall process: Hasegawa, T., *Racing the Enemy*, p. 237.

His Majesty's Sacred Judgment: Shimomura, H., *Shusenki* [Account of the End of the War], pp. 148–152.

Imperial Government's statement: Togo's Acceptance of Potsdam Proclamation on Behalf of the Emperor, Italy: General to Korea, Japan, President's Secretary's File, Truman Library, subject file 1940–53, Cabinet File: box 158.

Regime's last days: See Butow, R., *Japan's Decision to Surrender*, Hasegawa, T., *Racing the Enemy*, and Pacific War Research Society, *Japan's Longest Day*, for further narrative of these events.

Enraged staff officers: Hasegawa, T., *Racing the Enemy*, p. 243.

Sakomizu drafted Imperial Rescript: Hasegawa, T., *Racing the Enemy*, p. 244.

Hiroshi gazed at the stars: Pacific War Research Society, *Japan's Longest Day*, p. 153.

Anami to commit *seppuku*: Butow, R., *Japan's Decision to Surrender*, p. 219.

Anami's last words: Hasegawa, T., *Racing the Enemy*, p. 248.

Anami's *haiku*: Another version reads: 'Having received great favors / From His Majesty, / When dying / I have nothing to say.'

Emperor's implication about Japan's surrender: Butow, R., *Japan's Decision to Surrender*, p. 3.

Zenchiki's deep depression: Mitsue Fuiji (Hiraki),interview, Hiroshima, 2009.

Hiroshimans wept: Iwao Nakanishi, interview, Hiroshima, 2009.

Dr Hachiya and his staff's wrath: Hachiya, M., *Hiroshima Diary*, pp. 81, 83.

Governer urged maintaining pride: Japanese Atomic Bomb Report for the City of Hiroshima … Made by the Governor, Hiroshima Prefecture, 21 August 1945, Records of the Office of the Commanding General, Manhattan Project: Reports Pertaining to the Effects of the Atomic Bomb, NARA RG77, box 90.

Sobs of defiance: Kiyoko Mori, interview, Nagasaki, 2009.

Hirohito issued another rescript: quoted in Hasegawa, T., *Racing the Enemy*, p. 250.

Truman announced Japan's 'unconditional surrender' and celebrations were long and deep: General: Memorandum to Historical Files to September–October 1946, Historical File, 1924–53, Truman Library, box 188.

America's spiritual guide: Germany – Surrender, 8 May 1945 to Refugees, War, Japanese Surrender, 14 August 1945, President's Secretary's File, Historical File, Truman Library, box 195.

Day of prayer: Official program, press and radio conferences, September 1945, Ayers Papers, box 10.

Japanese and Allied representatives signed: General: Memorandum to Historical Files to September–October, 1946, Historical File, 1924–53, Truman Library, box 188.

Demonstration of who controlled Japan: President's Secretary's File, Truman Library (C788–793) four to five pages on address after signing of Japanese document of surrender, 1 September 1945.

CHAPTER 21 RECKONING

Upended graveyards: Butow, R., *Japan's Decision to Surrender*, p. 159.

Corpses towed ashore: Japanese Atomic Bomb Report for the City of Hiroshima … Made by the Governor, Hiroshima Prefecture, 21 August 1945, Records of the Office of the Commanding General, Manhattan Project: Reports Pertaining to the Effects of the Atomic Bomb, NARA RG77, box 90, 21 August 1945.

Hachiya wondered: Hachiya, M., *Hiroshima Diary*, p. 32.

Hiroshima cremation teams: Japanese Atomic Bomb Report for the City of Hiroshima … Made by the Governor, Hiroshima Prefecture, 21 August 1945, Records of the Office of the Commanding General, Manhattan Project: Reports Pertaining to the Effects of the Atomic Bomb, NARA RG77, box 90.

Casualties on day of bomb: Final Report of the Atomic Bomb Investigation Group at Hiroshima and Nagasaki, Records of the Office of the Commanding General, Manhattan Project: Reports Pertaining to the Effects of the Atomic Bomb, NARA RG77, box 90.

Dead sister's lunchbox: Keiko Nagai, interview, Hiroshima, 2009.

British scientific mission study: Report of the British Mission to Japan: An Investigation of the Effects of the Atomic Bombs Dropped at Hiroshima and Nagasaki, 29 January 1946, Records of the Office of the Commanding General, Manhattan Project: Reports Pertaining to the Effects of the Atomic Bomb, NARA RG77, box 90.

Trains resumed running: Records of the USSBS, Summary Report (Pacific War), Washington, DC, 1 July 1946, NARA RG243, box 93.

Schools 'completely burned' or 'totally destroyed': Japanese Atomic Bomb Report for the City of Hiroshima … Made by the Governor, Hiroshima Prefecture, 21 August 1945, Records of the Office of the Commanding General, Manhattan Project: Reports Pertaining to the Effects of the Atomic Bomb, NARA RG77, box 90; plus USSBS.

Hospitals completely wiped out: Records of the USSBS, The Effects of Atomic Bombs on Health and Medical Services in Hiroshima and Nagasaki, No. 12, Medical Division, March 1947, NARA RG243, box 94.

Doctors and nurses killed: Records of the USSBS, Japanese Atomic Bomb Report for the City of Hiroshima … Made by the Governor, Hiroshima Prefecture, 21 August 1945, Records of the Office of the Commanding General, Manhattan Project: Reports Pertaining to the Effects of the Atomic Bomb, NARA RG77, box 90.

Most prominent buildings destroyed: The Effects of Atomic Bombs on Health and Medical Services in Hiroshima and Nagasaki, No. 12, Medical Division, March 1947, NARA RG243, box 94.

Hiroshima telephone system: Japanese Atomic Bomb Report for the City of Hiroshima … Made by the Governor, Hiroshima Prefecture, 21 August 1945, Records of the Office of the Commanding General, Manhattan Project: Reports Pertaining to the Effects of the Atomic Bomb, NARA RG77, box 90.

Theatres, playhouses and brothels: Records of the USSBS, The Effects of Atomic Bombs on Health and Medical Services in Hiroshima and Nagasaki, No. 12, Medical Division, March 1947, NARA RG243, box 94.

Bomb dropped on Nagasaki missed its target: Summary Report (Pacific War), Washington, DC, 1 July 1946, NARA RG243, box 93.

Non-military areas worst affected: Final Report of the Atomic Bomb Investigation Group at Hiroshima and Nagasaki, Records of the Office of the Commanding General, Manhattan Project: Reports Pertaining to the Effects of the Atomic Bomb, NARA RG77, box 90.

Nagasaki's medical system: Records of the USSBS, The Effects of Atomic Bombs on Health and Medical Services in Hiroshima and Nagasaki, No. 12, Medical Division, March 1947, NARA RG243, box 94.

Dead and injured students: Kikuyo Nakamura, interview, Nagasaki, 2009.

Prisoners killed: Final Report of the Atomic Bomb Investigation Group at Hiroshima and Nagasaki, Records of the Office of the Commanding General, Manhattan Project: Reports Pertaining to the Effects of the Atomic Bomb, NARA RG77, box 90.

Explosion released several kinds of energy: Records of the USSBS, Summary Report (Pacific War), Washington, DC, 1 July 1946, NARA RG243, box 93.

Voluntary first aid teams: Japanese Atomic Bomb Report for the City of Hiroshima … Made by the Governor, Hiroshima Prefecture, 21 August

1945, Records of the Office of the Commanding General, Manhattan Project: Reports Pertaining to the Effects of the Atomic Bomb, NARA RG77, box 90.

Dr Hachiya's 'animal loneliness' and shame: Hachiya, M., *Hiroshima Diary*, p. 24, 52.

Dr Sasaki and Red Cross Hospital staff: Hersey, J., *Hiroshima*, pp. 33–34, 62.

Agonies and torment: Japan Broadcasting Corporation, see *Unforgettable Fire*.

Army opened stores: Japanese Atomic Bomb Report for the City of Hiroshima … Made by the Governor, Hiroshima Prefecture, 21 August 1945, Records of the Office of the Commanding General, Manhattan Project: Reports Pertaining to the Effects of the Atomic Bomb, NARA RG77, box 90.

People queued in open clearings: Records of the USSBS, Field Report Covering Air-Raid Protection and Allied Subjects in Nagasaki, Japan, Civilian Defense Division, March 1947, NARA RG243, box 94.

Local bureaucrats exploited opportunity: Hachiya, M., *Hiroshima Diary*, p. 144.

Looting rose: Hachiya, M., *Hiroshima Diary*, p. 117.

Press back at work: Japanese Atomic Bomb Report for the City of Hiroshima … Made by the Governor, Hiroshima Prefecture, 21 August 1945, Records of the Office of the Commanding General, Manhattan Project: Reports Pertaining to the Effects of the Atomic Bomb, NARA RG77, box 90.

Schoolchildren cleared their classrooms: Kiyoko Mori, interview, Hiroshima, 2009.

Orphans brought mementoes to class: Hayashi, K., 'The Empty Can', in Oe, K. (ed.), *Fire From the Ashes*, pp. 141–143.

Emiko Okada's illness: Emiko Okada, interview, Hiroshima, 2009.

Dr Hachiya studied the illness phenomenon: Hachiya, M., *Hiroshima Diary*, pp. 21, 36, 57, 97.

Soaring rates of miscarriage: Report of the British Mission to Japan: An Investigation of the Effects of the Atomic Bombs Dropped at Hiroshima and Nagasaki, 29 January 1946, Records of the Office of the Commanding General, Manhattan Project: Reports Pertaining to the Effects of the Atomic Bomb, NARA RG77, box 90.

Dr Hachiya's report on 'radiation sickness': Hachiya, M., *Hiroshima Diary*, pp. 125, 139–140.

Reports piqued American and British interest: Top Secret, Cable APCOM 5445 from General Farrell to O'Leary [Groves' assistant], 9 August 1945, Attack on Nagasaki, National Security Archive.

Number killed or wounded doubled: *The Times*, London, 23 August 1945.

29-year-old woman died after bruise: *The Times*, London, 30 August 1945.

Authority of Dr Oppenheimer re-invoked: Atomic Bomb – War Department, President's Secretary's File, Truman Library, Historical File, box 193.

Manhattan Project issued a memo: Final Report of the Atomic Bomb Investigation Group at Hiroshima and Nagasaki, Records of the Office of the Commanding General, Manhattan Project: Reports Pertaining to the Effects of the Atomic Bomb, NARA RG77, box 90.

First technical history of the bomb: Smyth, H.D., *Atomic Energy for Military Purposes*.

'Black rain': Reports and Other Records, 1928–47, NARA RG243, microfilm publication M1655, Roll 53.

'Toxic' memo contained precise medical guidance: Correspondence ('Top Secret') of the Manhattan Engineer District, 1942–46, Events Preceding and Following the Dropping of the First Atomic Bombs at Hiroshima and Nagasaki, NARA RG77, entry 1, roll 1, file 5, subfile 5b – Directives, Memorandums etc. to and from the Chief of Staff, Secretary of War.

Groves and Rea's phone conversation: Memorandum of Telephone Conversation Between General Groves and Lt Col. Rea, Oak Ridge Hospital, 9am, 28 August 1945, NARA, RG77, MED Records, Top Secret Documents, file no. 5b; see also Norris, R. S., *Racing for the Bomb*, pp. 339–441, and Bernstein, B.J., 'Reconsidering the "Atomic General": Leslie R. Groves', *Journal of Military History*, Vol. 67, No.3, pp. 907–908.

Farrell's scientific mission: Memo, Groves to Chief of Staff, 24 Aug 1945, Correspondence ('Top Secret') of the Manhattan Engineer District, 1942–46, Events Preceding and Following the Dropping of the First Atomic Bombs at Hiroshima and Nagasaki, NARA RG77, entry 1, roll 1, file 5, subfile 5b – Directives, Memorandums, to and from the Chief of Staff, Secretary of War.

Charles Ross invites reporters to test residual radiation levels at Trinity site: The Decision to Drop the Atomic Bomb on Japan, Truman Library, Harry S. Truman Student Research File (b file).

Burchett, first Western reporter to enter Hiroshima: Burchett, W.G., *Shadows of Hiroshima*, pp. 33–34.

'The Atomic Plague': *Daily Express*, 5 September 1945.

Burchett describes Hiroshima: Burchett, 'The First Nuclear War', in Bird, K. & Lifschultz, L. (eds), *Hiroshima's Shadow*, p. 69.

Laurence quoted Groves: *New York Times*, 9 September 1945.

'No Radioactivity in Hiroshima Ruin': *New York Times*, 12 September 1945.

Farrell's findings: Tinian Files, April–December 1945, box 17, envelope b, Confronting the Problem of Radiation Poisoning, documents 77a–b: General Farrell Surveys the Destruction, cable CAX 51813 from USS *Teton* to Commander in Chief Army Forces Pacific Administration, from Farrell to Groves, 10 September 1945, cable CAX 51948 from Commander in Chief Army Forces Pacific Advance Yokohoma Japan to Commander in Chief Army Forces Pacific Administration, 14 September 1945, National

Security Archive; Karl Compton's Report to President, 13 October 1945, Subject File 1940–53, Truman Library; Cabinet File, box 158.

MacArthur's press code: Sayle, M., in Bird, K. & Lifschultz, L. (eds), *Hiroshima's Shadow*, p. 46.

Reports of bomb disappeared from press: Braw, M., *The Atomic Bomb Suppressed*, pp. 5–6, 16–17.

Hundreds of foreign scientists anxious to study human exhibits of radiation disease: Top Secret, cable APCOM 5445 from General Farrell to O'Leary [Groves' assistant], 9 August 1945, Attack on Nagasaki National Security Archive.

Joint Commission for the Investigation of the Atomic Bomb in Japan: Atomic Bombings of Hiroshima & Nagasaki (Manhattan Engineer District) compiled by MED of the US Army under the direction of Groves. In time, the findings of the various groups of US scientific investigators were incorporated by Ashley Oughterson and Shields Warren into a book, *Medical Effects of the Atomic Bomb in Japan*, first published by the National Nuclear Energy Series, which contributed to a later general edition, *The Effects of Atomic Weapons*, produced in co-operation with the US Department of Defense and the US Atomic Energy Commission, under the direction of the Los Alamos Scientific Laboratory (McGraw-Hill, 1950).

Bomb's effects on humans: Final Report of the Atomic Bomb Investigation Group at Hiroshima and Nagasaki, Records of the Office of the Commanding General, Manhattan Project: Reports Pertaining to the Effects of the Atomic Bomb, NARA RG77, box 90.

Epilation findings: Japanese newspaper reports, 26 Sept 1945, of visit of US party led by Colonel Warren.

Japanese medical efforts severely hampered: Sugimine, H. & F. (eds), *Doctors' Testimonies of Hiroshima*, pp. 55, 56.

Agricultural and botanical findings: Hersey, J., *Hiroshima*, p. 91.

Residual radiation's stimulating effect on plant growth: Records of USSBS, Reports and Other Records, 1928–1947, NARA RG243, microfilm publication M1655, roll 53.

Ants and bugs resumed life cycles: Records of USSBS, Reports and Other Records, 1928–47, NARA RG243, microfilm publication M1655, roll 53.

CHAPTER 22 HIBAKUSHA

Seiko's ordeal: Seiko Ikeda, interview, Hiroshima, 2009.

James Forrestal's note to Truman: Richard B. Russell to Harry S. Truman, 7 August 1945: re Japan Should Be Beaten to Dust, Forced to Beg, telegram, White House Official File, Truman Papers.

Instructions were to experiment: Records of USSBS, The Effects of Atomic Bombs on Health and Medical Services in Hiroshima and Nagasaki, No. 12, Medical Division, March 1947, NARA RG243, box 94.

Presidential directive: White House Official File, Truman Papers, box 1527: Atomic Bomb.

To probe, prod and test: The ABCC Medical Research Program in Hiroshima, quoted in Osada, A. (ed.), *Children of Hiroshima*, p. 35.

Japanese doctors wary: Lifton, R.J., *Death in Life*, p. 345.

Insensitivity of foreign scientists: Lifton, R.J., *Death in Life*, p. 347.

ABCC valued the dead over the living: Burchett, W.G., *Shadows of Hiroshima*, pp. 59–61.

Experiments devoted to genetic effects: Jablon, S., *Atomic Bomb Radiation Dose Estimation at ABCC* (Technical Report 23–71); and Jablon, S., 'The Origin and Findings of the ABCC', lecture, February 1973.

Question of hereditary genetic mutation: quoted in Lifton, R.J., *Death in Life*, p. 117.

Increased rate of leukaemia: For the effects of nuclear weapons and radiation poisoning in humans, see three vast studies: Committee for the Compilation of Materials, *Hiroshima and Nagasaki*; Hiroshima International Council, *Effects of A-bomb Radiation on the Human Body*; and the (now dated) Glasstone, S. & Hirschfelder, J.O. (eds), *The Effects of Atomic Weapons*.

Hibakusha: Chisholm, A., *Faces of Hiroshima*, p. 24.

Keloid afflicted: Chisholm, A., *Faces of Hiroshima*, pp. 36–37.

Preventing 'contamination': Lifton, R.J., *Death in Life*, pp. 170–171.

Burakumin: Lifton, R.J., *Death in Life*, p. 268.

Preludes to an ocean of loneliness: Oe, K. *Hiroshima Notes*, p. 35.

Hibakusha's suicidal feelings: Ishida, T., *The Formation of the Atomic Bomb Experience*, Vol. 1, p. 1.

'Damned Untouchables': quoted in Inoue, M., 'The House of Hands', in Oe, K. (ed.), *Fire From the Ashes*, pp. 145, 168.

Goke-mura: The Force that Supports the Home Front (Japanese wartime brochure) p. 13.

Relief that Japan 'is not a Christian country': Oe, K., *Hiroshima Notes*, p. 84.

Hibakusha and second generation: Kikuyo Nakamura, interview, Nagasaki, 2009.

Stronger souls resisted condemnation: Chisholm, A., *Faces of Hiroshima*, p. 38.

Iwao's survival story: Iwao Nakanishi, interview, Hiroshima, 2009.

Boy in the torpedo tunnel: Tsuruji Matsuzoe, interview, Nagasaki, 2009.

She lived off the kindness of strangers and own wits: Tomiko Matsumoto, interview, Hiroshima, 2009.

Plastic surgery operations: Taeko Nakamae, interview, Hiroshima, 2009.

Compensation for war victims ceased: http://www.ne.jp/asahi/hidankyo/nihon/
rn_page/english/message.html.

Immediate medical treatment offered to *Lucky Dragon* crew: Herbert Passin,
ABCC academic, quoted in Chisholm, A., *Faces of Hiroshima*, p. 69.

Public outcry: Chisholm, A., *Faces of Hiroshima*, p. 69.

Citizen's movement against nuclear weapons: Committee for the Compilation
of Materials, *Hiroshima and Nagasaki*, p. 552.

'Hiroshima Maidens' on *This is Your Life*: quoted in Chisholm, A., *Faces of
Hiroshima*, p. 89.

Hiroko T. could eat through her mouth: Chisholm, A., *Faces of Hiroshima*, p. 105.

Medical complaint categories: The debate over the dangers of residual radiation
continues; see Lifton, R.J., *Death in Life*, p. 561, note.

Compensation fund planned: *Japan Times*, 7 August 2009.

Free of morbid self-pity: Chisholm, A., *Faces of Hiroshima*, p. 139.

Nursing Home for A-bomb survivors: Dr Nanao Kamada, interview, *Kurakake
Nozomi–en* [Nursing Home for A-bomb Survivors], Hiroshima 2009.

Child evacuees waited for parents: Committee for the Compilation of Materials,
Hiroshima and Nagasaki, pp. 435, 436.

Shoso's experience: Shoso Kawamoto, interview, Hiroshima, 2009.

Mitsue's family: Mitsue Fujii (Hiraki), interview, Hiroshima, 2009.

CHAPTER 23 WHY

Media caressed the bomb as the saviour of mankind: Alperovitz, G., *The Decision
to Use the Atomic Bomb*, p. 427.

Gallup poll responses: Morgan Gallop poll, August 1945.

Letters to the editor etc.: Ian Barton, Ross–on–Wye, to *The Times*, 10 August
1945; Vivien Cutting, Mavis Eurich and Olive Sampson, Southampton, to
The Times, 10 August 1945; Sir William Beveridge, Northumberland, to *The
Times*, 14 August 1945; George Bernard Shaw to *The Times*, 14 August
1945; and *The Times*, 14 August 1945.

Provincial firebrands echoed Truman: See Caron, G., *Fire of a Thousand Suns*,
pp. 257–261.

The Federal Council of Churches' disapproval: quoted in Conant, J. *109 East
Palace*, p. 284.

Appeal for understanding and forgiveness: 'America's Atomic Atrocity',
Christian Century, 29 August 1945.

Anglican Church shared Vatican's disgust: *The Times*, letters page, 18 August 1945.

Nuclear reckoning preoccupied scientists: Teller, E. & Brown, A., *The Legacy of
Hiroshima*, pp. 21–22.

Oppenheimer addressed the mesa's workforce: Oppenheimer, J. R., *Letters and
Recollections*, p. 311.

Truman and Oppenheimer's meeting: quoted in Lifton, R.J. & Mitchell, G., *Hiroshima in America*, p. 168.

Oppenheimer's collective sense of regret: Oppenheimer, J.R., *Uncommon Sense*, p. 113; Cockburn, S. & Ellyard, D., *Oliphant*, p. 125; Oppenheimer, J. R., *Letters and Recollections*, p. xix.

Oppenheimer's speech to ALAS: Oppenheimer, J. R., *Letters and Recollections*, pp. 315–325.

The rest were wistful dreams: Oppenheimer, J.R., *Uncommon Sense*, p. 186.

Of 4,756,705 American citizens: Goodchild, P., *J. Robert Oppenheimer*, p. 284.

Oliphant's guilt for his role: Cockburn, S. & Ellyard, D., *Oliphant*, pp. xiii, 124.

Emigré scientists' position: Sherwin, M., *A World Destroyed*, p. 118.

Conant's pride: Conant, J., *109 East Palace*, p. 228.

Conant's fear management: quoted in Conant, J., *109 East Palace*, p. 281.

Conant's arms race had limits: Conant, J., *109 East Palace*, pp. 468, 476.

Conant fretful over his role: Conant, J., *109 East Palace*, p. 752.

Stimson bemoaned his countrymen's indifference: quoted in Malloy, S. L., Atomic Tragedy, p. 643.

Stimson objected to targeting: quoted in Rhodes, R., *The Making of the Atomic Bomb*, p. 106.

USSBS speculation: USSBS; also quoted in Bernstein, B., *The Atomic Bomb*, p. 56.

USSBS conclusions criticised: See Gentile, G.P., 'Advocacy or Assessment? The USSBS of Germany and Japan', in Maddox, R.J. (ed.), *Hiroshima in History*, pp. 120–139.

America had no choice: Stimson, H.L., 'The Decision to Use the Bomb', originally published in *Harper's Magazine*, quoted in Baker, P.R., (ed.), *The Atomic Bomb*, p. 28.

The *Washington Post*'s critique: *Washington Post*, 28 January 1947.

No serious consideration to another course of action: Bernstein, B., *The Atomic Bomb*, pp. 114–115, p. 120.

Truman's estimation: Ayers Papers, Diary, 6 August 1945, Truman Library, box 19.

The President never lost any sleep: Truman to Klein, letter, date unknown, likely late 1950s, The Decision to Drop the Atomic Bomb on Japan, Truman Library, Harry S. Truman Student Research File (b file).

Hiroshima City Council's response to Truman: quoted in Alperovitz, G., *The Decision to Use the Atomic Bomb*, pp. 565–566.

Truman's claims to saved lives: Truman to Kupcinet, letter, 5 August 1963, The Decision to Drop the Atomic Bomb on Japan, Truman Library, Harry S. Truman Student Research File (b file).

Truman protested too much: quoted in Lifton, R.J. & Mitchell, G., *Hiroshima in America*, p. 176.

Spectacular figures: quoted in Alperovitz, G., *The Decision to Use the Atomic Bomb*, pp. 516–517; for range of Truman's estimates, see pp. 515–520, and Stimson, H.L., 'The Decision to Use the Bomb', quoted in Bernstein, B., *The Atomic Bomb*, p. 21.

Truman's several sources: Frank, R., *Downfall*, p. 133; see also Giangreco, D.M., 'A Score of Bloody Okinawas and Iwo Jimas', in Maddox, R.J. (ed.), *Hiroshima in History*, pp. 89–92.

Hoover's calculation derided: General George A. Lincoln to General Hull, 4 June 1945, American–British–Canadian Top Secret Correspondence, Records of the War Department General and Special Staffs, NARA, RG165, box 504.

Hoover's upper limit: Frank, R., *Downfall*, p. 133.

Hoover approved *planning* of Kyushu invasion: Skates, J.R., *The Invasion of Japan*, p. 77.

Truman chose none of these alternatives: Bernstein, B., 'Why was the bomb used?' in Bernstein, B., *The Atomic Bomb*, p. 109.

Truman denied Japanese given no warning: quoted in Alperovitz, G., *The Decision to Use the Atomic Bomb*, p. 55.

Byrnes told US News: Bernstein, B., *The Atomic Bomb*, pp. 20–21.

Byrnes undermined Truman's public position: quoted in Alperovitz, G., *The Decision to Use the Atomic Bomb*, p. 585.

No evidence to support blackest reading of Byrnes' role: Brown Papers, Diary, box 10, folder 16, letter, G. Bernard Noble, Dept of State, Historical Division, to Byrnes, 23 July 1954, Byrnes, James F., Potsdam: State Department, Inquiries for Publications, 1954.

Maddox's spirited attempt: See Maddox R.J. (ed.) *Hiroshima in History*, pp. 14–23.

Leahy was the most emphatic opponent: See Leahy, W., *I Was There*.

Halsey dismissed weapon as a 'toy': quoted in Alperovitz, G., *The Decision to Use the Atomic Bomb*, p. 445.

MacArthur implacably opposed: quoted in Hastings, M., *Nemesis*, p. 341.

Arnold and LeMay were dismissive: quoted in Hastings, M., *Nemesis*, pp. 343–344.

Gordon Daniels: quoted in Frank, R., *Downfall*, p. 7.

Eisenhower adamantly opposed: Eisenhower, D., *Mandate for Change, 1953–56*, p. 380.

Newsweek interview: *Newsweek*, 11 November 1963.

General Tomoyuki Yamashita's attitude: Records of USSBS, Reports and Other Records, 1928–47, NARA RG243 microfilm publication M1655, roll 53.

Contrast between two Japans: Mohan, U. & Tree, S., 'The Construction of Conventional Wisdom', in Bird, K. & Lifschultz, L. (Ed.), *Hiroshima's Shadow*.

Forrest Davis' observation: quoted in Mohan, U. & Tree, S., 'The Construction of Conventional Wisdom', in Bird, K. & Lifschultz, L. (Ed.), *Hiroshima's Shadow*, p. 153.

Understanding of the bomb as a danger to humankind: County Judge File: 3 December 1930 to 7 March 1951, President's Secretary's File, Longhand Notes File, 1930–55, Truman Library, box 281.

EPILOGUE: DEAD HEAT

Death sentence of America's next enemy: Memorandum, Major General Lauris Norstard to Major General Groves, 15 September 1945, Correspondence of the Manhattan Engineer District, 1942–1946, Stockpile, Storage and Military Characteristics, NARA RG77, entry 1, roll 1, file 3.

Groves approved the bomb production plan: Memorandum, Major General Lauris Norstard to Major General Groves, 15 September 1945, Correspondence of the Manhattan Engineer District, 1942–1946, Stockpile, Storage and Military Characteristics, NARA RG77, entry 1, roll 1, file 3.

Stimson warned of hostility if Russia freezed out of atomic secret: Documents Pertaining to the Atomic Bomb, the Atomic Bomb Collection, Truman Library, box 4.

White House asked media to self-censor: White House Official File, Truman Papers, box 1527: Atomic Bomb.

Hope of avoiding nuclear arms race eclipsed fear of Soviet Union: Bush to Truman, 25 September 1945 and Dean Acheson to Truman, 25 September 1945, Truman Library, Atomic Bomb: Cabinet Files.

Einstein's reply: Albert Einstein letter, Truman Library, Atomic Bomb: Cabinet Files.

The Joint Chiefs' conclusion: William Leahy, Joint Chiefs of Staff on Non-Proliferation, 23 October 1945, Truman Library, Atomic Bomb: Cabinet Files.

First act of nuclear diplomacy: Molotov and Byrnes, Brown Papers, box 68: folder 14.

Molotov's slip: quoted in Herken, G., *The Winning Weapon*, p. 49.

Brynes' concern: Letter, Lieutenant Colonel Calvert to Groves, Correspondence of the Manhattan Engineer District, 1942–46, Miscellaneous, NARA RG77, entry 1, roll 1, file 20.

Soviet Union turned its back: The Decision to Drop the Atomic Bomb on Japan, Truman Library, Harry S. Truman Student Research File (b file), Presidential Encyclopedia, p. 15.

Truman's lost confidence in Byrnes: Truman, H., *Memoirs of Harry S. Truman*, pp. 547, 550–552.

Origins of the coming arms race: quoted in Herken, G., *The Winning Weapon*, pp. 35–36, 94.

Pincher Plan: Herken, G., *The Winning Weapon*, pp. 219–224, 227–228, 248–250.

LeMay demanded rethink of US military policy: Curtis Lemay to NAA Banquet, 19 July 1946, Truman Library, Atomic Bomb: Press Releases, box 2.

B-36 heavy bombers: Inhabited Regions of World in Range of B-36's Atomic Bombs, 7 November 1946, War Dept, Bureau of Public Relations, Truman Library, Atomic Bomb.

Pre-emptive nuclear strike received imprimatur: Truman Library, Atomic Bomb: Atomic Testing: Crossroads, Press Release on Bikini and First Strike Policy, and 'The Evaluation of the Atomic Bomb as a Military Weapon', the Final Report of the Joint Chiefs of Staff Evaluation Board for Operation Crossroads, 30 June 1947.

Cold War arrived sooner than anticipated: Correspondence, Russia Drops Bomb, Ross Papers.

Groves' expectations upset: Correspondence, Russia Drops Bomb, Ross Papers.

CIA assessment of Russian nuclear threat: Truman Library, Atomic Bomb: Atomic Energy: Central Intelligence Agency, Joint Atomic Energy Intelligence Committee: Status of the Soviet Atomic Energy Program, 4 July 1950.

Nuclear fever gripped the American imagination: quoted in Boyer, P., *By the Bomb's Early Light*, p. 118.

Nuclear Armageddon propaganda: United States Civil Defense Booklet, 8 September 1950, Truman Library, Official File, box 1671. See also, Atomic Warfare – Manual of Basic Training, British Home Office, Civil Defence, Pamphlet No. 6, 1950.

Sumner Pike: Truman Library, Atomic Bomb: Atomic Energy: President's Directive [1 of 3] to Atomic Weapons: Procedures for Use, and 31 January 1950 [2 of 3] (C549–553) six pages on development of superbomb plus likely radioactivity, with letter to President from Sumner Pike, 7 December 1949, box 175.

Truman decided to proceed with hydrogen bomb: Truman Library, Atomic Bomb: Atomic Weapons: Thermonuclear, Truman Direction to Develop Thermonuclear Devices, 31 January 1950.

'Pandora's box 1': *Washington Post*, 2 January 1950.

Gore recommended: Truman Library, Atomic Bomb: 'A–Z', Favoring Use of Atomic Bomb in Korean Emergency, Al Gore, Tennessee, to Truman, 14 April 1951.

Church leaders were exponents: Protesting Use of Atomic Bomb in Korean
Emergency, Rev. Kenneth E. Eyler, Wesleyan Methodist Church, East
Michigan at Magnolia, to Truman, 29 November 1950; God's Way
Foundation to Truman, 5 December 1950, Truman Library, Atomic Bomb.

W.D. Westbrook: Protesting Use of Atomic Bomb in Korean Emergency, W.D.
Westbrook to Truman, 2 December 1950, Truman Library, Atomic Bomb.

Clara Bergamini: Protesting Use of Atomic Bomb in Korean Emergency, Clara
Bergamini, 30 November 1950, Truman Library, Atomic Bomb.

Senator Brien McMahon's letter: Brien McMahon via James Lay to President re
Requirement to Mass Produce Thermonuclear Weapons, 30 May 1952, and
President's reply, National Security Council – Atomic File, Truman Library,
President's Secretary's File, Subject File 1940–53.

Beginning of an arms race: *Washington Post*, 24 October 1952.

Cold War exceeded projections: quoted in Herken, G., *Counsels of War*, p. 83.

The Report from Iron Mountain: quoted in Herken, G., *Counsels of War*, p. 216.

Jonathan Schell's portentous vision: Schell, J.E., *The Fate of the Earth*, chapter 1.

Superpowers held their fire: For the most recent assessment of the global atomic
threat and its origins, see Baggott, J., *Atomic*, and Rhodes, R., *Twilight of the
Bombs*.

Paul Fussell offers the most eloquent: Fussell, P., *Thank God for the Atom Bomb
and Other Essays*, New York: Summit Books, 1998. First published as
'Hiroshima: A Soldier's View', *New Republic*, August 1981

APPENDICES

Carey Sublette: Chronology For The Origin Of Atomic Weapons: Nuclear
Weapons Frequently Asked Questions, Version 2.13: 15 May 1997, http://
nuclearweaponarchive.org/Nwfaq/Nfaq0.html; Tentative Chronology of
Part Played by Scientists in Decision to Use the Bomb Against Japan', 29
May 1957, Truman Library, Subject File, Ayers Papers, and Truman Papers,
US War Department, Public Relations Division, Press Section, Background
Information on Development of Atomic Energy Under Manhattan Project,
December 1946.

Casualty tables: From Hiroshima Shiyakusho, *Hiroshima Genbaku Sensaishi* Vol.
1; Hatano, S. & Watanuki, T. 'A-bomb Disaster Investigation Report', Vol. 1,
p. 621, quoted in Committee for the Compilation of Materials, *Hiroshima
and Nagasaki*; and Hiroshima International Council for Medical Care of the
Radiation-Exposed (Shigematsu, I. *et al.*), *Effects of A-bomb Radiation on the
Human Body*.

SELECT BIBLIOGRAPHY

BOOKS

Ackerman, E. (1953) *Japan's Natural Resources*. Chicago: University of Chicago Press

Aldrich, R. (2000) *Intelligence and the War Against Japan*. Cambridge: Cambridge University Press

Alperovitz, G. (1996) *The Decision to Use the Atomic Bomb*. New York: Vintage

Axell, A. & Kase, H. (2002) *Kamikaze: Japan's Suicide Gods*. London: Pearson Education

Baggott, J. (2009) *Atomic: The First War of Physics and the Secret History of the Atom Bomb: 1939–49*. London: Icon Books

Baker, P.R. (ed.) (1968) *The Atomic Bomb: The Great Decision*. New York: Holt, Rinehart and Winston

Barker, A.J. (1979) *Japanese Army Handbook*. London: Ian Allan

Bernstein, B. (1975) *The Atomic Bomb: The Critical Issues*. Boston: Little, Brown and Co.

Bird, K. & Lifschultz, L. (eds) (1998) *Hiroshima's Shadow: Writings on the Denial of History and the Smithsonian Controversy*. Stony Creek: Pamphleteer's Press

Bird, K. & Sherwin, M. (2005) *American Prometheus: The Triumph and Tragedy of J. Robert Oppenheimer*. New York: Alfred A. Knopf

Bix, H.P. (2000) *Hirohito and the Making of Modern Japan*. New York: HarperCollins

Bizony, P. (2008) *Atom*. London: Icon Books

Blackett, P.M.S. (1949) *Fear, War and the Bomb: Military and Political Consequences of Atomic Energy*. New York: McGraw-Hill

Blackett, P.M.S. (2007) *Studies of War: Nuclear and Conventional*. New York: McGraw-Hill

Boyer, P. (1994) *By the Bomb's Early Light: American Thought and Culture at the Dawn of the Atomic Age*. Chapel Hill: University of North Carolina Press

Braw, M. (1997) *The Atomic Bomb Suppressed: American Censorship in Occupied Japan (Asia and the Pacific)*. Armonk: M.E. Sharpe Inc.

Bundy, M. (1990) *Danger and Survival: Choices about the Bomb in the First Fifty Years*. New York: Vintage

Burchett, W.G. (1987) *Shadows of Hiroshima*. London: Verso Books

Burke-Gaffney, B. *et al.* (trans.) (2005). *The Light of Morning: Memoirs of the Nagasaki Atomic Bomb Survivors*. Nagasaki National Peace Memorial Hall for the Atomic Bomb Victims

Butow, R. (1954) *Japan's Decision to Surrender*. Stanford: Stanford University Press

Byrnes, J. (1947) *Speaking Frankly*. New York: Harper & Brothers

Byrnes, J. (1958) *All In One Lifetime*. New York: Harper & Brothers

Caron, G. (ed.) (1995) *Fire of a Thousand Suns*. USA: Robert Beard Books

Cheshire, L. (1996) *Where is God In All This?* London: Hyperion Books

Chinnock, F.W. (1970) *Nagasaki: The Forgotten Bomb*. London: Allen & Unwin

Chisholm, A. (1985) *Faces of Hiroshima: A Report*. New York: Jonathan Cape

Christman, A. (1998) *Target Hiroshima: Deak Parsons and the Creation of the Atomic Bomb*. Annapolis: US Naval Institute Press

Churchill, W. (1953) *The Second World War*, Vol. 6: *Triumph and Tragedy*. Boston: Houghton Mifflin

Clark, R.W. (1961) *The Birth of the Bomb*. UK: Horizon Press

Clark, R.W. (1981) *The Greatest Power on Earth: The Story of Nuclear Fission*. Fort William: Sidgwick and Jackson

Clarke, H.V. (1984) *Last Stop Nagasaki!* Sydney: Allen & Unwin

Cockburn, S. & Ellyard, D. (1981) *Oliphant: the Life and Times of Sir Mark Oliphant*. Adelaide: Axiom Books

Committee for the Compilation of Materials on Damage Caused by the Atomic Bombs in Hiroshima and Nagasaki (1981) *Hiroshima and Nagasaki: The Physical, Medical, and Social Effects of the Atomic Bombings* (Ishikawa, E. & Swain, D.L., trans.). New York: Basic Books

Conant, J. (2005) *109 East Palace: Robert Oppenheimer and the Secret City of Los Alamos*. New York: Simon & Schuster

Cook, H.T. & T.F. (1992) *Japan at War: An Oral History*. New York: The New Press

Coox, A. (1970) *Japan: The Final Agony*. New York: Ballantine Books

Craig, W. (1967) *The Fall of Japan*. New York: The Dial Press

de Seversky, A.P. (1942) *Victory Through Air Power*. New York: Garden City Publishing

Douhet, G. (1998) *The Command of the Air* (Ferrari, D., trans.). Washington: Air Force History and Museums Program.

Dollinger, H. (1968) *The Decline and Fall of Nazi Germany and Imperial Japan*. New York: Crown Publishers

Dower, J. (1993) *Japan in War and Peace: Selected Essays*. New York: The New Press

Drea, E.J. (1992) *MacArthur's ULTRA: Codebreaking and the War Against Japan, 1942–45*. Kansas: University Press of Kansas

Eisenhower, D. (1963) *Mandate for Change, 1953–56: The White House Years*. New York: Doubleday

Evans, G. & Kawaguchi, Y. (co-chairs) (2009) *Eliminating Nuclear Threats: A Practical Agenda for Global Policymakers*. Australia and Japan: Report of the International Commission on Nuclear Non-Proliferation and Disarmament

Feis, H. (1967) *The Atomic Bomb and the End of World War II*. Princeton: Princeton University Press

Fermi, L. (1954) *Atoms in the Family: My Life with Enrico Fermi*. Chicago: University of Chicago Press

Francillon, R. (1987) *Japanese Aircraft of the Pacific War*. Annapolis: US Naval Institute Press

Frank, R. (2001) *Downfall: The End of the Imperial Japanese Empire*. New York: Penguin Books

Gaimusho [Ministry of Foreign Affairs] (ed.) (1952) *Shusen Shiroku* [The Historical Records of the End of the War] Vol. 4, annotated by J. Eto (Higuchi, T., trans.), Tokyo: Shimbun Gekkansha

Garthoff, R.L. (1968) *Soviet Military Policy: A Historical Analysis*. New York: Frederick A. Praeger

Giangreco, D.M. (2009) *Hell to Pay: Operation DOWNFALL and the Invasion of Japan, 1945–1947*. Annapolis: US Naval Institute Press

Glantz, D.M. (1984) *August Storm: The Soviet 1945 Strategic Offensive in Manchuria*. Ann Arbor: University of Michigan Library

Glasstone, S. & Hirschfelder, J.O. (eds) (1950) *The Effects of Atomic Weapons: Prepared For and in Cooperation With the US Department of Defense and the US Atomic Energy Commission*. Washington, DC: US Government Printing Office

Glynn, P. (2009) *A Song for Nagasaki*. San Francisco: Ignatius Press

Goodchild, P. (1985) *J. Robert Oppenheimer: Shatterer of Worlds*. New York: Fromm International

Grew, J. (1944) *Ten Years in Japan*. London: Hammond, Hammond & Co.

Groueff, S. (1967) *Manhattan Project: The Untold Story of the Making of the Atomic Bomb*. Boston: Little, Brown

Groves, L. (1962) *Now It Can Be Told: The Story of the Manhattan Project*. New York: Da Capo Press

Hachiya, M. (1995) *Hiroshima Diary*. North Carolina: University of North Carolina Press

Ham, P. (2003) *Kokoda*. Sydney: HarperCollins

Hane, M. (2003) *Peasants, Rebels, Women and Outcastes: The Underside of Modern Japan*. Lanham: Rowman & Littlefield

Harriman, A. & Abel, E. (1976) *Special Envoy to Churchill and Stalin, 1941–1946*. New York: Hutchinson

Harris, A. (2005) *Bomber Offensive*. South Yorkshire: Pen and Sword Publishing

Hasegawa, T. (2005) *Racing the Enemy: Stalin, Truman, and the Surrender of Japan*. Cambridge, Mass.: Harvard University Press

Hastings, M. (1999) *Bomber Command*. London: Pan Macmillan

Hastings, M. (2008) *Nemesis*. London: HarperPerennial

Hayashi, S. (1979) *Kogun: The Japanese Army in the Pacific War*. Westport: Greenwood Publishing Group

Hearn, L. (1997) *Lafcadio Hearn's Japan: An Anthology of His Writings on the Country and Its People*. Vermont: Tuttle Publishing

Hein, L. & Selden, M. (1997) *Living with the Bomb: American and Japanese Cultural Conflicts in the Nuclear Age*. New York: East Gate

Herken, G. (1987) *Counsels of War*. New York: Oxford University Press

Herken, G. (1991) *The Winning Weapon: The Atomic Bomb in the Cold War, 1945–1950*. New York: Random House

Hersey, J. (1989) *Hiroshima*. New York: Vintage

Hershberg, J. (1995) *James B. Conant: Harvard to Hiroshima and the Making of the Nuclear Age*. Stanford: Stanford University Press

Hiroshima International Council for Medical Care of the Radiation-Exposed (ed.), with Shigematsu, I., Ito, C., Kamada, N., Akiyama, M. & Sasaki, H. (1995) *Effects of A-bomb Radiation on the Human Body*. (Harrison, B. trans.), Langhorne: Harwood Academic

Hiroshima Peace Memorial Museum (1994) *The Outline Of Atomic Bomb Damage In Hiroshima*. Hiroshima, Japan: Hiroshima Peace Memorial Museum

Hiroshima Shiyakusho, *Hiroshima Genbaku Sensaishi* [RHAWD] (1971), Vol. 1. Hiroshima: Hiroshima City

Holloway, D. (1994) *Stalin and the Bomb*. New Haven: Yale University Press

Hoshina, Z. (1975) *Daitoa Senso Hishi: Hoshino Zenshiro Kaiso-roku* [Secret History of the Greater East Asia War: Memoir of Zenshiro Hoshina] (Tajima, H., trans.). Tokyo, Japan: Hara-Shobo

Hosokawa, H. & Kamei, H. (eds) (1996) *The Diary of Yoko Moriwaki: A Junior High School Student of Hiroshima First Girls' School*. Tokyo: Heiwa Bunka

Ibuse, M. (1969) *Black Rain* (Bester, J., trans.). New York: Kodansha International

Ienaga, S. (1978) *The Pacific War*. New York: Pantheon Books

Ishida, T. (2004) *The Formation of the Atomic Bomb Experience: The Statistical Study, Japan*, Vol 1.

Itoh, T. (ed.) (2000) *Sokichi Takagi: Nikki to Joho* [Sokichi Takagi: Diary and Documents] (Tajima, H., trans.). Tokyo, Japan: Misuzu-Shobo

Jane, F. (1984) *The Imperial Japanese Navy*. London: Conway Maritime Press

Jansen, M. (2002) *The Making of Modern Japan*. Cambridge, Mass.: Harvard University Press

Japan Broadcasting Corporation (1977) *Unforgettable Fire: Pictures Drawn by Atomic Bomb Survivors*. New York: Pantheon Books

Japan Illustrated Encyclopaedia (1999) *Keys to the Japanese Heart and Soul*. Tokyo: Kodansha International Ltd

Jungk, R. (1985) *Children of the Ashes: The Story of a Rebirth*. London: Pelican Books

Kelly, C. (ed.) (2007) *The Manhattan Project: The Birth of the Atomic Bomb in the Words of Its Creators, Eyewitnesses and Historians*. New York: Black Dog & Leventhal Publishers

Kiyosawa, K. (1999) *A Diary of Darkness: The Wartime Diary of Kiyosawa Kiyoshi*. Princeton: Princeton University Press

Lamont, L. (1965) *Day of Trinity*. New York: Scribner

Lanouette, W. & Szilard, B. (1994) *Genius in the Shadows: A Biography of Leo Szilard, the Man Behind the Bomb*. Chicago: University of Chicago Press

Laurence, W. (1947) *Dawn Over Zero: The Story of the Atomic Bomb*. New York: Alfred E. Knopf

Lawren, W. (1988) *The General and the Bomb: A Biography of General Leslie R. Groves*. New York: Dodd Mead

Leahy, W. (1979) *I Was There*. New York: Arno Press

LeMay, C. (1986) *Superfortress: The Boeing B-29 and American Airpower in WWII*. Yardley: Westholme Publishing

Lifton, R.J. (1991) *Death in Life: Survivors of Hiroshima*. North Carolina: University of North Carolina Press

Lifton, R.J. & Mitchell, G. (1996) *Hiroshima in America*. New York: HarperPerennial

McCullough, D. (1992) *Truman*, Touchstone. New York: Simon & Schuster

Macpherson, W.J. (1987) *The Economic Development of Japan, c. 1868–1941*. London: Palgrave Macmillan

Maddox, R.J. (ed.) (2007) *Hiroshima in History: The Myths of Revisionism*. Columbia: University of Missouri Press

Malloy, S. L. (2008) *Atomic Tragedy: Henry L. Stimson and the Decision to Use the Bomb Against Japan*. Ithaca: Cornell University Press

Marshall, J. (1995) *To Have and Have Not: Southeast Asian Raw Materials and the Origins of the Pacific War*. Los Angeles, Berkeley: University of California Press

Mee, C. (1995) *Meeting at Potsdam*. New York: Franklin Square Press

Miller, M. (2005) *Plain Speaking: An Oral Biography of Harry S. Truman*. New York: Black Dog and Leventhal Publishers

Miscamble, W. (2007) *From Roosevelt to Truman: Potsdam, Hiroshima and the Cold War*. New York: Cambridge University Press

Mishima, Y. (1977) *Yukio Mishima on Hagakure: The Samurai Ethic and Modern Japan*. London: Souvenir Press

Mitchell, R. (1983) *Censorship in Imperial Japan*. Princeton: Princeton University Press

Moore, C. (1967) *The Japanese Mind: Essentials of Japanese Philosophy and Culture*. Hawaii: East-West Center Press

Morris-Suzuki, T. (1985) *Showa: An Inside History of Hirohito's Japan*. New York: Schocken

Murakami, H. (2002) *After the Quake*. London: The Harvill Press

Nitobe, I. (2001) *Bushido: The Soul of Japan*. Boston: Tuttle Publishing

Norris, R.S. (2002) *Racing for the Bomb: General Leslie R. Groves, The Manhattan Project's Indispensable Man*. Vermont: Steerforth Press

Oe, K. (1996) *Hiroshima Notes* (Swain, D. & Yonezawa, T., trans.). New York: Grove Press

Oe, K. (ed.) (2007) *Fire From the Ashes*: Short Stories from Hiroshima and Nagasaki. UK: Readers International

Ogura, T. (1982) *Letters from the End of the World: A Firsthand Account of the Bombing of Hiroshima* (Murakami, K. & Fujii, S., trans.). Tokyo: Kodansha International Ltd

Oppenheimer, J.R. (1956) *The Constitution of Matter*. Oregon: Oregon State System of Higher Education

Oppenheimer, J.R. (1980) *Letters and Recollections* (Kimball Smith, A. & Weiner, C., eds). Stanford: Stanford University Press

Oppenheimer, J.R. (1984) *Uncommon Sense*. Boston: Birkhäuser

Orwell, G. (2009) *Diaries* (Davison, P., ed.). London: Harvill Secker

Osada, A. (ed.) (1982) *Children of Hiroshima*. London: HarperCollins

The Pacific War Research Society (1980) *Japan's Longest Day*. New York: Kodansha International Ltd

Pais, A. (1994) *Niels Bohr's Times: In Physics, Philosophy and Polity*. New York: Oxford University Press

Pape, R. (1996) *Bombing to Win: Air Power and Coercion in War*. Ithaca and London: Cornell University Press

Rhodes, R. (1986) *The Making of the Atomic Bomb*. New York: Simon & Schuster

Rhodes, R. (2010) *Twilight of the Bombs: Recent Challenges, New Dangers, and the Prospects for a World Without Nuclear Weapons*. New York: Alfred A. Knopf

Roberts, A. & Guelff, R. (eds) (1982) *Documents on the Laws of War*. Oxford: Oxford University Press

Robertson, D. (1994) *Sly and Able: A Political Biography of James F. Byrnes*. New York: W.W. Norton & Company

Schell, J.E. (1982) *The Fate of the Earth*. London: Pan Books

Segrè, E. (1970) *Enrico Fermi, Physicist*. Chicago: The University of Chicago Press

Selden, K., Lifton, R.J., & Selden, M. (eds) (1997) *The Atomic Bomb: Voices from Hiroshima and Nagasaki*. New York: M.E. Sharpe

Serber, R. (1992) *The Los Alamos Primer: The First Lectures on How to Build an Atomic Bomb*. Berkeley: University of California Press

Sherwin, M. (2003) *A World Destroyed: Hiroshima and its Legacies*. Stanford: Stanford University Press

Shillony, B. (1981) *Politics and Culture in Wartime Japan*. New York: Oxford University Press

Sigal, L. (1988) *Fighting to a Finish: The Politics of War Termination in the United States and Japan, 1945*. Ithaca: Cornell University

Skates, J.R. (1988) *The Invasion of Japan: Alternative to the Bomb*. Columbia: University of South Carolina

Smyth, H.D. (1945) *Atomic Energy for Military Purposes; the Official Report on the Development of the Atomic Bomb under the Auspices of the United States Government, 1940–1945*. Princeton: Princeton University Press

Spector, R. (1985) *Eagle Against the Sun*. New York: Random House

Sugimine, H. & F. (eds) (2008) *Doctors' Testimonies of Hiroshima*. Kyoto: Kyoto Physicians' Association Appealing the Prevention of Nuclear War and the Abolition of Nuclear Weapons and International Human Network Global Project for International Understanding and Exchange Program

Sweeney, C., Antonucci, J. & Antonucci, M. (1997) *War's End: An Eyewitness Account of America's Last Atomic Mission*. New York: Avon Books

Takaki, R. (1995) *Hiroshima: Why America Dropped the Atomic Bomb*. Canada: Little, Brown

Takashi, N. (1994) *The Soil: A Portrait of Rural Life in Meiji Japan* (Waswo, A., trans.). Berkeley: University of California Press

Takeuchi, T. (1935) *War and Diplomacy in the Japanese Empire*. London: George Allen & Unwin

Tanaka, Y. (1997) *Hidden Horrors: Japanese War Crimes in World War II*. Colorado: Westview Press

Teller, E. & Brown, A. (1975) *The Legacy of Hiroshima*. Westport: Praeger Publishing

Teller, E. & Shoolery, J. (2001) *Memoirs*. Cambridge, Mass.: Perseus Publishing

Thorne, C. (1978) *Allies of a Kind: The United States, Britain and the War Against Japan, 1941–45*. London: Hamish Hamilton

Togo, S. (1956) *The Cause of Japan*. New York: Simon & Schuster

Toland, J. (1982) *Infamy: Pearl Harbor and its Aftermath*. New York: Doubleday

Toland, J. (2003) *The Rising Sun: The Decline and Fall of the Japanese Empire, 1936–1945*. New York: Modern Library Inc.

Truman, H.S. (1973) *Harry S. Truman*. London: Hamish Hamilton

Truman, H.S. (1983) *Dear Bess: The Letters from Harry to Bess Truman, 1910–1959*. Columbia: University of Missouri Press

Truman, H.S. (1986) *Memoirs of Harry S. Truman*. Vol. 1 *Year of Decisions*. Cambridge, Mass.: De Capo Press.

Yamamoto, T. (2001) *Bushido: The Way of the Samurai*, New York: Square One Publications

Tuchman, B. (1994) *The Guns of August*. New York: Random House

Turnbull, S.R. (2002) *The Samurai: A Military History*. London: Routledge Curzon

Van Der Rhoer, E. (2000) *Deadly Magic: A Personal Account of Communications Intelligence in World War II in the Pacific*. Bloomington: iUniverse

Vigor, P.H. (1983) *Soviet Blitzkrieg Theory*. London: The Macmillan Press

Volkogonov, D. (2000) *Stalin: Triumph and Tragedy*. London: Phoenix Press

Walker, S. (2005) *Shockwave: Countdown to Hiroshima*. New York: HarperCollins

Walzer, M. (1977) *Just and Unjust Wars*. London: Allen Lane

Webster, C. & Frankland, N. (2006) *The Strategic Air Offensive Against Germany 1939–1945*, Vol. III. East Sussex, UK: Naval and Military Press

Wells, H.G. [c1914] (2011) *The World Set Free*. Charleston: Createspace

Williams, R.C. & Cantelon, P.L. (eds) (1984) *The American Atom: A Documentary History of Nuclear Policies from the Discovery of Fission to the Present, 1939–1984*. Philadelphia, PA: University of Pennsylvania Press

Yamazaki, J. & Fleming, L. (1994) *Children of the Atomic Bomb: An American Physician's Memoir of Nagasaki, Hiroshima, and the Marshall Islands*. Durham, NC: Duke University Press

JOURNAL ARTICLES

Bernstein, B. J. (2003) 'Reconsidering the "Atomic General": Leslie R. Groves'. *The Journal of Military History*, Vol. 67, No. 3, pp. 883–920

Harris, A. (August 1944) 'The Organization of Mass Bombing Attacks'. *The Journal of the Royal United Service Institution*, No. 555

Hatano, S. & Watanuki, T. (1953) 'A-bomb Disaster Investigation Report – Fouth Investigation: Chiefly Concerned with Occurrence of Keloids due to Thermal Burns of Atomic Bomb', CRIABC, Vol. 1

Shimomura, H. (1948) 'The Second Sacred Judgment: August 14, 1945'. *Shusenki* [Account of the End of the War] (Higuchi, T., trans.), Tokyo: Kamakura Bunko

Shinji, T. (Autumn 1997) 'Listening to the Wishes of the Dead: In the Case of Dr Nagai Takashi', *Crossroads*, 5

NEWSPAPERS AND MAGAZINES

Asahi Shimbun
Associated Press
Bulletin of the Atomic Scientists
Chicago Daily Tribune
Christian Century
The Daily Express
Fujin Koron
International Herald Tribune
The Japan Times
Nature
Naturwissenschaften
New Republic
Newsweek
The New York Times
Stars and Stripes
Time Magazine
The Times (London)
Washington Post
Yomiuri Hochi

WEBPAGES

http://www.cbcj.catholic.jp/eng/ehistory/table02.htm
http://en.wikipedia.org/wiki/Giulio_Douhet
http://en.wikipedia.org/wiki/Dehousing#Contents_of_the_dehousing_paper
http://en.wikipedia.org/wiki/US_Department_of_Energy
http://www.lanl.gov/history/atomicbomb/trinity.shtml:
http://www.lanl.gov/history/atomicbomb/victory.shtml
http://www.ne.jp/asahi/hidankyo/nihon/rn_page/english/message.html
http://www.jewishvirtuallibrary.org/
http://nuclearweaponarchive.org/Nwfaq/Nfaq0.html

SELECTED INTERVIEWS

(of about 80 undertaken by the author in Hiroshima, Nagasaki and aboard the Peace Boat)
Nagasaki: Kazuhiro Hamaguchi, Teruo Ideguchi, Tsuruji Matsuzoe, Kiyoko Mori, Kikuyo Nakamura

Hiroshima: Takehiko Ena, Mitsue Fujii (Hiraki), Hiromi Hasai, Kohji Hosokawa, Seiko Ikeda, Takeshi Inokuchi, Yoshiko Kajimoto, Dr Nanao Kamada at *Kurakake Nozomi-en* (Nursing Home for A-bomb Survivors), Fumie Katayama, Yukio Katayama, Shoso Kawamoto, Kenji Kitagawa, Miyoko Matsubara, Takashi Morita, Taeko Nakamae, Tomiko Nakamura, Iwao Nakanishi, Emiko Okada, Miyoko Watanabe

2008: Kiyomi Igura, Nagasaki woman, interviewed by Yuta Hiramatsu, 2009

Personal statement, circa end 1945: Takehiko Ena, kamikaze navigator Hatsue Ueda, housewife, letter to a friend

BROCHURES, CATALOGUES, PAMPHLETS

Atomic Warfare – Manual of Basic Training, British Home Office, Civil Defence, Pamphlet No. 6, 1950

Bock, F., Commemorative Booklet for 50th Anniversary Reunion, 509th Composite Group, August 1995

Heisei ju-nendo dai nikai kikakuten – Jugo wo sasaeru chikara to natte – josei to senso [Special Exhibition No. 2, 1998 –The Force that Supports the Home Front – Women and the War], Hiroshima Peace Memorial Museum, 1999

Nagasaki Testimonial Society, 'A Journey to Nagasaki' (brochure)

Japanese Field Manual for the Decisive Battle for the Homeland (pamphlet)

The Minami Alumni 80th Anniversary Commemorative Magazine, Hiroshima Minami High School (formerly: Hiroshima Daiichi Girls' High School), 1 March 1982 issue

'To All Employees of E. I. DuPont de Nemours & Company', E. I. DuPont de Nemours & Company, Delaware, 24 August 1945 (internal memo)

ARCHIVES AND RECORDS

British Cabinet Papers, Winston Churchill, Minister of Defence, Secretarial papers, Complete Classes from CAB and PREM series, Public Records' Office, UK

Walter J. Brown Papers, Clemson University, Clemson, SC, MSS 243

James F. Byrnes Papers, Clemson University Library, Clemson, SC, MSS90 [Series 5]

James Forrestal Diaries, Naval Historical Center, Washington, DC

George C. Marshall Library, Lexington, VA

National Security Archive, George Washington University, Washington, DC

Henry Stimson Papers, Manuscripts and Archives, Yale University Library, New Haven, Connecticut

Henry A. Wallace Papers, University of Iowa Libraries

Harry S. Truman Library and Museum, Independence, Missouri
Atomic Bomb Collection
Harry S. Truman Student Research File
President's Secretary's Files
Papers of Eben A. Ayers
Papers of R. Gordon Arneson
Papers of Charles Ross
White House Official File
US National Archives and Records Administration (NARA)
RG77 Records of the Office of the Commanding General, Manhattan Project
RG165 Records of the War Department General and Special Staffs
RG218 Records of the Joint Chiefs of Staff, Central Decimal Files, 1942–45
RG243 Records of the US Strategic Bombing Survey
RG353 Records of Interdepartmental and Intradepartmental Committees, Secretary's Staff Meetings Minutes, 1944–47
RG457 'Magic' Diplomatic Summaries 1 January 1943–3 November 1945, boxes 1–19

ACKNOWLEDGEMENTS

I'D LIKE TO EXPRESS MY deep appreciation for the help offered by so many people during the research and writing of this book – in particular, the survivors of the atomic bombs and conventional air raids, who relived their experiences during hundreds of hours of interviews. The memories of several are intimately involved in this narrative: Mitsue Fujii (Hiraki), Hiroe Sato, Kohji Hosokawa, Seiko Ikeda, Teruo Ideguchi, Takeshi Inokuchi, Yoshiko Kajimoto, Tsuruji Matsuzoe, Iwao Nakanishi and Takehiko Ena.

HarperCollins has risen brilliantly to the task of publishing a book of this scale and scope, and I thank everyone who participated – chiefly Mary Rennie, Fiona Henderson, Christine Farmer, Shona Martyn and Michael Moynahan. Mary Rennie, Katie Stackhouse and John Mapps deserve special mention for their patient and exacting attention to detail; as do Matt Stanton, Rachel Dennis and Laurie Whiddon, respectively designer, publishing assistant and map illustrator, and typesetter Graeme Jones, designer Natalie Winter and picture researchers Linda Brainwood and Mika Kubo. Thanks also for the great support of literary agents Deborah Callaghan and Jane Burridge.

Several translators and interpreters were indispensable in rendering the Japanese – sometimes of an unusual form used at the time of the war, or of a difficult dialect – into modern English: Debbie Edwards, Meri Joyce, Mutsuko Yoshido, Keiko Ogura, Stephanie Oley, Alex Sayle and Kayo Shiraishi, who gave their time and expertise generously and I'm most grateful to them.

My gratitude also to the staff and volunteers of the 63rd Global Voyage of the Peace Boat, with whom I travelled from Sydney to Tokyo in January 2009. About 100 survivors of the atomic bombs were on this specially organised 'Global Voyage for a Nuclear-Free World: Peace Boat Hibakusha Project'. Especially helpful were the Peace Boat executive committee and organisers, chiefly Akira Kawasaki and Meri Joyce, as well as many dedicated staff and volunteers, namely: Sumiko Hatakeyama, Yuta Hiramatsu, Mariko Ishii, Mayuko Miyata, Izumi Shigaki, Yoko Takayama, Ray Ueno. Thanks also to the on-board film makers, Erika Bagnarello and Takashi Kunimoto.

Two organisations provided the tranquillity necessary for the completion of this history: the Scots College, Sydney, which made a quiet office available in Royle House, where I worked as a 'writer in residence' during 2010; and the Berghutte Ski Lodge, Thredbo, which I haunted during the southern-hemisphere summer of that year. My thanks to Dr Ian Lambert, principal of Scots, and the house master (then Peter Graham) and staff of Royle House; and to Michael Horton, Drew Blomfield and the Committee of Berghutte.

Of critical assistance in checking the physics was Associate Professor Reza Hashemi-Nezhad, Director of the Institute of Nuclear Science at the University of Sydney, who ran a close eye over the scientific chapters. Helpful to my research, also, were the indefatigable Jordan DeBor in America, and my ever tolerant mother, Shirley Ham, who took notes and gathered articles: thank you. I am similarly grateful to the Sayle family. In 2010, months before his death, Murray Sayle – the greatly admired Australian journalist whose essay 'Did the Bomb End the War?' consumed an entire issue of *The New Yorker* in 1995 – agreed to meet me and revisit the subject. He felt no less strongly that the bomb had been militarily unnecessary (a view shared by several US generals and admirals in 1945). His wife, Jennifer, generously took me through his extraordinary library; and his son, Alex, met me in Nagasaki to interpret discussions with atomic-bomb survivors; a job he did superbly for many hours.

Museums, archives, libraries were, as always, more than receptive to the insatiable demands of the subject. The Hiroshima Peace Memorial Museum and the Nagasaki Atomic Bomb Museum offered constant, untiring assistance during numerous visits. The staff of the Hiroshima Peace Museum were extraordinarily helpful in arranging interviews and interpreters, and I'd like to thank in particular Natsuki Okita and Chikage Sakamoto of the Museum's Outreach Division. In America, I'm grateful to the staff of the Harry S. Truman Library in Independence, Missouri – whose presidential grant was especially welcome; the National Archives in Washington, DC; and the James Byrnes Room in Clemson University, Clemson, South Carolina. Of great assistance, too, were Ednamaya E. (Lisa) Blevins, the public affairs specialist at the US Army White Sands Missile Range, who gave me a private tour of the Alamogordo atomic test site. Jeff Berger, Director, Communications and Government Affairs, Los Alamos National Laboratory, offered his time at short notice; as did Heather McClenahan, Assistant Director of the Los Alamos Historical Society.

My gratitude, also, to many people from different backgrounds who generously gave their time and knowledge: Nori Tohei and the staff of *Nihon Hidankyo* (the Japan Confederation of A- and H-bomb Sufferers Organisations); Noriko Honda, media and project officer, the Australian Embassy, Tokyo; Dr Nanao Kamada, president of *Kurakake Nozomi-en* (Nursing Home for A-bomb Survivors), Hiroshima; Kenji Kitagawa, Emeritus Professor of Hiroshima University and President of Hiroshima UNESCO; Steven Leeper, Chairperson, Board of Directors, Hiroshima Peace Culture Foundation; Professor Masaharu Hoshi, Department of Radiation Biophysics, Research Institute for Radiation Biology and Medicine, Hiroshima; Professor Martin Sherwin (as well, for his hospitality in Washington); Professor Norio Takahashi, Department of Genetics, and Dr Evan B. Douple, Associate Chief of Research, at the Radiation Effects Research Foundation, Hiroshima; Dr David B. Thomson,

Nuclear & Plasma Physics, Los Alamos Committee on Arms Control and International Security; Tatsuo Sekiguchi, Special Director, Nagasaki Broadcasting Company; Takayuki Shimomura, research scholar; Shinji Takahashi, former Professor of sociology, Nagasaki; Tomisha Taue, Mayor of Nagasaki; and Mari Yamauchi, Foreign Press Centre, Tokyo. Thanks also to the London *Sunday Times* for giving me the time off to complete this book.

Family members, friends and colleagues assisted in all sorts of ways – sometimes, unwittingly – and I'd like to mention a few: Georgia Arnott, Drew Blomfield, John Bullen, Janet Ham, Reg Carter, Louise Chapman, Don Featherstone, Mark Friezer, Juliet Herd, Michael Horton, Rob Jarrett, Tony Maniaty, Allan McKay, Justin Mclean, Graham Paterson, Tony Rees, David Reynolds, April Pressler, Emma and Steve Tolhurst, Andy and Renee Welsh, and Andrew Wiseman.

And I thank my parents, and brother and sister, who are ever interested and supportive; my beloved son, Ollie, whose unfolding perception of the world is a source of constant inspiration; and my beautiful wife, Marie, without whose tender interventions I could not have contemplated these terrible events with such intensity and for so long.

INDEX

radiation (*continued*)
  Joint Commission for the
    Investigation of the Atomic Bomb
    in Japan 428–30, 436
  neutron rays 317
  nuclear disintegration, product of 92,
    102
  research 90, 124, 125–6
    post-war 420–31, 435–9
    sharing of information 430–1
  types 90–1
radiation sickness
  'A-bomb disease' 338
  post-war acknowledgement 450
  reports 407, 417–26
    Japan, by 421, 422, 431
    United States' response 420–31, 447
  signs and symptoms 417–21, 422, 433
Radio Saipan 339–40
Radium Girls 90, 424
Rafferty, James 161
Ramer, Robert 60
Ramsey, Rear Admiral D.C. 172, 175
Ramsey, Dr Norman F. 290, 358
'Rape of Nanking' 12, 18, 269
  Nanking Massacre Memorial Museum
    444
Rea, Lieutenant Colonel Charles 407,
  422–4
Reagan, President Ronald 508
Red Cross hospitals 29, 274–5, 330, 410,
  413, 415, 428, 437
Rhodes, Richard 505
Rockwell, Theodore 128
Rongerik Island 447
Roosevelt, President Franklin D. 1–11,
  17, 57, 61, 72, 78, 82, 100, 103, 104,
  112, 193, 205, 206, 409, 475
Ross, Charlie 213, 304, 307, 352, 425
Rumania 212, 242
Russell, Senator Richard B. 193, 378–9
Russia (Union of Soviet Socialist
  Republics – USSR)
  Afghanistan, in 507
  atomic bomb targets 488, 489, 504,
    506
  atomic research and development
    129–30, 246, 463, 499–500
  'August Storm' 350

Russia Russia (Union of Soviet Socialist
  Republics – USSR) (*continued*)
  Cold War involvement *see* Cold War
  distrust of 2–3, 10, 77, 86, 114, 237,
    245, 386, 491
  Japan, relations with 180, 181, 184–9,
    252–6, 260, 264–6, 346–55, 382,
    474
    declaration of war 353–4, 358, 390,
      392
  Manchuria, invasion of 349–51, 352,
    358–9, 387–8, 474, 480, 481, 510
  Moscow Declaration of 1943 249
  nuclear weapon count 508
  Potsdam Declaration, signatory to
    260, 264, 346, 478
  Red Army 2, 10, 176–7, 182, 210, 212,
    216, 217, 350
  Red Star journal 214
  Soviet collapse 508
  Soviet–Japanese Neutrality Pact 7–8,
    177, 181, 184–5, 211
  T-34 350
  United States
    atomic bomb targets in 490
    intelligence report 77
    nuclear arms race 487, 491, 496,
      499–500, 506–8
    relations 11, 77–8, 84–6, 114, 130,
      164–5, 200, 204, 211, 212, 232,
      240, 246–50, 301, 355, 359, 386,
      387–8, 478–81, 487, 488–510
Russian Protocol 9
Russo-Japanese War 15, 16, 181
Rutherford, Lord Ernest 90–8, 102,
  103
Ryukyus, the 169, 277

Sachs, Alexander 103
Sagane, Ryokichi 361
Saipan 13, 21, 25, 23, 25, 57, 58
Sakhalin 16, 348, 350
Sakomizu, Chief Cabinet Secretary
  Hisatsune 262, 341–2, 344, 353–4,
  381–2, 398
samurai code (*bushido*) 20, 144
samurai tradition 33–4, 277, 344, 474
  poem 268
Sasaki, Kenichi 316

United States of America (*continued*)
National Defense Research
    Committee 108, 129
National Research Council of the
    National Academy of Sciences 436
naval blockade 22, 33, 174, 214, 342,
    343, 476, 481, 509
negotiations and peace talks with
    Japan 182–3, 188–9, 207–9, 252,
    256–7, 261, 302, 395
Nevada 497
nuclear weapons 488
    count 508
    public reaction 377–8, 459–62, 503–4
    research and development 497, 502
Oak Ridge 124–5, 126, 127, 128, 154,
    155, 422
'offensive defence' strategy 497–8
Office of Strategic Services 183
Office of War Mobilization 114, 175,
    206
Pacific Fleet Headquarters 303
Pearl Harbor 11, 17, 18, 41, 73, 191–2,
    206, 238, 244, 248, 290, 355, 363,
    368, 371, 376, 378, 379, 406, 459,
    476, 479, 487
Russia, relations with 11, 77–8, 84–6,
    114, 130, 164–5, 200, 204, 211,
    212, 232, 240, 246–50, 301, 355,
    359, 386, 387–8, 478–81, 487
    nuclear arms race 487, 491, 496,
    499–500, 506–8
SeaBees 285
surrender by Japan
    authorised signatories 405–6
    'unconditional surrender', Truman's
    announcement of 404
Uranium Committee 108
US Navy Bureau of Medicine and
    Surgery 428
US Strategic Bombing Strategy 61,
    63, 65
US Strategic Bombing Survey
    (USSBS) 61, 65, 144, 145, 175, 274,
    275, 409, 412, 413, 422, 429, 435,
    469–70, 471
Vietnam War, in 507
Walter Reed Hospital 500
Wendover, Utah 284–5, 287, 292

United States of America (*continued*)
Wesleyan Methodist Church of East
    Michigan 504
UN Charter 249
University of California 103
University of Chicago 115, 120, 121,
    162–3
    MetLab 121, 162, 229
University of Manchester 90
Uranium Committee 108
Urey, Harold 109, 110, 127
US Radium Corporation 90
USS *Augusta* (*Augie*) 210, 214–15, 236,
    301, 304, 306, 310
    Augusta Press 214, 302, 308
USS *Bunker Hill* 170
USS *Enterprise* 170
USS *Howarth* 170
USS *Indianapolis* 283, 284
USS *Missouri* 405, 406, 425

V-E Day 79
Van Kirk, Captain Theodore 'Dutch'
    291, 295, 296, 299
Vasilevsky, Marshal Aleksandr 388
Vietnam 507, 509
    Dien Bien Phu 506
von Neumann, John 120, 147
Vonnegut, Kurt 55
    Cat's Cradle 507

Wallace, Henry 82, 109, 110, 389
Walton, Ernest 97, 99
Warren, Shields 428
Warren, Colonel Stafford 125, 226, 428,
    430
Washington Post 471, 484, 502
Watanabe, Miyoko 67, 251, 259
Weckerling, Brigadier General John 188–9
Weil, George 121, 122
Weller, George 428
Wells, H.G. 92, 98, 103
Westbrook, W.D. 504
Westinghouse 124, 161
Whetham, Sir William Dampier 92
White, James 161
Wigner, Eugene 100, 121, 122
Wiley, Senator Alex 352
Wilson, Major John 289